Nutritional Neurosciences

Series Editor

Mohamed Essa, Sultan Qaboos University, Qaboos, Oman

This book series aims to publish volumes focusing on both basic and clinical research in the field of nutritional neuroscience with a focus on delineating the effect of nutrition on brain function and behavior. The books will examine the role of different nutrients, food agents and supplements (both macro and micro) on brain health, neurodevelopment, neurochemistry, and behaviour. The books will examine the influence of diet, including phytochemicals, antioxidants, dietary supplements, food additives, and other nutrients on the physiology and metabolism of neurons, neurotransmitters and their receptors, cognition, behavior, and hormonal regulations.

The books will also cover the influence of nutrients and dietary supplements on the management of neurological disorders. It details the mechanism of action of phytonutrients on signaling pathways linked with protein folding, aggregation, and neuroinflammation. The books published in the series will be useful for neuroscientists, nutritionists, neurologists, psychiatrists, and those interested in preventive medicine.

Wael Mohamed
Editor

Nutrition and Traumatic Brain Injury (TBI)

From Bench to Bedside

Springer

Editor
Wael Mohamed
Basic Medical Science Department, Kulliyyah of Medicine
International Islamic University Malaysia (IIUM)
Kuantan, Pahang, Malaysia

ISSN 2730-6712 ISSN 2730-6720 (electronic)
Nutritional Neurosciences
ISBN 978-981-97-6340-5 ISBN 978-981-97-6341-2 (eBook)
https://doi.org/10.1007/978-981-97-6341-2

This Springer imprint is published by the registered company Springer Nature Singapore Pte Ltd.
The registered company address is: 152 Beach Road, #21-01/04 Gateway East, Singapore 189721, Singapore

I express my heartfelt gratitude and deep appreciation as I dedicate this work to my wife, Dr. Rehab Ismaeil. Your unwavering support and unconditional love have been constant pillars throughout my academic journey and beyond. I am truly grateful for the love and inspiration you bring to my life, and I acknowledge that none of this would be possible without you.

Wael Mohamed
Pahang, Malaysia

Preface

> The physician who solely walks the corridors of academia may know the paths, but it is the one who also walks alongside patients that truly understands the journey.
> —Sir William Withey Gull (1816–1890)

Welcome to *Nutrition and Traumatic Brain Injury (TBI): From Bench to Bedside*. This book represents an ambitious endeavor to explore the intricate interplay between nutrition and traumatic brain injury (TBI), from the foundational research conducted in laboratories to the practical applications at the patient's bedside. Traumatic brain injury stands as one of the most pressing public health concerns globally, affecting millions of individuals each year. Whether caused by accidents, sports injuries, or military combat, the consequences of TBI can be profound, influencing cognitive function, emotional well-being, and overall quality of life. In the face of such challenges, understanding the role of nutrition in mitigating the effects of TBI emerges as a crucial frontier in both medical research and clinical practice.

This book is a collaborative effort, bringing together experts from diverse fields including neuroscience, nutrition science, clinical medicine, and rehabilitation. By synthesizing cutting-edge research findings with clinical insights and practical recommendations, our aim is to provide a comprehensive resource for researchers, healthcare professionals, caregivers, and individuals affected by TBI. The journey through these pages begins with a journey into the laboratory, where scientists delve into the molecular mechanisms underlying the relationship between nutrition and brain health. From exploring the impact of specific nutrients on neuronal function to unraveling the complex pathways of neuroinflammation and neuroprotection, these foundational insights lay the groundwork for understanding how dietary interventions can influence TBI outcomes.

Moving from the laboratory to the clinic, we examine the translation of research findings into clinical practice. Through case studies, clinical trials, and evidence-based guidelines, we explore the practical applications of nutritional strategies in the management and rehabilitation of TBI patients. From optimizing macronutrient balance to harnessing the potential of dietary supplements and functional foods, we uncover the multifaceted ways in which nutrition can support brain health and recovery. Yet, this book is not merely a compendium of scientific knowledge; it is also a testament to the resilience of the human spirit. Through the stories of TBI survivors and their caregivers, we gain a deeper understanding of the challenges they face and the triumphs they achieve. Their experiences remind us of the profound impact that compassionate care, informed by scientific understanding, can have on the journey to recovery.

As editor of this volume, I am deeply grateful to all the contributors who have generously shared their expertise, insights, and experiences. It is my hope that *Nutrition and TBI: From Bench to Bedside* will serve as a valuable resource, inspiring further research, informing clinical practice, and ultimately improving the lives of those affected by traumatic brain injury.

Finally, we thank all the authors for their significant efforts in contributing such excellent chapters for this new edition. We are also sincerely grateful to each author for their patience during the compilation and final editing of this book.

Kuantan, Pahang, Malaysia Wael Mohamed

Acknowledgments

I extend my sincere gratitude to all the authors who contributed their expertise and insights to this project. Their dedication and collaboration with the editor have made the journey of writing this book truly enjoyable. I also want to express my appreciation to those authors specializing in neuroscience and clinical research. Their years of experience and diligent work in various aspects of TBI have enriched this work immeasurably. The topics discussed herein hold significant value in the realms of nutrition, psychiatry, neurological disorders, and neurodegeneration. Lastly, I wish to thank my friends and colleagues for their unwavering support, encouragement, and inspiration throughout the duration of this project. Your belief in me fueled my determination to see this endeavor through. Thank you.

Kuantan, Pahang, Malaysia Wael Mohamed

Contents

Editor and Contributors

About the Editor

Wael Mohamed, BBCH (Egypt), MMSc, MD/PhD (USA) a Physician Neuroscientist, earned his PhD from PSU, USA, and presently holds the position of Professor Madya at IIUM Medical School, Malaysia. With over 200 speaking engagements both domestically and internationally, Dr. Mohamed is a seasoned lecturer. He has authored more than 150 peer-reviewed papers in the field of neuroscience/psychiatry, boasting an h-index of 23. Additionally, he serves as an editor for several international journals, having contributed to numerous journal special issues focusing on brain disorders. Furthermore, he is involved in editing several neuroscience books with several publishers. Dr. Mohamed has received research grants from esteemed national and international organizations including IBRO, ISN, MJF, STDF, FRGS, and INDO-ASEAN, totaling half a million US dollars in research funding. He is also the founder of the AfrAbia-PD-Genomic Consortium (AA-PD-GC).

Contributors

Basma Emad Aboulhoda Department of Anatomy and Embryology, Faculty of Medicine, Cairo University, Cairo, Egypt

Ahmad Afyouni Medical Learning Skills Academy, Beirut, Lebanon

Faculty of Medical Sciences, Lebanese University, Beirut, Lebanon

Rabia Ajana Biology, Ecology and Health Laboratory, Department of Biology, Faculty of Science, Abdelmalek Essaadi University, Tetouan, Morocco

Gideon S. Alex University of Port Harcourt, Port Harcourt, Rivers State, Nigeria

Driss Ait Ali Biological and Health Sciences Team, Higher Institute of Nursing Professions and Health Techniques, Ministry of Health, Ouarzazate, Morocco

Youssef Aoulad El Hadj Ali Department of Chemistry, Faculty of Science, University of Abdelmalek Essaâdi, Tetouan, Morocco

Mohamed M. Ammar Microbiology and Biochemistry Program, Faculty of Science, Benha University-Obour Campus, Benha, Egypt

Bouchra Anzieh Medical Learning Skills Academy, Beirut, Lebanon

Holy Spirit University of Kaslik, School of Medicine and Medical Sciences, Jouniyeh, Lebanon

Zahraa Awada Faculty of Medical Sciences, Lebanese University, Beirut, Lebanon

Medical Learning Skills Academy, Beirut, Lebanon

Abdelmonaim Azzouz Department of Chemistry, Faculty of Science, University of Abdelmalek Essaâdi, Tetouan, Morocco

Abdelali Bitar Laboratory of Anthropogenic, Biotechnology and Health, Team Physiopathology Nutritional, Neurosciences and Toxicology, Faculty of Sciences, Chouaib Doukkali University, El Jadida, Morocco

Ibtissam Chakir Biology, Ecology and Health Laboratory, Department of Biology, Faculty of Science, Abdelmalek Essaadi University, Tetouan, Morocco

Multidisciplinary Faculty, LPRD, Sultan Moulay Slimane University, Beni-Mellal, Morocco

Sakhr Dadda Department of Biology, Faculty of Sciences and Techniques, University Moulay Ismail, Errachidia, Morocco

Andrea Eid Medical Learning Skills Academy, Beirut, Lebanon

Holy Spirit University of Kaslik, School of Medicine and Medical Sciences, Jouniyeh, Lebanon

Hanane El Fatimi Mohammed VI University of Health Sciences, Casablanca, Morocco

Sana El Foutat Metabolics Platform, Biochemistry Laboratory, Faculty of Medicine, Cadi Ayyad University, Marrakech, Morocco

Abdelmoujoud El Haboussi Institute of Sports Sciences, Hassan 1er University, Settat, Morocco

Omar El Hiba Laboratory of Anthropogenic, Biotechnology and Health, Team Physiopathology Nutritional, Neurosciences and Toxicology, Faculty of Sciences, Chouaib Doukkali University, El Jadida, Morocco

Brahim El Houate Biological and Health Sciences Team, Higher Institute of Nursing Professions and Health Techniques, Ministry of Health, Ouarzazate, Morocco

Abdelaati El Khiat Biological and Health Sciences Team, Higher Institute of Nursing Professions and Health Techniques, Ministry of Health, Ouarzazate, Morocco

Laboratory of Clinical and Experimental Neurosciences and Environment, Faculty of Medicine and Pharmacy, Cadi Ayyad University, Marrakech, Morocco

Biological Engineering Laboratory, Faculty of Sciences and Techniques (FST), Sultan Moulay Slimane University, Beni Mellal, Morocco

Laboratory of Anthropogenic, Biotechnology and Health, Team Physiopathology Nutritional, Neurosciences and Toxicology, Faculty of Sciences, Chouaib Doukkali University, El Jadida, Morocco

Interdisciplinary Laboratory in Bio-Resources, Environment and Materials, Higher Normal School, Marrakech, Morocco

Institute of Sports Sciences, Hassan 1er University, Settat, Morocco

Mohamed El Koutbi Biological and Health Sciences Team, Higher Institute of Nursing Professions and Health Techniques, Ministry of Health, Ouarzazate, Morocco

Bilal El-Mansoury Laboratory of Anthropogenic, Biotechnology and Health, Team Physiopathology Nutritional, Neurosciences and Toxicology, Faculty of Sciences, Chouaib Doukkali University, El Jadida, Morocco

Enas El Saftawy Department of Medical Parasitology, Faculty of Medicine, Cairo University, Cairo, Egypt

Department of Medical Parasitology, Armed Forces College of Medicine, Cairo, Egypt

Kholoud Elsamman Biochemistry and Nutrition Department, Faculty of Women for Science, Ain Shams University, Cairo, Egypt

Nour Fakih Department of Natural Sciences, Lebanese American University, Beirut, Lebanon

Medical Learning Skills Academy, Beirut, Lebanon

Naima Fdil Metabolics Platform, Biochemistry Laboratory, Faculty of Medicine, Cadi Ayyad University, Marrakech, Morocco

Hagar F. Forsan Animal Production Research Institute (APRI), Agricultural Research Center (ARC), Dokki, Giza, Egypt

Jad Ghauch Faculty of Medicine and Medical Sciences, USEK, Jounieh, Lebanon

Medical Learning Skills Academy, Beirut, Lebanon

Asmaa Haj-Khlifa Faculty of Sciences Semlalia, Cadi Ayyad University, Marrakech, Morocco

Youssef Ait Hamdan Interdisciplinary Laboratory in Bio-Resources, Environment and Materials, Higher Normal School, Cadi Ayyad University, Marrakech, Morocco

Hiba Hamdar Medical Learning Skills Academy, Beirut, Lebanon

Plovdiv Medical University, Plovdiv, Bulgaria

Rewida R. Hamed Chemistry Department, Faculty of Science, Helwan University, Helwan, Egypt

Mahmoud Hammad Medical Learning Skills Academy, Beirut, Lebanon

Faculty of Medicine, American University of Beirut, Beirut, Lebanon

Nada Ibrahim Hendi Faculty of Medicine, Ain Shams University, Cairo, Egypt

Rabii Hilali Biology, Ecology and Health Laboratory, Department of Biology, Faculty of Science, Abdelmalek Essaadi University, Tetouan, Morocco

Hussein Hmedi Medical Learning Skills Academy, Beirut, Lebanon

Faculty of Medical Sciences, Lebanese University, Beirut, Lebanon

Katrine Ibrahim Medical Learning Skills Academy, Beirut, Lebanon

Faculty of Medical Sciences, Lebanese University, Beirut, Lebanon

Amadi Ogonda Ihunwo Faculty of Health Sciences, School of Anatomical Sciences, University of the Witwatersrand, Johannesburg, South Africa

Ahmad Jaber Medical Learning Skills Academy, Beirut, Lebanon

Faculty of Medicine, Lebanese American University, Dbeyieh, Lebanon

Jana Jaber Faculty of Medical Sciences, Lebanese University, Beirut, Lebanon

Medical Learning Skills Academy, Beirut, Lebanon

Arumugam Jayakumar Miller School of Medicine, University of Miami, Miami, FL, USA

Safaa Joumaa Faculty of Medical Sciences, Lebanese University, Beirut, Lebanon

Medical Learning Skills Academy, Beirut, Lebanon

Haidar Kanso Medical Learning Skills Academy, Beirut, Lebanon

Faculty of Medicine, Damascus University, Damascus, Syria

Kamal Kaoutar Biological and Health Sciences Team, Higher Institute of Nursing Professions and Health Techniques, Ministry of Health, Ouarzazate, Morocco

Ayoub Kasrati Biology, Ecology and Health Lab, Department of Biology, Faculty of Science, Abdelmalek Essaadi University, Tetouan, Morocco

Department of Pharmacy, Faculty of Medicine and Pharmacy, Hassan 1st University, Settat, Morocco

Zeinab Kazan Faculty of Medical Sciences, Lebanese University, Beirut, Lebanon

Medical Learning Skills Academy, Beirut, Lebanon

Mohamed Khaled Faculty of Medicine, Alzexandria University, Alexandria, Egypt

Firas Kobeissy Department of Neurobiology and Neuroscience Institute, Morehouse School of Medicine (MSM), Center for Neurotrauma, Multiomics and Biomarkers, Atlanta, GA, USA

Younsse Korchyou Biological and Health Sciences Team, Higher Institute of Nursing Professions and Health Techniques, Ministry of Health, Ouarzazate, Morocco

Jawad Laadraoui Faculty of Science Aïn Chock, Hassan II University, Casablanca, Morocco

Nahed S. Lamloum Department of Biochemistry, Faculty of Science, Beni-Suef University, Beni-Suef, Egypt

Adejoke Elizabeth Memudu Anatomy Department, Neuroscience Unit, Faculty of Basic Medical Sciences, Edo State University Uzairue, Auchi, Edo State, Nigeria

Wael Mohamed Department of Basic Medical Science, Kulliyyah of Medicine, International Islamic University Malaysia, Kuantan, Pahang, Malaysia

Clinical Pharmacology Department, Menoufia Medical School, Menoufia University, Menoufia, Egypt

Layal Msheik Faculty of Medical Sciences, Lebanese University, Beirut, Lebanon

Medical Learning Skills Academy, Beirut, Lebanon

Hajar Ouaki Laboratory of Biostatistics, Clinical Research and Epidemiology (LBRCE), Faculty of Medicine and Pharmacy, Mohammed V University, Rabat, Morocco

Dia Eddine Oudghiri Biology, Ecology and Health Laboratory, Department of Biology, Faculty of Science, Abdelmalek Essaadi University, Tetouan, Morocco

Biology Higher Institute of Nursing Professions and Health Techniques (ISPITS), Tetouan, Morocco

Fatima Rachidi Laboratory of Biostatistics, Clinical Research and Epidemiology (LBRCE), Faculty of Medicine and Pharmacy, Mohammed V University, Rabat, Morocco

Biology Department, Higher School of Teachers (ENS), Mohammed V University in Rabat, Rabat, Morocco

El Hassan Sakar Biology, Ecology and Health Lab, Department of Biology, Faculty of Science, Abdelmalek Essaadi University, Tetouan, Morocco

Nour Shaheen Faculty of Medicine, Alzexandria University, Alexandria, Egypt

Rana I. Soliman Biotechnology and Biomolecular Chemistry Department, Faculty of Science, Cairo University, Giza, Egypt

Nourhan S. Sultan Biotechnology and Biomolecular Chemistry Department, Faculty of Science, Cairo University, Giza, Egypt

Batoul Taher Faculty of Medical Sciences, Lebanese University, Beirut, Lebanon

Medical Learning Skills Academy, Beirut, Lebanon

Lahcen Tamegart Biology, Ecology and Health Laboratory, Department of Biology, Faculty of Science, Abdelmalek Essaadi University, Tetouan, Morocco

See Jia Whei General Medicine, Sriwijaya University, Palembang, Indonesia

Mohamed Y. Zaky Molecular Physiology Division, Zoology Department, Faculty of Science, Beni-Suef University, Beni-Suef, Egypt

Part I
Understanding Traumatic Brain Injury (TBI)

Chapter 1
Defining and Biomechanics of TBI

Mohamed M. Ammar, Nourhan S. Sultan, Rewida R. Hamed, and Rana I. Soliman

Abstract Globally, traumatic brain injury (TBI) is a significant public health concern. Its annual impact in certain countries amounts to 2% of the population. It causes approximately two million visits to emergency rooms, of which 15% necessitate hospitalization. Despite therapeutic advancements, substantial long-term neurological deficits persist in 35% of survivors, and mortality rates remain high at 20% within this group. It is believed that this delay in achieving better outcomes is partially attributable to the limited implementation of standard treatment protocols. This chapter's objectives are to delve deeper into the field's history, categories, symptoms, biomechanics, and promising future directions.

Keywords Traumatic brain injury (TBI) · Pathophysiology · Primary injury · Secondary injury · Biomechanics

1.1 Introduction

Traumatic brain injury (TBI) is a significant contributor to both morbidity and mortality, with an estimated ten million cases occurring annually worldwide and over 1.4 million cases in the United States (Langlois et al. 2006). The implementation of an effective trauma system and adherence to well-researched trauma guidelines

M. M. Ammar
Microbiology and Biochemistry Program, Faculty of Science, Benha University-Obour Campus, Benha, Egypt
e-mail: mohammed401517@fsc.bu.edu.eg

N. S. Sultan (✉) · R. I. Soliman
Biotechnology and Biomolecular Chemistry Department, Faculty of Science, Cairo University, Giza, Egypt
e-mail: nourhansabri@gstd.sci.cu.edu.eg; 202030051@std.sci.cu.edu.eg

R. R. Hamed
Chemistry Department, Faculty of Science, Helwan University, Helwan, Egypt
e-mail: ruwaida.ramadan.hamid702@science.helwan.edu.eg

W. Mohamed (ed.), *Nutrition and Traumatic Brain Injury (TBI)*, Nutritional Neurosciences, https://doi.org/10.1007/978-981-97-6341-2_1

have both contributed to enhanced outcomes following traumatic brain injuries (TBIs) (Stocchetti and Maas Andrew 2014). Additionally, our comprehension of the pathophysiology of TBI has evolved, which has led to the refinement of acute clinical management for patients with TBI. Furthermore, developments in critical care have significantly contributed to improved outcomes for TBI patients (Khellaf et al. 2019).

In this chapter, we describe the background of this disease, its pathophysiology, the animal model used for TBI, biomechanics, prospects, and the treatment of this disease.

1.2 History of TBI

Ancient beliefs about head injuries may be prehistoric (High et al. 2005). Battleground tomb skulls with holes bored above fractures show that trepanation might have alleviated TBI (Catanese 2007). Ancient Mesopotamians recognized that head injuries might cause convulsions, loss of consciousness, and impairment of vision, hearing, or speech (Scurlock and Andersen 2005). Edwin Smith's Papyrus classified brain traumas as well as signs by presentation and tractability (Sanchez and Burridge 2007). Hippocrates and other ancient Greek physicians believed the brain was the center of the mind after witnessing head injuries (Binder 1986).

For cranial injuries, the Middle Ages, in addition to Renaissance specialists, continued trepanation (Binder 1986). Doctors also classified head trauma signs during the Middle Ages, and the term "concussion" gained increased usage (Zillmer et al. 2006). Berengario da Carpi meticulously documented concussion symptoms for the first time in the sixteenth century (Binder 1986).

Intracranial pressure, with no damage to the skull, was initially postulated to be the underlying cause of pathology following TBI in the eighteenth century. As a treatment, skull opening to alleviate tension was suggested after this concept was validated at the close of the nineteenth century (Catanese 2007).

Psychosis was identified as a syndrome associated with traumatic brain injury (TBI) in the nineteenth century (Arciniegas et al. 2003). Post-concussion syndrome was initially attributed to psychological factors rather than brain tissue disruption, which sparked a debate (Binder 1986). The debate persists to this day.

Phineas Gage was struck by a long iron rod directly in the head, which led to the destruction of either or both of his frontal lobes in the accident. He may have been the first to report personality changes following a brain injury. Since then, there have been numerous reports of personality changes following a brain injury (Cummings 1993; Eslinger and Damasio 1985; Kim 2002; Mattson and Levin 1990; McDonald et al. 2003; Milders et al. 2003; Stone et al. 1998).

Technological advances in the twentieth century enhanced treatment and diagnosis, including imaging techniques such as CT and MRI. In addition, diffusion tensor imaging is in the twenty-first century. The advent of intracranial pressure monitoring in the 1950s is widely acknowledged as the inception of the "modern era" in head injury research (Marshall 2000). Rehabilitation was uncommon, and TBI mortality was high prior to the twentieth century; however, advancements in health services through World War I enabled rehabilitation and decreased fatalities (High et al. 2005). The first TBI rehabilitation facilities probably occurred during World War I (High et al. 2005). The extensive number of TBIs caused by explosives during World War I provided researchers with valuable information regarding brain function localization (Jones et al. 2007). Soldiers returning from war frequently sustain blast-related injuries; research indicates that the symptoms of these TBIs are essentially identical to those of TBIs caused by direct hits to the brain (Belanger et al. 2009).

Since the 1970s, when knowledge of TBI as an important care issue emerged, significant advances have been made in the study of brain trauma. These advances include the identification of primary and secondary brain injuries (Catanese 2007). Standardized guidelines for TBI treatment were established and widely distributed during the 1990s. These guidelines included protocols pertaining to various aspects, such as medication administration and intracranial pressure management (Marshall 2000). TBI survival has been enhanced by research since the early 1990s; that time was dubbed the "Decade of the Brain" due to developments in neurological studies (Catanese 2007).

1.3 Definition

Traumatic brain injury (TBI) is defined as an external mechanical force that inflicts a nondegenerative, noncongenital insult to the brain, which may result in temporary or permanent impairment of physical, mental, and cognitive functions, accompanied by a reduced or altered state of consciousness (Burke 2008). The number of males who suffer brain injuries is twice that of females. There are two age groups with the highest incidence rates: 15 to 24 years old and 75 years and older (Valera et al. 2021). Brain injuries are frequently induced by violent activities, sports, motor vehicle collisions, and falls (Ahmed et al. 2017).

1.4 Types of TBI

TBIs can be classified into the following types:

*1.4.1 Concussion**

This mild TBI with no physical impairment often follows a nonpenetrating TBI. Acceleration/deceleration forces from head impacts generally cause it. An individual's mental state can range from perplexity to loss of consciousness. A conventional CT or MRI scan cannot diagnose this. Diffusion tensor and functional MRI may detect concussions early (Laker 2011; Voss et al. 2015).

Second impact syndrome: After a concussion, an athlete may play without fully recuperating and suffer another injury, which can lead to malignant cerebral edema within a short period (McLendon et al. 2016).

Chronic traumatic encephalopathy (CTE): It develops after repeated mild TBI. Sportspeople often have psychological disorders, suicidal thoughts, attention difficulties, and issues with memory and executive function (Lenihan and Jordan 2015).

1.4.2 Extra-axial Hematoma

Subdural hematomas (SDH) and epidural hematomas (EDH) are both classified as extra-axial hematomas (Aromatario et al. 2021). EDH is typically acute and caused by hemorrhage from the middle meningeal artery and its branches or a fracture. Acute or chronic SDH may occur as a result of bruising in a bridging vein (Kulesza et al. 2021).

1.4.3 Contusion

Brain contusions may manifest as a coup or contrecoup (Rodríguez 2016). Contrary to coup contusions, which occur at the impact region, contrecoup lesions generally manifest in the basis-frontal and anterior temporal lobes on the contralateral side of the impact site (Adatia et al. 2021).

1.4.4 Traumatic Subarachnoid Hemorrhage (SAH)

Trauma is the leading cause of subarachnoid hemorrhage, which is produced by the rupture of small capillaries and subsequent blood ingress into the subarachnoid space. SAH, which is a consequence of aneurysmal rupture, manifests in the basal cisterns, whereas it frequently occurs over the convexity (Gutierrez et al. 2004).

1.4.5 Diffuse Axonal Injury (DAI)

Any shearing, stretching, or twisting injury to the axons of neurons may cause this, which may be the cause of mild-to-moderate traumatic brain injury. This effect is mostly observed at the interface between gray and white matter, where axons from neurons enter white matter that is less fluid-filled, thick, and fatty (myelinated). Stretching of the neural axon by such shearing pressures can result in axonal swelling, elevated permeability, calcium influx, separation, and axonal death. This is due to the cytoskeleton being damaged. Autopsies usually reveal diffuse laminar necrosis (Georges and Das 2023; Saboori and Walker 2019).

1.5 Pathophysiology of TBI

TBI occurs when the brain is initially subjected to an external mechanical force that impacts the skull. The damage process encompasses both immediate and delayed stages. During the initial stage, there is a notable occurrence of brain tissue degradation, encompassing neurons, glia, and endothelial cells that constitute the blood-brain barrier. During the delayed phase, there is a release of "toxins" from cells that have been harmed, which subsequently initiates cascades in adjacent cells, so intensifying the initial injury (Freire et al. 2023; Reis et al. 2017).

The brain undergoes cellular and molecular alterations subsequent to TBI, encompassing an inflammatory reaction, excitotoxicity, and oxidative stress. These modifications give rise to modifications in cerebral tissue, impacting the nervous system and leading to temporary or enduring functional impairment. Gaining a comprehensive understanding of these alterations is essential for enhancing the provision of care, rehabilitation, and overall well-being for those afflicted by TBI while simultaneously mitigating the economic consequences associated with this condition (Freire et al. 2023).

TBI has a profound effect on both the functioning and structure of the brain, resulting in enduring morbidity and mortality on a global scale. The disruption of tissue equilibrium caused by the impairment of the blood-brain barrier is a crucial element in the detrimental occurrences linked to TBI. Brain function and structure impairment following a TBI can lead to significant restrictions on individuals' ability to work, impacting their health, social security systems, and general quality of life (Freire et al. 2023).

A diverse array of symptoms can manifest in individuals with TBI, impacting several dimensions of their cognitive, physical, and emotional well-being. Several prevalent symptoms include cognitive disabilities such as common cognitive symptoms of TBI include memory loss, disorientation, and challenges with concentration and attention. These limitations can have a substantial influence on an individual's capacity to carry out routine activities and uphold their autonomy. Common physical signs of TBI include headaches, dizziness, and balance issues. The severity of these symptoms can be incapacitating and may necessitate medical intervention for effective management. Common emotional changes associated with TBI include irritability, depression, and anxiety. These modifications have the potential to adversely affect an individual's overall well-being and interpersonal connections, necessitating the provision of psychological assistance (Fife and Kalra 2015). Figure 1.1 illustrates the pathophysiology of TBI.

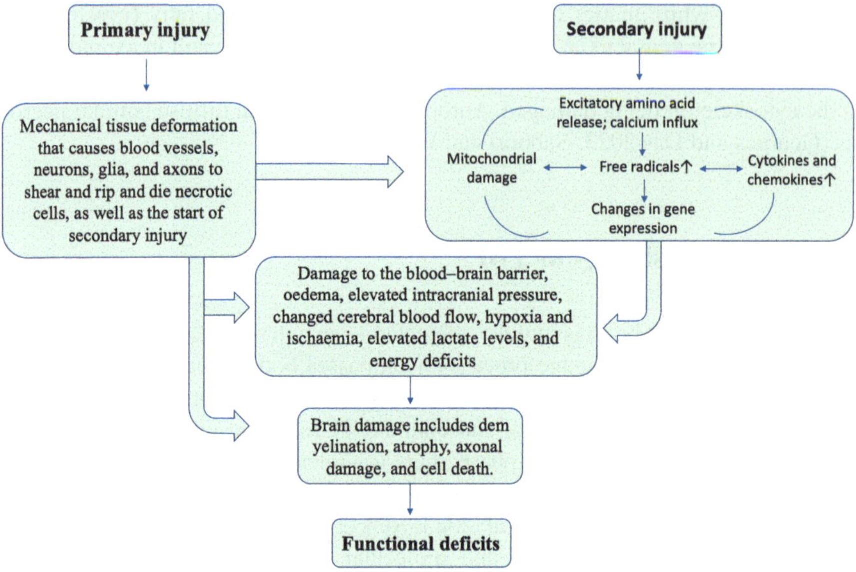

Fig. 1.1 Simplified pathophysiology of TBI

1.5.1 Primary Injury

Within the realm of TBI, the primary injury refers to the initial damage that directs its influence on the brain tissue. The possible outcomes include direct damage to the brain tissue, such as bruises, cuts, or bleeding within the skull. Additionally, it is important to note that TBI has the potential to result in cranial fractures or intracranial hemorrhaging, hence exacerbating the severity of the injury. The occurrence of diffuse axonal injury is a significant consequence of the initial injury, which arises from abrupt acceleration or deceleration forces that induce the stretching and fracture of the nerve fibers in the brain. Acquiring a comprehensive understanding of the mechanisms and consequences associated with the initial injury in cases of TBI is crucial in order to effectively diagnose, treat, and manage individuals affected by this illness. Gaining insight into the mechanisms of harm in TBI is essential for formulating efficacious treatment strategies and enhancing patient results. Healthcare workers must possess a comprehensive understanding of the diverse range of potential consequences that may occur as a result of TBI, including cognitive deficits, emotional disorders, and physical limitations (Block et al. 2022; Sen et al. 2020).

1.5.2 Secondary Injury

Secondary injury in the context of TBI refers to the subsequent harm that occurs following the first impact and primary injury. It encompasses a sequence of occurrences that can exacerbate the initial harm and lead to further neurological problems. The secondary-injury pathways in TBI involve a range of mechanisms, such as inflammation, excitotoxicity, oxidative stress, and the formation of edema. A comprehensive understanding and effective control of secondary-injury pathways are crucial in the treatment of TBI to minimize additional harm to brain tissue and improve patient outcomes. Healthcare practitioners have the opportunity to mitigate the level of damage and enhance the overall prognosis for persons with TBI by focusing on these specific pathways. The statement mentioned above underscores the significance of timely intervention and thorough treatment approaches in mitigating the consequences of secondary damage on cognitive abilities (Mishra et al. 2016; Paul Andre et al. 2022).

1.6 TBI Models in Animals

Replicating clinically relevant and standardized patterns of brain injury that are highly linked with motor and cognitive disability is the ultimate goal of employing TBI models in the laboratory. It is possible to ascertain the appropriate tissue damage by adjusting specific parameters related to the injury's intensity. The researcher will be able to correlate changes in behavior, histology, and molecules across time by varying these variables. Rats with substantial traumatic injuries seldom experience the coma-like state commonly seen in individuals with diffuse traumatic brain injury (DBI). However, they do experience noticeable tissue degeneration and motor impairment. This is a drawback of these popular models, even if a moderate TBI matches a human brain concussion (Gennarelli et al. 1982; Smith et al. 2000). In order to make certain that treatments other than the actual trauma, like surgery and anesthesia, won't affect the outcome parameters, the researcher will mix the TBI groups with animals who were sham-operated as controls. It has been demonstrated that some aesthetics, such as isoflurane, may actually protect damaged brain tissue (Statler et al. 2006).

Many factors can lead to traumatic brain injury in animals. Static loading and progressive, gradual pressures exerted on the head are two examples of these. These forces typically only cause major injuries once they are experienced over a lengthy distance. As an alternative, the most typical mechanism is a quick dynamic loading, which might happen when the head collides with an object or as an impulsive loading brought on by the head stopping abruptly in mid-motion without necessarily colliding with anything. Because of the inertia it creates, the latter can result in translational, rotational, and angular movements that apply shearing forces to the structural elements of the brain.

One of the main objectives of using animal models in research is to comprehend the essential clinical challenges resulting from TBI, among many other research concerns. Following an initial trauma, there is continuous and prolonged neurodegeneration that manifests as increased permeability of the cerebral endothelium, loss of viable tissue due to cell death, activating and developing blood leukocytes into the brain's parenchyma, and enlargement of lesions. These are the chronological phases that follow brain injury. Extensive documentation of these descriptions can be found elsewhere.

Edema, or increased mass brain volume, is the macroscopic outcome of cell swelling and vascular fluid leakage into the brain. This localized constriction of neighboring arteries may result in further ischemia insults. Complex biochemical processes, such as the release of excitotoxic neurotransmitters, immunological mediators, oxygen or nitrogen radicals, and metabolic crises, cause these drastic morphological changes (Marcoux et al. 2008; McLntosh et al. 1996; Morganti-Kossmann et al. 2008).

1.7 Biomechanics of TBI

A biomechanical reconstruction of a brain injury caused by trauma involving soft tissue destruction is presented. A variational biomechanical simulation of delicate cellular tissues has been employed to foresee elastic deformation-induced damage to axonal cells and bubble damage. The compound's action is divided into two distinct components: elastoplastic and viscous, involving rate effects, shearing and permeable adaptability, and boundary viscoelasticity. The mechanical harm done to brain tissue can be characterized as dimensional (compression or tension) or shear-type. The study comprises the use of finite element simulations of neurological impairment, with a concentration on frontal and oblique head impacts with external objects. Damage to the tissue is going to be confined, extended, intense, and reversible or irreversible. The work's future directions, including the relationship between mechanical injury and physiologic dysfunction in the brain, as well as the connection to important medical and scientific worries, are discussed. Biomechanical modeling of traumatic brain injuries involves the development of difficult constitutive equations that take into consideration massive types, both rate and time effects, and effective decay models. Modern mechanical scholarship focuses on hyperelastic or finite viscoelastic formulations that adjust for very small perturbations that depart from optimum temperatures (El Sayed et al. 2008). As a result, establishing a relationship between neurological impairment and mechanical dosage allows for the deduction of brain injury processes and associated impact tolerance. The information gathered is crucial for preventing and diagnosing brain injuries. This study provides an overview of our most recent head injury research, which has focused on establishing brain damage biomechanics through the development of a thorough and validated mathematical framework. To accomplish our goal, we created an elaborate three-dimensional in-nature computational head human model, finite aspect swine head theories, and positive gravity accelerometer sensors that were used for assessing skull kinematics and mental pressure. So far, this research has yielded a predicted and partially validated mathematical model of the skull, demonstrating a clear correlation between brain mechanical response and experimentally observed damage (King et al. 1995). The relationship between mechanical forces and neuropathology is crucial to comprehending traumatic brain damage. Large shear loads over interaction produce the destruction of white material processes, which leads to synaptic trauma, which has been shown to be an excellent predictor of long-term medical consequences.

Nevertheless, the association underlying the biomechanical forces and sequences of white matter wreckage attributed to persisting propagated MRI irregularities is not totally understood. This affects its ability to accurately predict the nature of harm to the head and design proper protection. Recognizing the association between biomechanics power patterns and injury to the axonal site is important for awaiting the implications of numerous kinds of head injuries and building brain protection innovations (such as balloons and helmets) that prevent axonal injury. White matter networks are very susceptible to mechanical trauma via various methods (Donat et al. 2021). Injuries may trigger micro- and macro-failures. For biomechanical injury risk factors, it must be taken into account to look at how a circumstance either led to a mechanical load that was above what can be tolerated under typical circumstances (Mechanism 1) or minimized the degree of acceptance throughout time to a level at which an ordinary mechanical load cannot be anymore understood or the usual function becomes compromised (Mechanism 2). The technique could also satisfy both parameters for harm, substantial loads, and the possibility of endurance lowering (acute or prolonged-term) by nanoscale trauma (Mechanism 3) (Tierney 2021). Biomechanical research that incorporates computational and real-world rebuilding has contributed to the knowledge of the qualities that distinguish both minor and serious types of TBI. Additional studies on everyday life TBI from falls and slips offer information on the mechanisms driving this sort of brain injury. However, these researchers were not assessing whether they were appropriate for biomechanical measures associated with the medical diagnosis of TBI for a number of emotions but rather focusing on shocks or falls from upright posture.

As a consequence, there needs to be more clarity on the potential constraints of these measures for electrical power greater than upright landings (Post et al. 2017). Micromechanical actions of neuro-axonal components happen consecutively in tandem, featuring harm and "relaxation" periods in multiple regions of the brain. An original combined computing paradigm is developed, coupling brain-scale kinematics with micromechanical destruction of axonal and synaptic components. The macroscale model encompasses burst-wave interactions, the effects of blast waves on human fatness, and force waves spreading throughout the human body, which includes brain or body hemodynamic interactions such as neck rotation, movement in the case of translocation, and contact with the floor. The mathematical mesh will be built, which includes both the place of the explosion for computational fluid hydrology (CFD) examination of the blast wave as well as the mind and body for utilizing the finite element method, or FEM, in order to influence auto mechanics calculations (Post et al. 2017).

1.8 Future Prospective and Potential Therapeutics

A major attempt has been made to develop neuroprotective agents. The primary goal of the next section is to address TBI. Pathogenesis as well as known and prospective new therapeutic options for managing TBI-related inflammatory events and behavioral impairments. TBI management focuses on prevention and gets closer, specifically methods for organizing and targeting specific mechanisms linked to the complex secondary-injury pathway. In the past, neurotoxic arrivals have been heavily influenced by a Eurocentric orientation, with changes in neuronal-based damage pathways serving as the primary or sole concentration of preventative interventions. As a result, it is essential to identify injuries that affect the endothelium, microglia, astroglia, oligodendroglia, and progenitor cells. Recent protective effects on neuron efforts define this multifaceted structure and interaction, concentrating on treatments that encourage nonneuronal cell regrowth and optimum performance while stopping the root cause pathway of neuron cell destruction (Crupi et al. 2020). Because previous injuries in TBI frequently result in acute injuries and the death of necrotic cells, which are expected to be reversible, medical treatments focus on regulating the injury site and guarding it against additional harm. As already mentioned, secondary wounds can be triggered by any number of dangerous factors that develop over time. This opens the door to medical intervention in occurrences that might end in the death of neurons and glial cells at the point of injury, such as ongoing inflammatory processes, excitotoxicity, oxidative harm, and cell death from apoptosis. A lot of studies have been accomplished in pursuit of more knowledge of the underlying causes of later-stage brain injuries in order to produce improved rehabilitation tactics that target many stages (Ng and Lee 2019). Recognizing TBI pathology will facilitate the formulation of efficient drugs or combinations of medications that focus on multiple secondary-injury pathways but remain safe during the best treatment intervals. This method has the potential to improve neuroprotective treatment translation from lab to bedside. Some noteworthy therapy techniques that have moved on to the clinical research stage for brain damage include tranexamic acid, cytidine 5′-diphosphocholine (CDP)-choline, methylphenidate, NA-1(Tat-NR2B9c), cannabidiol (CBD), and harmless remote ischemic conditioning (RIC). However, effective medicines for TBI remain scarce, with therapy primarily comprising emergency operations, intracranial pressure (ICP) maintenance, and rehabilitation therapies. TBI's diverse nature and complex pathophysiology necessitate a mix of therapies to alleviate subsequent injury and post-traumatic impairments (Jarrahi et al. 2020).

1.9 Conclusion

TBI is a significant health concern and challenge. There are numerous promising research avenues to enhance neurointensive care protocols. These avenues establish empirical support for surgical intervention and apply prospective neuroprotection pharmacotherapies to human pathology. Different types of pathology mean that a single treatment cannot affect all of them. This is why personalized medicine and combination therapy are so important in TBI. It may explain why there has been a considerable number of failures to establish a benefit in this particular pathology, contrary to traditional randomized control trials.

References

Adatia K, Newcombe VFJ, Menon DK (2021) Contusion progression following traumatic brain injury: a review of clinical and radiological predictors, and influence on outcome. Neurocrit Care 34(1):312–324. https://doi.org/10.1007/s12028-020-00994-4

Ahmed S, Venigalla H, Mekala HM, Dar S, Hassan M, Ayub S (2017) Traumatic brain injury and neuropsychiatric complications. Indian J Psychol Med 39(2):114–121. https://doi.org/10.4103/0253-7176.203129

Arciniegas DB, Harris SN, Brousseau KM (2003) Psychosis following traumatic brain injury. Int Rev Psychiatry 15(4):328–340. https://doi.org/10.1080/09540260310001606719

Aromatario M, Torsello A, D'Errico S, Bertozzi G, Sessa F, Cipolloni L, Baldari B (2021) Traumatic epidural and subdural hematoma: epidemiology, outcome, and dating. Medicina 57(2). https://doi.org/10.3390/medicina57020125

Belanger HG, Kretzmer T, Yoash-Gantz R, Pickett T, Tupler LA (2009) Cognitive sequelae of blast-related versus other mechanisms of brain trauma. J Int Neuropsychol Soc 15(1):1–8. https://doi.org/10.1017/S1355617708090036

Binder LM (1986) Persisting symptoms after mild head injury: a review of the postconcussive syndrome. J Clin Exp Neuropsychol 8(4):323–346. https://doi.org/10.1080/01688638608401325

Block H, Hunter SC, Bellon M, George S (2022) Implementing a behavior management approach in the hospital setting for individuals with challenging behaviors during acute traumatic brain injury. Brain Inj 36(9):1176–1186. https://doi.org/10.1080/02699052.2022.2110941

Burke D (2008) Traumatic brain injury. In: Frontera WR, Silver JK, Rizzo TD (eds) Essentials of physical medicine and rehabilitation, 2nd edn. W.B. Saunders, Philadelphia, pp 913–918

Catanese C (2007) Traumatic brain injury: methods for clinical and forensic neuropsychiatric assessment, CRC Press

Crupi R, Cordaro M, Cuzzocrea S, Impellizzeri D (2020) Management of traumatic brain injury: from present to future. Antioxidants 9(4). https://doi.org/10.3390/antiox9040297

Cummings JL (1993) Frontal-subcortical circuits and human behavior. Arch Neurol 50(8):873–880. https://doi.org/10.1001/archneur.1993.00540080076020

Donat CK, Yanez Lopez M, Sastre M, Baxan N, Goldfinger M, Seeamber R, Ghajari M et al (2021) From biomechanics to pathology: predicting axonal injury from patterns of strain after traumatic brain injury. Brain 144(1):70–91. https://doi.org/10.1093/brain/awaa336

El Sayed T, Mota A, Fraternali F, Ortiz M (2008) Biomechanics of traumatic brain injury. Comput Methods Appl Mech Eng 197(51):4692–4701. https://doi.org/10.1016/j.cma.2008.06.006

Eslinger PJ, Damasio AR (1985) Severe disturbance of higher cognition after bilateral frontal lobe ablation. Neurology 35(12):1731. https://doi.org/10.1212/WNL.35.12.1731

Fife TD, Kalra D (2015) Persistent vertigo and dizziness after mild traumatic brain injury. Ann N Y Acad Sci 1343(1):97–105. https://doi.org/10.1111/nyas.12678
Freire MAM, Rocha GS, Bittencourt LO, Falcao D, Lima RR, Cavalcanti JRLP (2023) Cellular and molecular pathophysiology of traumatic brain injury: what have we learned so far? Biology 12(8). https://doi.org/10.3390/biology12081139
Gennarelli TA, Thibault LE, Adams JH, Graham DI, Thompson CJ, Marcincin RP (1982) Diffuse axonal injury and traumatic coma in the primate. Ann Neurol 12(6):564–574. https://doi.org/10.1002/ana.410120611
Georges A, Das MJ (2023) Traumatic brain injury. StatPearls Publishing, Treasure Island (FL)
Gutierrez G, Reines H, Wulf-Gutierrez ME (2004) Clinical review: hemorrhagic shock. Crit Care 8(5):373–381. https://doi.org/10.1186/cc2851
High WM Jr, Sander AM, Struchen MA, Hart KA (2005) Rehabilitation for traumatic brain injury. Oxford University Press
Jarrahi A, Braun M, Ahluwalia M, Gupta RV, Wilson M, Munie S, Vaibhav K et al (2020) Revisiting traumatic brain injury: from molecular mechanisms to therapeutic interventions. Biomedicines 8(10). https://doi.org/10.3390/biomedicines8100389
Jones E, Fear NT, Wessely S (2007) Shell shock and mild traumatic brain injury: a historical review. Am J Psychiatry 164(11):1641–1645. https://doi.org/10.1176/appi.ajp.2007.07071180
Khellaf A, Khan DZ, Helmy A (2019) Recent advances in traumatic brain injury. J Neurol 266(11):2878–2889. https://doi.org/10.1007/s00415-019-09541-4
Kim E (2002) Agitation, aggression, and disinhibition syndromes after traumatic brain injury. NeuroRehabilitation 17:297–310. https://doi.org/10.3233/NRE-2002-17404
King AI, Ruan JS, Zhou C, Hardy WN, Khalil TB (1995) Recent advances in biomechanics of brain injury research: a review. J Neurotrauma 12(4):651–658. https://doi.org/10.1089/neu.1995.12.651
Kulesza B, Mazurek M, Rams Ł, Nogalski A (2021) Acute epidural and subdural hematomas after head injury: clinical distinguishing features. Indian J Surg 83(1):96–104. https://doi.org/10.1007/s12262-020-02304-w
Laker SR (2011) Epidemiology of concussion and mild traumatic brain injury. PM&R 3(10S2):S354–S358. https://doi.org/10.1016/j.pmrj.2011.07.017
Langlois J, Rutland-Brown W, Wald M (2006) The epidemiology and impact of traumatic brain injury: a brief overview. J Head Trauma Rehabil 21:375–378. https://doi.org/10.1097/00001199-200609000-00001
Lenihan MW, Jordan BD (2015) The clinical presentation of chronic traumatic encephalopathy. Curr Neurol Neurosci Rep 15(5):23. https://doi.org/10.1007/s11910-015-0541-5
Marcoux J, McArthur DA, Miller C, Glenn TC, Villablanca P, Martin NA, Vespa PM et al (2008) Persistent metabolic crisis as measured by elevated cerebral microdialysis lactate-pyruvate ratio predicts chronic frontal lobe brain atrophy after traumatic brain injury*. Crit Care Med 36(10):2871–2877
Marshall LF (2000) Head injury: recent past, present, and future. Neurosurgery 47(3):546–561
Mattson AJ, Levin HS (1990) Frontal lobe dysfunction following closed head injury. A review of the literature. J Nerv Ment Dis 178(5):282–291
McDonald S, Flanagan S, Rollins J, Kinch J (2003) TASIT: a new clinical tool for assessing social perception after traumatic brain injury. J Head Trauma Rehabil 18(3):219–238
McLendon LA, Kralik SF, Grayson PA, Golomb MR (2016) The controversial second impact syndrome: a review of the literature. Pediatr Neurol 62:9–17. https://doi.org/10.1016/j.pediatrneurol.2016.03.009
McLntosh TK, Smith DH, Meaney DF, Kotapka MJ (1996) Neuropathological sequelae of traumatic brain injury: relationship to neurochemical and biomechanical. Lab Investig 74(2):315
Milders M, Fuchs S, Crawford JR (2003) Neuropsychological impairments and changes in emotional and social behaviour following severe traumatic brain injury. J Clin Exp Neuropsychol 25(2):157–172. https://doi.org/10.1076/jcen.25.2.157.13642

Mishra V, Skotak M, Schuetz H, Heller A, Haorah J, Chandra N (2016) Primary blast causes mild, moderate, severe and lethal TBI with increasing blast overpressures: experimental rat injury model. Sci Rep 6(1):26992. https://doi.org/10.1038/srep26992

Morganti-Kossmann C, Satgunaseelan L, Bye N, Kossmann T (2008) Modulation of immune response by head injury. Injury 38:1392–1400. https://doi.org/10.1016/j.injury.2007.10.005

Ng SY, Lee AYW (2019) Traumatic brain injuries: pathophysiology and potential therapeutic targets. Front Cell Neurosci 13

Paul Andre P-B, William Thomas W, Graeme Philip H, Gordon MM (2022) Primary repair of multiligament knee injury with InternalBrace ligament augmentation. BMJ Case Rep 15(1):e247173. https://doi.org/10.1136/bcr-2021-247173

Post A, Taylor K, Hoshizaki TB, Brien S, Cusimano MD, Marshall S (2017) A biomechanical analysis of traumatic brain injury for slips and falls from height. Trauma 21(1):27–34. https://doi.org/10.1177/1460408617721564

Reis C, Gospodarev V, Reis H, Wilkinson M, Gaio J, Araujo C, Zhang JH et al (2017) Traumatic brain injury and stem cell: pathophysiology and update on recent treatment modalities. Stem Cells Int 2017:6392592. https://doi.org/10.1155/2017/6392592

Rodríguez AF (2016) Traumatic brain injuries. In: Hoffmann Nunes R, Abello AL, Castillo M (eds) Critical findings in neuroradiology. Springer International Publishing, Cham, pp 211–217

Saboori P, Walker G (2019) Brain injury and impact characteristics. Ann Biomed Eng 47(9):1982–1992. https://doi.org/10.1007/s10439-019-02199-z

Sanchez GM, Burridge AL (2007) Decision making in head injury management in the Edwin Smith papyrus. Neurosurg Focus 23(1):1–9

Scurlock JA, Andersen B (2005) Diagnoses in Assyrian and Babylonian medicine: ancient sources, translations, and modern medical analyses. University of Illinois Press

Sen T, Saha P, Gupta R, Foley LM, Jiang T, Abakumova OS, Sen N et al (2020) Aberrant ER stress induced neuronal-IFNβ elicits white matter injury due to microglial activation and T-cell infiltration after TBI. J Neurosci 40(2):424–446

Smith D, Nonaka M, Miller R, Leoni M, Chen X-H, Alsop D, Meaney D (2000) Immediate coma following inertial brain injury dependent on axonal damage in the brainstem. J Neurosurg 93:315–322. https://doi.org/10.3171/jns.2000.93.2.0315

Statler KD, Alexander H, Vagni V, Holubkov R, Dixon CE, Clark RSB, Kochanek PM et al (2006) Isoflurane exerts neuroprotective actions at or near the time of severe traumatic brain injury. Brain Res 1076(1):216–224. https://doi.org/10.1016/j.brainres.2005.12.106

Stocchetti N, Maas Andrew IR (2014) Traumatic intracranial hypertension. N Engl J Med 370(22):2121–2130. https://doi.org/10.1056/NEJMra1208708

Stone VE, Baron-Cohen S, Knight RT (1998) Frontal lobe contributions to theory of mind. J Cogn Neurosci 10(5):640–656. https://doi.org/10.1162/089892998562942

Tierney G (2021) Concussion biomechanics, head acceleration exposure and brain injury criteria in sport: a review. Sports Biomech 1–29. https://doi.org/10.1080/14763141.2021.2016929

Valera EM, Joseph A-LC, Snedaker K, Breiding MJ, Robertson CL, Colantonio A, Bellgowan PSF et al (2021) Understanding traumatic brain injury in females: a state-of-the-art summary and future directions. J Head Trauma Rehabil 36(1):E1–E17

Voss JD, Connolly J, Schwab KA, Scher AI (2015) Update on the epidemiology of concussion/mild traumatic brain injury. Curr Pain Headache Rep 19(7):32. https://doi.org/10.1007/s11916-015-0506-z

Zillmer E, Schneider J, Tinker J, Kaminaris C (2006) A history of sports-related concussions. Sports neuropsychology: Assessment and management of traumatic brain injury, 17–42

Chapter 2
Probing Traumatic Brain Injury Pathophysiology: Lessons from the Zebrafish Model

Wael Mohamed and Firas Kobeissy

Abstract Traumatic brain injury (TBI) is a major public health issue worldwide, as its intricate pathophysiology makes it difficult to develop effective treatments. This chapter explores the use of the zebrafish (ZF) model as a powerful tool for studying the complexities of TBI pathophysiology. Zebrafish provide distinctive benefits, such as genetic manipulability, optical clarity, and swift embryonic growth, rendering them an optimal model organism for investigating neurotrauma. We investigate the molecular, cellular, and behavioural alterations observed in zebrafish after inducing traumatic brain injury (TBI), providing insights into the crucial pathways implicated in the response to injury, inflammation, neurodegeneration, and potential regeneration processes. In addition, we explore how knowledge acquired from studying zebrafish can contribute to translational research endeavours focused on creating innovative therapeutic approaches for traumatic brain injury (TBI) in humans. This chapter provides a thorough summary of the zebrafish model's contributions to our knowledge of TBI pathophysiology. It also emphasises the potential of this model to improve clinical management and treatment outcomes for patients with traumatic brain injuries.

Keywords TBI · Zebrafish · Larvae · Laser · Weight drop

W. Mohamed (✉)
Basic Medical Science Department, Kulliyyah of Medicine, International Islamic University Malaysia (IIUM), Kuantan, Pahang, Malaysia

Clinical Pharmacology Department, Menoufia Medical School, Menoufia University, Shebeen El-Kom, Egypt

F. Kobeissy
Department of Neurobiology and Neuroscience Institute Morehouse School of Medicine (MSM), Center for Neurotrauma, Multiomics and Biomarkers, Atlanta, GA, USA

W. Mohamed (ed.), *Nutrition and Traumatic Brain Injury (TBI)*, Nutritional Neurosciences, https://doi.org/10.1007/978-981-97-6341-2_2

2.1 Introduction

Traumatic brain injury (TBI) is a major health problem worldwide, yet we still don't have effective therapies since we don't fully grasp its complicated pathogenesis. TBI is a major issue in public health, with few treatment options and a poor understanding of its underlying causes. TBI is a significant contributor to death and disability on a global scale (Hay et al. 2016; Nguyen et al. 2016). Additionally, it is a significant contributing factor to the development of neurodegeneration and dementia, including chronic traumatic encephalopathy (CTE) (Gardner and Yaffe 2015).

TBIs are a significant health issue in the United States, accounting for 30% of all recorded fatalities attributable to injuries. In 2010, there were about 2.5 million documented occurrences of TBI in the United States. These cases included both isolated injuries and injuries that occurred with other forms of trauma (Taylor et al. 2017). It is believed that these estimates are far lower than the actual number of injuries since many injuries are probably not being recorded and/or people are not seeking medical attention (Corrigan et al. 2010). There has been a growing amount of research dedicated to studying the frequency, underlying mechanisms, and healing process of TBIs (Taylor et al. 2017). Consequently, TBIs are now being approached as a medical condition rather than just a traumatic incident (Masel and DeWitt 2010), which includes mild TBIs (mTBI), generally referred to as concussions. The first symptoms of a mild traumatic brain injury (mTBI) might include loss of consciousness, amnesia, headaches, and nausea (Len and Neary 2011). However, in 30% of instances, these symptoms can continue and develop into post-concussive syndrome (PCS; Lewine et al. 2007). Long-term effects of the condition include cognitive and memory problems (Vanderploeg et al. 2005) and motor limitations (De Beaumont et al. 2007). From a pathological perspective, injuries may be categorised as either primary or secondary. Primary injuries occur immediately after the impact and might include fractures, as well as elevated pressure and bleeding. Secondary injuries occur subsequent to the traumatic incident and may result in disturbances in the functioning of bodily systems at the molecular and cellular levels. This may manifest as issues with the release and absorption of neurotransmitters, scarring caused by astrocytes at the location of the damage, inflammation, and the death of neuronal and glial cells via necrosis and apoptosis (Mckee and Daneshvar 2015). TBI may occur due to direct physical injuries, abrupt changes in brain acceleration and deceleration, or shock wave effects, such as pressure waves caused by explosive bursts (Cruz-Haces et al. 2017). However, the main processes involved in both TBI and Chronic Traumatic Encephalopathy (CTE) have many similarities. These include the widespread presence of hyperphosphorylated Tau pathology, degeneration of axons, and loss of neurons (Hay et al. 2016; Ojo et al. 2016). The processes by which physical damage leads to the development of progressive Tau disease are yet unknown and suggest potential targets for therapeutic intervention. Advancements in this area are hindered by the limited availability of appropriate models. Inflicting physical damage on a cell culture is challenging and

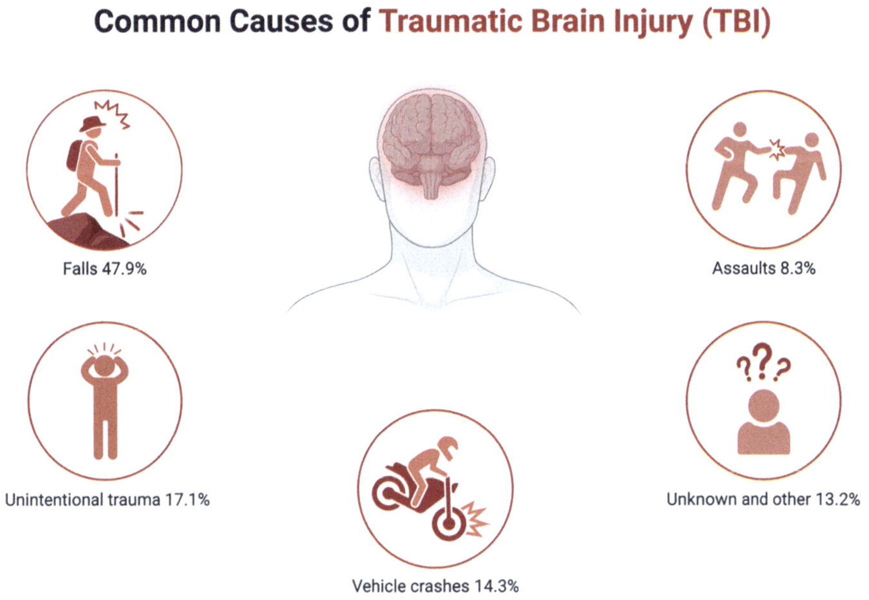

Fig. 2.1 Common causes of TBI

does not accurately reflect the intricate biopathology that encompasses several elements of brain function.

This chapter consolidates significant discoveries from studies utilising zebrafish (ZF) models of traumatic brain injury (TBI) and examines their impact on enhancing our comprehension of TBI pathogenesis and the development of therapeutic interventions (Fig. 2.1).

2.2 Zebrafish in TBI Research: A Model of Hope and Insight

The zebrafish (Danio rerio) has become a remarkable model organism for studying traumatic brain injury (TBI), providing distinct benefits for understanding the causes of harm and evaluating prospective treatments. Zebrafish (Danio rerio) are a promising model organism for researching TBI because they may be easily manipulated genetically, they mature quickly, and their neuroanatomy is comparable to that of mammals. Zebrafish TBI models provide a means to accurately manipulate injury parameters, enabling the creation of localised or widespread brain injuries using several methods, such as mechanical impact, exposure to high-pressure waves, and chemical assault (Burgess et al. 2010; McGinn et al. 2018). These models simulate important features of human TBI, such as damage to neurons, inflammation in the brain, disruption of the blood-brain barrier, and cognitive deficits (Shah et al. 2018; McGinn et al. 2020). The genetic tractability of zebrafish is crucial because it allows for specific manipulation of molecular pathways that are involved in the development of TBI. This manipulation helps in identifying new targets for therapeutic interventions (Sun et al. 2020).

Adult animals possess restricted neuroregenerative capacities after a central nervous system (CNS) lesion (Lieschke and Currie 2007). Zebrafish have the unique capacity to regenerate and repair brain tissue throughout their adult lives, as shown by März et al. (2011), Kishimoto et al. (2012), and Skaggs et al. (2014). Although many crucial processes of zebrafish neuroregeneration have been discovered (Kishimoto et al. 2012), there has been no comprehensive analysis of the whole transcriptome of adult zebrafish during neuroregeneration. Presently, zebrafish TBI research uses stab or lesion models to cause damage (Kishimoto et al. 2012; Skaggs et al. 2014). While both models are innovative for studying specific lesions in the zebrafish brain, they may be classified as moderate or severe models of TBI since they both entail the penetration of the epidermal layer and the blood-brain barrier (Saatman et al. 2008). Zebrafish embryos and larvae have a high level of optical transparency, which enables the observation of cellular and molecular reactions to injury in real time. This ability provides valuable information about dynamic processes, including neuroinflammation, neurogenesis, and axonal regeneration (Chen et al. 2009; Monteiro et al. 2021). In addition, the development of imaging techniques, such as two-photon microscopy and functional MRI, allows for detailed observation of both structural and functional alterations in the damaged zebrafish brain (Dreosti et al. 2015; Ahrens et al. 2013). The combination of zebrafish traumatic brain injury (TBI) models with high-throughput screening methods enables the efficient assessment of pharmacological substances and genetic modifiers for their capacity to protect the nervous system (Rao et al. 2021). In addition, drug discovery platforms that utilise zebrafish provide cost-effective and morally acceptable alternatives to conventional mammalian models. This facilitates the faster translation of pre-clinical research findings into clinical applications (Patten et al. 2017).

To summarise, zebrafish models of TBI offer a robust framework for studying the intricate pathways that contribute to TBI pathophysiology and assessing innovative therapeutic approaches. Further multidisciplinary research that utilises the advantages of the zebrafish model shows significant potential for enhancing outcomes for individuals with TBI.

2.3 Zebrafish Larvae: A Window into Traumatic Brain Injury Mechanisms

Traumatic brain injuries are a prominent source of mortality and impairment in younger individuals, as well as a significant predisposing factor for neurodegenerative disorders and dementia in older folks. They can arise from direct physical injuries, whiplash, or shockwaves, such as those generated by explosions (Cruz-Haces et al. 2017). However, traumatic experiences can also have repercussions that extend beyond the acute loss of life and damage to neurons. Specifically, they have the capacity to interfere with a protein called Tau, which typically preserves the integrity of numerous neuronal cells. During this occurrence, an atypical and excessively phosphorylated form of Tau builds up in cells and disseminates throughout the central nervous system by transforming healthy Tau proteins into detrimental variation (Johnson et al. 2013; Ojo et al. 2016; Zanier et al. 2018). The accumulation of Tau protein is a characteristic feature of diseases known as tauopathies, such as Alzheimer's disease and a degenerative brain disorder observed in athletes who frequently sustain head injuries. Scientists require animal models that are conveniently accessible and allow for the observation and manipulation of the proliferation of aberrant Tau proteins following brain damage. Rodent models, such as rats and mice, are available; nevertheless, they are costly and not optimal for visualising internal brain processes. In a recent publication in eLife, Ted Allison from the University of Alberta and his colleagues, including Hadeel Alyenbaawi from Alberta and Majmaah University as the first author, along with other researchers from Alberta and Pittsburgh, present novel zebrafish larvae models for tauopathies and traumatic brain injuries (Alyenbaawi et al. 2021). The initial model consists of transgenic zebrafish that have been genetically modified to express a 'Tau-green fluorescence protein (GFP) reporter' protein. This modification allows for direct observation of the propagation of the aberrant protein. In order to obtain this outcome, Alyenbaawi et al. genetically modified the animals to express a modified version of Tau protein fused with green fluorescence protein (GFP) in their neurons. Due to their transparency, the larvae allow for easy visualisation of their nervous system and the fluorescent Tau. Subsequently, the fish were subjected to the injection of aberrant mice Tau proteins, resulting in the aggregation of reporter Tau into mobile 'puncta', which are characteristic little dots observed in tauopathies. A greater number of puncta were detected when extracts from brains affected by Tau-linked disorders were injected into the larvae, as opposed to the injection of normal proteins. Alyenbaawi et al. developed a cost-effective and straightforward zebrafish model to study traumatic brain injury. The larvae were inserted into a sealed syringe, and a weight was then released onto the plunger, generating a shockwave that imitated the effects of blast injuries in people. Engaging in this routine for three consecutive days results in situations that resemble those encountered in repetitive sports injuries. The shockwave treatment in the Tau-GFP reporter larvae resulted in the formation of fluorescent puncta in the brain and spinal cord, which is in line with the development of Tau pathologies associated with traumatic brain injuries (Fig. 2.2). Previous studies have demonstrated that healthy mice developed a condition associated with Tau when they were given brain extracts from other mice that had suffered traumatic brain injuries (Zanier et al. 2018). More than 50% of individuals who have experienced a traumatic brain injury, particularly those who have been exposed to a blast injury, develop epileptic seizures. These seizures can trigger or worsen the advancement of illnesses associated with Tau protein. Seizure-like behaviours were observed in zebrafish larvae following traumatic brain injury, and the strength of these seizures was found to be directly proportional to the extent of

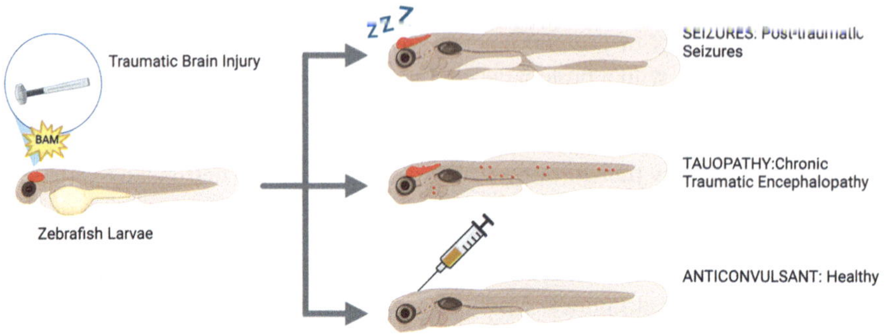

Fig. 2.2 TBI causes seizures and Tau-linked disorders in zebrafish larvae. Zebrafish larvae experience brain damage (left). Without therapy, many have seizures and develop a Tau-linked disease that causes protein accumulation and neuron death (top right). Treatment with anticonvulsants may prevent larvae against seizures and Tau-linked disease (bottom right)

aberrant Tau distribution. The drugs that either facilitated or inhibited seizures had a corresponding effect on the severity of the Tau-linked condition. This indicates that anticonvulsant medications may be beneficial in the treatment of brain injuries in a clinical setting (Fig. 2.2). Alyenbaawi et al. thoroughly acknowledged the constraints of their models, noting, for example that the Tau-GFP reporter might propagate in larvae even in the absence of abnormal Tau proteins. This could be due to the comparatively elevated levels of the Tau reporter in transgenic animals, highlighting the significance of regulating the expression levels of the transgene. With the exception of a recent model that utilised ultrasound, there have been limited ways thus far for simulating traumatic brain injury in zebrafish (Cho et al. 2020). Despite the fact that larvae have a more immediately observable central nervous system, a high throughput, and an ethical advantage compared to adults, this was particularly true for them. The models created by Alyenbaawi and his colleagues are a valuable contribution to our understanding of the mechanisms related to traumatic brain injury.

The larval zebrafish model, which has huge sample numbers, may help reduce some of the variability in the results. On a more positive note, the differences between individuals and the random occurrence of tauopathy in a model that can process a large amount of data might be used to get a new understanding of the spontaneous development of Alzheimer's disease or other tauopathies that are not caused by genetics. However, future research will also need the analysis of the biochemical and biophysical properties of human Tau inclusions in zebrafish, in comparison to patients or rodent models. In relation to these TBI techniques, further enhancements may yet have the potential to enhance the uniformity of the damage and decrease the apparent differences seen across people. The observed variability is present, but it is partially counterbalanced by the ability to obtain large sample sizes. These TBI methods provide significant benefits in terms of genetic and drug accessibility in high-throughput formats, while still maintaining the necessary complexity of in vivo conditions to study the causes and treatments of diseases. Additionally, it is necessary to determine whether the mechanisms we have discovered are present in all types of TBI. This model addresses a gap in the research by providing a unique 'closed-head' TBI model, which is different from most animal models that involve removing parts of the skull before causing brain injury. This model is particularly applicable to brain damage occurring in the human foetus, such as during vehicle crashes or domestic violence, due to its consideration of the developmental stage and aqueous medium. Furthermore, it is likely that the damage described here affects not just the brain but also other tissues, such as the spinal cord. This factor should be taken into account while analysing the data. Additional research is required to understand the impact of our explosion damage on a smaller scale, such as the brain of a larva measuring less than 500 mm. Therefore, the information obtained may have varying degrees of applicability. Ultimately, this model examines only larval fish. Although there are several logistical and ethical benefits, this approach restricts investigation to immediate impacts that take place during a span of few days. Findings from this research, after being improved and confirmed using the in vivo zebrafish larva model, should be examined in rat models. This is necessary since assessing the long-term effectiveness of therapies for these progressive late-onset dementias is similarly time-consuming in rodent models.

2.4 ZF-TBI Model: Unravelling Injury Mechanisms Through Weight-Drop Technique

The present models for TBI in zebrafish only induce mild or severe TBIs. Maheras et al. (2018) created a model for mild traumatic brain injury (mTBI) by modifying an existing mouse mTBI model produced by Mychasiuk et al. (2014). The modification included employing a weight-drop technique. A specialised apparatus specifically constructed for adult ZF was used (Maheras et al. 2018). The weight-drop model of mild traumatic brain injury (mTBI) in adult zebrafish. This innovative paradigm enables the induction of non-invasive, widespread damage to the brain of a mature zebrafish. The fish are rendered unconscious using a 0.02% solution of tricaine-S (MS-222), and then positioned on the foam cradle. They are then subjected to a hit with a 0.33-gram ball bearing, which exerts a force of 0.0032 Newtons and delivers an impact energy of 35 millijoules. The guiding tube is positioned precisely 1 cm above the zebrafish skull, guaranteeing that the impact occurs at a certain spot between the fish's eyes.

Given the widespread occurrence of mTBIs that result in cognitive impairments lasting both in the short and long term, it is crucial to create a cost-effective and easily reproducible mTBI model that can be used in any laboratory. This work confirms the effectiveness of a new adult zebrafish mTBI model, which shows a significant injury response at 3 dpi and a significant neuroregeneration response at 21 dpi. The model needs minimum equipment and may be easily implemented in a normal zebrafish aquatic housing system. For the efficient and successful setup, just a standardised ball bearing, clamped tube equipment, and foam block cradle are necessary to consistently deliver mTBIs in a satisfactory manner. This versatile model may also be modified to study the consequences of other traumatic brain injuries, including several, less severe impacts that are typical of chronic traumatic encephalopathy (Mouzon et al. 2014; Petraglia et al. 2014). In addition, this unique model includes a behavioural test that can accurately assess spatial memory impairments in fish with mTBI, a phenomenon that has been reported in previous animal models of mTBI as well as in human patients (Luo et al. 2017). Zebrafish are a valuable disease model because of the significant similarity between zebrafish and human brains. Approximately, 70% of human genes have a clear counterpart in zebrafish (Howe et al. 2013). The creation of this mTBI model is particularly noteworthy since it utilises zebrafish, which have a remarkably comparable genome to humans. The genetic similarity between zebrafish and humans has significant implications for using zebrafish neuroregeneration in the human brain. Moreover, the striking resemblance, particularly in the genetic makeup related to diseases, between humans and zebrafish offers scientists significant opportunities for study and future advancements in pharmaceuticals (Lieschke and Currie 2007).

2.5 ZF-TBI: Exploring Traumatic Brain Injury Through Laser-Based Modelling

The zebrafish (Danio rerio) is a widely used and innovative model organism in neuroscience research (Bashirzade et al. 2022; de Abreu et al. 2022; Kodera and Matsui 2022), and can also be utilised for researching the pathobiology of TBI. These fish can be used to model the symptoms and pathogenic mechanisms of TBI, as well as examine the molecular diseases connected to TBI, which can contribute to the development of its therapy. Currently, there are multiple well-established zebrafish models of TBI. These models involve applying a mechanical blow to the skull, similar to what is done in rats. This leads to a peak of the inflammatory response on day 3 after the injury and neuroregeneration on day 21 post-insult (Maheras et al. 2018). Exploiting the potential of this influential model species, we have devised an innovative technique for inducing TBI in adult zebrafish using non-invasive laser stimulation. Lasers are becoming essential instruments in biomedicine because of their accuracy and ability to generate high-density energy fluxes, which may be used to precisely target and damage cells for various biological investigations (Harreguy et al. 2020). Zebrafish larvae serve as a useful model for conducting laser experiments to evaluate the functional connections among hindbrain neurons inside specific segments (Liu and Fetcho 1999), study visuomotor behaviour (Roeser and Baier 2003), investigate regeneration following laser axotomy (Hu et al. 2018), and conduct research on spinal cord injuries (Dehnisch Ellstrom et al. 2019).

Due to the diverse nature of TBI, it is not feasible to replicate the complete range of TBI pathophysiology in animals. However, researchers have created many animal models to study certain pathophysiological mechanisms and outcomes of TBI, as well as prospective targets for therapy. Various components of TBI have been well replicated in animal models, encompassing both mammals and zebrafish. In the traditional sense, an animal model is deemed valid if it exhibits similarities to the human condition in terms of symptoms (face validity), response to treatments or worsening effects (predictive validity), pathological mechanisms, and theoretical understanding of dysfunctional processes (construct validity) (Belzung and Lemoine 2011). TBI models often encompass focal, diffuse, or mixed forms of injury. Diffuse brain damage is defined by a sudden acceleration-deceleration of the brain, typically without a direct hit to the head. Animals can have focal damage when a specific portion of the exposed dura or the closed skull is impacted, resulting in laceration, contusion, and hematoma (Shultz et al. 2017). Zebrafish TBI models typically consist of penetrating injuries to the brain tissues, ranging from moderate to severe. These injuries can involve the penetration of the epidermal layer and the blood-brain barrier. Nevertheless, these models fail to accurately replicate closed-head TBI. Two new models of closed-head TBI have recently been created and studied in adult zebrafish. One study used a standard mouse weight-drop technique to induce non-penetrating, diffuse moderate traumatic brain injury in adult zebrafish (Maheras et al. 2018). Another experimental model of TBI in adult zebrafish utilises a specific technique called targeted pulsed high-intensity focused ultrasound (McCutcheon et al. 2017). Both of these models accurately replicate a diffuse kind of TBI that impacts the entire brain tissue. Because of the small dimensions of the zebrafish head, it is difficult to develop a system that enables precise selection of the angle, force, and site of impact for region-specific analysis. In response to this requirement, we introduce an innovative model of focused TBI caused by laser, which does not penetrate the tissue. Heating is the primary cause of the main effects observed in biological tissues when exposed to laser light. Initially, a radiation quantum is absorbed in the vibrational-rotational band of the molecule, leading to its transition to an excited state. Subsequently, as a result of inelastic interactions with the surrounding environment, the molecule experiences a rise in its kinetic energy, leading to the occurrence of localised hyperthermia. The pathological alterations that occur in a tissue mostly rely on the intensity of radiation, duration of exposure, and the extent of penetration into the tissue. When biological tissues are heated by lasers, they can exhibit several responses. These include no change in temperature (<40 °C), temporary hyperthermia (42–50 °C), protein denaturation and cell necrosis (>60 °C), tissue removal through ablation (100 °C), carbonisation (150 °C), and melting (>300 °C). These effects can occur concurrently, ranging from carbonisation on the surface to hyperthermia at a depth of several millimetres. Therefore, in order to prevent unwanted thermal harm, it is crucial to carefully choose the characteristics of laser radiation, such as the period and duration of pulses, as well as the pulse energy. Another benefit of laser radiation is its ability to function at a distance, as transparent tissues are not susceptible to its harmful effects. Zebrafish strains exhibit significant variation in colour and transparency. Some strains have translucent head sections, which allow laser radiation to selectively injure the brain while leaving the surface tissues, such as skin and skull, unaffected. This feature makes zebrafish a suitable model for controlled and non-invasive TBI studies. In this study, we utilised laser irradiation generated by a laser diode with a power output of 500 mW and a wavelength of 405 nm.

Unlike mammals, zebrafish possess the extraordinary capacity to regenerate and restore their brain following an injury, mostly because of the abundance of neurogenic niches in their brain (Kizil et al. 2012). In mammals, the afflicted area mostly undergoes a glial proliferative process, whereas in zebrafish, proliferation largely occurs in neurogenic zones. Subsequently, neuronal precursors migrate from these zones to the injured area. Contrary to mammals, fish do not experience significant gliosis (overgrowth of glial cells in the affected area), and this does not cause obvious pathological effects. However, zebrafish radial glial cells have the ability to transform into neuroblasts. The zebrafish's ability to rapidly and nearly fully regenerate its neurological functions and structure is of great interest. Zebrafish models of TBI are particularly well-suited for investigating the mechanisms of neurorepair and identifying prospective targets for regenerative medicine (Cacialli 2021; Diotel et al. 2013).

The findings of Tikhonova et al. (2022) indicate that the newly established and distinctive model of laser-induced focused non-penetrating TBI in zebrafish accurately reproduces mild TBI. The laser radiation induces a closed-head brain damage that is specifically suited for subsequent sophisticated and region-specific studies. The benefits of this method include the ability to precisely target and control the extent of damage to nerve tissue by adjusting the laser intensity or time of exposure to the laser radiation. This neurotrauma model successfully replicated behavioural deficits, accurately mimicking the reactions reported in mammalian TBI models and clinical practice. This indicates that the model has face validity. The results indicated a satisfactory level of variability in the animals' reactions to the experimental exposure. Therefore, this experimental animal model can serve as a valuable tool for understanding the neurobiological pathways that are responsible for the effects of laser-induced TBI. The molecular changes observed in the brains of zebrafish wounded by lasers align with existing understanding of how traumatic brain injuries develop, providing evidence that the model accurately represents the condition. The findings indicate that Brain-derived neurotrophic factor (BDNF) plays a crucial role in enhancing the survival and functional restoration of fully developed neurons in the affected region during the initial phase of laser-induced mild TBI and throughout the subsequent recovery following the primary injury. In summary, their findings validate the therapeutic promise of BDNF or its mimetics for mild TBI and propose the manipulation of neuroinflammatory response as an additional viable strategy in TBI treatment.

2.6 Ultrasound-Aided Zebrafish Models: Probing Traumatic Brain Injury Dynamics

Several animal models have been created to comprehend the pathophysiology of TBI and assist in identifying new therapeutic approaches. These animal models have been employed to clarify the molecular characteristics of trauma, as well as to examine the enduring effects and neurological vulnerability to neurodegenerative illnesses (Ziong et al. 2013; Levin and Robertson 2013). The most often utilised model organisms are rats, mice, and swine, and a range of assays have been devised to measure the outcome (Ziong et al. 2013). Nevertheless, the assessment of medicinal compounds using these models continues to be a tedious and time-consuming procedure. The extensive pre-clinical development period is a considerable obstacle in promptly identifying chemical entities or signalling pathways for the treatment of TBI. Comprehensive outcome study of a single pharmacological item can require several years for evaluation to be completed. In vitro models have some capacity for high-throughput evaluation or therapies, but they are restricted in their ability to replicate the anatomical and cellular intricacies found in whole animal preparations. Although in vitro models have been valuable in dissecting precise mechanisms of cellular damage following trauma and providing a certain level of anatomical preservation (e.g. organotypic cultures), they do not completely capture the cell-cell interactions that occur after TBI. The zebrafish has the potential to serve as a model organism that can bridge the gap and overcome the technical and time-related challenges involved with pre-clinical drug validation. Zebrafish are particularly well-suited for studying neurological illnesses due to their highly developed central nervous systems, which have brain regions that are similar to those seen in mammals (Laird and Robberecht 2011). Live imaging and documenting of injury progression are greatly facilitated by transgenic lines, which enable the production of transparent fish or fish with specific and easily identifiable fluorescent reporters that are relevant to neurotrauma (Kalueff et al. 2014; Stewart et al. 2014). The adult zebrafish exhibits homology in the expression and function of crucial receptors involved in synaptic neurotransmission, such as α-amino-3-hydroxyl-5-methyl-4-isoxazole-propionate (AMPA) and N-methyl-D-aspartate (NMDA) receptors, which are known to play a significant role in the excitotoxic pathways associated with TBI. Zebrafish exhibit significant resemblance to mammals in terms of sequence homology and vulnerability of cytoskeletal proteins, including the heavy neurofilament chain and alphaII-spectrin (Miracle et al. 2009). In addition, there are other cognitive and behavioural outcome measures that may be quickly assessed using automated image collection and processing technologies in order to establish correlations with molecular discoveries (Stewart et al. 2014). Therefore, zebrafish have great significance and potential for being used as a model organism to enhance the effectiveness of translational studies on TBI. Ultrasound is a modality that uses energy and has several applications in medicine and biology for both diagnosis and treatment. High-intensity focused ultrasound (HIFU) is an emerging form of therapeutic ultrasound that has great potential in targeted, low-impact therapy methods. The coordinates (Kennedy et al. 2014; Tavakkoli and Sanghvi 2011) represent a point in a two-dimensional space. HIFU beams can be produced externally using specialised ultrasound transducers and directed towards a specific area within the body. HIFU energy, based on its exposure conditions, has the ability to cause thermal and/or mechanical impacts in tissue. Pulsed high-intensity focused ultrasound (pHIFU) is a unique form of ultrasound therapy that uses brief bursts of high-intensity ultrasound energy to mainly generate mechanical effects in tissue, such as stress forces and cavitation (O'Neill and Li 2012). Presently, the sole documented models of brain damage in zebrafish pertain to penetrating injuries inflicted upon brain tissues (Kishimoto et al. 2012). These models fail to accurately replicate closed-head traumatic brain damage.

TBI is a multifaceted illness process rather than a single pathophysiological event (Kolias et al. 2013). The importance of a TBI model lies in its construct, face, and predictive validity, which are crucial for ensuring that the model system has objective translational value (van der Staay 2006). Face validity refers to the extent to which a model appears to accurately represent a specific circumstance, whereas construct validity indicates that the model is based on a solid theoretical foundation (Varga et al. 2010). Predictive validity requires that the animal model accurately represents the human condition it is meant to imitate. Predictive validity can be quantified by assessing both the reliability and relevance of a given measure (Stewart and Kalueff 2013). Typically, an animal model is deemed valid if it closely resembles the human state in terms of the causes, physiological changes, symptoms, and response to treatments. Through our zebrafish model, we show that the adult zebrafish injury model has similarities with the mammalian rodent system, which is generally recognised for its build, face, and predictive validity. Due to the diverse nature of the injury, it is not feasible to replicate the complete range of TBI pathophysiology in an animal model. Therefore, several elements of TBI have been effectively replicated in various animal models (Ziong et al. 2013; Levin and Robertson 2013). The focused ultrasonic stimulus was employed to generate a type of closed-head, brain-specific trauma. This method of injury proved capable of replicating molecular alterations and related cognitive deficits with a significant level of resemblance to mammalian models of TBI. The adult zebrafish TBI model has predictive validity, which is the crucial capacity of animal models to accurately and consistently represent the human situation. The adult damage system we used showed consistent results in terms of outcomes and had a stronger capacity for processing compared to rodent systems. The model as a whole exhibits significant relevance to the study of mammalian pathophysiology by utilising clearly defined metrics of cellular and behavioural outcomes. Presumably, this model can be utilised to find possible therapeutic substances and targets linked to TBI. Although this system has its benefits, it cannot fully substitute for more extensive investigations conducted on larger mammals. The use of pHIFU ultrasonic technology to target specific areas of injury is presently not theoretically viable due to the size restrictions of both the pHIFU wave and the zebrafish brain. Additionally, the tissue samples obtained from certain brain regions are insufficient in size to provide research that is specific to those regions. Therefore, alterations in protein expression may be affected by signal-to-noise artefacts, which might obscure the true extent of changes in protein expression within the traumatised regions. Moreover, our current study solely focused on investigating the immediate impact of injury on the final result. The existing knowledge of whether zebrafish display long-lasting neurological impairments is lacking, as well as the understanding of how these timeframes align with pathological processes in mammals. Furthermore, the observed prolonged death rate in a subset of the group with normal body temperature injury 5 and 10 days after the damage indicates that additional injury mechanisms are still worsening the initial reaction to the injury. These findings indicate that zebrafish have comparable pathophysiological characteristics to mammals in terms of the evolution of secondary harm following initial injury. The time sequence of events and the thresholds for survival during brain trauma are currently unknown, particularly in relation to how they differ between different species, such as zebrafish and mammals. The current approach is a fast and cost-effective method for producing empirical information on the effectiveness of treatment before investing in a whole study with mammals. The ease of drug delivery, whether through water application or focused microinjections, enhances the value of this model as a translational bridge between in vitro and in vivo systems. It simplifies the evaluation process while yet keeping a certain level of anatomical complexity by utilising a whole organism. Although there are some drawbacks, the continuously expanding collection of transgenic reporter lines makes the model suitable for several potential applications in studying the pathophysiology of TBI. Recovery times following an injury were found to rise in a manner that was dependent on both the intensity and duration of the applied pressure. These findings indicate that our model has the ability to replicate certain characteristics of mild TBI (McCutcheon et al. 2017). The categorisation of damage severity in animal models, however, is still somewhat unclear and lacks standardisation across different injury platforms. There is currently no universally accepted scoring system for assessing the severity of injuries based on a rapid neurological assessment in animal models (Ziong et al. 2013) The zebrafish pHIFU injury model in adult individuals exhibits alterations in the expression of proteins in the white matter. The presence of axonal cytoskeletal proteins and activated caspase 3 indicated a typical response to stress in rodent models of non-penetrating TBI. NF160, a neurofilament of moderate size, plays a crucial role in providing structural support and tensile strength to axons. It has been observed that NF160, along with heavy and light neurofilaments, tends to collect in the swollen areas of axons caused by the pathological changes resulting from traumatic brain injury (TBI) (McCutcheon et al. 2017). Microtubule β-III tubulin is mostly found in the cytoskeletons of neurons and plays a crucial role in the growth and transportation of axons (Siedler et al. 2014). Secondary injury leads to the breakdown of microtubules, causing impaired intra-axonal transport, as well as swelling and separation of the axons (Sahuquillo and Vilalta 2007). In earlier studies, we have seen elevated levels of NF200 in less severe damage models (Park et al. 2013). This alteration in expression is additionally corroborated in our zebrafish injury model. It is noteworthy that there was a prior occurrence of intense neurofilament and microtubule expression before the activation of caspase-3 and the build-up of β-APP. The data suggest that there is a gradual development of mild damage mechanisms in the zebrafish. Further research is needed to confirm these phenomena in rodent models.

2.7 Conclusions and Perspectives

This summary provides an overview of the current studies using zebrafish as a model for studying TBI. The key findings from studies on TBI in zebrafish demonstrate the capacity to deliberately cause controlled damage using several methods, including mechanical impact, exposure to high-pressure waves, and chemical assault. These models accurately reproduce important characteristics of human TBI, such as the death of neurons, inflammation, and cognitive impairments. They offer vital knowledge on the processes of injury and potential treatments. Moreover, the ability of zebrafish larvae to allow light to pass through them makes it possible to observe cellular and molecular reactions to injury in real time. This makes it easier to investigate dynamic processes like neuroinflammation and axonal regeneration. Zebrafish larvae can be used in high-throughput screening methods to quickly evaluate the effects of pharmacological substances and genetic modifiers on TBI outcomes. The combination of modern imaging techniques, like two-photon microscopy and functional MRI, with zebrafish models of TBI provides exceptional accuracy in both space and time, allowing for a detailed understanding of the changes in injury patterns and the reorganisation of neuronal circuits after the injury.

Zebrafish models of traumatic brain injury (TBI) have both advantages and disadvantages. These models are useful for researchers to study different aspects of TBI pathophysiology, neuroplasticity, and therapeutic approaches. Each model possesses distinct advantages and limits, rendering them appropriate for addressing various research inquiries and experimental goals. The selection of a model is contingent upon various factors, including the required degree of injury control, the requirement for accurate localisation, the accessibility of imaging tools, and the intricacy of the TBI pathology under investigation. Researchers should meticulously evaluate these aspects while choosing the best suitable model for their investigations. In summary, zebrafish models of traumatic brain injury (TBI) offer significant knowledge on the causes of injury and possible treatment approaches. This contributes to our comprehension of TBI pathophysiology and the advancement of treatment options.

In summary, zebrafish serve as a great supplementary model system for studying TBI. They offer distinct benefits in terms of revealing core mechanisms and discovering new treatment targets to alleviate TBI-related neurological impairments. Further collaboration across many fields of study, using the zebrafish model, shows potential for enhancing our knowledge of TBI pathogenesis and creating successful therapies to enhance patient outcomes (Table 2.1).

Table 2.1 Tabulated comparison of different zebrafish models for traumatic brain injury (TBI), including the weight-drop model, laser model, larvae model, open head model, and chemical induction model

Model	Description	Advantages	Disadvantages
Weight-drop model	Involves inducing TBI by delivering a controlled mechanical impact to the zebrafish head	Allows for precise control over the severity and location of the injury	Requires specialised equipment for precise impact delivery
Laser model	Utilises laser ablation to induce focal brain injury in zebrafish larvae	Enables precise and reproducible induction of focal brain lesions	Limited to inducing focal injuries and may not replicate diffuse brain trauma observed in some TBI cases
Larvae model	Utilises zebrafish larvae for studying TBI, often induced by various methods, including mechanical or chemical means	Larvae offer optical transparency for in vivo imaging and observation of brain injury and repair processes	May not fully recapitulate TBI pathology observed in adult zebrafish or humans
Open head model	Involves direct exposure and manipulation of zebrafish brain tissue to induce TBI, often through surgical procedures, such as partial craniotomy	Allows for precise targeting of specific brain regions and pathways for studying TBI	Invasive surgical procedures may introduce confounding factors and affect zebrafish viability or behaviour
Chemical induction model	Induces TBI in zebrafish through exposure to neurotoxic compounds or oxidative stress inducers, mimicking pathological processes associated with brain injury	Provides a less invasive method for inducing TBI compared to mechanical or surgical models	May lack specificity in replicating the complex pathophysiology of traumatic brain injury observed in humans

References

Ahrens MB, Orger MB, Robson DN et al (2013) Whole-brain functional imaging at cellular resolution using light-sheet microscopy. Nat Methods 10(5):413–420

Alyenbaawi H, Kanyo R, Locskai LF, Kamali-Jamil R, DuVal MG, Bai Q, Wille H, Burton AE (2021) Seizures are a druggable mechanistic link between TBI and subsequent tauopathy. elife 10:e58744

Bashirzade AA, Zabegalov KN, Volgin AD, Belova AS, Demin KA, de Abreu MS, Babchenko VY, Bashirzade KA, Yenkoyan KB, Tikhonova MA et al (2022) Modeling neurodegenerative disorders in zebrafish. Neurosci Biobehav Rev 138:104679

Belzung C, Lemoine M (2011) Criteria of validity for animal models of psychiatric disorders: focus on anxiety disorders and depression. Biol Mood Anxiety Disorders 1:9

Burgess HA, Johnson SL, Granato M (2010) Unidirectional startle responses and disrupted left-right co-ordination of motor behaviors in robo3 mutant zebrafish. Genes Brain Behav 9(5):500–511

Cacialli P (2021) Neurotrophins time point intervention after traumatic brain injury: from zebrafish to human. Int J Mol Sci 22:1585

Chen YC, Priyadarshini M, Panula P (2009) Complementary developmental expression of the two tyrosine hydroxylase transcripts in zebrafish. Histochem Cell Biol 132(3):375–381
Cho SJ, Park E, Telliyan T, Baker A, Reid AY (2020) Zebrafish model of posttraumatic epilepsy. Epilepsia 61:1774–1785
Corrigan JD, Selassie AW, Orman JAL (2010) The epidemiology of traumatic brain injury. J Head Trauma Rehab 25:72–80
Cruz-Haces M, Tang J, Acosta G, Fernandez J, Shi R (2017) Pathological correlations between traumatic brain injury and chronic neurodegenerative diseases. Transl Neurodegeneration 6:20. https://doi.org/10.1186/s40035-017-0088-2
de Abreu MS, Giacomini A, Demin KA, Petersen EV, Kalueff AV (2022) On the value of zebrafish outbred strains in neurobehavioral research. Lab Anim 51:5–6
De Beaumont L, Lassonde M, Leclerc S, Théoret H (2007) Long-term and cumulative effects of sports concussion on motor cortex inhibition. Neurosurgery 61(2):329–336
Dehnisch Ellstrom I, Spulber S, Hultin S, Norlin N, Ceccatelli S, Hultling C, Uhlen P (2019) Spinal cord injury in zebrafish induced by near-infrared femtosecond laser pulses. J Neurosci Methods 311:259–266
Diotel N, Vaillant C, Gabbero C, Mironov S, Fostier A, Gueguen MM, Anglade I, Kah O, Pellegrini E (2013) Effects of estradiol in adult neurogenesis and brain repair in zebrafish. Horm Behav 63:193–207
Dreosti E, Lopes G, Kampff AR, Wilson SW (2015) Automated light response measures from the larval zebrafish using a high-throughput screening platform. BMC Neurosci 16:1–15
Gardner RC, Yaffe K (2015) Epidemiology of mild traumatic brain injury and neurodegenerative disease. Mol Cell Neurosci 66:75–80
Harreguy MB, Marfil V, Grooms NWF, Gabel CV, Chung SH, Haspel G (2020) Ytterbium-doped fibre femtosecond laser offers robust operation with deep and precise microsurgery of C. elegans neurons. Sci Rep 10:4545
Hay J, Johnson VE, Smith DH, Stewart W (2016) Chronic traumatic encephalopathy: the neuropathological legacy of traumatic brain injury. Annu Rev Pathol 11:21–45
Howe K, Clark MD, Torroja CF et al (2013) The zebrafish reference genome sequence and its relationship to the human genome. Nature 496:498–503
Hu BB, Chen M, Huang RC, Huang YB, Xu Y, Yin W, Li L, Hu B (2018) In vivo imaging of Mauthner axon regeneration, remyelination and synapses re-establishment after laser axotomy in zebrafish larvae. Exp Neurol 300:67–73
Johnson VE, Stewart JE, Begbie FD et al (2013) Inflammation and white matter degeneration persist for years after a single traumatic brain injury. Brain 136:28–42
Kalueff AV, Echevarria DJ, Stewart AM (2014) Gaining translational momentum: more zebrafish models for neuroscience research. Prog Neuro-Psychopharmacol Biol Psychiatry 55:1–6
Kennedy JE, Ter Haar GR, Cranston D (2014) High intensity focused ultrasound: surgery of the future? Br J Radiol
Kishimoto N, Shimizu K, Sawamoto K (2012) Neuronal regeneration in a zebrafish model of adult brain injury. Dis Model Mech 5:200–209
Kizil C, Kaslin J, Kroehne V, Brand M (2012) Adult neurogenesis and brain regeneration in zebrafish. Dev Neurobiol 72:429–461
Kodera K, Matsui H (2022) Zebrafish, medaka and turquoise killifish for understanding human neurodegenerative/neurodevelopmental disorders. Int J Mol Sci 23(4):1399
Kolias AG, Guilfoyle MR, Helmy A, Allanson J, Hutchinson PJ (2013) Traumatic brain injury in adults. Pract Neurol 13:228–235
Laird AS, Robberecht W (2011) Modeling neurodegenerative diseases in zebrafish embryos. Methods Mol Biol 793:167–184
Len TK, Neary JP (2011) Cerebrovascular pathophysiology following mild traumatic brain injury. Clin Physiol Funct Imaging 31:85–93
Levin HS, Robertson CS (2013) Mild traumatic brain injury in translation. J Neurotrauma 30:610–617

Lewine JD, Davis JT, Bigler ED, Thoma R, Hill D, Funke M et al (2007) Objective documentation of traumatic brain injury subsequent to mild head trauma. J Head Trauma Rehab 22:141–155
Lieschke GJ, Currie PD (2007) Animal models of human disease: zebrafish swim into view. Nat Rev Genet 8:353–367
Liu KS, Fetcho JR (1999) Laser ablations reveal functional relationships of segmental hindbrain neurons in zebrafish. Neuron 23:325–335
Luo Y, Zou H, Wu Y, Cai F, Zhang S, Song W (2017) Mild traumatic brain injury induces memory deficits with alteration of gene expression profile. Sci Rep 7:1077
Maheras AL, Dix B, Carmo OMS, Young AE, Gill VN, Sun JL, Booker AR, Thomason HA, Ibrahim AE, Stanislaw L, Dallego JC, Ngo CN, Chen A, Fortini BK, Spence RD (2018) Genetic pathways of neuroregeneration in a novel mild traumatic brain injury model in adult zebrafish. eNeuro 5(1):ENEURO.0208-17.2017
März M, Schmidt R, Rastegar S, Strähle U (2011) Regenerative response following stab injury in the adult zebrafish telencephalon. Dev Dyn 240:2221–2231
Masel BE, DeWitt DS (2010) Traumatic brain injury: a disease process, not an event. J Neurotrauma 27:1529–1540
McCutcheon V, Park E, Liu E, Sobhebidari P, Tavakkoli J, Wen XY, Baker AJ (2017) A novel model of traumatic brain injury in adult zebrafish demonstrates response to injury and treatment comparable with mammalian models. J Neurotrauma 34:1382–1393
McGinn TE, Mitchell DM, Meighan PC et al (2018) Zebrafish swim bladder nematode infection as a model for assessing anthelmintic activity. Zebrafish 15(6):548–556
McGinn TE, Mitchell DM, Meighan PC et al (2020) Zebrafish swim bladder nematode infection as a model for assessing anthelmintic activity. Zebrafish 15(6):548–556
Mckee AC, Daneshvar DH (2015) The neuropathology of traumatic brain injury. Handb Clin Neurol 127:45–66
Miracle A, Denslow ND, Kroll KJ, Liu MC, Wang KK (2009) Spillway-induced salmon head injury triggers the generation of brain alphaII-spectrin breakdown product biomarkers similar to mammalian traumatic brain injury. PLoS One 4(2):e4491
Monteiro J, Aires R, Becker JD et al (2021) Lsd1 inhibition attenuates tissue damage in a zebrafish model of spinal cord injury. Front Cell Dev Biol 9:645888
Mouzon BC, Bachmeier C, Ferro A et al (2014) Chronic neuropathological and neurobehavioral changes in a repetitive mild traumatic brain injury model. Ann Neurol 75:241–254
Mychasiuk R, Farran A, Angoa-Perez M, Briggs D, Kuhn D, Esser MJ (2014) A novel model of mild traumatic brain injury for juvenile rats. J Vis Exp 94:1–7
Nguyen R, Fiest KM, McChesney J, Kwon CS, Jette N, Frolkis AD, Atta C, Mah S, Dhaliwal H, Reid A, Pringsheim T, Dykeman J, Gallagher C (2016) The international incidence of traumatic brain injury: a systematic review and meta-analysis. Can J Neurol Sci 43:774–785
O'Neill BE, Li KC (2012) Pulsed high-intensity focused ultrasound. Springer, Cham, pp 2181–2186
Ojo JO, Mouzon B, Algamal M, Leary P, Lynch C, Abdullah L, Evans J, Mullan M, Bachmeier C, Stewart W, Crawford F (2016) Chronic repetitive mild traumatic brain injury results in reduced cerebral blood flow, axonal injury, gliosis, and increased T-Tau and tau oligomers. J Neuropathol Exp Neurol 75:636–655
Park E, Eisen R, Kinio A, Baker AJ (2013) Electrophysiological white matter dysfunction and association with neurobehavioral deficits following low-level primary blast trauma. Neurobiol Dis 52:150–159
Patten SA, Armstrong GA, Lissouba A et al (2017) Fishing for causes and cures of motor neuron disorders. Dis Model Mech 10(4):507–518
Petraglia AL, Plog BA, Dayawansa S, Dashnaw ML, Czerniecka K, Walker CT et al (2014) The pathophysiology underlying repetitive mild traumatic brain injury in a novel mouse model of chronic traumatic encephalopathy. Surg Neurol Int 5:184
Rao SS, McDonald ML, Vittorio CC et al (2021) A zebrafish high throughput screening model for development of novel therapies for Parkinson's disease. Sci Rep 11(1):2091

Roeser T, Baier H (2003) Visuomotor behaviors in larval zebrafish after GFP-guided laser ablation of the optic tectum. J Neurosci 23:3726–3734

Saatman KE, Duhaime A-C, Bullock R et al (2008) Classification of traumatic brain injury for targeted therapies. J Neurotrauma 25(7):719–738

Sahuquillo J, Vilalta A (2007) Cooling the injured brain: how does moderate hypothermia influence the pathophysiology of traumatic brain injury. Curr Pharm Des 13:2310–2322

Shah AN, Davey CF, Whitebirch AC et al (2018) Rapid reverse genetic screening using CRISPR in zebrafish. Nat Methods 12(6):535–540

Shultz SR, McDonald SJ, Vonder Haar C et al (2017) The potential for animal models to provide insight into mild traumatic brain injury: translational challenges and strategies. Neurosci Biobehav Rev 76:396–414

Siedler DG, Chuah MI, Kirkcaldie MT, Vickers JC, King AE (2014) Diffuse axonal injury in brain trauma: insights from alterations in neurofilaments. Front Cell Neurosci 8:429

Skaggs K, Goldman D, Parent JM (2014) Excitotoxic brain injury in adult zebrafish stimulates neurogenesis and long-distance neuronal integration. Glia 62:2061–2079

Stewart AM, Kalueff AV (2013) Developing better and more valid animal models of brain disorders. Behav Brain Res 10:49–58

Stewart AM, Braubach O, Spitsbergen J, Gerlai R, Kalueff AV (2014) Zebrafish models for translational neuroscience research: from tank to bedside. Trends Neurosci 37:264–278

Sun L, Guo J, Zhang Z et al (2020) Engineered zebrafish as a model for understanding the role of neuroinflammation in Parkinson's disease. Zebrafish 17(5):308–317

Tavakkoli J, Sanghvi NT (2011) Ultrasound-guided HIFU and thermal ablation. In: Frenkel V (ed) Therapeutic ultrasound: mechanisms to applications. Nova Science Publishers, Hauppauge, NY, pp 137–161

Taylor CA, Bell JM, Breiding MJ, Xu L (2017) Traumatic brain injury-related emergency department visits, hospitalizations, and deaths - United States, 2007 and 2013. MMWR Surveill Summ 66:1–16

Tikhonova MA, Maslov NA, Bashirzade AA et al (2022) A novel laser-based zebrafish model for studying traumatic brain injury and its molecular targets. Pharmaceutics 14:1751

van der Staay FJ (2006) Animal models of behavioral dysfunctions: basic concepts and classifications, and an evaluation strategy. Brain Res Rev 52:131–159

Vanderploeg RD, Curtiss G, Belanger HG (2005) Long-term neuropsychological outcomes following mild traumatic brain injury. J Int Neuropsychol Soc 11:228–236

Varga OE, Hansen AK, Sandoe P, Olsson IA (2010) Validating animal models for preclinical research: a scientific and ethical discussion. Altern Lab Anim 38:245–248

Zanier ER, Bertani I, Sammali E, Pischiutta F, Chiaravalloti MA, Vegliante G, Masone A, Corbelli A, Smith DH, Menon DK et al (2018) Induction of a transmissible tau pathology by traumatic brain injury. Brain 127:2685–2699

Ziong Y, Mahmood A, Chopp M (2013) Animal models of traumatic brain injury. Nat Rev Neurosci 14:128–142

Chapter 3
Understanding Pathophysiological Changes of Traumatic Brain Injury

Mohamed Y. Zaky and Nahed S. Lamloum

Abstract Traumatic brain injury (TBI) has a disastrous effect on the lives and health of those who experience it, as well as on those who care for them. Besides the inherent complexities of individuals' conditions, there are notable repercussions for both health and social security systems due to TBI. This is primarily because individuals with TBI often experience a considerable decrease in their ability to work. Therefore, conducting research to better understand the physiological changes resulting from TBI is crucial. The present book chapter offers a concise summary of the classification of traumatic brain injuries and the underlying pathophysiological mechanisms involved.

Keywords Traumatic brain injury (TBI) · Physiological changes · Pathophysiological mechanisms · Health systems · Social security

3.1 Introduction

The brain is the most intricate and intriguing structure in the human body. Although there is a growing body of knowledge about this organ, much remains unknown, particularly regarding tissue and cell responses to mechanical, chemical, or physiological perturbations. Traumatic brain injury (TBI) is a common and serious event that can have long-term effects on brain function (Freire et al. 2023). TBI arises when an external mechanical force results in a modified state of consciousness (Joshua et al. 2022). Primary damage occurs when brain tissue is distorted due to injury, as well as direct harm to blood vessels, neurons, axons, and glial cells.

M. Y. Zaky
Molecular Physiology Division, Zoology Department, Faculty of Science, Beni-Suef University, Beni-Suef, Egypt
e-mail: mohamedzaki448@science.bsu.edu.eg

N. S. Lamloum (✉)
Department of Biochemistry, Faculty of Science, Beni-Suef University, Beni-Suef, Egypt

W. Mohamed (ed.), *Nutrition and Traumatic Brain Injury (TBI)*, Nutritional Neurosciences, https://doi.org/10.1007/978-981-97-6341-2_3

Secondary damage appeared due to a group of events that occur over time because of hypoxia, ischemia, edema, hydrocephalus, and elevated intracranial pressure (ICP), all of which are associated with the initial injury. Primary and secondary injuries are not different, and they can occur concurrently, resulting in temporary or permanent neurologic impairments (Joshua et al. 2022). In this context, our current work aims to integrate the existing understanding of TBI pathophysiology, with a focus on brain edema, excitotoxicity, oxidative stress, mitochondrial dysfunction, neuroinflammation, and apoptotic cell death processes.

3.2 Classification of Traumatic Brain Injuries

TBI is an extremely challenging condition to categorize owing to its high degree of heterogeneity, the challenge of accurately reproducing and scaling in animal models, and other reasons (Naumenko et al. 2023). A key factor in the classification of TBI is the physical mechanism of injury, which distinguishes between penetrating and closed wounds. Both scenarios have radically different injury biomechanics and penetrating trauma increases the risk of infection (Deshetty and Periyasamy 2023). Healthcare providers can employ the physical mechanism classification to promptly detect or anticipate vascular dysfunction at the outset of a patient's care (Hawryluk and Manley 2015). Many primary brain injury mechanisms (such as compression, rotation, displacement, rupture, and so forth) can be categorized according to their association with acceleration-deceleration forces or contact trauma (Naumenko et al. 2023). Since multiple molecular cascades can continue to function for weeks following the initial impact, it is believed that progressive adverse effects develop following the primary injury. Considering that these secondary injury mechanisms appear later, they may be treated or prevented (Hawryluk and Manley 2015).

Numerous clinical changes, such as cerebral edema, contusions, and diffuse axonal injury, are indicative of TBI (Foerschner et al. 2020). Pathoanatomical lesions come in two varieties: diffuse and focal. When specific skull tissues and planes are taken into consideration, these lesions may also be taken into consideration. For medical professionals, pathological classification is essential for creating emergency treatment plans (Naumenko et al. 2023). Computed tomography (CT) data are now frequently used in clinical practice for premorbid pathological classification. Although widely utilized, the Marshall Score System is not without its shortcomings. It is one-dimensional, missing subtype recognition for intracranial lesions, and primarily based on two features: radiological signs of elevated ICP and the necessity for surgery. A more recent classification technique that enables the evaluation of distinct CT components is the Rotterdam scale. Additionally, it assesses the clinical and anatomical aspects, such as epidural hematomas and traumatic subarachnoid hemorrhage, for their predictive value (Rafay et al. 2020).

Severity is often used to categorize TBI, which is essentially a symptom-based classification scheme. This is a common use for the Glasgow Coma Scale (GCS), a scoring system that is widely used in medicine. Ocular, verbal, and motor scales are the three component points that make up the scale (range 3–15), which may be used to rapidly assess brain damage degree. Individuals with GCS scores of 9–12, 9–15, or 9 are categorized as having mild, moderate, or severe TBI, respectively. There are notable differences in both the prognosis and the clinical course of the disease between these groups. Despite being incredibly accurate, the GCS damage assessment is challenging to apply successfully in the care of children (Naumenko et al. 2023). The latest addition, the FOUR score scale, addresses several GCS limitations, including the challenge of verbally scoring intubated patients and directly assessing the brainstem's condition (Hawryluk and Manley 2015).

3.3 Pathophysiology of Traumatic Brain Injury

Delays and prolonged secondary damage that might last for hours or years can result from the physiological, molecular, and metabolic processes that occur after a primary injury. Secondary injuries arise from various mechanisms (Fig. 3.1).

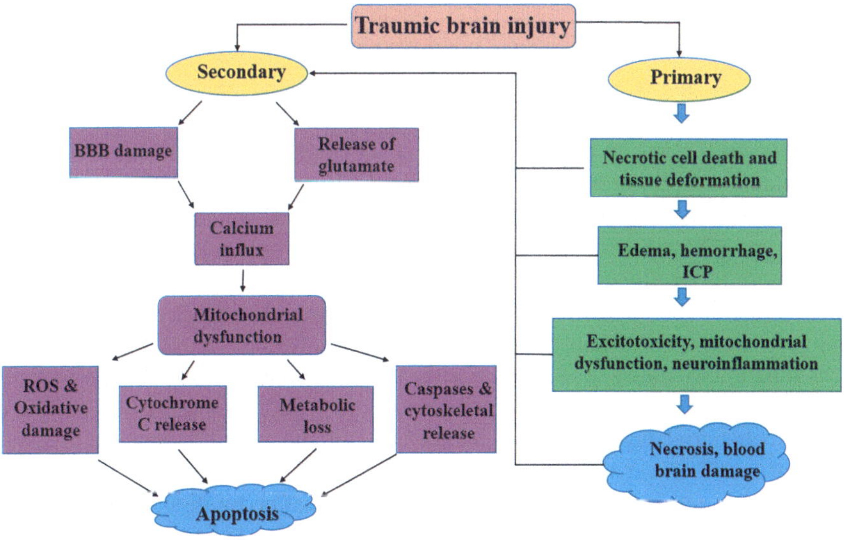

Fig. 3.1 The pathophysiological mechanism underlying primary and secondary brain injury

3.3.1 Brain Edema

The development of brain edema, which is defined as an elevated water content in the intracellular and/or extracellular spaces of the brain, leading to organ swelling (Zusman et al. 2020), is one of the crucial events that occur after TBI (Konar et al. 2019). Brain edema is traditionally categorized as either cytotoxic or vasogenic, with both types occurring hours after TBI (Arora et al. 2023).

Vasogenic edema is a form of cerebral edema defined by the buildup of extravascular fluid in the extracellular space because of increased blood–brain barrier (BBB) permeability is brought on by cerebral blood vessel damage, which leads to plasma extravasation and the formation of edema (Freire et al. 2023). A frequent side effect of several brain disorders, including infections, brain tumors, acute ischemic stroke, and TBI, is vascular edema (Dalby et al. 2020). Vasogenic edema is the result of a complicated chain of events. A cerebral blood vessel injury sets off an inflammatory response that releases growth factors, interleukins (ILs), and cytokines, among other chemical mediators (Freire et al. 2023). Increased intracranial pressure, the primary effect of vasogenic edema, can result in serious neurological symptoms and even potentially fatal conditions (Jha et al. 2019). The clinical assessment of symptoms and imaging tests, such as computed tomography (CT) and magnetic resonance imaging (MRI), are the foundation for the diagnosis of vasogenic edema (Greco et al. 2023). These imaging methods can detect cerebral edema and assist in identifying its underlying cause. The goal of vasogenic edema treatment is to lower intracranial pressure while managing symptoms. This frequently entails taking diuretic drugs, like mannitol, which aid in lowering the amount of fluid in the brain (Halstead and Geocadin 2019). Furthermore, osmotherapy and decompressive craniectomy are also used to treat cerebral edema (Addis et al. 2023).

On the other hand, cytotoxic edema is a kind of cerebral edema that is defined by the buildup of fluid inside brain cells as a result of modifications to cell membrane function (Jha et al. 2019). Contrary to vasogenic edema, which is caused by fluid leaking out of blood vessels, cytotoxic edema is caused by swelling of the cells and an increase in the amount of fluid inside them (Winkler et al. 2016). The primary cause of cytotoxic edema is the disturbance of ion homeostasis, specifically the failure of the Na^+/K^+ pump, which results in intracellular sodium (Na^+) and water accumulation (Winkler et al. 2016). This causes cellular swelling, compromised energy metabolism, and impaired cellular function. Furthermore, oxidative stress and mitochondrial dysfunction play a role in the initiation and advancement of cytotoxic edema (Freire et al. 2023).

As seen in rats in a model of TBI caused by cranial perforation, the membrane protein aquaporin 4 (AQP4), which is in charge of cellular water transport, exhibits variations in its expression based on the type of edema to form in the event of edema formation following a TBI. Whereas AQP4 was upregulated in cytotoxic edema, it was downregulated in vasogenic edema (Zhang et al. 2015). In this regard, it seems that AQP4 gene silencing or attenuation is useful in decreasing the development of edema following TBI (Fan et al. 2023). Significant effects on brain function and neuronal viability are associated with cytotoxic edema. Increased intracellular volume can lead to altered neural signaling pathways, cellular damage, and impaired neurotransmission (Jha et al. 2019).

Advanced imaging methods like diffusion-weighted imaging (DWI) and MRI are essential for diagnosing cytotoxic edema because they can show restricted diffusion and cellular swelling in the affected brain regions (Čivrný et al. 2023). The aim of cytotoxic edema treatment is to reduce cellular swelling and treat the underlying ailment. Optimizing cerebral perfusion, lowering intracellular Na^+ levels, and encouraging cellular recovery are a few examples of management techniques (Zusman et al. 2020). Furthermore, targeted treatments utilizing the vasopressin V1a receptor antagonist conivaptan and aquaporin molecule regulation appear promising in reducing cytotoxic edema (Kitchen et al. 2020).

3.3.2 Excitotoxicity

BBB disruption brought on by TBI causes an overabundance of neurotransmitters to be released, which are normally reliable for reabsorbing glutamate, to malfunction (Thapa et al. 2021). Glutamate interacts with both ionotropic and metabotropic glutamate receptors, as well as its various metabolites, to activate them. Among the ionotropic glutamate receptors facilitating the entry of calcium, potassium, and sodium ions into the membrane to induce depolarization are the *N*-methyl-d-aspartate (NMDA) and α-amino-3-hydroxy-5-methyl-4-isoxazolepropionic acid (AMPA) receptors (Brustovetsky et al. 2010). These receptors are overexpressed in TBI settings due to excessive glutamate release (Meldrum 2000). GluN2B has been demonstrated to be involved in mediating the excitotoxic response and to be present in the synaptic cytosol. Excess intracellular Ca^{2+} activates numerous downstream signaling molecules. Cell death results from the activation of apoptotic proteins, such as caspases, calcineurin, and calpain, caused by an excess of Ca^{2+} in the cytosol (Thapa et al. 2021). Accumulated reactive oxygen species (ROS) also compromise mitochondrial activity. Cells perish from oxidative stress due to excitatory neurotransmitters, and excitotoxicity causes cell death. The shear and strain stresses caused by brain injury immediately after trauma increase glutamate-independent excitotoxicity by activating NMDA receptors (Thapa et al. 2021). A study additionally showcased the mechanosensitivity of NMDA receptors to specific subunits and signaling pathways involved in modulating NMDA receptors when subjected to mechanical stimuli. Findings indicated that the GluN2B subunit facilitates the response to mechanical stimuli (Fig. 3.2) (Singh et al. 2012).

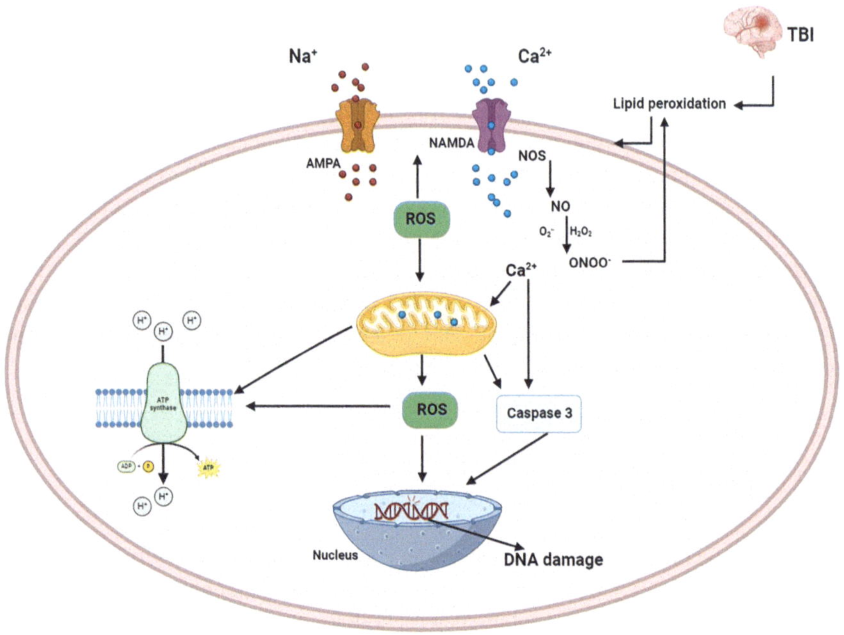

Fig. 3.2 An overview of the typical processes connected to excitotoxicity after TBI

3.3.3 Oxidative Stress

Oxidative metabolism stands as a significant metabolic advancement for aerobic organisms, as it yields 36 ATP molecules for each glucose molecule oxidized in the presence of oxygen (Reiter et al. 2017). The complexes of the respiratory chain exchange electrons through the mitochondrial electron transfer chain (ETC), which creates adenosine triphosphate (ATP), the cell's energy source. While some electrons escape during this process, which is crucial for cell survival, nearby O_2 molecules are reduced to superox-ide anion radical ($O_2^{\bullet-}$). About 1%–3% of O_2 is converted to $O_2^{\bullet-}$ under normal circumstances, where the oxygen level is physiological and the ETC is operating at maximum efficiency. However, in circumstances where the elements and operation of the ETC are disrupted, such as CNS injuries, greater electron leakage results in increased production of reactive oxygen species (ROS) (Reiter et al. 2017). O_2-derived molecules, or oxygen-free radicals, are unstable and extremely reactive due to unpaired electron(s) in the outer orbital layer. For essential molecules inside cells, free radicals, and other ROS (Fig. 3.2). Superoxide radical $O_2^{\bullet-}$, considered a primary ROS, interacts with other molecules within the cell. This interaction can trigger chemical reactions that generate new, highly reactive species known as 'secondary ROS,' thereby amplifying oxidative stress within the cellular environment. When $O_2^{\bullet-}$ reacts with other molecules after acting as the primary ROS, secondary ROS are created (Naseem and Parvez 2014). The body uses an antioxidant defense system to combat these harmful factors (Hakiminia et al. 2022). Oxidative stress is a damaging condition that occurs when ROS are overproduced or the antioxidant defense mechanism is ineffective (Naseem and Parvez 2014). When O_2 loses a single electron, $O_2^{\bullet-}$ is formed, which is a relatively weak reactant. $O_2^{\bullet-}$ is not extremely hazardous to neighboring molecules, but when it is transformed to more toxic agents, such as OH and ONOO, the damage is greater. Furthermore, SOD dismutase $O_2^{\bullet-}$ to generate H_2O_2. This non-radical ROS has a high lipophilicity and can easily pass through lipid membranes. H_2O_2 is more harmful because it reacts with transition metals (e.g., iron) to generate $^{\bullet}$OH via the Fenton reaction, which oxidizes Fe^{2+} to ferric iron (Fe^{3+}) (Reiter et al. 2017).

According to certain reports, the brain's $O2^{\bullet-}$ increases in the initial minutes and hours following brain damage (Kontos and Povlishock 1986). The brain uses a lot of oxygen (20% of inspired oxygen is consumed by it). Additionally, iron is a transition metal and the brain contains tissue that is rich in polyunsaturated fatty acids (PUFAs). This organ, however, has low levels of antioxidant factors. All things considered, this suggests that, among other organs, the brain is the most vulnerable to oxidative damage (Carretero et al. 2009). Ferritin and transferrin bind to Fe^{3+} efficiently at neutral pH, preserving iron in non-catalytic form. In an acidic environment, such as damaged brain tissue, both proteins lose their affinity and release their iron. Iron then participates in the reactions of O_2 radicals as a catalyst. Catalytic iron can also be obtained through trauma-induced bleeding. It seems that iron, which is released from Hb, is the leading cause of Hb-related oxidative damage (Hakiminia et al. 2022). $O2^{\bullet-}$ is created when the Fe^{2+} undergoes autoxidation. Fe^{2+} also reacts with H_2O_2 to produce $^{\bullet}$OH, as was previously mentioned (Hall et al. 2010).

Injured neuronal cells experience oxidative damage primarily in the form of lipid peroxidation (LP), which is a damaging process in TBI pathology (Hall et al. 2016). Lipid-rich structures and cell membranes are highly vulnerable to ROS attack. Reactive radicals cause disruptions to the fluidity, permeability, enzymatic activity, and cell membrane structure when they steal electrons from lipid molecules like arachidonic acid. This ultimately leads to cell damage. Particularly, the most prevalent are polyunsaturated fats (PUFAs), which contain multiple double bonds and offer hydrogen as an electron source. LP evolves in three stages: (a) By removing a hydrogen atom from a polyunsaturated fat molecule, a free radical is quenched, and initiation takes place. As a result, PUFA is changed into an alkyl radical or lipid. (b) The generated alkyl radical is oxidized due to its instability, resulting in the production of a lipid peroxyl radical. This radical then combines with another PUFA to generate a new alkyl lipid radical and a lipid hydroperoxide. We call this stage "propagation." (c) Termination is the last stage, which produces non-radical end products, mainly 2-propenal (acrolein) and 4-hydroxynonenal (4-HNE). These stable compounds are created when a lipid radical reacts with either a radical scavenger or another radical. These final products cause further formation of free radicals and damage to neuronal proteins, making them highly neurotoxic (Hall et al. 2010). All of these events point to ROS as the primary cause of the pathophysiologic processes following CNS trauma. An increase in intracranial pressure (ICP), disruption of cerebral perfusion, ischemia, brain edema, overproduction of inflammatory and toxic substances, mitochondrial damage, and energy failure are all caused by ROS (Najem et al. 2018). Interestingly, ROS cannot be directly quantified; instead, oxidative stress is determined by measuring the byproducts of ROS invasion of biologic molecules. F2-isoprostanes, 4-HNE, malondialdehyde (MDA), protein carbonyl, and 8-hydroxy-2'-deoxyguanosine (8-OHdG) are significant markers of oxidative stress (Hakiminia et al. 2022).

3.3.4 Mitochondrial Dysfunction

Organelles inside cells that are membrane-bound are called mitochondria. They are separated into two membranes: the inner membrane is nearly impermeable, while the outer membrane allows ions and small molecules to pass through. The K^+ ATPase channels, the Na^+/Ca^{++} exchanger, the Ca^{++} uniporter, and other ion channels and transporters are present in the inner membrane, which also contains the protein complexes that make up the electron transport chain (ETC). The synthesis of ATP is one of the most significant functions of mitochondria, which is why the term "powerhouse" of the cell is frequently applied to them. In a typical vertebrate cell, almost 95% of the ATP demand is generated by mitochondria (Mustafa and Alshboul 2013). In particular, neurons appear to be heavily reliant on healthy mitochondria to meet the metabolic requirements of membrane excitability, which is necessary for impulse transmission, neurotransmission, and other functions like plasticity (Yang et al. 2023). Neural cells need a lot of ATPs to maintain ionic gradients and neurotransmission; to meet this demand, oxidative phosphorylation in the mitochondria is used. As a result, aberrant mitochondrial function is frequently linked to neurodegeneration, and mitochondrial function is essential for neural survival (Celsi et al. 2009). Following TBI, mitochondrial damage is a common pathology. Electron microscopy experiments have demonstrated that in mitochondria isolated from brain trauma, the cristae membranes are destroyed (Diao et al. 2024). Furthermore, damaged mitochondria have enlarged and show indications of membrane discontinuity. It was demonstrated that high levels of lipid peroxidation (LP) were present in conjunction with structural disintegration in mitochondria using antibodies against particular markers of oxidative damage. Research on isolated mitochondria from rodents has demonstrated that elevated levels of 4-HNE and 3-nitrotyrosine (3-NT), indicators of LP and protein nitration, are induced by controlled cortical impact TBI. Proteomic analytical data have verified that post-traumatic oxidative damage targets the majority of the proteins within the ETC complexes, which are essential for ATP synthesis (Opii et al. 2007). This occurs at the same time that oxidative metabolism is compromised, and ATP stores are depleted as a result of TBI. It goes without saying that the severe alterations in energy metabolism that result from TBI are caused by mitochondrial dysfunction (Hakiminia et al. 2022). Neuronal cells appear to experience mitochondrial dysfunction as a result of glutamate neurotoxicity (Tiwari et al. 2023). Due to *N*-methyld-aspartate (NMDA) receptor activation, glutamate release after TBI causes a massive intracellular Ca^{++} accumulation. This is followed by an excess of Ca^{++} in the mitochondria, which puts them at risk of additional damage (Gao et al. 2023). Following a TBI, this Ca^{++} disruption exacerbates mitochondrial dysfunction and energy failure (Pandya et al. 2023). Furthermore, it has been demonstrated that the ensuing Ca^{++} load in the mitochondria causes the production of ROS, reactive nitrogen species (RNS), and their extremely reactive free radicals. As a result, these free radicals primarily originate from and target Ca^{++}-loaded neural mitochondria. They also cause protein modifications and mitochondrial membrane LP, which irreversibly impair mito-chondrial functions like ion transport, oxidative phosphorylation, and respiration, as well as reduce the amount of Ca^{++} that the mitochondria can buffer (Mustafa and Alshboul 2013). The most disastrous effects of the Ca^{++} load on mitochondria, however, are the stimulation of the mitochondrial permeability transition pore (mPTP) in the inner mitochondrial membrane and the consequent dumping of the matrix Ca^{++} pool back into the cytoplasm (Atlante et al. 2022). Solutes with molecular weights of less than 1500 Daltons can effortlessly pass through the inner mitochon-drial membrane when the permeability of the membrane rapidly increases (mPTP) (Arora et al. 2023). In addition to contributing to delayed Ca^{++} dysregulation, this mitochondrial breakdown increases the rise in cytosolic Ca^{++} (Zhao and Pan 2021). It seems that the creation of the mPTP is essential for the Ca^{++}-induced production of free radicals by mitochondria. It is accompanied by a sharp collapse of the potential of the mitochondrial membrane, which will prevent mitochondria from producing ATP and could cause ATP synthase to reverse its normal function (Hansson et al. 2008). The general dysfunction of mitochondria is linked to the death of neurons after TBI. It is proposed that the mode of cell death necrotic versus apoptotic after glutamate neurotoxicity is determined by the state of mitochondrial functions (Mustafa and Alshboul 2013).

Apoptosis appears to be facilitated by oxidative stress factors that are brought on by the creation of mPTP after injury by increasing the release of cytochrome c from mitochondria. Research utilizing mPTP inhibitors, such as cyclosporine A and its non-immunosuppressive counterpart NIM811, has demonstrated that in controlled cortical impact-TBI models, these drugs protect mitochondrial function and improve neurodegeneration, have shown that mPTP formation and mitochondrial dysfunction contribute to post-TBI neurodegeneration (Mustafa and Alshboul 2013).

3.3.5 Neuroinflammation

Pro-inflammatory substances include nitric oxide, ROS, interleukin (IL)-1β, interferon-γ (IFNγ), tumor necrosis factor-α (TNFα), and nitric oxide (Kumar and Khanum 2012). Notably, pro-inflammatory chemicals like IL-4, IL-13, prostaglandins, and neutrophils are also released. Numerous CNS microglia activation phenotypes have been identified by earlier research. Pro-inflammatory substances lead to increased levels of pro-inflammatory cytokines and ROS by enhancing the M1 microglia phenotype (Kumar and Khanum 2012).

It has been demonstrated that those anti-inflammatory drugs activate the M2 phenotype of microglia, supporting angiogenesis, tissue repair, neurite growth and extension, and the inhibition of deleterious immune responses (Yu et al. 2023).

Although research indicates that both of these phenotypes can be activated simultaneously, it is still unclear what processes cause one to predominate over the other (Kumar and Loane 2012). Furthermore, because it oversimplifies the complexity of the CNS microglia response in vivo, the M1/M2 classification has generated a lot of controversy in the literature (Sulhan et al. 2020).

In numerous of the previously discussed processes, oxidative stress resulting from produced ROS contributes to TBI neuroinflammation. The BBB can be compromised by membrane lipid peroxidation, which is brought on by ROS and their generated superoxide radical (Hall et al. 2010). By reacting with superoxide, numerous neuroinflammatory processes generate nitric oxide, a vasodilatory substance that aids in the production of a range of free radicals (Sulhan et al. 2020). The CNS is particularly vulnerable to free radical-induced peroxidation because of its high amounts of membrane lipids that are prone to peroxidation (Hall et al. 2010). Furthermore, numerous investigations have demonstrated that exposure to ROS promotes both tight junction protein degradation and BBB permeability (Sulhan et al. 2020).

3.4 Conclusion

After experiencing TBI, brain edema, the inflammatory response, oxidative stress, metabolic disturbances, and oxidative stress all play crucial roles in contributing to tissue damage and cognitive impairment. Considering the aforementioned, initiatives to lessen secondary brain injury seek to stop or slow down harmful processes while enhancing the environment for neuronal healing. In the treatment of TBI, it is critical to comprehend and address secondary brain injury because prompt identification and reduction of secondary injury may enhance patient outcomes and encourage neurological recovery.

References

Addis A, Baggiani M, Citerio G (2023) Intracranial pressure monitoring and management in aneurysmal subarachnoid hemorrhage. Neurocrit Care:1–11

Arora K, Vats V, Kaushik N et al (2023) A systematic review on traumatic brain injury pathophysiology and role of herbal medicines in its management. Curr Neuropharmacol 21:2487–2504

Atlante A, Amadoro G, Latina V, Valenti D (2022) Therapeutic potential of targeting mitochondria for Alzheimer's disease treatment. J Clin Med 11:6742

Brustovetsky T, Bolshakov A, Brustovetsky N (2010) Calpain activation and Na+/Ca2+ exchanger degradation occur downstream of calcium deregulation in hippocampal neurons exposed to excitotoxic glutamate. J Neurosci Res 88:1317–1328

Carretero M, Escames G, López LC et al (2009) Long-term melatonin administration protects brain mitochondria from aging. J Pineal Res 47:192–200

Celsi F, Pizzo P, Brini M et al (2009) Mitochondria, calcium and cell death: a deadly triad in neurodegeneration. Biochim Biophys Acta (BBA)-Bioenergetics 1787:335–344

Čivrný J, Tomáš D, Černá M (2023) MRI of cerebral oedema in ischaemic stroke and its current use in routine clinical practice. Neuroradiology:1–11

Dalby T, Wohl E, Dinsmore M et al (2020) Pathophysiology of cerebral edema—a comprehensive review. J Neuroanaesth Crit Care 8:163–172

Deshetty UM, Periyasamy P (2023) Potential biomarkers in experimental animal models for traumatic brain injury. J Clin Med 12:3923

Diao C, Yang Z, Hu Q et al (2024) Celastrol alleviates mitochondrial oxidative stress and brain injury after intracerebral hemorrhage by promoting OPA1-dependent mitochondrial fusion. Neuroscience 536:79–91

Fan Y, Yang Y, Lin K et al (2023) The effect of AQP4 on tau protein aggregation in neurodegeneration and persistent neuroinflammation after cerebral microinfarcts. Open Med 18:20230800

Foerschner L, Kanz K-G, Biberthaler P, Bogner-Flatz V (2020) Clinical risk factors of traumatic brain injury. Biomarkers Trauma Brain Inj, pp 323–331

Freire MAM, Rocha GS, Bittencourt LO et al (2023) Cellular and molecular pathophysiology of traumatic brain injury: what have we learned so far? Biology (Basel) 12:1139

Gao Y, Liu N, Chen J et al (2023) Neuropharmacological insight into preventive intervention in posttraumatic epilepsy based on regulating glutamate homeostasis. CNS Neurosci Ther 29:2430–2444

Greco A, Meomartino L, Gnudi G et al (2023) Imaging techniques in veterinary medicine. Part II: computed tomography, magnetic resonance imaging, nuclear medicine. Eur J Radiol Open 10:100467

Hakiminia B, Alikiaii B, Khorvash F, Mousavi S (2022) Oxidative stress and mitochondrial dysfunction following traumatic brain injury: from mechanistic view to targeted therapeutic opportunities. Fundam Clin Pharmacol 36:612–662. https://doi.org/10.1111/fcp.12767

Hall ED, Vaishnav RA, Mustafa AG (2010) Antioxidant therapies for traumatic brain injury. Neurotherapeutics 7:51–61

Hall ED, Wang JA, Bosken JM, Singh IN (2016) Lipid peroxidation in brain or spinal cord mitochondria after injury. J Bioenerg Biomembr 48:169–174

Halstead MR, Geocadin RG (2019) The medical management of cerebral edema: past, present, and future therapies. Neurotherapeutics 16:1133–1148

Hansson MJ, Månsson R, Morota S et al (2008) Calcium-induced generation of reactive oxygen species in brain mitochondria is mediated by permeability transition. Free Radic Biol Med 45:284–294

Hawryluk GWJ, Manley GT (2015) Classification of traumatic brain injury: past, present, and future. Handb Clin Neurol 127:15–21

Jha RM, Kochanek PM, Simard JM (2019) Pathophysiology and treatment of cerebral edema in traumatic brain injury. Neuropharmacology 145:230–246

Joshua AM, Vijaya Kumar K, Naidu RC (2022) Traumatic brain injury. Physiotherapy for adult neurological conditions. Springer, pp 423–493

Kitchen P, Salman MM, Halsey AM et al (2020) Targeting aquaporin-4 subcellular localization to treat central nervous system edema. Cell 181:784–799

Konar SK, Shukla D, Agrawal A (2019) Posttraumatic brain edema: pathophysiology, management, and current concept. Apollo Med 16:2–7

Kontos HA, Povlishock JT (1986) Oxygen radicals in brain injury. Cent Nerv Syst Trauma 3:257–263

Kumar GP, Khanum F (2012) Neuroprotective potential of phytochemicals. Pharmacogn Rev 6:81–90. https://doi.org/10.4103/0973-7847.99898

Kumar A, Loane DJ (2012) Neuroinflammation after traumatic brain injury: opportunities for therapeutic intervention. Brain Behav Immun 26:1191–1201

Meldrum BS (2000) Glutamate as a neurotransmitter in the brain: review of physiology and pathology. J Nutr 130:1007S–1015S

Mustafa AG, Alshboul OA (2013) Pathophysiology of traumatic brain injury. Neurosci J 18:222–235

Najem D, Rennie K, Ribecco-Lutkiewicz M et al (2018) Traumatic brain injury: classification, models, and markers. Biochem Cell Biol 96:391–406
Naseem M, Parvez S (2014) Role of melatonin in traumatic brain injury and spinal cord injury. Sci World J 2014:1–13
Naumenko Y, Yuryshinetz I, Zabenko Y, Pivneva T (2023) Mild traumatic brain injury as a pathological process. Heliyon
Opii WO, Nukala VN, Sultana R et al (2007) Proteomic identification of oxidized mitochondrial proteins following experimental traumatic brain injury. J Neurotrauma 24:772–789
Pandya JD, Musyaju S, Modi HR et al (2023) Comprehensive evaluation of mitochondrial redox profile, calcium dynamics, membrane integrity and apoptosis markers in a preclinical model of severe penetrating traumatic brain injury. Free Radic Biol Med 198:44–58
Rafay M, Gulzar F, Iqbal N, Sharif S (2020) Prognostic computed tomography scores in traumatic brain injury. Clin Neurol Neurosurg 195:105957
Reiter RJ, Rosales-Corral S, Tan DX et al (2017) Melatonin as a mitochondria-targeted antioxidant: one of evolution's best ideas. Cell Mol Life Sci 74:3863–3881
Singh P, Doshi S, Spaethling JM et al (2012) N-methyl-D-aspartate receptor mechanosensitivity is governed by C terminus of NR2B subunit. J Biol Chem 287:4348–4359
Sulhan S, Lyon KA, Shapiro LA, Huang JH (2020) Neuroinflammation and blood–brain barrier disruption following traumatic brain injury: pathophysiology and potential therapeutic targets. J Neurosci Res 98:19–28
Thapa K, Khan H, Singh TG, Kaur A (2021) Traumatic brain injury: mechanistic insight on pathophysiology and potential therapeutic targets. J Mol Neurosci 71:1725–1742
Tiwari KD, Sharma G, Prakash MM et al (2023) Effects of high glutamate concentrations on mitochondria of human neuroblastoma SH-SY5Y cells. Annales Pharmaceutiques Françaises, vol 81. Elsevier, pp 457–465
Winkler EA, Minter D, Yue JK, Manley GT (2016) Cerebral edema in traumatic brain injury: pathophysiology and prospective therapeutic targets. Neurosurg Clin 27:473–488
Yang S, Park JH, Lu H-C (2023) Axonal energy metabolism, and the effects in aging and neurodegenerative diseases. Mol Neurodegener 18:49
Yu W, Gong E, Liu B et al (2023) Hydrogel-mediated drug delivery for treating stroke. Chinese Chem Lett 108205:108205
Zhang C, Chen J, Lu H (2015) Expression of aquaporin-4 and pathological characteristics of brain injury in a rat model of traumatic brain injury. Mol Med Rep 12:7351–7357
Zhao H, Pan X (2021) Mitochondrial Ca2+ and cell cycle regulation. Int Rev Cell Mol Biol 362:171–207
Zusman BE, Kochanek PM, Jha RM (2020) Cerebral edema in traumatic brain injury: a historical framework for current therapy. Curr Treat Options Neurol 22:1–28

Chapter 4
Nutritional Changes in Traumatic Brain Injury

Jad Ghauch, Katrine Ibrahim, Ahmad Afyouni, Hussein Hmedi, Mahmoud Hammad, Haidar Kanso, and Hiba Hamdar

Abstract Traumatic brain injury (TBI) stands as a prevalent cause of mortality, intimately linked to complications that significantly influence clinical outcomes. Hyperglycemia, increased catabolism, protein wasting, and increased energy demand emerge as key factors contributing to a poor prognosis following TBI. Gastrointestinal dysfunction, which is common in severe TBI patients, highlights the importance of effective nutritional support. As a result, adequate nutrient management is critical in maintaining cellular homeostasis, lowering mortality, and reducing infectious complications. Early intervention with insulin to mitigate the negative effects of hyperglycemia, such as oxidative stress, inflammatory activation, tissue acidosis, endothelial dysfunction, and blood-brain barrier permeability, as well as protein supplementation and vitamins, are critical in promoting brain damage recovery, enhancing immune function, and preserving body mass. Early understanding and implementation of these nutritional strategies in TBI management can have a significant impact on patient outcomes.

J. Ghauch
Faculty of Medicine and Medical Sciences, USEK, Jounieh, Lebanon

Medical Learning Skills Academy, Beirut, Lebanon

K. Ibrahim · A. Afyouni · H. Hmedi
Medical Learning Skills Academy, Beirut, Lebanon

Faculty of Medical Sciences, Lebanese University, Beirut, Lebanon

M. Hammad
Medical Learning Skills Academy, Beirut, Lebanon

Faculty of Medicine, American University of Beirut, Beirut, Lebanon

H. Kanso
Medical Learning Skills Academy, Beirut, Lebanon

Faculty of Medicine, Damascus University, Damascus, Syria

H. Hamdar (✉)
Medical Learning Skills Academy, Beirut, Lebanon

W. Mohamed (ed.), *Nutrition and Traumatic Brain Injury (TBI)*, Nutritional Neurosciences, https://doi.org/10.1007/978-981-97-6341-2_4

Keywords TBI · Nutrients · Hyperglycemia · Gastric emptying · Energy expenditure · ROS · Inflammation

4.1 Introduction

Traumatic brain injury (TBI) is the leading cause of death among patients (Wong and Langley 2016), and it is associated with severe complications that have been shown to play a causal role in promoting poor clinical outcomes (Shi et al. 2016). Hyperglycemia, increased catabolism, protein wasting, and increased energy demand are among these consequences (Jeremitsky et al. 2003; Tavarez et al. 2021). In this regard, it is critical to maintain adequate and appropriate nutrient levels in order to maintain cellular homeostasis and reduce mortality and the incidence of infectious complications (Bjørklund and Chirumbolo 2017). Following a TBI, a stress response is elicited, which frequently includes elevated blood sugar levels (hyperglycemia) (Jeremitsky et al. 2005), resulting in cerebral ischemia and hypoxia (Zhao et al. 2011). Hence, early insulin injection-based hyperglycemia management will reduce its harmful effects, which include increased oxidative stress, inflammatory cascade activation, tissue acidosis, endothelial dysfunction, and increased blood-brain barrier permeability (Kavanagh and McCowen 2010). Protein supplementation, which is thought to be a mere source of calories, can also help recover from brain damage, support the immune system, and preserve body mass (McClave et al. 2016). Additionally, gastrointestinal dysfunction is a possibility in patients with severe traumatic brain injury (Haddad and Arabi 2012). Enteral feeding was well tolerated because it stimulates all functions of the gastrointestinal tract, maintains the integrity of the immunological gut barrier, and lowers the risk of sepsis complications. Previous research has indicated that malnutrition is linked to a higher death rate in patients with severe traumatic brain injury (TBI) (Rapp et al. 1983). Besides, early and effective nutritional support may improve insulin resistance and the prognosis of patients (Heyland et al. 1998). Furthermore, vitamins are essential for promoting neurological recovery and lessening the effects of traumatic brain injury (TBI) by aiding in the healing process (Lee et al. 2019).

4.2 Discussion

Traumatic brain injury frequently causes hypermetabolism and increased catabolism, resulting in a variety of gastrointestinal and nutritional complications and malnutrition in affected individuals (Lee and Oh 2022). Comprehending the dynamic alterations in the body and brain's nutrition is essential to comprehending the extensive consequences of traumatic brain injury and devising appropriate treatment strategies. According to the World Health Organization, a person's nutritional status is determined by how well their intake, absorption, and utilization of nutrients interact with their unique physiological and pathological status (Ciancarelli et al. 2022). However, following a traumatic brain injury, a hypermetabolic state is generated, and there is an alteration in micro- and macronutrients (Lucke-Wold et al. 2018), which is caused by an oxidative stress mechanism that initiates the release of free radicals and systemic neuroinflammation (Werner and Engelhard 2007), which can disrupt normal physiological processes, affecting nutrient absorption, utilization, and storage (Lucke-Wold et al. 2014). Intracellular reactive oxygen species levels have been identified as a critical factor in TBI patients (Fesharaki-Zadeh 2022). These levels can directly damage cellular structures and biomolecules, such as proteins, lipids, and DNA, resulting in cellular dysfunction and affecting the transport of nutrients across the blood-brain barrier by compromising cell membrane integrity (Carvalho and Moreira 2018).

Unlocking the pathological pathway of reactive oxygen species (ROS) and inflammation in traumatic brain injury.

Traumatic brain injury causes the release of free radicals, such as superoxide anions (ROS) and reactive nitrogen species (RNS) (Ciancarelli et al. 2012), which in turn activates neuronal necrosis and apoptosis (Siotto et al. 2022). Free radical scavengers like glutathione peroxidase (GPx), copper and zinc, and Cu/Zn superoxide dismutase (Sies 1986), which have short half-lives and thus restrict brain bioavailability, are frequently employed to counteract the elevated levels of ROS (Ciancarelli et al. 2015). Thus, a nutritional deficiency of minerals like copper, calcium, zinc, iron, selenium, and magnesium may cause the inappropriate function of these scavengers, contributing to the development of inflammation and the long-term cytotoxic effects of free radicals (Mirończuk et al. 2021). In particular, deficiencies in magnesium may exacerbate inflammation and oxidative stress, while deficiencies in zinc may reduce body mass index and impair appetite and food intake (Veronese and Barbagallo 2021; Baltaci and Mogulkoc 2012). Thus, research findings have demonstrated the importance of proper nutrient intake in reducing oxidative stress (Ciancarelli et al. 2022). However, studies have shown that oxidative stress and inflammation brought on by traumatic brain injury interact with one another by triggering transcription factors and proinflammatory gene expression that releases inflammatory cytokines (Siotto et al. 2020). Following an acute traumatic injury, oxidative stress contributes to inflammation by generating a multitude of cytokines, such as tumor necrosis factor-α (TNF-α) and interleukin-1β (IL-1β) (Abdul-Muneer et al. 2015). This results in disruption of the blood-brain barrier, edema, and neuronal loss (Soares et al. 1995). Inflammation, in conjunction with oxidative stress, may cause dysfunction in other organs and systems, such as the intestinal barrier, appetite, and satiety-regulating systems, precipitating and exacerbating malnutrition (Spychala et al. 2018). Therefore, as a result, consuming nutrients with anti-free radical properties may improve the gut microbiome, which acts to normalize the acid-base balance and reduce the production of proinflammatory cytokines (Spychala et al. 2018), as research has long demonstrated the link between interleukin 8 and malnourishment (Pourhassan et al. 2021), and increased levels of tumor necrosis factor (TNF) (Takele et al. 2016). Furthermore, it's noteworthy that undernourishment following traumatic brain injury may impede clinical recovery.

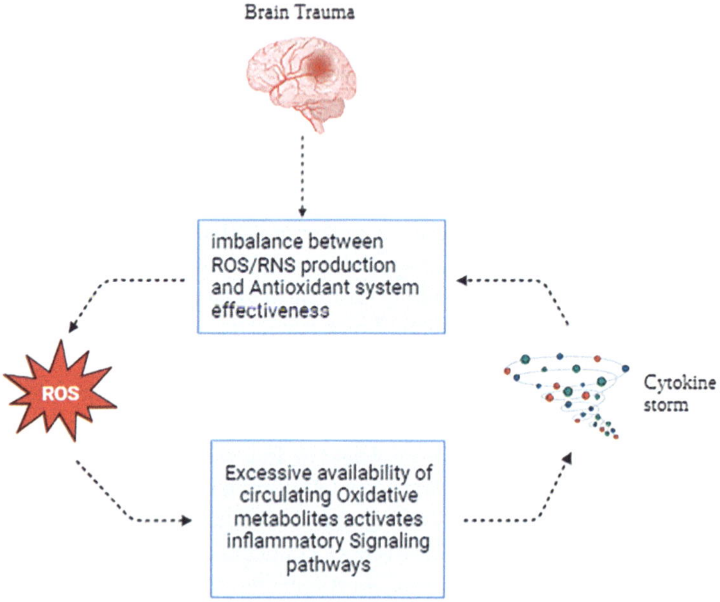

Fig. 4.1 The pathological relationship between reactive oxidative stress and inflammation following brain trauma

As a result, the interplay between oxidative stress and inflammation brought on by traumatic brain injury may play a significant and modifiable role in these patients' malnourishment and, ultimately, in their overall prognosis (Ciancarelli et al. 2022) (Fig. 4.1).

4.3 Brain Metabolism in Traumatic Brain Injury

One of the primary concerns in primary traumatic brain injury is the patient's control over their glucose level and metabolism (Moro et al. 2013), as trauma typically increases catecholamine secretion, which opposes insulin and inflammatory mediators (Lee and Oh 2022). As a result of this hormonal shift following a TBI, corticosteroids and cytokines are secreted more frequently, which causes hypermetabolism and fluctuations in energy (Faisy et al. 2003). It has been established by studies that hyperglycemia of more than 200 mg/dl (11 mmol/l) is the primary factor causing brain injury and the most common symptom in brain-injured patients (Moro et al. 2013; Matovu et al. 2021). It also exacerbates nerve damage, which ultimately results in the death of nerve cells (Hermanides et al. 2018).

Acute post-injury hyperglycemia has been shown to have a deleterious effect on traumatic brain injury patient outcomes, particularly during the first 48 h (Bergsneider et al. 1997; Hutchinson et al. 2009). Although there are many different causes of hyperglycemia (Godoy et al. 2016), one of the main mechanisms is the induction of neuronal, hormonal, and autonomic response, which releases a large amount of cortisol, catecholamine, and cytokines, which in turn stimulates the production of hepatic glucose and, eventually, insulin resistance (Godoy et al. 2010). Some factors that contribute to this problem are high-calorie diets, intravenous fluids that contain dextrose, vasoactive amines like dopamine and norepinephrine, steroids, insulin resistance, and underdiagnosed diabetes (Godoy et al. 2010). Following a traumatic brain injury, hyperglycemia has significant implications for clinical outcomes and mortality (Shi et al. 2016), including increased oxidative stress, activation of the inflammatory cascade, tissue acidosis, and endothelial dysfunction (Tomlinson and Gardiner 2008). However, at the systemic level, hyperglycemia can cause an increase in blood osmolarity and hypovolemia, making people more susceptible to sepsis and organ dysfunction (Kavanagh and McCowen 2010).

Hypoglycemia, defined as a blood glucose level of less than 70 mg/dl (Moghissi et al. 2009), on the other hand, was found to have a negative impact on TBI patient outcomes (Oddo et al. 2008a). Hypoglycemia causes mitochondrial dysfunction via the production of oxygen-free radicals, changes in transmembrane ion gradients, glutamate-mediated excitotoxicity, and activation of the apoptotic pathway (Godoy et al. 2010; Carre et al. 2013; Suh et al. 2007). In critically ill patients, moderate hypoglycemia (41–70 mg/dl) (NICE-SUGAR Study Investigators et al. 2012), can lead to neuroglycopenia, a condition in which serum glucose levels in the "normal range" (80–100 mg/dl) are associated with insufficient levels of glucose in the brain to meet its demands (Vespa et al. 2005, 2006).

Hyperglycemia and hypoglycemia are two abnormal metabolic states that have a negative impact on the outcome of traumatic brain injury (Graffagnino et al. 2010). As a result of the direct relationship between blood glucose level and prognosis (Oddo et al. 2008b), glycemic control should be done by insulin injections, conventional glycemic control (CGC) that aims to keep the glucose level at 220 mg/dL, and intensive glycemic control (IGC) that keeps the glucose level between 80 and 110 mg/dL (Coester et al. 2010; Bilotta et al. 2008). The earlier glycemic control is initiated, the better the outcome (Godoy et al. 2009); however, it is critical to provide adequate calories to support the increased metabolic demands during traumatic brain injury to avoid hypoglycemia (Krenitsky 2011).

4.4 Variation in Energy Expenditure

In traumatic brain injury, energy dysfunction is a well-known pathological pathway phenomenon (Magistretti and Pellerin 1999). Patients in the acute phase of TBI would exhibit a hypermetabolic state characterized by catecholamine secretion and strongly associated with increased morbidity and weight loss (Faisy et al. 2003; Kurtz and Rocha 2020). This increased physiological response alters energy expenditure, specifically measured resting energy expenditure (mREE) (Morbitzer et al. 2020). Lee et al.'s study emphasizes the uniqueness of TBI patients' energy expenditure and calorie requirements, highlighting the impact of factors like paralysis on mREE (Lee and Oh 2022). Patients who are paralyzed or immobilized may have mREE values that range between 100% and 125% of expected values, highlighting the dynamic nature of energy demands in this population (Lee and Oh 2022).

Moreover, Masha et al.'s study highlights the complex relationship between TBI patients' energy expenditure and the development of malnutrition, acknowledging the influence of multiple factors (Masha'al 2016). The study by Jalloh et al. (2015) clarifies the complex relationship between the pathophysiology of TBI, changes in glucose metabolism, and adjustments in energy absorption into the brain. Together, these discoveries advance our knowledge of the mechanisms underlying variations in energy expenditure in patients with traumatic brain injury. In the meantime, patients with acute brain injury must consume the right amount of calories (Morbitzer et al. 2020) as inadequate calorie intake can result in poor outcomes and complications like electrolyte imbalance, impaired organ function, and inability to wean off of mechanical ventilation (Brandi et al. 1997).

4.5 Delayed Gastric Emptying and Intolerance to Enteral Nutrition

Gastrointestinal dysfunction, including gastroparesis or delayed gastric emptying that results in feeding intolerance, is one of the most common consequences of traumatic brain injury (Fu et al. 2020; Ott et al. 1991). Normal gastrointestinal environment changes following a traumatic brain injury (TBI) result in dysbiosis of bacteria, which in turn causes a systemic inflammatory response (Patel et al. 2016). This response is mediated by cytokines and chemokines, primarily interleukines IL-1β, IL-6, IL-8, IL-10, and TNF-α, which damage sensory visceral neurons and cause autonomic dysfunction (Zhu et al. 2018a). Although the exact mechanism underlying the alteration of the gut microbiome following a moderate-to-severe traumatic brain injury (TBI) remains unclear, brain damage is caused by a feedback loop that is triggered by TBI and results in gut inflammation. This loop increases the permeability of the blood-brain barrier, which permits immune cells to enter the brain and increases the burden of TBI on the central nervous system (Hanscom et al. 2021). In addition, the onset of gastric hypokinesia, or delayed gastric emptying, may take place 1–2 weeks following the injury, contingent on its severity (Krakau et al. 2007). However, in TBI patients, enteral nutrition may assume this delay in emptying (Gore et al. 1981). Individuals who have traumatic brain injury (TBI) may become intolerant to enteral nutrition (EN) for a variety of reasons, including elevated intracranial pressure, damage to the autonomic nervous system, surgical procedures, and intubation (Abdelmalik et al. 2017).

Patients with TBI should be placed at a 45° head-of-bed elevation to reduce the risk of aspiration pneumonia and gastric reflux (Kattelmann et al. 2006). To prevent reflux and meet caloric needs in smaller volumes, concentrated enteral formulas with higher calorie content are recommended (Lee and Oh 2022). Prokinetic medications, such as metoclopramide or erythromycin, can also be used to promote gastrointestinal motility in patients at risk of swallowing aspiration (McClave et al. 2016). Early EN initiation (within 24–48 h) versus delayed EN in patients with TBI has been shown to be associated with lower mortality, lower infection rates, and shorter hospital stays (Lee and Oh 2022). Furthermore, by decreasing infection rates, such as those of pneumonia and central line infections, and decreasing length of stays in intensive care units, EN offers benefits over parenteral nutrition (PN) (McClave et al. 2016). To lessen the negative effects of an energy imbalance and guarantee the patient receives vital nutrients and energy support, the European Society for Clinical Nutrition and Metabolism recommends PN supplementation if malnutrition lasts for 4–7 days (Singer et al. 2019).

4.6 Impact of Nutrition on TBI Recovery

One of the most important aspects to consider during the healing process of TBI patients is their nutritional health (Ford and Ouma 2021). Nutritional targets, which may play a role in the mechanism of brain injury (Lucke-Wold et al. 2018), may reduce the secondary effects of TBI, such as ROS, inflammation, and neuronal cell apoptosis (Begum et al. 2013). Table 4.1 discusses commonly used dietary supplements.

Highlights of the topics discussed herein and recent recommendations for nutritional therapy.

Highlights of the issues and recommendations for nutrition therapy in patients with TBI are listed in the Table 4.2.

Table 4.1 The essential supplements required in traumatic brain injury patients and their effects

Supplements	Effects
1. Omega-3 fatty acids	Anti-inflammatory effects (Smith-Ryan et al. 2020)
2. Vitamins C and E, along with other antioxidants	Combat oxidative stress associated with TBI (Venegoni et al. 2017)
3. *N*-acetylcysteine	Precursor to glutathione has shown promise in reducing neuroinflammation post-TBI (Khayyat et al. 2016)
4. Curcumin	Anti-inflammatory and antioxidant properties, curcumin, derived from turmeric, might have neuroprotective effects in TBI (Salehi et al. 2019)
5. Vitamin D	Support neuroprotection and neurogenesis, potentially benefiting TBI recovery (Jung et al. 2022)
6. Iron and vitamin B12	Supports cognitive function and energy (Daglas and Adlard 2018; Wu et al. 2019)
7. Coenzyme Q10	Mitochondrial support, antioxidant (Allan et al. 2019)

Table 4.2 The recommendations for nutrition therapy in patients with TBI

Protein support	Protein is more than just a source of calories; it aids in brain damage recovery, immune function enhancement, and body mass preservation (Smith-Ryan et al. 2020). Protein supplementation can be beneficial for patients who do not receive enough EN, and it is critical to monitor protein intake on a regular basis (Lee and Oh 2022). In patients with TBI, protein should account for 15%–20% of total calories, with at least 2 g/kg body weight administered daily (Quintard and Ichai 2019)
Glycemic control	TBI causes a stress response, which frequently includes elevated blood sugar levels (hyperglycemia), which can worsen neurological outcomes in the presence of cerebral ischemia and hypoxia (Kurtz and Rocha 2020). TBI patients must keep their glucose levels between 8 and 11 mmol/l (Oddo et al. 2008a). It is important to note, however, that IGC therapy was associated with a significantly increased risk of hypoglycemia episodes, which should be carefully managed and factored into the treatment strategy (Zhu et al. 2018b)
Determination of energy expenditure	If indirect calorimetry is not available, the ASPEN-SCCM guideline recommends using a published predictive equation or a simple weight-based eq. (25–30 kcal/kg/day) to determine energy requirements in critically ill patients (quality of evidence: Expert consensus) (McClave et al. 2016)
Enteral nutrition	Early EN initiation (within 24–48 h) has been shown to be associated with lower mortality, lower infection rates, and shorter hospital stays in patients with TBI when compared to delayed EN (Singer et al. 2019)
Selection of nutrients	The main goal of the nutritional approach is to establish the necessary calories and protein to prevent the loss of lean body mass, particularly skeletal muscle mass, and to maintain nitrogen balance (Markovic et al. 2021). As a result, special formulas enriched with nutrients such as glutamine, omega-3 fatty acids, nucleotides, and arginine have been proposed to improve neuroprotection from secondary brain injuries (Smith-Ryan et al. 2020)

4.7 Conclusion

Traumatic brain injury is a serious condition that can result in long-term physical and psychological disabilities, as well as death because it is associated with hypermetabolism and variations in energy expenditure, as well as gastrointestinal dysfunction. Nutrition is critical in TBI patients for maintaining cellular homeostasis and lowering mortality and the incidence of infectious complications. As a result, providing specific nutritional supplements and infusions soon after a mild traumatic brain injury can help improve patients' overall recovery. Starting with adequate protein intake is important for tissue repair, recovery, and metabolic regulation, as well as proper glycemic control because TBI frequently triggers a stress response that includes elevated blood sugar levels (hyperglycemia), which can exacerbate neurological outcomes in the presence of cerebral ischemia and hypoxia. Last but not least, the administration of micronutrients (minerals, vitamins, and trace elements) aids in the restoration of cellular balance, as well as the reduction of free radical production, inflammation, and damage in TBI patients. Finally, more research strategies are needed to improve the outcomes of these patients.

References

Abdelmalik PA, Dempsey S, Ziai W (2017) Nutritional and bioenergetic considerations in critically ill patients with acute neurological injury. Neurocrit Care 27(2):276–286. https://doi.org/10.1007/s12028-016-0336-9

Abdul-Muneer PM, Chandra N, Haorah J (2015) Interactions of oxidative stress and neurovascular inflammation in the pathogenesis of traumatic brain injury. Mol Neurobiol 51(3):966–979. https://doi.org/10.1007/s12035-014-8752-3

Allan K, Hayes K, Thomas M, Barnard K (2019) Coenzyme Q10 supplementation in traumatic brain injury: a scoping review protocol. JBI Database System Rev Implement Rep 17(9):1901–1908. https://doi.org/10.11124/JBISRIR-2017-003984

Baltaci AK, Mogulkoc R (2012) Leptin and zinc relation: in regulation of food intake and immunity. Indian J Endocrinol Metab 16(Suppl 3):S611–S616. https://doi.org/10.4103/2230-8210.105579

Begum G, Harvey L, Dixon CE, Sun D (2013) ER stress and effects of DHA as an ER stress inhibitor. Transl Stroke Res 4(6):635–642. https://doi.org/10.1007/s12975-013-0282-1

Bergsneider M, Hovda DA, Shalmon E, Kelly DF, Vespa PM, Martin NA, Phelps ME, McArthur DL, Caron MJ, Kraus JF, Becker DP (1997) Cerebral hyperglycolysis following severe traumatic brain injury in humans: a positron emission tomography study. J Neurosurg 86(2):241–251. https://doi.org/10.3171/jns.1997.86.2.0241

Bilotta F, Caramia R, Cernak I, Paoloni FP, Doronzio A, Cuzzone V, Santoro A, Rosa G (2008) Intensive insulin therapy after severe traumatic brain injury: a randomized clinical trial. Neurocrit Care 9(2):159–166. https://doi.org/10.1007/s12028-008-9084-9

Bjørklund G, Chirumbolo S (2017) Role of oxidative stress and antioxidants in daily nutrition and human health. Nutrition (Burbank, Los Angeles County, Calif) 33:311–321. https://doi.org/10.1016/j.nut.2016.07.018

Brandi LS, Bertolini R, Calafà M (1997) Indirect calorimetry in critically ill patients: clinical applications and practical advice. Nutrition (Burbank, Los Angeles County, Calif.) 13(4):349–358. https://doi.org/10.1016/s0899-9007(97)83059-6

Carre E, Ogier M, Boret H, Montcriol A, Bourdon L, Jean-Jacques R (2013) Metabolic crisis in severely head-injured patients: is ischemia just the tip of the iceberg? Front Neurol 4:146. https://doi.org/10.3389/fneur.2013.00146

Carvalho C, Moreira PI (2018) Oxidative stress: a major player in cerebrovascular alterations associated to neurodegenerative events. Front Physiol 9:806. https://doi.org/10.3389/fphys.2018.00806

Ciancarelli I, Di Massimo C, De Amicis D, Carolei A, Tozzi Ciancarelli MG (2012) Evidence of redox unbalance in post-acute ischemic stroke patients. Curr Neurovasc Res 9(2):85–90. https://doi.org/10.2174/156720212800410885

Ciancarelli I, Di Massimo C, De Amicis D, Pistarini C, Tozzi Ciancarelli MG (2015) Uric acid and Cu/Zn superoxide dismutase: potential strategies and biomarkers in functional recovery of post-acute ischemic stroke patients after intensive neurorehabilitation. Curr Neurovasc Res 12(2):120–127. https://doi.org/10.2174/1567202612666150311104900

Ciancarelli I, Morone G, Iosa M, Cerasa A, Calabrò RS, Iolascon G, Gimigliano F, Tonin P, Tozzi Ciancarelli MG (2022) Influence of oxidative stress and inflammation on nutritional status and neural plasticity: new perspectives on post-stroke neurorehabilitative outcome. Nutrients 15(1):108. https://doi.org/10.3390/nu15010108

Coester A, Neumann CR, Schmidt MI (2010) Intensive insulin therapy in severe traumatic brain injury: a randomized trial. J Trauma 68(4):904–911. https://doi.org/10.1097/TA.0b013e3181c9afc2

Daglas M, Adlard PA (2018) The involvement of iron in traumatic brain injury and neurodegenerative disease. Front Neurosci 12:981. https://doi.org/10.3389/fnins.2018.00981

Faisy C, Guerot E, Diehl JL, Labrousse J, Fagon JY (2003) Assessment of resting energy expenditure in mechanically ventilated patients. Am J Clin Nutr 78(2):241–249. https://doi.org/10.1093/ajcn/78.2.241

Fesharaki-Zadeh A (2022) Oxidative stress in traumatic brain injury. Int J Mol Sci 23(21):13000. https://doi.org/10.3390/ijms232113000
Ford LM, Ouma JRA (2021) A descriptive study of malnutrition in traumatic brain injury patients. Panam J Trauma Crit Care Emerg Surg 10(3):107–112. https://doi.org/10.5005/jp-journals-10030-1359
Fu W, Shi N, Wan Y, Mei F, Qiu B, Bao Y, Zhang Y, Hao J, He J, Peng X (2020) Risk factors of acute gastrointestinal failure in critically ill patients with traumatic brain injury. J Craniofac Surg 31(2):e176–e179. https://doi.org/10.1097/SCS.0000000000006130
Godoy DA, Piñero GR, Svampa S, Papa F, Di Napoli M (2009) Early hyperglycemia and intravenous insulin-the rationale and management of hyperglycemia for spontaneous intracerebral hemorrhage patients: is time for change? Neurocrit Care 10(1):150–153. https://doi.org/10.1007/s12028-008-9098-3
Godoy DA, Di Napoli M, Rabinstein AA (2010) Treating hyperglycemia in neurocritical patients: benefits and perils. Neurocrit Care 13(3):425–438. https://doi.org/10.1007/s12028-010-9404-8
Godoy DA, Behrouz R, Di Napoli M (2016) Glucose control in acute brain injury: does it matter? Curr Opin Crit Care 22(2):120–127. https://doi.org/10.1097/MCC.0000000000000292
Gore RM, Mintzer RA, Calenoff L (1981) Gastrointestinal complications of spinal cord injury. Spine 6(6):538–544. https://doi.org/10.1097/00007632-198111000-00002
Graffagnino C, Gurram AR, Kolls B, Olson DM (2010) Intensive insulin therapy in the neurocritical care setting is associated with poor clinical outcomes. Neurocrit Care 13(3):307–312. https://doi.org/10.1007/s12028-010-9469-4
Haddad SH, Arabi YM (2012) Critical care management of severe traumatic brain injury in adults. Scand J Trauma Resusc Emerg Med 20:12. https://doi.org/10.1186/1757-7241-20-12
Hanscom M, Loane DJ, Shea-Donohue T (2021) Brain-gut axis dysfunction in the pathogenesis of traumatic brain injury. J Clin Invest 131(12):e143777. https://doi.org/10.1172/JCI143777
Hermanides J, Plummer MP, Finnis M, Deane AM, Coles JP, Menon DK (2018) Glycaemic control targets after traumatic brain injury: a systematic review and meta-analysis. Crit Care 22(1):11. https://doi.org/10.1186/s13054-017-1883-y
Heyland DK, MacDonald S, Keefe L, Drover JW (1998) Total parenteral nutrition in the critically ill patient: a meta-analysis. JAMA 280(23):2013–2019. https://doi.org/10.1001/jama.280.23.2013
Hutchinson PJ, O'Connell MT, Seal A, Nortje J, Timofeev I, Al-Rawi PG, Coles JP, Fryer TD, Menon DK, Pickard JD, Carpenter KL (2009) A combined microdialysis and FDG-PET study of glucose metabolism in head injury. Acta Neurochir 151(1):51–61. https://doi.org/10.1007/s00701-008-0169-1
Jalloh I, Carpenter KLH, Helmy A et al (2015) Glucose metabolism following human traumatic brain injury: methods of assessment and pathophysiological findings. Metab Brain Dis 30:615–632. https://doi.org/10.1007/s11011-014-9628-y
Jeremitsky E, Omert L, Dunham CM, Protetch J, Rodriguez A (2003) Harbingers of poor outcome the day after severe brain injury: hypothermia, hypoxia, and hypoperfusion. J Trauma 54(2):312–319. https://doi.org/10.1097/01.TA.0000037876.37236.D6
Jeremitsky E, Omert LA, Dunham CM, Wilberger J, Rodriguez A (2005) The impact of hyperglycemia on patients with severe brain injury. J Trauma 58(1):47–50. https://doi.org/10.1097/01.ta.0000135158.42242.b1
Jung E, Ro YS, Park JH, Moon SB, Lee SGW, Park GJ, Ryu HH, Shin SD (2022) Vitamin D deficiency and prognosis after traumatic brain injury with intracranial injury: a multi-center observational study. J Neurotrauma 39(19–20):1408–1416. https://doi.org/10.1089/neu.2022.0053
Kattelmann KK, Hise M, Russell M, Charney P, Stokes M, Compher C (2006) Preliminary evidence for a medical nutrition therapy protocol: enteral feedings for critically ill patients. J Am Diet Assoc 106(8):1226–1241. https://doi.org/10.1016/j.jada.2006.05.320
Kavanagh BP, McCowen KC (2010) Clinical practice. Glycemic control in the ICU. N Engl J Med 363(26):2540–2546. https://doi.org/10.1056/NEJMcp1001115

Khayyat A, Tobwala S, Hart M, Ercal N (2016) N-acetylcysteine amide, a promising antidote for acetaminophen toxicity. Toxicol Lett 241:133–142. https://doi.org/10.1016/j.toxlet.2015.11.008
Krakau K, Hansson A, Karlsson T, de Boussard CN, Tengvar C, Borg J (2007) Nutritional treatment of patients with severe traumatic brain injury during the first six months after injury. Nutrition (Burbank, Los Angeles County, Calif.) 23(4):308–317. https://doi.org/10.1016/j.nut.2007.01.010
Krenitsky J (2011) Glucose control in the intensive care unit: a nutrition support perspective. Nutr Clin Pract 26(1):31–43. https://doi.org/10.1177/0884533610392237
Kurtz P, Rocha EEM (2020) Nutrition therapy, glucose control, and brain metabolism in traumatic brain injury: a multimodal monitoring approach. Front Neurosci 14:190. https://doi.org/10.3389/fnins.2020.00190
Lee HY, Oh BM (2022) Nutrition management in patients with traumatic brain injury: a narrative review. Brain NeuroRehabilit 15(1):e4. https://doi.org/10.12786/bn.2022.15.e4
Lee JM, Jeong SW, Kim MY, Park JB, Kim MS (2019) The effect of vitamin D supplementation in patients with acute traumatic brain injury. World Neurosurg 126:e1421–e1426. https://doi.org/10.1016/j.wneu.2019.02.244
Lucke-Wold BP, Turner RC, Logsdon AF, Bailes JE, Huber JD, Rosen CL (2014) Linking traumatic brain injury to chronic traumatic encephalopathy: identification of potential mechanisms leading to neurofibrillary tangle development. J Neurotrauma 31(13):1129–1138. https://doi.org/10.1089/neu.2013.3303
Lucke-Wold BP, Logsdon AF, Nguyen L, Eltanahay A, Turner RC, Bonasso P, Knotts C, Moeck A, Maroon JC, Bailes JE, Rosen CL (2018) Supplements, nutrition, and alternative therapies for the treatment of traumatic brain injury. Nutr Neurosci 21(2):79–91. https://doi.org/10.1080/1028415X.2016.1236174
Magistretti PJ, Pellerin L (1999) Cellular mechanisms of brain energy metabolism and their relevance to functional brain imaging. Philos Trans R Soc Lond Ser B Biol Sci 354(1387):1155–1163. https://doi.org/10.1098/rstb.1999.0471
Markovic SJ, Fitzgerald M, Peiffer JJ, Scott BR, Rainey-Smith SR, Sohrabi HR, Brown BM (2021) The impact of exercise, sleep, and diet on neurocognitive recovery from mild traumatic brain injury in older adults: a narrative review. Ageing Res Rev 68:101322. https://doi.org/10.1016/j.arr.2021.101322
Masha'al DA (2016) The change in nutritional status in traumatic brain injury patients: a retrospective descriptive study. USF Tampa Graduate Theses and Dissertations. https://digitalcommons.usf.edu/etd/6316
Matovu P, Kirya M, Galukande M, Kiryabwire J, Mukisa J, Ocen W, Lowery Wilson M, Abio A, Lule H (2021) Hyperglycemia in severe traumatic brain injury patients and its association with thirty-day mortality: a prospective observational cohort study in Uganda. PeerJ 9:e10589. https://doi.org/10.7717/peerj.10589
McClave SA, Taylor BE, Martindale RG, Warren MM, Johnson DR, Braunschweig C, McCarthy MS, Davanos E, Rice TW, Cresci GA, Gervasio JM, Sacks GS, Roberts PR, Compher C, Society of Critical Care Medicine, & American Society for Parenteral and Enteral Nutrition (2016) Guidelines for the provision and assessment of nutrition support therapy in the adult critically ill patient: society of critical care medicine (SCCM) and American Society for Parenteral and Enteral Nutrition (A.S.P.E.N.). JPEN 40(2):159–211. https://doi.org/10.1177/0148607115621863
Mirończuk A, Kapica-Topczewska K, Socha K, Soroczyńska J, Jamiołkowski J, Kułakowska A, Kochanowicz J (2021) Selenium, copper, zinc concentrations and cu/Zn, cu/se molar ratios in the serum of patients with acute ischemic stroke in northeastern Poland-a new insight into stroke pathophysiology. Nutrients 13(7):2139. https://doi.org/10.3390/nu13072139
Moghissi ES, Korytkowski MT, DiNardo M, Einhorn D, Hellman R, Hirsch IB, Inzucchi SE, Ismail-Beigi F, Kirkman MS, Umpierrez GE, American Association of Clinical Endocrinologists, & American Diabetes Association (2009) American Association of Clinical Endocrinologists and

American Diabetes Association consensus statement on inpatient glycemic control. Diabetes Care 32(6):1119–1131. https://doi.org/10.2337/dc09-9029

Morbitzer KA, Wilson WS, Chaben AC, Darby A, Dehne KA, Brown ER, Rhoney DH, Jordan JD (2020) Energy expenditure in critically ill adult patients with acute brain injury: indirect calorimetry vs. predictive equations. Front Neurol 10:1426. https://doi.org/10.3389/fneur.2019.01426

Moro N, Ghavim S, Harris NG, Hovda DA, Sutton RL (2013) Glucose administration after traumatic brain injury improves cerebral metabolism and reduces secondary neuronal injury. Brain Res 1535:124–136. https://doi.org/10.1016/j.brainres.2013.08.044

NICE-SUGAR Study Investigators, Finfer S, Liu B, Chittock DR, Norton R, Myburgh JA, McArthur C, Mitchell I, Foster D, Dhingra V, Henderson WR, Ronco JJ, Bellomo R, Cook D, McDonald E, Dodek P, Hébert PC, Heyland DK, Robinson BG (2012) Hypoglycemia and risk of death in critically ill patients. N Engl J Med 367(12):1108–1118. https://doi.org/10.1056/NEJMoa1204942

Oddo M, Schmidt JM, Carrera E, Badjatia N, Connolly ES, Presciutti M, Ostapkovich ND, Levine JM, Le Roux P, Mayer SA (2008a) Impact of tight glycemic control on cerebral glucose metabolism after severe brain injury: a microdialysis study. Crit Care Med 36(12):3233–3238. https://doi.org/10.1097/CCM.0b013e31818f4026

Oddo M, Schmidt JM, Mayer SA, Chioléro RL (2008b) Glucose control after severe brain injury. Curr Opin Clin Nutr Metab Care 11(2):134–139. https://doi.org/10.1097/MCO.0b013e3282f37b43

Ott L, Young B, Phillips R, McClain C, Adams L, Dempsey R, Tibbs P, Ryo UY (1991) Altered gastric emptying in the head-injured patient: relationship to feeding intolerance. J Neurosurg 74(5):738–742. https://doi.org/10.3171/jns.1991.74.5.0738

Patel JJ, Rosenthal MD, Miller KR, Martindale RG (2016) The gut in trauma. Curr Opin Crit Care 22(4):339–346. https://doi.org/10.1097/MCC.0000000000000331

Pourhassan M, Babel N, Sieske L, Westhoff TH, Wirth R (2021) Longitudinal changes of cytokines and appetite in older hospitalized patients. Nutrients 13(8):2508. https://doi.org/10.3390/nu13082508

Quintard H, Ichai C (2019) Nutritional and metabolic supplementation for the injured brain: an update. Curr Opin Crit Care 25(2):126–131. https://doi.org/10.1097/MCC.0000000000000588

Rapp RP, Young B, Twyman D, Bivins BA, Haack D, Tibbs PA, Bean JR (1983) The favorable effect of early parenteral feeding on survival in head-injured patients. J Neurosurg 58(6):906–912. https://doi.org/10.3171/jns.1983.58.6.0906

Salehi B, Stojanović-Radić Z, Matejić J, Sharifi-Rad M, Anil Kumar NV, Martins N, Sharifi-Rad J (2019) The therapeutic potential of curcumin: a review of clinical trials. Eur J Med Chem 163:527–545. https://doi.org/10.1016/j.ejmech.2018.12.016

Shi J, Dong B, Mao Y, Guan W, Cao J, Zhu R, Wang S (2016) Review: traumatic brain injury and hyperglycemia, a potentially modifiable risk factor. Oncotarget 7(43):71052–71061. https://doi.org/10.18632/oncotarget.11958

Sies H (1986) Biochemistry of oxidative stress. Angew Chem Int Ed 25:1058–1071. https://doi.org/10.1002/anie.198610581

Singer P, Blaser AR, Berger MM, Alhazzani W, Calder PC, Casaer MP, Hiesmayr M, Mayer K, Montejo JC, Pichard C, Preiser JC, van Zanten ARH, Oczkowski S, Szczeklik W, Bischoff SC (2019) ESPEN guideline on clinical nutrition in the intensive care unit. Clin Nutr (Edinburgh, Scotland) 38(1):48–79. https://doi.org/10.1016/j.clnu.2018.08.037

Siotto M, Santoro M, Aprile I (2020) Vitamin D and rehabilitation after stroke: status of art. Appl Sci 10(6):1973. https://doi.org/10.3390/app10061973

Siotto M, Germanotta M, Santoro M, Canali R, Pascali S, Insalaco S, Cipollini V, Papadopoulou D, Antonacci E, Aprile I (2022) Oxidative stress status in post stroke patients: sex differences. Healthcare (Basel, Switzerland) 10(5):869. https://doi.org/10.3390/healthcare10050869

Smith-Ryan AE, Hirsch KR, Saylor HE, Gould LM, Blue MNM (2020) Nutritional considerations and strategies to facilitate injury recovery and rehabilitation. J Athl Train 55(9):918–930. https://doi.org/10.4085/1062-6050-550-19

Soares HD, Hicks RR, Smith D, McIntosh TK (1995) Inflammatory leukocytic recruitment and diffuse neuronal degeneration are separate pathological processes resulting from traumatic brain injury. J Neurosci 15(12):8223–8233. https://doi.org/10.1523/JNEUROSCI.15-12-08223.1995

Spychala MS, Venna VR, Jandzinski M, Doran SJ, Durgan DJ, Ganesh BP, Ajami NJ, Putluri N, Graf J, Bryan RM, McCullough LD (2018) Age-related changes in the gut microbiota influence systemic inflammation and stroke outcome. Ann Neurol 84(1):23–36. https://doi.org/10.1002/ana.25250

Suh SW, Hamby AM, Swanson RA (2007) Hypoglycemia, brain energetics, and hypoglycemic neuronal death. Glia 55(12):1280–1286. https://doi.org/10.1002/glia.20440

Takele Y, Adem E, Getahun M, Tajebe F, Kiflie A, Hailu A, Raynes J, Mengesha B, Ayele TA, Shkedy Z, Lemma M, Diro E, Toulza F, Modolell M, Munder M, Müller I, Kropf P (2016) Malnutrition in healthy individuals results in increased mixed cytokine profiles, altered neutrophil subsets and function. PLoS One 11(8):e0157919. https://doi.org/10.1371/journal.pone.0157919

Tavarez T, Roehl K, Koffman L (2021) Nutrition in the neurocritical care unit: a new frontier. Curr Treat Options Neurol 23(5):16. https://doi.org/10.1007/s11940-021-00670-8

Tomlinson DR, Gardiner NJ (2008) Glucose neurotoxicity. Nat Rev Neurosci 9(1):36–45. https://doi.org/10.1038/nrn2294

Venegoni W, Shen Q, Thimmesch AR, Bell M, Hiebert JB, Pierce JD (2017) The use of antioxidants in the treatment of traumatic brain injury. J Adv Nurs 73(6):1331–1338. https://doi.org/10.1111/jan.13259

Veronese N, Barbagallo M (2021) Magnesium and micro-elements in older persons. Nutrients 13(3):847. https://doi.org/10.3390/nu13030847

Vespa P, Bergsneider M, Hattori N, Wu HM, Huang SC, Martin NA, Glenn TC, McArthur DL, Hovda DA (2005) Metabolic crisis without brain ischemia is common after traumatic brain injury: a combined microdialysis and positron emission tomography study. J Cerebral Blood Flow Metab 25(6):763–774. https://doi.org/10.1038/sj.jcbfm.9600073

Vespa P, Boonyaputthikul R, McArthur DL, Miller C, Etchepare M, Bergsneider M, Glenn T, Martin N, Hovda D (2006) Intensive insulin therapy reduces microdialysis glucose values without altering glucose utilization or improving the lactate/pyruvate ratio after traumatic brain injury. Crit Care Med 34(3):850–856. https://doi.org/10.1097/01.CCM.0000201875.12245.6F

Werner C, Engelhard K (2007) Pathophysiology of traumatic brain injury. Br J Anaesth 99(1):4–9. https://doi.org/10.1093/bja/aem131

Wong VS, Langley B (2016) Epigenetic changes following traumatic brain injury and their implications for outcome, recovery and therapy. Neurosci Lett 625:26–33. https://doi.org/10.1016/j.neulet.2016.04.009

Wu F, Xu K, Liu L, Zhang K, Xia L, Zhang M, Teng C, Tong H, He Y, Xue Y, Zhang H, Chen D, Hu A (2019) Vitamin B_{12} enhances nerve repair and improves functional recovery after traumatic brain injury by inhibiting ER stress-induced neuron injury. Front Pharmacol 10:406. https://doi.org/10.3389/fphar.2019.00406

Zhao QJ, Zhang XG, Wang LX (2011) Mild hypothermia therapy reduces blood glucose and lactate and improves neurologic outcomes in patients with severe traumatic brain injury. J Crit Care 26(3):311–315. https://doi.org/10.1016/j.jcrc.2010.08.014

Zhu CS, Grandhi R, Patterson TT, Nicholson SE (2018a) A review of traumatic brain injury and the gut microbiome: insights into novel mechanisms of secondary brain injury and promising targets for neuroprotection. Brain Sci 8(6):113. https://doi.org/10.3390/brainsci8060113

Zhu C, Chen J, Pan J, Qiu Z, Xu T (2018b) Therapeutic effect of intensive glycemic control therapy in patients with traumatic brain injury: a systematic review and meta-analysis of randomized controlled trials. Medicine 97(30):e11671. https://doi.org/10.1097/MD.0000000000011671

Chapter 5
Resilience and Nutri-protection Against Traumatic Brain Injury Prior and Post-injury

Bilal El-Mansoury, Hanane El Fatimi, Asmaa Haj-Khlifa, Jawad Laadraoui, Abdelaati El Khiat, Abdelali Bitar, Arumugam Jayakumar, and Omar El Hiba

Abstract Traumatic brain injury (TBI) is defined as a physiological disturbance of brain function or a structural impairment caused by trauma as a result of an outside force. Across all age groups and in all nations, TBI is a major worldwide health burden and public health challenge with a growing economic impact on society. Until now, there are no effective treatments for this condition. However, recently nutrition has proven an important role in the treatment of TBI-related injuries and can also have a role in the prevention of the disease. Indeed, patients with TBI displayed increased catabolism This causes protein waste and raises the need for energy and on the other hand undernutrition is linked to worse neurologic outcomes and a higher death rate. To enhance patients' prognoses after TBI, prompt and effi-

B. El-Mansoury (✉) · A. Bitar · O. El Hiba
Laboratory of Anthropogenic, Biotechnology and Health, Team physiopathology Nutritional, Neurosciences and Toxicology, Faculty of Sciences, Chouaib Doukkali University, El Jadida, Morocco
e-mail: el-mansouri.b@ucd.ac.ma

H. El Fatimi
Mohammed VI University of Health Sciences, Casablanca, Morocco

A. Haj-Khlifa
Faculty of Science Semlalia, Cadi Ayyad University, Marrakech, Morocco

J. Laadraoui
Faculty of Science Aïn Chock, Hassan II University, Casablanca, Morocco

A. El Khiat
Laboratory of Anthropogenic, Biotechnology and Health, Team physiopathology Nutritional, Neurosciences and Toxicology, Faculty of Sciences, Chouaib Doukkali University, El Jadida, Morocco

Higher Institute of Nursing Professions and Health Techniques, Ministry of Health, Ouarzazate, Morocco

A. Jayakumar
Miller School of Medicine, University of Miami, Miami, FL, USA

W. Mohamed (ed.), *Nutrition and Traumatic Brain Injury (TBI)*, Nutritional Neurosciences, https://doi.org/10.1007/978-981-97-6341-2_5

cient nutritional therapy is especially important. An overview of nutrition's significance in TBI post-injury and even pre-injury treatment is given in this chapter.

Keywords Traumatic brain injury · Pathophysiology · Nutrition · Treatment · Resilience · Protection

5.1 Introduction

Traumatic brain injury (TBI) is a frequent brain damage that affects millions of people worldwide. TBI has significant social and economic impacts (Maas et al. 2017; WHO 2006). Epidemiological previsions until 2030 suggest that the frequency of impairment associated with TBI will be 2–3 times higher than that of neurological impairments resulting from cerebrovascular illnesses or Alzheimer's disease (Pietro et al. 2020). Hence, it is a major global health concern with no current FDA-approved medications for its treatment (Ismail et al. 2020).

TBI is defined as altered brain function or brain structure damage caused by an external force (Menon et al. 2010). It is well known that TBI encompasses primary and secondary damages. "Focal" anatomical lesions, such as contusion, laceration, diffuse axonal injury, and intracranial hemorrhage, are caused by primary damage (mechanical), which happens at the scene of the injury. This type of injury is only responsive to preventive measures, not therapeutic ones. A second wave of long-lasting pathological alterations known as the secondary injury mechanism arises in response to the usually transient primary damage mechanisms.

TBI is still a significant worldwide health burden and public health concern for individuals of all ages in all countries, despite recent advancements in trauma research and ongoing collaborative interdisciplinary studies intended to address this issue and improve patient outcomes.

New targets and strategies are required to alleviate the long-term effects of TBI. Research has indicated that nutrition may serve as a therapeutic intervention for TBI as well as a preventative (resilience) measure (Yu et al. 2019). It is interesting to note that today's extremely harmful eating habits, which include excessive processed food consumption, inadequate dietary *n*-3 polyunsaturated fatty acids (PUFA) intakes (especially DHA deficiencies), and insufficient dietary sulfur, may be the cause of an increased susceptibility to neurological injury following TBI. This chapter provides important data on the role of nutrition in the treatment of TBI post-injury and even its role prior to injury.

5.2 Definition and Epidemiology of Traumatic Brain Injury

TBI is described as a physiological disturbance of brain function or a traumatized structural injury brought on by an external force (Defense 2015). TBI can be defined as closed (nonpenetrating) or open (penetrating) (Cifu and Lew 2017; Taylor et al. 2017). This profoundly impacts brain cell homeostasis and resulting in either transient or chronic brain damage, neuromotor impairments, neurocognitive deficiencies, or mental illnesses (Maas et al. 2008; Mtbi 2004). This condition has a major influence on both civilians and military personnel (Menon et al. 2010). Furthermore, a shock, blow, or bump to the head can result in a TBI, as can a penetrating head injury that inhibits brain activity normally. These findings are reported by the Centers for Disease Control and Prevention (CDC).

It is highly established that TBI incidence is estimated to be 939 per 100,000 worldwide, with falls, vehicle accidents, wars, and sports being the main causes (Dewan et al. 2018; Wojcik et al. 2010). Based on estimations, TBI deaths account for 90% of mortality in developing countries, with a global fatality incidence ranging from 7% to 23% (El-Menyar et al. 2017; Johnson and Griswold 2017). Additionally, TBI imposes a financial burden on societies where its cost approaches 400 billion dollars a year (Maas et al. 2017). Furthermore, epidemiological studies conducted up until 2030 show that the incidence of patients with TBI-related disability is two to three times higher than that of patients with neurological impairments brought on by cerebrovascular or Alzheimer's disease. Despite recent developments in trauma research and ongoing efforts by collaborative interdisciplinary studies to address this issue and improve patient outcomes, TBI continues to pose a significant global health burden and public health challenge for patients of all ages and income levels worldwide (Pietro et al. 2020).

5.3 Types of TBI

5.3.1 Diffuse Axonal Injury

Diffuse axonal injury (DAI) is a common type of TBI that is characterized by a clinical spectrum of gradually growing brain axonal damage together with decreasing injury severity. The most severe cases of DAI result in a protracted traumatic coma that is not connected to ischemia, increased intracranial pressure, or mass lesions. When DAI is less severe, it is linked to concussive syndromes. Pathological studies of DAI reveal extensive but varied microscopic axon destruction in the brainstem, cerebral and cerebellar hemispheres, and white matter. The brain stem and corpus callosum are particularly vulnerable to damage. Thus, these areas are susceptible to varying degrees of severe traumatic axonal damage. In severe cases of DAI, hemorrhage affects specifically some regions of the brain (i.e., the corpus callosum and dorsolateral quadrants of the rostral brainstem), which are helpful indicators for the radiologic and postmortem identification of DAI, but in milder cases, no macroscopic abnormalities may be discovered (Blumbergs et al. 1994; Graham et al. 1997; Reilly and Bullock 1997).

5.3.2 *Contusions*

The appearance of contusions, which are typical symptoms of TBI, indicates that a head injury has occurred. The majority of animal models of TBI have shown some degree of hemorrhage, despite the fact that distinct brain deformation models produce it at the gray-white matter interface rather than in the cortical mantle (Cortez et al. 1989; Dixon et al. 1991; McIntosh et al. 1989). Moreover, damage to small blood vessels causes contusions, which are focal surface injuries with hemorrhages that frequently lie perpendicular to the cortical surface. Additionally, contusions frequently impact gyral crests and spread to a wedge-shaped necrotic area that affects the white matter subjacent to the lesion. This is because the initial localized lesion generates ongoing bleeding, ischemic necrosis, and vasogenic edema, all of which lead to the lesion growing. Since contusions are primarily impact-related phenomena, they frequently appear beneath the impact site (coup contusions). In impact acceleration models of closed head injury, they are frequently linked to skull fractures (fracture contusions). Contrecoup contusions can also be detected roughly opposite the impact site, whereas herniation contusions happen near the margins of brain hernias. However, deadly diffuse brain injuries may have minor or no contusions. Physical disturbance of the cerebral parenchyma causes laceration and contusion, which are two aspects of a continuous spectrum of injury at the brain's surface (Cortez et al. 1989; Dixon et al. 1991; McIntosh et al. 1989).

5.3.3 *Concussion (Mild TBI)*

Mild TBI (mTBI) refers to an acute neurophysiologic brain dysfunction brought on by impact contact forces or abrupt acceleration or deceleration that results in a brief shift in awareness and/or anterograde (and possibly retrograde) amnesia. The most troublesome diagnostic criteria pertain to the least severe level of alteration of consciousness, which is typically defined as the subjective experience of feeling dazed or disoriented and/or unable to account for events that occurred seconds or minutes following the injury.

5.3.4 *Hematomas*

5.3.4.1 Epidural Hematomas

An epidural hematoma (EDH) is extra-axial blood accumulation in the space that may exist between the dura mater's outer layer and the skull.

The dura is contained by the lateral sutures, particularly the coronal sutures. This is regarded as a serious neurosurgical emergency and is frequently brought on by TBI. The illness has the potential to be fatal and may leave individuals permanently disabled if appropriate clinical attention is not received (Irie et al. 2011). A successful conclusion depends on an early diagnosis and evacuation. Intracranial hematomas (ICH), including EDH, are thought to occur in 0.2% of instances of mild TBI, 3% of cases of moderate TBI, and 25–45% of cases of severe TBI (Thurman and Guerrero 1999). Due to the dura mater's tight attachment to the inside of the skull, EDH, which accounts for 1–9% of all TBIs, are significantly less common in the elderly (Hamlat et al. 2005). Age is one of the important variables that significantly affects how well individuals with EDH respond to treatment, among many other parameters with varying prognostic significance (Jones et al. 1993).

5.3.4.2 Subdural Hematomas

Subdural hematomas (SDHs) are typically brought on by a rupture of the veins that run through the subdural region and connect the superior surface of the cerebral hemispheres to the sagittal sinus. SDHs, which typically involve the entire hemisphere, are more prevalent than extradural hematomas because of the unrestricted blood flow across the subdural region (Yokobori et al. 2016).

SDHs are more widespread than extradural hematomas and typically cover the entire hemisphere due to the free flow of blood throughout the subdural region. According to studies, the death rate for traumatic SDH can range from 30 to 90%; however, people who undergo neurosurgical intervention within 4 h of the injury appear to have lower death rates (Seelig et al. 1981).

5.3.4.3 Subarachnoid Hemorrhage

Subarachnoid hemorrhage (SAH) is the result of blood spilling into the subarachnoid space, which envelops the spinal cord and brain. TBI is the most frequent cause of SAH. The majority of the remaining instances are brought on by a blood artery that bursts spontaneously. Furthermore, aneurysmal, nonaneurysmal, and perimesencephalic sources can be distinguished as the causes of spontaneous SAH (Abraham and Chang 2016). Consequently, severe headache, nausea, vomiting, neck discomfort, nuchal stiffness, and photophobia are among the signs of SAH. In 20–30% of cases, the initial bleeding is deadly. Seizures, hydrocephalus, cerebral vasospasm-related delayed cerebral ischemia, and rebleeding are among the complications of SAH. Strategies that quickly reduce intracranial pressure raise the risk of rebleeding. It is important to note that SAH is a severe acute neurological illness that affects more than 30,000 Americans annually (Go et al. 2014; Longstreth et al. 1994). Despite improvements in medical and surgical care, SAH continues to be a significant contributor to premature mortality, representing 27% of all stroke-related potential years of life lost prior to age 65 (Johnston et al. 1998). In a study, it was reported that SAH had a 57% mortality rate after 6 months and a 43% risk of death right after ictus (Bonita and Thomson 1985). In addition, the most common causes of death following SAH were identified by the International Cooperative Aneurysm Study, which was carried out in the 1980s and indicated vasospasm, direct consequences of the first hemorrhage, and rebleeding (Kassell et al. 1990). According to a study, SAH had a 57% mortality rate after six months and a 43% risk of death right after ictus (Bonita and Thomson 1985). In addition, the most common causes of death following SAH were identified by the International Cooperative Aneurysm Study, which was carried out in the 1980s and indicated vasospasm, direct consequences of the first hemorrhage, and rebleeding (Kassell et al. 1990). More recently, poor outcomes following SAH have been linked to intraventricular hemorrhage, worldwide cerebral edema, and medical problems (Bonita and Thomson 1985). Furthermore, vasospasm, the immediate consequences of the first hemorrhage, and rebleeding were found to be the most common causes of death following SAH in the International Cooperative Aneurysm Study, which was carried out in the 1980s (Kassell et al. 1990). It has been determined that medical problems, intraventricular hemorrhage, and worldwide cerebral edema all contribute to a poor outcomes following SAH (Helbok et al. 2011; Wartenberg et al. 2006).

5.3.4.4 Intraparenchymal Hemorrhage

Intraparenchymal hemorrhage (IPH) is a common occurrence that can be fatal. It happens as a result of a brain parenchymal blood vessel burst. Broadly speaking, primary IPH, which makes up 78–88% of instances of IPH, refers to rupture of tiny arteries or arterioles that have been damaged. This is most frequently caused by either cerebral amyloid angiopathy (CAA) or hypertension. In spite of this, coagulopathy, cerebral venous thrombosis, moyamoya, vasculitis, tumors, hemorrhagic conversion of ischemic stroke, rupture of a mycotic aneurysm, or vascular malformations like arteriovenous malformations (AVM), arteriovenous fistulas, or cavernous malformations can all result in secondary IPH (Gross et al. 2019).

5.3.4.5 Intraventricular Hemorrhage

Intraventricular hemorrhage (IVH) is characterized by an influx of blood into the ventricles of the brain. In the USA, it affects around 12,000 premature infants annually and has a very bad prognosis (Ballabh 2010; Strahle et al. 2012). Furthermore, over the past 20 years, there has been an increase in the incidence of severe IVH. Hence, posthemorrhagic hydrocephalus (PHH), which affects 48% of patients with grades 3 and 4 IVH, is a consequence of high-grade IVH, in addition to the severe neurological abnormalities linked to the condition (Garton et al. 2016).

It is highly established that the infection caused by the IVH is heterogeneous, complicated, and multifactorial. Hemorrhage can occur due to an inherent fragility in the vascular system of the germinal matrix, and a change in the cerebral blood flow can lead to a rupture of the vascular system. Indeed, the disruption of homeostasis mechanisms may accentuate hemorrhage if plaquette or coagulation issues are linked. Vaginal accouchement, a low Apgar score, severe respiratory distress syndrome, pneumothorax, hypoxia, hypercapnia, convulsions, persistence of the arterial canal, infection, and other factors appear to primarily increase cerebral blood flow fluctuations and serve as significant risk factors for the development of IVH (Ballabh 2014).

5.4 Skull Fracture

A skull fracture (SF) can serve as an objective measure of the severity of head trauma and primary brain injury since they are caused by forces powerful enough to shatter the skull bone.

Thus, we postulated that SF may have an impact on patients' overall prognosis and mortality from TBI. Previous fundamental research on TBI in mice has shown that SF is linked to the release of inflammatory cytokines and a bad prognosis (Zvejniece et al. 2020).

In addition, it has been noted that individuals with SF are more likely to experience cerebral lesions that need to be surgically repaired (Faried et al. 2019; Munoz-Sánchez et al. 2009). The efficacy of SF as a predictor of mortality is unknown, despite the general consensus that SF raises the risk of complications. Only a small number of studies have indicated a relationship between SF and mortality (Fabbri et al. 2008; Tseng et al. 2011), and the value of SF as a predictor of mortality is unclear. The main types of TBI are depicted in Fig. 5.1.

5.5 Pathophysiology of TBI

It is commonly recognized that tissue lesions follow an initial brain injury. Subsequent injuries ensue, which lead to modifications in cellular function and the dissemination of injury through several mechanisms such as neuroinflammation, excitotoxicity, excessive production of free radicals, and compromise of the integrity of the blood-brain barrier (BBB), altered calcium homeostasis, ischemic injury, formation of edema, and intracranial hypertension (Greve and Zink 2009; Zink et al. 2010; Dixon 2017).

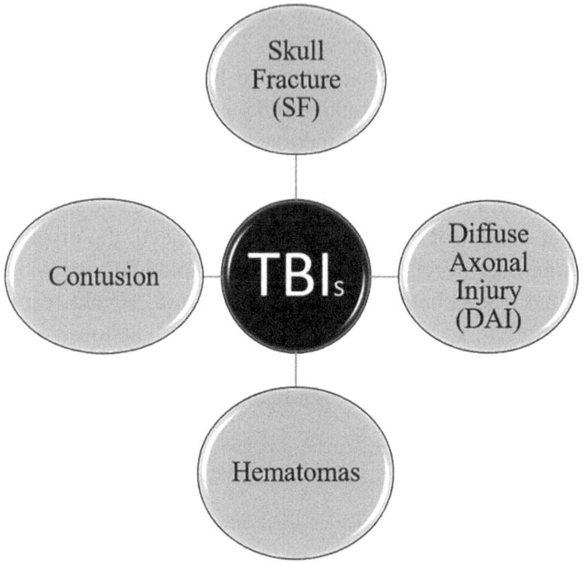

Fig. 5.1 The four main types of traumatic brain injury. Skull fracture, diffuse axonal injury, hematomas, and contusions

After TBI, the brain displays the typical signs of inflammation (Louveau et al. 2015). Indeed, traumatic insults trigger a complex inflammatory response in the brain that includes both the activation of CNS immune cells and peripheral immune cells infiltrating the brain (via a redesigned BBB). Numerous chemokines, adhesion molecules, and inflammatory cytokines mediate these activities along with reactive oxygen and nitrogen species, complement factors, and other mediators (Mathew et al. 1994; Rock and Kono 2008). These cytokines (as well as associated chemokines) increase cerebrovascular permeability, attract peripheral immune cells, and stimulate excitotoxicity to mediate inflammation (McGinn and Povlishock 2016). An increasing body of research suggests that the modifications of the BBB's function caused by brain damage may have a major impact on the injury development, the response to treatment, and neuronal healing (Zink et al. 2010). Reactive oxygen species (ROS), proinflammatory cytokines, vascular endothelial growth factor, and matrix metalloproteinases (MMPs) are among the factors that have been connected to this BBB leakage (Zink et al. 2010). Brain swelling and increased intracranial pressure (ICP) are caused by cerebral edema, which develops concurrently with these inflammatory processes. These effects are harmful following TBI (McGinn and Povlishock 2016).

On the other hand, excessive excitatory neurotransmission plays an important role in the pathophysiologic events that follow TBI. Excitatory amino acid neurotransmitters, especially glutamate, are released in large quantities following TBI (Werner and Engelhard 2007), leading to excitotoxicity. Indeed, it has been showed that extracellular glutamate increases rapidly and transiently in the minutes that follow a brain insult; the extent and duration of this rise are correlated with the severity of the lesion (Chamoun et al. 2010; Faden et al. 1989; Katayama et al. 1990; Koura et al. 1998). In addition, excitotoxicity leads to increased production of ROS when the body's natural antioxidant defenses, such as glutathione peroxidase, catalase, and superoxide dismutase, are saturated leading to oxidative stress. This leads to increased protein oxidation and peroxidation of cellular and vascular structures along with DNA damage (Werner and Engelhard 2007).

5.6 Nutri-protection Against TBI Prior to Injury

The most effective way to prevent the negative effects of TBI, whether minor or severe, in the short or long term, is to prevent the injury in the first place (Lewis 2016). Nutrient methods have been advocated as an adjuvant to other therapies after injury and as a means of promoting resilience before TBI due to the research that shows diet can influence brain function in both healthy and wounded populations (Erdman et al. 2011). Additionally, the nutrition status of TBI patients pre- and post-injury could be also important to assess whether specific nutrient supplements would improve their health outcomes (Erdman et al. 2011). For instance, nutritional supplementation has been proposed as a preventative and/or therapeutic approach to mitigate the negative consequences associated with sports-related concussions and subconcussive injuries (Walrand et al. 2021). Although treating concussions related to sports has received a lot of attention, there is currently significant interest in prevention before impact.

Several experimental studies have been conducted using nutrition against TBI prior to injury. In fact, a number of supplements, including sulfur amino acids, minerals, PUFA, vitamins, and antioxidants, have demonstrated encouraging outcomes as concussion recovery or prevention aids in animal research as well as in human studies following TBI (Walrand et al. 2021). Therefore, increased consumption of PUFA, particularly docosahexaenoic acid (DHA) or certain amino acids like cysteine, as well as dietary fortification with micronutrients in particular, the minerals involved in brain resiliency, such as magnesium, zinc, and sulfur are some potential nutritional treatments utilized to help avoid concussion and related symptoms in sports (Walrand et al. 2021).

Additionally, Yu et al. (2019) used the controlled cortical impact rodent model to study the benefits of a preventative diet rich in fruits and vegetables on TBI recovery. It was demonstrated that mice fed the diets exhibited less inflammation, greater neuronal density in the hippocampus, smaller lesions, and improved neuromotor performance. These findings might suggest that consuming a diet high in fruits and vegetables either raw or powdered prevents the harmful consequences of TBI (Yu et al. 2019).

Furthermore, other studies have sustained the neuroprotective effect of CDP-choline against TBI (Başkaya et al. 2000; Dempsey and Rao 2003; Dixon et al. 1997). These studies demonstrated that CDP-choline significantly reduced cortical contusion magnitude, enhanced neurological recovery, and prevented TBI-induced hippocampal neuronal death (Erdman et al. 2011). Moreover, pre-injury dietary supplementation with fish oil effectively reduces elevations in post-injury protein oxidation, leading to the stabilization of multiple molecular mediators of learning and memory, mitochondrial calcium homeostasis, cellular energy homeostasis, and improved cognitive performance as demonstrated by studies in rodent models of TBI (Wu et al. 2007). Consequently, nutritional supplementation may be beneficial both prior and following TBI (Turnagöl et al. 2022).

On the other hand, magnesium therapy may be able to assist in the restoration of the cell's ability to do these tasks since after a TBI, intracellular magnesium concentrations drop dramatically. Interestingly, it was also found magnesium pretreatment can stop the decrease in intracellular free magnesium following trauma (McIntosh et al. 1987; Vink et al. 1988). In addition, magnesium therapy was reported to lower injury-induced cognitive impairment in the Morris water maze by drastically reducing the loss of ipsilateral hippocampal cells in the brain after fluid percussion injury (Enomoto et al. 2005).

Overall, nutrition appears to have an effect on outcomes within minutes to hours following an injury (Erdman et al. 2011). However, the search for effective nutrition therapies to boost TBI resilience is still fraught with uncertainties and difficulties. Thus, nutritional supplementation before TBI may be advantageous, and further research in this field may be profitable (Turnagöl et al. 2022).

5.7 Nutrition Against TBI Post-injury

TBI and its secondary morbidity usually induce a state of malnutrition. In TBI patients, malnutrition is associated with more infectious complications, poorer neurological outcomes, reduced immune defenses, and a higher mortality rate. It is, therefore, essential to offer suitable dietary treatment in order to avoid additional brain injury and to encourage the best possible recovery. A healthy diet and post-injury supplementation with nutrient compounds, including vitamins, have also been shown to improve functional recovery after TBI (Scrimgeour and Condlin 2014).

Zinc (Zn) is essential for brain health. However, an excess or lack of zinc can worsen neurological diseases and cause brain damage (Li et al. 2022). Prior research has demonstrated that individuals with TBI exhibit markedly reduced serum Zn levels in addition to elevated urine zinc excretion. Young and colleagues carried out a study in post-TBI patients for 3 months using intravenous Zn supplementation to compensate for Zn losses. Findings revealed that the Zn-supplemented group's mortality rate was half that of the control group (Young et al. 1996). Furthermore, a recent study confirmed the effectiveness of zinc supplementation in TBI patients, as evidenced by noticeably increased plasma zinc concentrations, improvements in the Glasgow Coma Scale, as well as a reduction in the duration of hospitalization (Khazdouz et al. 2018).

A rat study showed that dietary zinc therapy following TBI significantly improved the rats' ability to learn and remember spatial information (Cope et al. 2012). Additionally, studies employing chronic zinc supplementation before injury in rats demonstrated a greater protective effect against cognitive and behavioral deficits than those observing zinc as a treatment, suggesting that chronic zinc supplementation may be used in populations at high risk of TBI to avoid unfavorable outcomes (Levenson 2020).

The brain needs PUFA, namely DHA and eicosapentaenoic acid (EPA) to function well. Supplementing with omega-3 has been shown to normalize levels of synapsin-1, cyclic element-binding protein adenosine monophosphate (cAMP) (Kumar et al. 2014), brain-derived neurotrophic factor (BDNF), and cAMP, as well as to increase memory and learning skills and reduce oxidative damage (Kumar et al. 2014). There may be neuroprotective benefits to dietary supplements containing DHA and/or EPA, according to a number of preclinical studies (Bailes and Mills 2010; Mills et al. 2011; Wu et al. 2004). Wu and colleagues recently showed that DHA supplementation raises the amount of DHA in TBI rats' brains, which may help maintain membrane integrity and enhance cognitive performance. Moreover, the consumption of DHA-rich foods led to a decrease in β-amyloid precursor protein, which is a sign of axonal damage. In addition, a reduction in apoptotic markers and an improvement in memory were also observed after 30 days (Wu et al. 2011). Moreover, Zhang and colleagues have also demonstrated that omega-3-rich fish oil preserves the polarity and amount of AQP4, protecting the function of the lymphatic system, and controls the integrity of the blood-brain barrier by preventing the loss of tight junction components brought on by head trauma in mice (Zhang et al. 2020). In TBI, omega-3 supplementation increases neuronal cell survival, apoptosis reduction, synaptic plasticity, neurotransmission, and oxidative stress reduction (Kumar et al. 2014). All these beneficial effects suggest the importance of considering omega-3 supplementation as a therapeutic option for brain injury survivors.

Vitamin D is crucial for the normal development of the brain. It has been shown that vitamin D reduces the inflammatory response after brain trauma, attenuating neuronal damage and improving CNS functional outcomes. Previous studies have found that vitamin D administration can attenuate inflammation and neuronal damage following TBI in animal models, especially when administered in combination with progesterone (Aminmansour et al. 2012; Tang et al. 2015). A study involving human patients with TBI found that progesterone and vitamin D together significantly improved outcomes after TBI, compared with progesterone alone or placebo. In addition, it has been shown that vitamin D deficiency can increase the inflammatory response after TBI. Furthermore, vitamin D lowers the rate of free radical scavenging and reduces oxidation by increasing glutathione, a powerful intracellular antioxidant (Saadatmand et al. 2021). Sharma and colleagues showed that the amount of consciousness and length of mechanical ventilation in the early stages of severe TBI were positively impacted by vitamin D administration during the acute phase of injury. (Sharma et al. 2020). Vitamin D can also enhance regeneration of the axonal and neuronal cytoskeleton and protection of microtubules, thus promoting axogenesis and increasing the diameter of axons injured following brain trauma.

In addition, vitamin D upregulates neurotrophic growth factors and other markers involved in neuronal development, survival, and function. In addition to its CNS neuroprotective effects, vitamin D could also improve the outcome of TBI through its antimicrobial factors. According to Lee and colleagues, vitamin D supplementation can improve long-term performance and cognitive outcomes in individuals with mild-to-moderate TBI who showed considerable vitamin D insufficiency during the acute phase of injury (Lee et al. 2019). Given its significant pharmacological function, vitamin D therapy administered early on may be seen as a sensible intervention for the treatment of TBI.

Vitamin E and its derived molecules are a strong protective agent against the long-term comorbidity of TBI. Impaired BDNF production and reduced cognitive performance have been observed due to increased levels of oxidative stress following TBI. However, dietary supplementation with vitamin E significantly normalized levels of oxidative stress and counteracted cognitive decline in animals with TBI. Additionally, vitamin E significantly stabilized levels of BDNF, which is highly important for the maintenance of synaptic plasticity and cognition (Wu et al. 2010). Moreover, the experiment conducted by Aigou and colleagues suggested that dietary vitamin E supplementation is able to protect against massive TBI by upregulating SIR2 and SOD (Wu et al. 2010). Pretreatment with α-tocopherol, an active form of vitamin E, has been shown to protect against lipid peroxidation in TBI in guinea pigs, while in a rat model of TBI, post-injury therapy decreased cellular inflammatory markers and enhanced cognitive and motor deficits. In human studies, vitamin E treatment improved Glasgow Scale scores and decreased mortality in individuals with severe TBI compared to other groups (Denniss and Barker 2023; Razmkon et al. 2011).

Thanks to its anti-inflammatory and antioxidant effects, curcumin may prevent the brain from injury induced by oxidative stress. It can also cross the BBB and therefore exhibits neuroprotection in certain models of CNS diseases. Studies indicate that curcumin has a major and perhaps neuroprotective impact on TBI (Khayatan et al. 2022). TBI has been shown to affect cognitive performance, and reduced BDNF levels and its downstream effectors in synapsin, CREB, and synaptic plasticity appear to be related to this. Wu and coworkers found that by reducing the negative effects of oxidative stress, curcumin administration restores normal levels of BDNF protein following TBI (Wu et al. 2006). By blocking the TLR-4, NF-κB, and NLRP-3 pathways as well as the BDNF/TrkB, Nrf2, and PI3K/AKT signaling pathways, curcumin can control key signaling pathways and molecules that regulate neuroinflammation in TBI. Preclinical and clinical studies looking at curcumin's impact on TBI support these findings (Farkhondeh et al. 2020).

Further evidence that curcumin treatment enhances hippocampus function has been shown in a rat model of TBI. A decrease in microglial activation in several hippocampal regions and a return to normal astrocyte shape were shown to be the causes of the curcumin-induced improvements in cognitive function, according to a study by Sun et al. Thus, by promoting neurogenesis and decreasing neuroinflammation, curcumin can lessen the harm caused by TBI (Sun et al. 2020). The study by Samini et al. evaluated curcumin's ability to reduce TBI-induced lesions in rats.

The results showed that administration of curcumin prior to injury produced a significant decrease in the volume of cortical lesions associated with TBI. Additional evidence of decreased muscle coordination and balance resulting from TBI was found in the altered rotarod performance and inclined plane tests subsequent to TBI in rats. Nevertheless, curcumin pretreatment enhanced the rats' coordination, indicating that curcumin can notably avert changes in sensory-motor function. Curcumin has a strong neuroprotective potential in TBI, as confirmed by these results.

Nevertheless, a high-fat, high-sucrose (HFS) diet has been shown to exacerbate secondary TBI lesions by enhancing the oxidative and inflammatory processes in the cortex and hippocampus, and reducing compensatory hippocampal cell proliferation, leading to reduced brain recovery after TBI. HFS consumption has been associated with decreased hippocampal expression of proteins required to facilitate neuronal function and synaptic plasticity, notably BDNF, CREB, and synapsin I (Samini et al. 2013). Moreover, Hoane et al. showed that learning and memory impairments were aggravated in TBI mice after 8 weeks of high-fat diet (HFD) feeding. At the same time, neuromuscular strength and motor coordination were impaired by HFD and TBI, implying that HFD is an important risk factor for aggravated pot-TBI injuries (Hoane et al. 2011). Ibeh and colleagues demonstrated that HFD-TBI mice produced more ROS. Similarly, TBI mice fed a high-fat diet had repressed antioxidant defenses, which accounted for the increasing oxidative stress (Stanley et al. 2023). On the other hand, calorie restriction and reduced nutritional intake have been associated with longer lifespans, enhanced brain health, decreased neurodegeneration, and increased synaptic plasticity. Therefore, the greatest protection for the nervous system would likely come from a diet high in vitamins, fiber, phytochemicals, healthy fats, and foods low in added sugars, refined carbs, saturated fats, and overly processed foods. Moreover, a healthy lifestyle that includes exercise, rest, and a balanced diet may also be very important in fostering neuroprotective or, conversely, pathogenic processes that may affect the severity of neurocognitive deficits brought on by TBI (Markovic et al. 2021).

5.8 Conclusion

Nutrition plays a crucial part in the care and healing of TBI patients. Proper nutrition can significantly impact various aspects of TBI management, including cognitive function, physical recovery, and overall well-being. Through a combination of macronutrients and micronutrients, tailored nutrition plans can help support brain healing, reduce inflammation, and mitigate secondary complications. Additionally, nutritional interventions can protect against TBI prior to injury. Collaboration between healthcare professionals is crucial to develop and implement comprehensive nutrition plans that optimize recovery and quality of life for patients recovering from TBI. Further research and clinical studies are warranted to better understand the specific nutritional requirements and optimal interventions for TBI patients, ultimately improving their long-term outcomes and quality of life.

References

Abraham MK, Chang WTW (2016) Subarachnoid hemorrhage. Emerg Med Clin North Am 34(4):901–916. https://doi.org/10.1016/j.emc.2016.06.011

Aminmansour B, Nikbakht H, Ghorbani A, Rezvani M, Rahmani P, Torkashvand M, Nourian M, Moradi M (2012) Comparison of the administration of progesterone versus progesterone and vitamin D in improvement of outcomes in patients with traumatic brain injury: a randomized clinical trial with placebo group. Adv Biomed Res 1:58

Bailes JE, Mills JD (2010) Docosahexaenoic acid reduces traumatic axonal injury in a rodent head injury model. J Neurotrauma 27(9):1617–1624

Ballabh P (2010) Intraventricular hemorrhage in premature infants: mechanism of disease. Pediatr Res 67(1):1–8

Ballabh P (2014) Pathogenesis and prevention of intraventricular hemorrhage. Clin Perinatol 41:47–67. https://doi.org/10.1016/j.clp.2013.09.007

Başkaya MK, Doğan A, Rao AM, Dempsey RJ (2000) Neuroprotective effects of citicoline on brain edema and blood—brain barrier breakdown after traumatic brain injury. J Neurosurg 92(3):448–452

Blumbergs PC, Scott G, Manavis J, Wainwright H, Simpson DA, McLean AJ (1994) Stalning af amyloid percursor protein to study axonal damage in mild head injury. Lancet 344(8929):1055–1056

Bonita R, Thomson S (1985) Subarachnoid hemorrhage: epidemiology, diagnosis, management, and outcome. Stroke 16(4):591–594

Chamoun R, Suki D, Gopinath SP, Goodman JC, Robertson C (2010) Role of extracellular glutamate measured by cerebral microdialysis in severe traumatic brain injury. J Neurosurg 113(3):564–570

Cifu DX, Lew HL (2017) Braddom's rehabilitation care: a clinical handbook E-book. Elsevier Health Sciences

Cope EC, Morris DR, Scrimgeour AG, Levenson CW (2012) Use of zinc as a treatment for traumatic brain injury in the rat: effects on cognitive and behavioral outcomes. Neurorehabil Neural Repair 26(7):907–913

Cortez SC, McIntosh TK, Noble LJ (1989) Experimental fluid percussion brain injury: vascular disruption and neuronal and glial alterations. Brain Res 482(2):271–282

Defense A (2015) Traumatic brain injury: updated definition and reporting. Department of Defense, Washington, DC

Dempsey RJ, Rao VLR (2003) Cytidinediphosphocholine treatment to decrease traumatic brain injury—induced hippocampal neuronal death, cortical contusion volume, and neurological dysfunction in rats. J Neurosurg 98(4):867–873

Denniss RJ, Barker LA (2023) Brain trauma and the secondary cascade in humans: review of the potential role of vitamins in reparative processes and functional outcome. Behav Sci 13(5):388

Dewan MC, Rattani A, Gupta S, Baticulon RE, Hung Y-C, Punchak M, Agrawal A, Adeleye AO, Shrime MG, Rubiano AM (2018) Estimating the global incidence of traumatic brain injury. J Neurosurg 130(4):1080–1097

Di Pietro V, Yakoub KM, Caruso G, Lazzarino G, Signoretti S, Barbey AK, Tavazzi B, Lazzarino G, Belli A, Amorini AM (2020) Antioxidant therapies in traumatic brain injury. Antioxidants 9(3). https://doi.org/10.3390/antiox9030260

Dixon KJ (2017) Pathophysiology of traumatic brain injury. Phys Med Rehabil Clin N Am 28(2):215–225. https://doi.org/10.1016/j.pmr.2016.12.001

Dixon CE, Clifton GL, Lighthall JW, Yaghmai AA, Hayes RL (1991) A controlled cortical impact model of traumatic brain injury in the rat. J Neurosci Methods 39(3):253–262

Dixon CE, Ma X, Marion DW (1997) Effects of CDP-choline treatment on neurobehavioral deficits after TBI and on hippocampal and neocortical acetlycholine release. J Neurotrauma 14(3):161–169

El-Menyar A, Mekkodathil A, Al-Thani H, Consunji R, Latifi R (2017) Incidence, demographics, and outcome of traumatic brain injury in the Middle East: a systematic review. World Neurosurg 107:6–21
Enomoto T, Osugi T, Satoh H, McIntosh TK, Nabeshima T (2005) Pre-injury magnesium treatment prevents traumatic brain injury–induced hippocampal ERK activation, neuronal loss, and cognitive dysfunction in the radial-arm maze test. J Neurotrauma 22(7):783–792
Erdman J, Oria M, Pillsbury L (2011) Improving acute and subacute health outcomes in military personnel. In: Nutrition and traumatic brain injury
Fabbri A, Servadei F, Marchesini G, Stein SC, Vandelli A (2008) Early predictors of unfavourable outcome in subjects with moderate head injury in the emergency department. J Neurol Neurosurg Psychiatry 79(5):567–573
Faden AI, Demediuk P, Panter SS, Vink R (1989) The role of excitatory amino acids and NMDA receptors in traumatic brain injury. Science 244(4906):798–800
Faried A, Halim D, Widjaya IA, Badri RF, Sulaiman SF, Arifin MZ (2019) Correlation between the skull base fracture and the incidence of intracranial hemorrhage in patients with traumatic brain injury. Chin J Traumatol 22(05):286–289
Farkhondeh T, Samarghandian S, Roshanravan B, Peivasteh-Roudsari L (2020) Impact of curcumin on traumatic brain injury and involved molecular signaling pathways. Recent Pat Food Nutr Agric 11(2):137–144
Garton T, Keep RF, Wilkinson DA, Strahle JM (2016) Intraventricular hemorrhage: the role of blood components in secondary intraventricular hemorrhage : the role of blood components in secondary injury and hydrocephalus. Transl Stroke Res 7:447–451. https://doi.org/10.1007/s12975-016-0480-8
Go AS, Mozaffarian D, Roger VL, Benjamin EJ, Berry JD, Blaha MJ, Dai S, Ford ES, Fox CS, Franco S (2014) Heart disease and stroke statistics—2014 update: a report from the American Heart Association. Circulation 129(3):e28–e292
Graham DI, Lantos PL, Greenfield JG (1997) Greenfield's neuropathology
Greve MW, Zink BJ (2009) Pathophysiology of traumatic brain injury. Mt Sinai J Med 15–21. https://doi.org/10.1002/MSJ
Gross BA, Jankowitz BT, Friedlander RM (2019) Cerebral Intraparenchymal hemorrhage: a review. JAMA 321(13):1295–1303. https://doi.org/10.1001/jama.2019.2413
Hamlat A, Mazzon A, Adn M, Morandi X, Riffaud L, Guegan Y, Brassier G (2005) Intracranial epidural haematomas in elderly patients: observations in 14 patients. Acta Neurochir 147:1055–1060
Helbok R, Ko S-B, Schmidt JM, Kurtz P, Fernandez L, Choi HA, Connolly ES, Lee K, Badjatia N, Mayer SA (2011) Global cerebral edema and brain metabolism after subarachnoid hemorrhage. Stroke 42(6):1534–1539
Hoane MR, Swan AA, Heck SE (2011) The effects of a high-fat sucrose diet on functional outcome following cortical contusion injury in the rat. Behav Brain Res 223(1):119–124
Irie F, Le Brocque R, Kenardy J, Bellamy N, Tetsworth K, Pollard C (2011) Epidemiology of traumatic epidural hematoma in young age. J Trauma -Injury Infection Crit Care 71(4):847–853. https://doi.org/10.1097/TA.0b013e3182032c9a
Ismail H, Shakkour Z, Tabet M, Abdelhady S, Kobaisi A, Abedi R, Nasrallah L, Pintus G, Al-Dhaheri Y, Mondello S, El-Khoury R, Eid AH, Kobeissy F, Salameh J (2020) Traumatic brain injury: oxidative stress and novel anti-oxidants such as mitoquinone and edaravone. Antioxidants 9(10):1–18. https://doi.org/10.3390/antiox9100943
Johnson WD, Griswold DP (2017) Traumatic brain injury: a global challenge. Lancet Neurol 16(12):949–950
Johnston SC, Selvin S, Gress DR (1998) The burden, trends, and demographics of mortality from subarachnoid hemorrhage. Neurology 50(5):1413–1418
Jones NR, Molloy CJ, Kloeden CN, North JB, Simpson DA (1993) Extradural haematoma: trends in outcome over 35 years. Br J Neurosurg 7(5):465–471

Kassell NF, Torner JC, Haley EC, Jane JA, Adams HP, Kongable GL (1990) The international cooperative studyon the timing of aneurysm surgery: part 1: overall management results. J Neurosurg 73(1):18–36

Katayama Y, Becker DP, Tamura T, Hovda DA (1990) Massive increases in extracellular potassium and the indiscriminate release of glutamate following concussive brain injury. J Neurosurg 73(6):889–900

Khayatan D, Razavi SM, Arab ZN, Niknejad AH, Nouri K, Momtaz S, Gumpricht E, Jamialahmadi T, Abdolghaffari AH, Barreto GE (2022) Protective effects of curcumin against traumatic brain injury. Biomed Pharmacother 154:113621

Khazdouz M, Mazidi M, Ehsaei M, Ferns G, Kengne AP, Norouzy A-R (2018) Impact of zinc supplementation on the clinical outcomes of patients with severe head trauma: a double-blind randomized clinical trial. J Dietary Suppl 15(1):1–10

Koura SS, Doppenberg EMR, Marmarou A, Choi S, Young HF, Bullock R (1998) Relationship between excitatory amino acid release and outcome after severe human head injury. Intracranial pressure and Neuromonitoring in brain injury: proceedings of the tenth international ICP symposium, Williamsburg, Virginia, May 25–29, 1997, 244–246

Kumar PR, Essa MM, Al-Adawi S, Dradekh G, Memon MA, Akbar M, Manivasagam T (2014) Omega-3 fatty acids could alleviate the risks of traumatic brain injury–a mini review. J Tradit Complement Med 4(2):89–92

Lee JM, Jeong SW, Kim MY, Park JB, Kim MS (2019) The effect of vitamin D supplementation in acute traumatic brain injury patients. World Neurosurg 126:e1421–e1426

Levenson CW (2020) Zinc and traumatic brain injury: from chelation to supplementation. Med Sci (Basel, Switzerland) 8(3). https://doi.org/10.3390/medsci8030036

Lewis MD (2016) Concussions, traumatic brain injury, and the innovative use of omega-3s. J Am Coll Nutr 35(5):469–475. https://doi.org/10.1080/07315724.2016.1150796

Li Z, Liu Y, Wei R, Yong VW, Xue M (2022) The important role of zinc in neurological diseases. Biomol Ther 13(1):28

Longstreth WT, Nelson LM, Koepsell TD, van Belle G (1994) Subarachnoid hemorrhage and hormonal factors in women: a population-based case-control study. Ann Intern Med 121(3):168–173

Louveau A, Smirnov I, Keyes TJ, Eccles JD, Rouhani SJ, Peske JD, Derecki NC, Castle D, Mandell JW, Lee KS (2015) Structural and functional features of central nervous system lymphatic vessels. Nature 523(7560):337–341

Maas AIR, Stocchetti N, Bullock R (2008) Moderate and severe traumatic brain injury in adults. Lancet Neurol 7(8):728–741

Maas AIR, Menon DK, Adelson PD, Andelic N, Bell MJ, Belli A, Bragge P, Brazinova A, Büki A, Chesnut RM (2017) Traumatic brain injury: integrated approaches to improve prevention, clinical care, and research. Lancet Neurol 16(12):987–1048

Markovic SJ, Fitzgerald M, Peiffer JJ, Scott BR, Rainey-Smith SR, Sohrabi HR, Brown BM (2021) The impact of exercise, sleep, and diet on neurocognitive recovery from mild traumatic brain injury in older adults: a narrative review. Ageing Res Rev 68:101322

Mathew P, Graham DI, Bullock R, Maxwell W, McCulloch J, Teasdale G (1994) Focal brain injury: histological evidence of delayed inflammatory response in a new rodent model of focal cortical injury. Brain Edema IX: Proceedings of the Ninth International Symposium Tokyo, May 16–19, 1993, 428–430

McGinn MJ, Povlishock JT (2016) Pathophysiology of traumatic brain injury. Neurosurg Clin N Am 27(4):397–407. https://doi.org/10.1016/j.nec.2016.06.002

McIntosh TK, Vink R, Noble L, Yamakami I, Fernyak S, Soares H, Faden AL (1989) Traumatic brain injury in the rat: characterization of a lateral fluid-percussion model. Neuroscience 28(1):233–244

McIntosh TK, Faden AI, Bendall MR, Vink R (1987) Traumatic brain injury in the rat: alterations in brain lactate and pH as characterized by 1H and 31P nuclear magnetic resonance. J Neurochem 49(5):1530–1540

Menon DK, Schwab K, Wright DW, Maas AI (2010) Position statement: definition of traumatic brain injury. Arch Phys Med Rehabil 91(11):1637–1640

Mills JD, Bailes JE, Sedney CL, Hutchins H, Sears B (2011) Omega-3 fatty acid supplementation and reduction of traumatic axonal injury in a rodent head injury model. J Neurosurg 114(1):77–84

Mtbi C (2004) Methodological issues and research recommendations for mild traumatic brain injury: the WHO Collaborating Centre Task Force on Mild Traumatic Brain Injury. J Rehabil Med 43:113–125

Munoz-Sánchez MA, Murillo-Cabezas F, Cayuela-Domínguez A, Rincón-Ferrari MD, Amaya-Villar R, León-Carrión J (2009) Skull fracture, with or without clinical signs, in mTBI is an independent risk marker for neurosurgically relevant intracranial lesion: a cohort study. Brain Inj 23(1):39–44

Organization WH (2006) Neurological disorders: public health challenges. World Health Organization

Razmkon A, Sadidi A, Sherafat-Kazemzadeh E, Mehrafshan A, Jamali M, Malekpour B, Saghafinia M (2011) Administration of vitamin C and vitamin E in severe head injury: a randomized double-blind controlled trial. Neurosurgery 58:133–137

Reilly P, Bullock R (1997) Head injury: pathophysiology and management of severe closed injury. Chapman Hall

Rock KL, Kono H (2008) The inflammatory response to cell death. Annu Rev Pathol Mech Dis 3:99–126

Saadatmand K, Khan S, Hassan Q, Hautamaki R, Ashouri R, Lua J, Doré S (2021) Benefits of vitamin D supplementation to attenuate TBI secondary injury? Transl Neurosci 12(1):533–544

Samini F, Samarghandian S, Borji A, Mohammadi G (2013) Curcumin pretreatment attenuates brain lesion size and improves neurological function following traumatic brain injury in the rat. Pharmacol Biochem Behav 110:238–244

Scrimgeour AG, Condlin ML (2014) Nutritional treatment for traumatic brain injury. J Neurotrauma 31(11):989–999. https://doi.org/10.1089/NEU.2013.3234

Seelig JM, Becker DP, Miller JD, Greenberg RP, Ward JD, Choi SC (1981) Traumatic acute subdural hematoma: major mortality reduction in comatose patients treated within four hours. N Engl J Med 304(25):1511–1518

Sharma S, Kumar A, Choudhary A, Sharma S, Khurana L, Sharma N, Kumar V, Bisht A (2020) Neuroprotective role of oral vitamin D supplementation on consciousness and inflammatory biomarkers in determining severity outcome in acute traumatic brain injury patients: a double-blind randomized clinical trial. Clin Drug Investig 40:327–334

Stanley I, Bakkar N-MZ, Ahmad F, Nwaiwu J, Barsa C, Mekhjian S, Reslan MA, Eid AH, Harati H, Nabha S (2023) High fat diet exacerbates long term metabolic, neuropathological, and behavioral derangements in an experimental mouse model of traumatic brain injury

Strahle J, Garton HJL, Maher CO, Muraszko KM, Keep RF, Xi G (2012) Mechanisms of hydrocephalus after neonatal and adult intraventricular hemorrhage. Transl Stroke Res 3(Suppl 1):25–38

Sun G, Miao Z, Ye Y, Zhao P, Fan L, Bao Z, Tu Y, Li C, Chao H, Xu X (2020) Curcumin alleviates neuroinflammation, enhances hippocampal neurogenesis, and improves spatial memory after traumatic brain injury. Brain Res Bull 162:84–93

Tang H, Hua F, Wang J, Yousuf S, Atif F, Sayeed I, Stein DG (2015) Progesterone and vitamin D combination therapy modulates inflammatory response after traumatic brain injury. Brain Inj 29(10):1165–1174

Taylor CA, Bell JM, Breiding MJ, Xu L (2017) Traumatic brain injury–related emergency department visits, hospitalizations, and deaths—United States, 2007 and 2013. MMWR Surveill Summ 66(9):1–16

Thurman D, Guerrero J (1999) Trends in hospitalization associated with traumatic brain injury. JAMA 282(10):954–957

Tseng W-C, Shih H-M, Su Y-C, Chen H-W, Hsiao K-Y, Chen I-C (2011) The association between skull bone fractures and outcomes in patients with severe traumatic brain injury. J Trauma Acute Care Surg 71(6):1611–1614

Turnagöl HH, Koşar, Şükran N, Güzel Y, Aktitiz S, Atakan MM (2022) Nutritional considerations for injury prevention and recovery in combat sports. Nutrients 14(1). https://doi.org/10.3390/nu14010053

Vink R, McIntosh TK, Demediuk P, Weiner MW, Faden AI (1988) Decline in intracellular free Mg2+ is associated with irreversible tissue injury after brain trauma. J Biol Chem 263(2):757–761

Walrand S, Gaulmin R, Aubin R, Sapin V, Coste A, Abbot M (2021) Nutritional factors in sport-related concussion. Neurochirurgie 67(3):255–258. https://doi.org/10.1016/j.neuchi.2021.02.001

Wartenberg KE, Schmidt JM, Claassen J, Temes RE, Frontera JA, Ostapkovich N, Parra A, Connolly ES, Mayer SA (2006) Impact of medical complications on outcome after subarachnoid hemorrhage. Crit Care Med 34(3):617–623

Werner C, Engelhard K (2007) Pathophysiology of traumatic brain injury. Br J Anaesth 99(1):4–9. https://doi.org/10.1093/bja/aem131

Wojcik BE, Stein CR, Bagg K, Humphrey RJ, Orosco J (2010) Traumatic brain injury hospitalizations of US army soldiers deployed to Afghanistan and Iraq. Am J Prev Med 38(1):S108–S116

Wu A, Ying Z, Gomez-Pinilla F (2004) Dietary omega-3 fatty acids normalize BDNF levels, reduce oxidative damage, and counteract learning disability after traumatic brain injury in rats. J Neurotrauma 21(10):1457–1467

Wu A, Ying Z, Gomez-Pinilla F (2006) Dietary curcumin counteracts the outcome of traumatic brain injury on oxidative stress, synaptic plasticity, and cognition. Exp Neurol 197(2):309–317

Wu A, Ying ZHE, Gomez-Pinilla F (2007) Omega-3 fatty acids supplementation restores mechanisms that maintain brain homeostasis in traumatic brain injury. J Neurotrauma 24(10):1587–1595

Wu A, Ying Z, Gomez-Pinilla F (2010) Vitamin E protects against oxidative damage and learning disability after mild traumatic brain injury in rats. Neurorehabil Neural Repair 24(3):290–298

Wu A, Ying Z, Gomez-Pinilla F (2011) The salutary effects of DHA dietary supplementation on cognition, neuroplasticity, and membrane homeostasis after brain trauma. J Neurotrauma 28(10):2113–2122. https://doi.org/10.1089/NEU.2011.1872

Yokobori S, Yamaguchi M, Igarashi Y, Hironaka K, Onda H, Kuwamoto K, Araki T, Fuse A, Yokota H (2016) Outcome and refractory factor of intensive treatment for geriatric traumatic brain injury: analysis of 1165 cases registered in the Japan Neurotrauma Data Bank. World Neurosurg 86:127–133

Young B, Ott L, Kasarskis E, Rapp R, Moles KAY, Dempsey RJ, Tibbs PA, Kryscio R, McClain C (1996) Zinc supplementation is associated with improved neurologic recovery rate and visceral protein levels of patients with severe closed head injury. J Neurotrauma 13(1):25–34

Yu J, Zhu H, Taheri S, Monday W, Perry S, Kindy MS (2019) Reduced neuroinflammation and improved functional recovery after traumatic brain injury by prophylactic diet supplementation in mice. Nutrients 11(2). https://doi.org/10.3390/nu11020299

Zhang E, Wan X, Yang L, Wang D, Chen Z, Chen Y, Liu M, Zhang G, Wu J, Han H (2020) Omega-3 polyunsaturated fatty acids alleviate traumatic brain injury by regulating the glymphatic pathway in mice. Front Neurol 11:707

Zink BJ, Szmydynger-Chodobska J, Chodobski A (2010) Emerging concepts in the pathophysiology of traumatic brain injury. Psychiatr Clin N Am 33(4):741–756. https://doi.org/10.1016/j.psc.2010.08.005

Zvejniece L, Stelfa G, Vavers E, Kupats E, Kuka J, Svalbe B, Zvejniece B, Albert-Weissenberger C, Sirén A-L, Plesnila N (2020) Skull fractures induce neuroinflammation and worsen outcomes after closed head injury in mice. J Neurotrauma 37(2):295–304

Chapter 6
Antioxidants and Traumatic Brain Injury

Layal Msheik, Batoul Taher, Zeinab Kazan, Safaa Joumaa, Nour Fakih, and Hiba Hamdar

Abstract This chapter investigates the relationship between oxidative stress and traumatic brain injury (TBI) pathophysiology, with a focus on reactive oxygen species (ROS) and free radicals. It describes antioxidants' protective potential, classifying them as enzymatic or nonenzymatic, and delineating their modes of action in scavenging, repairing, and avoiding oxidative damage. Neuroprotective antioxidants such as *N*-acetylcysteine, Alpha Lipoic Acid (ALA), vitamins C and E, glutathione, and melatonin are mentioned. The abstract emphasizes their involvement in cellular component stabilization, neural damage prevention, and cell survival. TBI's effect on the blood–brain barrier (BBB) and the ability of antioxidants to preserve BBB integrity via anti-inflammatory effects are investigated. Endogenous antioxidant mechanisms are enhanced, and exogenous antioxidants are used to boost these defenses. The chapter dives into antioxidants' control of apoptosis pathways, avoiding excessive cell death and assisting in tissue healing. Furthermore, it investigates research indicating antioxidants' protective benefits against cognitive decline and neurological abnormalities linked with TBI, stressing their potential for brain plasticity and cognitive function recovery. Finally, the notion of combining antioxidants with other medications and adapting interventions to specific patient needs is reviewed, providing a thorough overview of antioxidants in TBI and prospective research options.

Keywords Traumatic brain injury · Antioxidant · Reactive oxygen species · Vitamins · Inflammation · Free radicals

L. Msheik · B. Taher · Z. Kazan · S. Joumaa
Faculty of Medical Sciences, Lebanese University, Beirut, Lebanon

Medical Learning Skills Academy, Beirut, Lebanon

N. Fakih
Medical Learning Skills Academy, Beirut, Lebanon

Department of Natural Sciences, Lebanese American University, Beirut, Lebanon

H. Hamdar (✉)
Medical Learning Skills Academy, Beirut, Lebanon

W. Mohamed (ed.), *Nutrition and Traumatic Brain Injury (TBI)*, Nutritional Neurosciences, https://doi.org/10.1007/978-981-97-6341-2_6

6.1 Introduction

Traumatic brain injury (TBI) is a complex and devastating medical condition that affects millions of people worldwide often resulting in long-term physical, cognitive, and emotional impairments. According to the Centers for Disease Control and Prevention (CDC), a traumatic brain injury (TBI) results from either a forceful impact, a blow, or a strong jolt to the head, or it can occur when something penetrates the head, leading to an abnormal disruption in the brain's normal functioning. These traumatic brain injuries can be broadly categorized as either closed injuries (nonpenetrating) or open injuries (penetrating) (Surveillance Report of Traumatic Brain Injury-related Emergency Department Visits, Hospitalizations, and Deaths 2014). Annually, about 1.7 million individuals suffer traumatic brain injuries, which tragically lead to approximately 52,000 deaths. A substantial number of those affected, approximately 124,000 individuals, end up with long-term disabilities as a result of these injuries (Fesharaki-Zadeh 2022). The financial burden of these disabilities is substantial, with an estimated annual cost of $56 billion, underscoring a significant socioeconomic challenge (Wang et al. 2016). Traumatic brain injury (TBI) has a complex and heterogeneous pathophysiological etiology that manifests on multiple scales. TBI can lead to diverse pathological changes, including axonal shearing, or diffuse axonal injury (DAI), neurovascular compartment damage, neuronal network disruption, as well as apoptotic and necrotic cellular damage, all contributing to the multifaceted nature of TBI's impact on the brain (Fesharaki-Zadeh 2022). The post-injury neuroinflammatory response encompasses the release of proinflammatory cytokines and the formation of free radicals, both of which are key components of the immune and oxidative stress responses in the injured brain (Shi et al. 2019).

The injury's pathophysiology is a complex interplay of various mechanisms, including the initial mechanical damage and a cascade of secondary injury processes that unfold over time. Among these secondary processes, oxidative stress has emerged as a critical player in the progression of TBI-induced damage (Schimmel et al. 2017). Oxidative stress results from the disruption of the balance between the production of highly reactive molecules known as free radicals and the body's ability to neutralize them using antioxidants. Free radicals are highly reactive and capable of causing cellular and tissue damage by initiating a process known as oxidation (Lobo et al. 2010). Oxidative stress plays a key role in the ongoing damage in the brain when it comes to TBI. It has been linked to cellular damage, an increase in inflammation, and disruption of the blood–brain barrier (BBB), all of which are secondary injury processes (Fesharaki-Zadeh 2022).

The acknowledgment of the crucial role of oxidative stress in TBI has sparked interest in antioxidant therapy as a potential strategy to mitigate its deleterious consequences (Ashok et al. 2022). Antioxidants are a variety of compounds found naturally in the body or obtained through food sources or supplementation. Their major function is to transfer electrons to free radicals, effectively neutralizing them and preventing further harm to cells. Many antioxidants include anti-inflammatory and neuroprotective actions in addition to direct antioxidant activities, making them particularly appropriate for TBI treatments. Furthermore, antioxidants can influence processes such as inflammation and cell survival by modulating multiple cellular signaling pathways and gene expression (Scheff and Ansari 2017).

The diverse array of antioxidants offers a promising avenue for intervention in TBI. The spectrum of antioxidants studied in the context of TBI is extensive and includes vitamins (such as C, E, and A), polyphenols (found in foods like tea and fruits), antioxidant enzymes (like superoxide dismutase and catalase), omega-3 fatty acids (commonly found in fish oil), and synthetic antioxidants (like *N*-acetylcysteine) (Farooqui and Farooqui 2017). These antioxidants have demonstrated the potential to reduce oxidative damage, inflammation, and cell death, ultimately promoting recovery and improving outcomes in TBI patients (Di Pietro et al. 2020). However, translating antioxidant therapies from the laboratory to clinical practice is a complex and challenging process (Ashok et al. 2022). Clinical studies in TBI frequently confront challenges due to the heterogeneity of patient groups, varied injury severity, and complex interactions with different therapy modalities (Maas et al. 2004). Nonetheless, the potential for antioxidants to improve outcomes in TBI patients is an enthralling avenue of research that offers hope and commitment for enhanced therapeutic strategies in the future (Di Pietro et al. 2020).

In this chapter, we will investigate the complex interplay between antioxidants and TBI, focusing on the mechanisms driving oxidative stress, its role in TBI pathophysiology, and the secondary cell death pathways that are involved. Moreover, by getting a comprehensive understanding of the role of antioxidants in TBI, we intend to shed light on the promising potential they possess for improving TBI outcomes and opening up new perspectives for TBI management.

6.2 Discussion

6.2.1 Oxidative Stress's Role in TBI Pathology

The pathophysiological etiology of traumatic brain injury (TBI) is diverse and manifests at several levels. TBI can lead to several pathological alterations, such as axonal shearing, often referred to as diffuse axonal injury (DAI), damage to the neurovascular compartment, disruption of neuronal networks, and cellular damage characterized by apoptosis and necrosis (Cordaro et al. 2021; Acosta et al. 2017). The consequences of these alternations are profound. To illustrate, oxidative damage in TBI patients plays a significant role in disrupting critical cellular components and processes such as lipids, proteins, and nucleic acids (Chan 2001). Moreover, the post-injury neuroinflammatory response encompasses the release of proinflammatory cytokines and the formation of free radicals, both of which are key components of the immune and oxidative stress responses in the injured brain (Shi et al. 2019). Oxidative stress is a pivotal factor in TBI, marked by the generation of reactive oxygen species (ROS), such as superoxide and hydrogen peroxide, and reactive nitrogen species (RNS), including peroxynitrites (PN). This imbalance typically results from the interplay between the biochemical pathways that produce ROS and those responsible for their removal, including the enzymatic and nonenzymatic antioxidant cellular defense mechanisms. For instance, oxidative enzymes, particularly NADPH oxidase (NOX) and NOX2 are overproduced in oxidative damage in TBI (Fesharaki-Zadeh 2022). Furthermore, oxidative stress following traumatic brain injury (TBI) has been substantiated by both preclinical and human studies (Jamjoom et al. 2021). This oxidative stress has been observed to endure for up to 6 months post-TBI, as demonstrated in studies like Licastro et al. (2016). Reactive oxygen species (ROS) have damaging effects on nucleic acids, leading to DNA fragmentation, which includes both single- and double-stranded breaks, with the latter being irreversible and manifesting a few hours after the brain injury. Additionally, it can result in the modification and loss of DNA bases, further compromising genetic integrity (Mendes Arent et al. 2014). Moreover, ROS harm cell membranes by interacting with polyunsaturated fatty acids, leading to the formation of lipoperoxyl radicals. This process, known as lipid peroxidation, can disrupt the integrity of cell membranes, increase permeability, and decrease ATPase activity, potentially compromising their function and structural integrity, leading to further injury (Fesharaki-Zadeh 2022; Thapa et al. 2021).

Traumatic brain injuries disrupt normal brain cell function through direct, rotational, and shear forces, with these forces present across all injury severities. Rotational forces can harm the brain's white matter tracts, potentially causing diffuse axonal injury, which can be evaluated using a specialized MRI technique called diffusion tensor imaging. Additionally, axonal injury leads to local swelling, which hinders signal transmission, while traumatic injury can alter cerebral blood flow, initially causing reduced cerebral blood flow (CBF) followed by unresponsive vasodilation, possibly due to nitric oxide release in the tissue (Capizzi et al. 2020) (Fig. 6.1).

6.2.2 *Antioxidants: General Overview*

Free radicals are created by the body's natural cell metabolism as well as by external factors (such as pollution, radiation, and medications). These free radicals are normally neutralized by various antioxidant systems in our bodies. The disruption of equilibrium between free radicals and antioxidant systems can result in the occurrence of oxidative stress (OxS), which is responsible for the impairment of biomolecules and the subsequent development of various chronic degenerative diseases (Sies 2015).

A. Endogenous Antioxidants

   (a) Enzymatic:

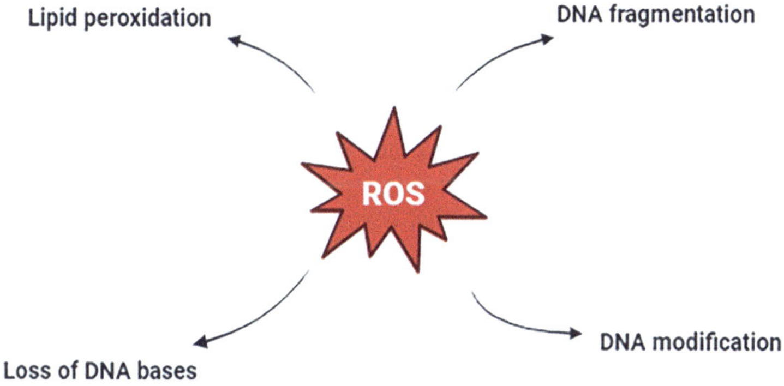

Fig. 6.1 Effects of ROS

Superoxide dismutases (SODs) catalase the dismutation of highly reactive superoxide radicals (O_2•−) into hydrogen peroxide (H_2O_2) and molecular oxygen. While, catalase (CAT) can transform hydrogen peroxide (H_2O_2) into water and molecular oxygen. Glutathione peroxidase (GPx) facilitates the reduction of hydrogen peroxide (H_2O_2) or organic peroxide (ROOH) to water or alcohol. The reaction under consideration involves the utilization of reduced glutathione (GSH), which undergoes oxidation to become oxidized glutathione (GSSG) (Aguilar et al. 2016).

(b) Nonenzymatic:

The reduced form of glutathione (GSH) is the active form that can directly protect against reactive oxygen/nitrogen species, in addition to functioning as a coenzyme for detoxication enzymes such as GPx mentioned above (Aguilar et al. 2016). Finally, melatonin can directly detoxify reactive oxygen and nitrogen species and indirectly stimulate antioxidant enzymes while suppressing the activity of prooxidant enzymes. In addition to that, it suppresses the formation of toxic hydroxyl radicals, reducing oxidative stress in the process (Reiter et al. 2016).

B. Exogenous Antioxidants

Exogenous antioxidants are antioxidants that come from external sources, such as dietary intake or supplements. They possess the capacity to exert a noteworthy influence on enhancing the body's inherent antioxidant systems, fortifying cellular protection against oxidative stress through various mechanisms, each of which will be exemplified (Zingg 2019).

Direct Scavenging of Free Radicals: Vitamin E can act as a lipid soluble antioxidant by collecting reactive oxygen and nitrogen species, protecting the cell membrane and preventing its destabilization and therefore cell dysfunction (Bouayed and Bohn 2010).

Regeneration of Antioxidants: Vitamin C can regenerate vitamin E by undertaking an ox-red reaction, causing the reduction of vitamin E radicals formed by scavenging free radicals (Niki 1987).

Enhanced Enzymatic Activity: Selenium (Se) acts as a cofactor and is an essential component of many enzymes, such as GSH, that fulfill their role as antioxidant enzymes (Sedighi et al. 2014).

Chelation of metals: The process of chelation involving metals is observed in curcumin, which exhibits the ability to interact with metals within the body. This binding action effectively inhibits the catalytic activity of metals, thereby limiting the occurrence of harmful processes that develop free radicals (Mary et al. 2018).

Antioxidants can also be found in cellular signaling pathways that control the body's response to oxidative stress. Nuclear factor erythroid 2-related factor 2 (Nrf2), for example, is a transcription factor that, when activated by antioxidants, can increase the production of several antioxidant and detoxification enzymes, enhancing the cell's ability to combat oxidative damage (Lobo et al. 2010). Finally, certain antioxidants, such as vitamin C and glutathione, have the ability to replenish other antioxidants. Vitamin C, for example, may renew vitamin E, allowing it to continue defending cell membranes from lipid peroxidation (Salehi et al. 2018; Lü et al. 2010) (Table 6.1).

6.2.3 Major Antioxidants and Their Role in Traumatic Brain Injury

The *N*-acetylcysteine (NAC) antioxidant action is owing to its ability to operate as a reduced glutathione (GSH) precursor; GSH is a well-known direct antioxidant as well as a substrate for various antioxidant enzymes. Furthermore, under some settings where endogenous Cys and GSH are depleted, NAC can operate as a direct antioxidant for some oxidant species such as NO_2 and HO_X. Its antioxidant action

Table 6.1 Types of endogenous antioxidants and their main characteristics with examples (Irato and Santovito 2021)

Type of endogenous antioxidants	Enzymatic	Nonenzymatic
Characteristics	• Catalyzing specific chemical reactions that help neutralize free radicals	• Neutralize free radicals and oxidative stress • Obtained from the diet or produced within the body
Examples	• Alpha-dioxygenase • Ascorbate peroxidase • Catalase • Dehydroascorbate reductases • Glutathione peroxidase • Glutathione reductase • Glutathione-S-transferase • NADPH oxidase • Peroxiredoxin • Superoxide dismutase	• Ascorbate • Fibrin • Glutathione • Melatonin • Mycothiol • Phenolics • Serum albumin

could possibly be attributed to its ability to break down thiolated proteins, releasing free thiols as well as reduced proteins, which have significant direct antioxidant activity in some situations (Aldini et al. 2018).

There are no therapies shown to improve outcome after severe traumatic brain injury (TBI). However, clinical trial shows that *N*-acetylcysteine (NAC) and the synergistic adjuvant probenecid combination shows protective effect (Clark et al. 2017).

Delayed traumatic brain injury includes neuroinflammation, oxidative damage, apoptosis, and glutamate toxicity. NAC inhances brain glutathione, prevents lipid oxidation, and protects mitochondria (Lawless and Bergold 2022). Extensive research has been conducted on the antioxidant benefits of Alpha Lipoic Acid (ALA), including its ability to serve as a metal-chelating agent, scavenge free radicals, regenerate endogenous antioxidants like glutathione and vitamins C and E, and repair oxidized damage. It has the power to raise glutathione levels within cells, which chelate and help the body eliminate a wide range of toxins, including harmful metals. For instance, research indicated that delivery of ALA resulted in a reduction of iron ions in epithelial cells. This alteration was linked to increased cell resistance to a hydrogen peroxide challenge, indicating that ALA directly affects the decrease of oxidative stress (Salehi et al. 2019).

Furthermore, vitamin E, a powerful antioxidant, and its related compounds present an appealing possibility for long-term comorbidity linked with TBI (Dobrovolny et al. 2018). Vitamin E is a vital free radical-scavenging antioxidant for humans; nevertheless, it is only effective against free radical-mediated oxidation and not against non-radical oxidants (Niki and Noguchi 2021).

Low serum vitamin E level was associated with poor prognosis at 1 and 6 months after TBI with intracranial injury. In clinical settings, early antioxidant supplementation, including vitamin E in critically ill patients, including TBI patients, was associated with decrease in organ failure and hospital stay. Vitamin E supplementation showed a significant reduction in mortality and improvement of long-term functional outcomes of TBI patients (Park et al. 2022) (Fig. 6.2).

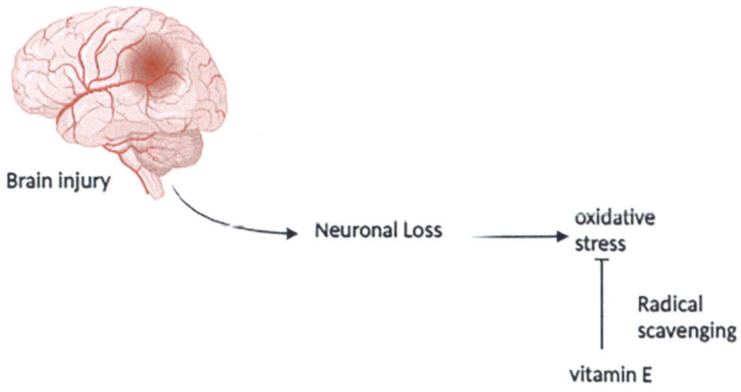

Fig. 6.2 Protective role of vitamin E against oxidative stress

Ascorbate works as an electron donor in redox chemistry, making it a powerful reducing agent and free radical scavenger, capable of neutralizing O_2•, OH•, O_3, NO_2•, and ROO• (Tyuryaeva and Lyublinskaya 2023). These results have been emphasized by a systemic review that shows that in adult patients with TBI, antioxidant therapy such as amino acids, vitamins C and E, progesterone, *N*-acetylcysteine, and endogenous may be safe and effective supplementary therapies (Shen et al. 2016).

On the other hand, transcription factor NF-E2-related factor 2 (Nrf2) defends against oxidative stress, neuroinflammation, and neuronal death; it regulates antioxidant proteins such as oxygenase-1 (HO-1), NAD(P)H quinone dehydrogenase 1 (NQO1), and glutathione peroxidase (GPx). TBI pathogenesis is exacerbated by Nrf2 knockout (KO), whereas Nrf2 activator therapy reduces neuronal apoptosis and neuroinflammatory responses by lowering oxidative damage (Wu et al. 2022).

Glutathione (GSH) is the major low-molecular-mass thiol within cells and one of the most abundant antioxidants in body fluids. Two GSH molecules contribute one electron each and are transformed into glutathione disulfide (GSSG) under prooxidant circumstances. The GSH/GSSG molar ratio is regarded as a reliable indicator of oxidative stress and disease risk (Giustarini et al. 2016).

However, a study found that reactive oxygen species (ROS) produced by repetitive head injury are largely responsible for increased cortical cell death and that therapeutic administration of the antioxidant glutathione significantly reduces this cell death, preserves the glia limitans, and prevents myelomonocytic cells from entering the brain parenchyma (Mason et al. 2021).

Melatonin reduces oxidative stress through a variety of mechanisms, including direct detoxification of reactive oxygen and reactive nitrogen species and indirect stimulation of antioxidant enzymes while decreasing the activity of prooxidant enzymes. Furthermore, melatonin chelates transition metals, which lowers the generation of the highly harmful hydroxyl radical, resulting in a reduction in oxidative stress. Its large concentrations in mitochondria most likely help it survive oxidative damage and cellular death (Reiter et al. 2016). Treatment with ramelteon (a melatonin MT1/MT2 receptor agonist) appears to inhibit oxidative stress and inflammation through the Nrf2-antioxidant response element pathway and may be a possible chronotherapeutic approach for the management of TBI (Wang et al. 2019).

Finally, phenolic acids are mainly found in fruits, olives, tea, and cereals (Manach et al. 2004). Many mechanisms allow phenolic acids to exert neuroprotection: they can reduce prooxidant and proinflammatory mediator levels; increased antioxidant and anti-inflammatory mediator production and/or activity; strengthen the antioxidant machinery; rescue and/or increase neurotrophic factor release; prevent and/or counteract protein aggregation; increase the production and function of antiapoptotic mediators and reduce that of proapoptotic mediators; activate pro-survival signaling pathways while also blocking cell death signaling pathways; and reduce the activity of reactive astrocytes and microglia (Caruso et al. 2022) (Fig. 6.3).

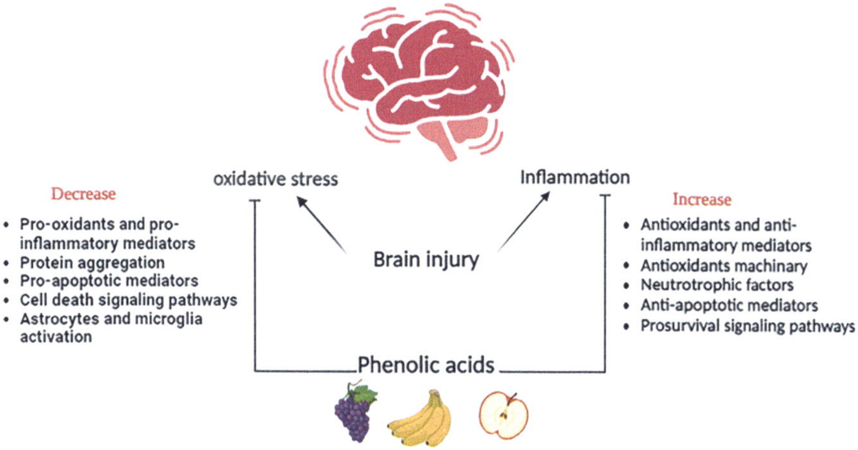

Fig. 6.3 Neuroprotective role of phenolic acids

6.2.4 Protection of Neuronal Cells

Traumatic brain injury (TBI) causes a complex, generally detrimental secondary inflammatory response in the brain. The severity of the inflammation determines the degree of harm. Extravasation of neutrophils, blood–brain barrier disruption, activation of astrocytes and microglia, migration of leucocytes and phagocytes, and cytokine and chemokine synthesis all occur immediately after primary trauma. All of these cause oxidative stress. Reactive oxygen species (ROS) and reactive nitrogen species (RNS) overpower the antioxidative response, causing irreversible cellular damage by reacting with proteins, lipids, carbohydrates, and nucleic acids (Günther et al. 2018).

Furthermore, during TBI, the increased production of ROS and RNS after TBI is caused by severe excitotoxicity and a lack of endogenous antioxidant mechanisms. These reactive species enhance oxidative stress and are associated with the formation of superoxidized cellular and vascular structures, oxidized protein, cleaved DNA, and an altered mitochondrial electron transport chain. Thus, these can cause astrocytes to release proinflammatory factors such as interleukin (IL)-6, monocyte chemoattractant protein (MCP)-1, and matrix metallopeptidase 9 (MMP-9), resulting in BBB compromise and neuroinflammation. Endothelial cells, which are enriched in mitochondria for the normal performance of ATP-dependent active transporters, can thus fail in the presence of elevated oxidative stress and decreased mitochondrial functions. Moreover, oxidative stress and energy deprivation cause cellular ion channel malfunction, which leads to neuronal depolarization and excitatory neurotransmitter aggregation, which aggravates neuronal depolarization and increases hazardous calcium levels. All these mechanisms are sufficient to trigger early or late apoptosis, immediate cell necrosis, and delayed cell necrosis (Zhou et al. 2020). However, patients with good recovery experience less oxidative stress in the acute phase after TBI or have better redox function (Visser et al. 2022).

Free radical scavengers, antioxidant system activators, inhibitors of free radical-producing enzymes, and antioxidant enzymes are examples of antioxidant therapies. These treatments have shown promise in reducing edema, protecting the blood–brain barrier (BBB), reducing contusion volume, and decreasing inflammation. Additionally, numerous antioxidant therapies supported better functional recovery after TBI in terms of sensory, motor, and cognitive abilities (Davis and Vemuganti 2022). According to another study, mild TBI causes early mitochondrial dysfunction, which may determine how vulnerable cells are to recurrent head impact. Therefore, treatments that focus on mitochondrial dysfunction may enhance outcomes following repeated mild TBI (Hubbard et al. 2019).

Furthermore, TBI-induced oxidative damage causes mitochondrial electron transport chain and membrane potential impairments, resulting in an increase in free radical production and cell death. The results of studies on mitochondria-targeted treatments, such as antioxidants, are promising (Hakiminia et al. 2022). Antioxidant therapy has helped patients recovering from TBI regain cognitive and functional abilities, and they represent a promising therapeutic approach (Fesharaki-Zadeh 2022). However, ascorbic acid, an antioxidant, has been used to reverse the detrimental cellular effect that can unintentionally increase ROS (Kim and Priefer 2020).

6.2.5 Antioxidants and Blood–Brain Barrier

BBB functions in maintaining brain homeostasis by ensuring a protected and regulated influx of oxygen and nutrients while safeguarding the fragile cerebral tissue by blocking the entry of bloodborne pathogens, peripheral immune cells, and circulating factors (Chodobski et al. 2011; Keaney and Campbell 2015). This highly selective semipermeable barrier isolates the brain parenchyma from circulating blood and ensures regulated influx-efflux processes while also acting as a solute exchange barrier between blood and brain (Cash and Theus 2020). BBB is composed of specialized CNS endothelial cells distinguished from others by the presence of BBB-specific transporter and receptor proteins to control metabolite entry and exit across cells (transcellular transport) and high electrical resistance tight junctions (TJs) that stick together the endothelial cells to limit movement between adjacent cells (paracellular transport) (Keaney and Campbell 2015). BBB endothelial cells are regulated by cells of "neurovascular unit NVU"; this includes pericytes, microglia, astrocytes, and neurons that help maintain the balance and homeostasis through direct cell-to-cell contact or paracrine soluble factors (Keaney and Campbell 2015; Cash and Theus 2020).

Recent studies have shifted their interest to study beyond BBB to NVU (Chodobski et al. 2011). Astrocytes play a role in maintaining BBB integrity by forming astrocytic end feet around endothelial cells (Karve et al. 2016) and whose processes are in close contact with cerebrovascular endothelium of parenchymal blood microvessels, and thus are critical for normal function of the BBB (Chodobski et al. 2011). In addition, endothelial–astrocyte interactions play an important role in maintaining the homeostasis of the brain microenvironment through its aquaporins-4 water channels as these channels improve the redistribution of water in the brain (Li et al. 2022). Microglia, on the other hand, are specialized immune cells of the brain with phagocytic and antigen-presenting capabilities (Karve et al. 2016) that function in close paracrine manner with the brain endothelium to help protect against toxins and injuries (Chodobski et al. 2011). It is also activated in CNS disorders like Alzheimer's disease (AD) and multiple sclerosis where BBB dysfunction and neuroinflammation usually occur. Pericytes are perivascular cells whose processes ensheath CNS vessel walls and have an important role in regulating capillary diameter, cerebral blood flow, and extracellular matrix protein secretion and levels (Keaney and Campbell 2015).

TBI compromises the integrity of the blood–brain barrier (BBB) early in the injury, occurring within hours of injury (Cash and Theus 2020). Rendering it dysfunctional, this damage could persist for many months or years and is considered to be a major risk factor for high mortality and morbidity (Cash and Theus 2020; Feng et al. 2023). BBB dysfunction is thought to be caused by two biological processes; the first being the loss of tight junction proteins of the barrier and thus an increase in paracellular transport of molecules and cells that are usually restricted, for instance, immune cells like neutrophils that will penetrate the brain parenchyma and exacerbate the inflammatory process. The second process is the transcytosis of larger molecules and proteins like albumin and fibrinogen across endothelial cells, highlighting that those serum proteins are normally restricted from crossing the BBB and entering the brain (Cash and Theus 2020). Importantly, astrocytes are among the first responder brain cells to TBI, and the mechanical forces of the primary injury trigger reactive astrocytes or astrogliosis (Dinet et al. 2019). Both astrocytes and microglia can rapidly respond to injury by increasing the production of cytokines and multiple factors that may have a profound effect on BBB function; this includes TGF-beta, glutamate, ROS, MMP, and vascular endothelial growth factor (VEGF) A, all of which are implicated in BBB alteration after TBI (Chodobski et al. 2011; Dinet et al. 2019). Moreover, protein matrix metallopeptidase 9 (MMP-9), which is extensively produced by reactive astrocytes in response to the primary injury (Dinet et al. 2019) will digest tight junctions between endothelial cells causing more and more entry of immune cells and factors that will further alter the permeability of the BBB (Schimmel et al. 2017; Cash and Theus 2020). In addition, BBB damage triggers astrocytic aquaporin-4 upregulation, and this upregulation leads to astrocytic swelling, which further increases brain edema and adversely affects the injured brain (Li et al. 2022; Dinet et al. 2019).

TBI and other neurodegenerative illnesses cause oxidative damage to the blood–brain barrier, which is protected by its own system of endogenous antioxidants (Freeman and Keller 2012). One of the endogenous antioxidants is glutathione GSH, a BBB transporter that mediates the entry of endogenous substrates required for cellular response to stress (Ronaldson and Davis 2015; Abdullahi et al. 2018).

Obviously, maintaining high levels of GSH is essential for the protection of cells against oxidative stress (Ronaldson and Davis 2015). Nevertheless, under severe oxidative stress due to TBI, GSH is oxidized to produce glutathione disulfide (GSSG), lowering the GSH/GSSG ratio and causing evident cellular loss of GSH from BBB endothelial cells (Abdullahi et al. 2018). Knowing that decreased cellular concentration of GSH is detrimental to the cells (Ronaldson and Davis 2015), one therapeutic way of fighting against oxidative stress is to target the BBB transporters that mediate the GSH efflux and consequently its loss. One of those transporters is the Multidrug resistance protein (MRPs) family specifically MRP1, MRP2, and MRP4 (Abdullahi et al. 2018). Under oxidative stress, MRPs expression is upregulated resulting in active efflux of GSH and hence an alteration in the redox balance and an increased risk of cell damage (Ronaldson and Davis 2015). In summary, targeting MRP transporters during oxidative stress may have protective and therapeutic advantages to BBB protection (Abdullahi et al. 2018). Other endogenous antioxidants include uric acid and coenzyme Q10, but those are secondary antioxidants as their primary role is different (Di Pietro et al. 2020).

Definitely, the ROS increases rapidly after TBI and the endogenous antioxidant defense system of the brain is not sufficient to prevent damage (Feng et al. 2023), so it is fundamental to supply antioxidants from a nutritional diet like vitamins C and E, phenolic acids, and curcumin. Dietary antioxidants are exogenous antioxidants not produced by the body so we depend on our regular diet to reach adequate concentrations of antioxidants. Here we highlight the importance of antioxidant-rich diet and its importance in protecting against oxidative stress in TBI and other neurodegenerative diseases (Di Pietro et al. 2020). Exogenous antioxidants presented important in vitro benefits on oxidative stress with little data related to their effect on BBB. However, their therapeutic application in vivo remains limited (Pinto et al. 2020). Indeed, and despite the fact that preclinical studies have reported relative success for antioxidant therapy in human diseases associated with oxidative stress, clinical trials proved limited benefits (Di Pietro et al. 2020; Benfeito et al. 2013).

One of the major causes behind it is the non-permeability of the BBB to these antioxidants, and thus their inability to cross the BBB and enter the brain (Feng et al. 2023; Pinto et al. 2020); other causes include the lack of standardized, clearly effective pharmacological approaches of TBI patients (Di Pietro et al. 2020). New strategies have emerged to surpass this limitation and are divided into invasive and noninvasive ones. The invasive is, as its name implies, the direct administration of the drug into the brain. Although more efficient, it has several setbacks that favor noninvasive strategies usage like the cost and the risks it holds as a technique. In contrast, noninvasive techniques were more accepted and are being advanced to improve the bioavailability of dietary antioxidants in the brain. Some of the techniques include increasing the lipophilicity of the administered drug by either adding lipophilic moieties or its conjugation with signaling molecules that target its therapeutic action. Other techniques include delivering antioxidants by loading them with vectors that aim to deliver them in a specifically targeted way. In addition to using vectors, new ways of administration are explored, for instance, the intranasal route has been reported to circumvent BBB and the hepatic first-pass effect, and as a result lowers the dosage and subsequently the side effects (Pinto et al. 2020).

6.2.6 Enhancement of Endogenous Antioxidant Systems

It should be noted that the administration of exogenous antioxidants to cells may retard the absorption of endogenous antioxidants; hence, maintaining the overall "cell antioxidant potential" at a constant level (Poljsak et al. 2013). Therefore, it is possible that antioxidant supplements can enhance the organism's ability to mitigate oxidative stress, which cannot be effectively addressed by the body's own antioxidant defense (Pisoschi and Pop 2015). The topic surrounding synthetic antioxidants has been a subject of debate. One example of this is the assertion that the recommended daily allowance of vitamins C and E may not adequately address the effects of oxidative stress. However, antioxidant supplementation usage proves its effectiveness if the initial oxidative stress level is above normal or above the individual's set point of regulation (Cutler and Mattson 2003). Thus, antioxidant supplements may reduce the increased levels of oxidative stress that cannot be inactivated by endogenous sources (Poljsak et al. 2013).

6.2.7 Modulation of Apoptosis and Cell Death

ROS and the consequent cellular redox shift in the brain are capable of triggering cell death by damaging critical cellular components and activating several apoptosis pathways (Kannan and Jain 2000; Klein and Ackerman 2003). Furthermore, oxidative stress may cause reentry into the cell cycle in some terminally developed neurons. Following DNA replication, these neurons die (Klein and Ackerman 2003). The first mechanism to be illuminated is the activation of the extrinsic pathway by apoptosis via death receptor activation on the cell membrane. Caspases will be activated as a result of this (Redza-Dutordoir and Averill-Bates 2016). The mitochondrial route triggered apoptosis is another significant mechanism (Kannan and Jain 2000). Proteins are released between the mitochondrial inner and outer membranes under the control of B-cell lymphoma 2 (BCL-2) family members. These proteins trigger apoptosis by activating caspase (Wang and Youle 2009). Unfortunately, ROS are mostly produced in the mitochondria, causing malfunction and initiating apoptosis (Liu et al. 2019). As a result of TBI, energy metabolism is disrupted, resulting in ROS production from mitochondria and apoptosis. Apoptotic cell death has been observed in contusional brain lesions during the immediate post-traumatic phase of TBI, as well as in regions distant from the location immediately affected days to weeks following trauma. TBI is a dynamic disorder that causes neuronal degeneration over time and is most likely caused by neurodegenerative processes years after the event (Akamatsu and Hanafy 2020).

Oxidative stress and mitochondrial malfunction were discovered to affect all brain cells, mostly triggering death in neural stem cells (Cho et al. 2020). Enzymatic and dietary antioxidants have antiapoptotic and antioxidative stress actions (Kahl et al. 2004; Watson et al. 2000). Enzymatic antioxidants such as mnSOD protect cells from apoptosis caused by certain radicals such as superoxide anion radicals or H_2O_2, blocking both apoptosis via the receptor-mediated apoptotic pathway and cell death via the mitochondrial route (Kahl et al. 2004). Vitamin E, as a dietary antioxidant, protects striatal neurons against apoptosis (Watson et al. 2000). Glutathione levels in the brain are a predictor of oxidative stress protection. When this antioxidant was removed from brain cells, oxidative stress-induced apoptosis increased dramatically (Franco and Cidlowski 2009). Another example of antioxidants' capacity to inhibit brain cell death is the B-cell lymphoma 2 (BCL-2) protein, which possesses antioxidant and antiapoptotic activities (Hockenbery et al. 1993). When glutathione is depleted in brain cells, the expression of BCL-2 prevents the formation of reactive oxygen species and lipid peroxides, sparing the cells against ROS-induced death (Kane et al. 1993). Furthermore, antioxidants can minimize cell death, and prevent or postpone the emergence of ultrastructural apoptotic patterns, DNA fragmentation, and caspase activation (Salucci et al. 2014).

TBI, as previously stated, causes the death of brain cells via ROS production from the mitochondria; antioxidants have a high potential to counteract this.

6.2.8 Neuroprotection and Cognitive Function

High-level cognitive operations rely on dispersed brain networks and highly regulated neurotransmitter systems, which are disturbed by injury, resulting in location-specific impairments (Stocchetti and Zanier 2016). The primary clinical manifestation of all forms of TBI is cognitive dysfunction, as there is a dose-response association between the severity of the TBI and the cognitive deficits up to 6 months or more post-injury (Filley and Kelly 2018). Following a TBI, cognitive deficits include impaired memory and attention, being easily distracted, having trouble focusing, seeing, and understanding spatial relationships, having trouble communicating with others, and having problems with executive functions (Barman et al. 2016).

To fight oxidative stress, which as mentioned previously is a major contributor to the secondary nervous tissue damage associated with TBI, dietary antioxidants are introduced to the patients' diets (Dobrovolny et al. 2018). Phenolic acids exert neuroprotective and cognition-enhancing effects as they can intervene and increase the release of neurotrophic factors, prevent or counteract protein aggregation, increase the expression of antiapoptotic agents, decrease the levels of prooxidants and proinflammatory mediators, increase the production and the activity of anti-inflammatory mediators, and enhance the antioxidant machinery (Caruso et al. 2022). Flavonoids, which are characterized by low toxicity, possess neuroprotective activity, and promote synaptogenesis and neurogenesis by inhibiting oxidative stress and neuroinflammation, thus showing promising effects in improving cognitive functions regardless of age (84). Bensalem J et al. demonstrated that the administration of flavonoids found in grape and berry extracts considerably enhanced episodic and exploratory memory, particularly in the subgroup with the most significant deficits (Bensalem et al. 2019). Ascorbic acid is a powerful water-soluble antioxidant linked to improved cognitive performance, including improved vocabulary, recognition, and free recall, as well as a slower rate of cognitive aging (Harrison et al. 2014). When combined with vitamin C, which restores the activity of oxidized vitamin E,, a significant component of cell membranes, is utilized as a booster of cognitive dysfunction after traumatic brain injury (TBI) (Khalili et al. 2022).

Antioxidant therapy is one of the potential treatments that are being explored to target known changes in neuroplasticity after traumatic brain injury (Sophie Su et al. 2016). Polyphenol-rich diets enhance cognitive functions by supporting survival, differentiation, and neuronal function improvement and regeneration (Teng and Chen 2019). By directly binding to their receptors and increasing the production of neurotrophic factors like Brain-derived neurotrophic factor (BDNF), polyphenols activate critical intracellular signaling cascades that are responsible for stress tolerance, neuronal survival, and effects like those of antidepressants (Arias-Sánchez et al. 2023). Additionally, learning and memory formation in the brain are significantly influenced by these active pathways (Zhang et al. 2020). Furthermore, they alter the gut microbiota's constituents, which influences the synthesis of neurotransmitters and neuropeptides that improve brain activity (Zhang et al. 2022). Likewise, polyunsaturated fatty acids, mainly eicosapentaenoic acid (EPA) and docosahexaenoic acid (DHA), found in fish oils, prevent cognitive decline by increasing nerve membrane neuroplasticity, promoting synaptogenesis, modulating signal transduction pathways in neuronal cells, and attenuating inflammatory processes (Franzoni et al. 2021).

6.2.9 Combination Therapies and Personalized Approaches

Targets for treatment interventions to enhance recovery outcomes are provided by a better knowledge of the neurophysiological effects of traumatic brain injury (Romeu-Mejia et al. 2019). Numerous pharmacological treatments are available that effectively reduce inflammation (Kou and VandeVord 2014). Glucocorticoids, Non-steroidal anti-inflammatory drugs (NSAIDS), and tumour necrosis factor (TNF)-α inhibitors are among these medications; the latter is particularly utilized to prevent TNF-α's proinflammatory impact, which appears quickly following traumatic injury (Bergold 2016). *N*-acetyl cysteine is an additional anti-inflammatory drug that is also classified as an antioxidant. It can reverse TBI-induced mitochondrial dysfunctions and has an antiapoptotic action (Du et al. 2013). To retrain motor skills after a neurologic injury, rehabilitation specialists choose and create a treatment plan based on the principles of motor learning (Jasey and Ward 2019). In addition to speech, occupational, and physical therapy, noninvasive transcranial brain stimulation is a relatively new therapeutic method that encourages brain remodeling within the affected tracts (Filley and Kelly 2018). Exercise suggestions that are specifically tailored are an effective intervention for treating the functional deficits and symptom burden that often follow traumatic brain injury (Fernandes et al. 2017), as physical activity influences neurogenesis and enhances patients' memory and attention (Cichon et al. 2020). Additional secondary symptoms of traumatic brain injury (TBI), such as headache complaints, sleep problems, and/or mood disorders, are treated with prescription drugs (Plantier et al. 2016). Dietary changes must be made after traumatic brain injury (TBI) to mitigate the negative consequences of the subsequent damage (McGeown et al. 2021). Because of their chemical properties, antioxidants are consequently considered a valuable adjuvant for the treatment of traumatic brain damage (Di Pietro et al. 2020).

There is no one-size-fits-all treatment strategy in TBI patients, as evidenced by the heterogeneity of TBI pathophysiology and the variation in patient responses to therapy (Di Pietro et al. 2020). This emphasizes how crucial it is to create tailored therapy methods and precision prognostication to regulate the secondary damage processes that are the source of increasing morbidity and mortality (Poljsak et al. 2013). Functional medicine specialists will work individually with each patient to establish a customized treatment plan that targets problem areas and closely monitor progress to make adjustments as needed. Dietitians typically start by treating patients' diets, giving them a balanced diet that addresses deficiencies and gives them the antioxidants and probiotics they need to promote brain health (Richer 2017). However, the dosage, length of time, mode of administration, and subsequent bioavailability of the provided antioxidants all affect how successful the antioxidant treatment is (McKeegan et al. 2021). Excess concentrations of antioxidants may have the opposite effect, increasing oxidative stress by acting as prooxidants (Sotler et al. 2019). The factor that requires oxidative stress to be measured before their administration remains to be demonstrated in clinical practice (Zeiler et al. 2021). Furthermore, sociodemographic factors have a significant impact on the bioavailability of antioxidants, which alters the guidelines for various age and sex groups (Beydoun et al. 2015). This suggests that to tailor and maximize patient treatment regimens, such as antioxidant therapy, further clinical studies are required (Khalili et al. 2017).

6.3 Conclusion

In conclusion, antioxidants have tremendous promise for reducing oxidative stress associated with traumatic brain damage. *N*-acetylcysteine, ALA, and vitamins C and E all show promise in terms of neuroprotection and cognitive function maintenance. Antioxidants' varied roles in regulating cellular components and boosting endogenous defenses highlight their therapeutic significance. Future studies should look at tailored applications and combination medicines to improve TBI treatment effectiveness.

References

Abdullahi W, Tripathi D, Ronaldson PT (2018) Blood-brain barrier dysfunction in ischemic stroke: targeting tight junctions and transporters for vascular protection. Am J Physiol Cell Physiol 315(3):C343–C356. https://doi.org/10.1152/ajpcell.00095.2018

Acosta SA, Tajiri N, Sanberg PR, Kaneko Y, Borlongan CV (2017) Increased amyloid precursor protein and tau expression manifests as key secondary cell death in chronic traumatic brain injury. J Cell Physiol 232:665–677

Aguilar TAF, Navarro BCH, Pérez JAM (2016) Endogenous antioxidants: a review of their role in oxidative stress. INTECH. https://doi.org/10.5772/65715

Akamatsu Y, Hanafy KA (2020) Cell death and recovery in traumatic brain injury. Neurotherapeutics 17(2):446–456. https://doi.org/10.1007/s13311-020-00840-7

Aldini G, Altomare A, Baron G et al (2018) N-acetylcysteine as an antioxidant and disulphide breaking agent: the reasons why. Free Radic Res 52(7):751–762. https://doi.org/10.1080/10715762.2018.1468564

Arias-Sánchez RA, Torner L, Fenton, Navarro B (2023) Polyphenols and neurodegenerative diseases: potential effects and mechanisms of neuroprotection. Molecules 28(14):5415. https://doi.org/10.3390/molecules28145415

Ashok A, Andrabi SS, Mansoor S, Kuang Y, Kwon BK, Labhasetwar V (2022) Antioxidant therapy in oxidative stress-induced neurodegenerative diseases: role of nanoparticle-based drug delivery systems in clinical translation. Antioxidants (Basel, Switzerland) 11(2):408. https://doi.org/10.3390/antiox11020408

Barman A, Chatterjee A, Bhide R (2016) Cognitive impairment and rehabilitation strategies after traumatic brain injury. Indian J Psychol Med 38(3):172–181. https://doi.org/10.4103/0253-7176.183086

Benfeito S, Oliveira C, Soares P et al (2013) Antioxidant therapy: still in search of the 'magic bullet'. Mitochondrion 13(5):427–435. https://doi.org/10.1016/j.mito.2012.12.002

Bensalem J, Dudonné S, Etchamendy N et al (2019) Polyphenols from grape and blueberry improve episodic memory in healthy elderly with lower level of memory performance: a Bicentric double-blind, randomized, placebo-controlled clinical study. J Gerontol A Biol Sci Med Sci 74(7):996–1007. https://doi.org/10.1093/gerona/gly166

Bergold PJ (2016) Treatment of traumatic brain injury with anti-inflammatory drugs. Exp Neurol 275(Pt 3):367–380. https://doi.org/10.1016/j.expneurol.2015.05.024

Beydoun MA, Fanelli-Kuczmarski MT, Kitner-Triolo MH et al (2015) Dietary antioxidant intake and its association with cognitive function in an ethnically diverse sample of US adults. Psychosom Med 77(1):68–82. https://doi.org/10.1097/PSY.0000000000000129

Bouayed J, Bohn T (2010) Exogenous antioxidants--double-edged swords in cellular redox state: health beneficial effects at physiologic doses versus deleterious effects at high doses. Oxidative Med Cell Longev 3(4):228–237. https://doi.org/10.4161/oxim.3.4.12858

Capizzi A, Woo J, Verduzco-Gutierrez M (2020) Traumatic brain injury: an overview of epidemiology, pathophysiology, and medical management. Med Clin North Am 104(2):213–238. https://doi.org/10.1016/j.mcna.2019.11.001

Caruso G, Godos J, Privitera A, Lanza G, Castellano S, Chillemi A, Bruni O, Ferri R, Caraci F, Grosso G (2022) Phenolic acids and prevention of cognitive decline: polyphenols with a neuroprotective role in cognitive disorders and Alzheimer's disease. Nutrients 14(4):819. https://doi.org/10.3390/nu14040819

Cash A, Theus MH (2020) Mechanisms of blood-brain barrier dysfunction in traumatic brain injury. Int J Mol Sci 21(9):3344. Published 2020 May 8. https://doi.org/10.3390/ijms21093344

Chan PH (2001) Reactive oxygen radicals in signaling and damage in the ischemic brain. J Cereb Blood Flow Metab 21(1):2–14. https://doi.org/10.1097/00004647-200101000-00002

Cho JH, Lee S, Jeon H, Kim AH, Lee W, Lee Y, Yang S, Yun J, Jung YS, Lee J (2020) Tetrabromobisphenol A-induced apoptosis in neural stem cells through oxidative stress

and mitochondrial dysfunction. Neurotox Res 38(1):74–85. https://doi.org/10.1007/s12640-020-00179-z
Chodobski A, Zink BJ, Szmydynger-Chodobska J (2011) Blood-brain barrier pathophysiology in traumatic brain injury. Transl Stroke Res 2(4):492–516. https://doi.org/10.1007/s12975-011-0125-x
Cichon N, Saluk-Bijak J, Gorniak L, Przyslo L, Bijak M (2020) Flavonoids as a natural enhancer of neuroplasticity-an overview of the mechanism of neurorestorative action. Antioxidants (Basel) 9(11):1035. Published 2020 Oct 23. https://doi.org/10.3390/antiox9111035
Clark RSB, Empey PE, Bayır H, et al (2017) Phase I randomized clinical trial of N-acetylcysteine in combination with an adjuvant probenecid for treatment of severe traumatic brain injury in children. PLoS One 12(7):e0180280. Published 2017 Jul 7. https://doi.org/10.1371/journal.pone.0180280
Cordaro M, Trovato Salinaro A, Siracusa R, D'Amico R, Impellizzeri D, Scuto M, Ontario ML, Crea R, Cuzzocrea S, Di Paola R et al (2021) Hidrox® roles in neuroprotection: biochemical links between traumatic brain injury and Alzheimer's disease. Antioxidants 10:818
Cutler RG, Mattson MP (2003) Measuring oxidative stress and interpreting its relevance in humans. In: Cuttler RG, Rodriguez H (eds) Critical reviews of oxidative stress and aging e advances in basic science, diagnostic and intervention, vol 1. World Scientific Publishing Co, New Jersey, p 131. (chapter 8)
Davis CK, Vemuganti R (2022) Antioxidant therapies in traumatic brain injury. Neurochem Int 152:105255. https://doi.org/10.1016/j.neuint.2021.105255
Di Pietro V, Yakoub KM, Caruso G, Lazzarino G, Signoretti S, Barbey AK, Tavazzi B, Lazzarino G, Belli A, Amorini AM (2020) Antioxidant therapies in traumatic brain injury. Antioxidants (Basel, Switzerland) 9(3):260. https://doi.org/10.3390/antiox9030260
Dinet V, Petry KG, Badaut J (2019) Brain-immune interactions and neuroinflammation after traumatic brain injury. Front Neurosci 13:1178. https://doi.org/10.3389/fnins.2019.01178
Dobrovolny J, Smrcka M, Bienertova-Vasku J (2018) Therapeutic potential of vitamin E and its derivatives in traumatic brain injury-associated dementia. Neurol Sci 39(6):989–998. https://doi.org/10.1007/s10072-018-3398-y
Du X, Ewert DL, Cheng W et al (2013) Effects of antioxidant treatment on blast-induced brain injury. PLoS One 8(11):e80138. Published 2013 Nov 5. https://doi.org/10.1371/journal.pone.0080138
Farooqui T, Farooqui AA (eds) (2017) Role of the Mediterranean diet in the brain and neurodegenerative diseases. Academic
Feng J, Zheng Y, Guo M, Ares I, Martínez M, Lopez-Torres B, Martinez MA et al (2023) Oxidative stress, the blood–brain barrier and neurodegenerative diseases: the critical beneficial role of dietary antioxidants. Acta Pharm Sin B 13:3988–4024
Fernandes J, Arida RM, Gomez-Pinilla F (2017) Physical exercise as an epigenetic modulator of brain plasticity and cognition. Neurosci Biobehav Rev 80:443–456. https://doi.org/10.1016/j.neubiorev.2017.06.012
Fesharaki-Zadeh A (2022) Oxidative stress in traumatic brain injury. Int J Mol Sci 23(21):13000. https://doi.org/10.3390/ijms232113000
Filley CM, Kelly JP (2018) White matter and cognition in traumatic brain injury. J Alzheimers Dis 65(2):345–362. https://doi.org/10.3233/JAD-180287
Franco R, Cidlowski JA (2009) Apoptosis and glutathione: beyond an antioxidant. Cell Death Differ 16(10):1303–1314. https://doi.org/10.1038/cdd.2009.107
Franzoni F, Scarfò G, Guidotti S, Fusi J, Asomov M, Pruneti C (2021) Oxidative stress and cognitive decline: the neuroprotective role of natural antioxidants. Front Neurosci 15:729757. Published 2021 Oct 13. https://doi.org/10.3389/fnins.2021.729757
Freeman LR, Keller JN (2012) Oxidative stress and cerebral endothelial cells: regulation of the blood-brain-barrier and antioxidant based interventions. Biochim Biophys Acta 1822(5):822–829. https://doi.org/10.1016/j.bbadis.2011.12.009

Giustarini D, Tsikas D, Colombo G et al (2016) Pitfalls in the analysis of the physiological antioxidant glutathione (GSH) and its disulfide (GSSG) in biological samples: an elephant in the room. J Chromatogr B Analyt Technol Biomed Life Sci 1019:21–28. https://doi.org/10.1016/j.jchromb.2016.02.015

Günther M, Al Nimer F, Piehl F, Risling M, Mathiesen T (2018) Susceptibility to oxidative stress is determined by genetic background in neuronal cell cultures. eNeuro 5(2):ENEURO.0335-17.2018. Published 2018 Mar 19. https://doi.org/10.1523/ENEURO.0335-17.2018

Hakiminia B, Alikiaii B, Khorvash F, Mousavi S (2022) Oxidative stress and mitochondrial dysfunction following traumatic brain injury: from mechanistic view to targeted therapeutic opportunities. Fundam Clin Pharmacol 36(4):612–662. https://doi.org/10.1111/fcp.12767

Harrison FE, Bowman GL, Polidori MC (2014) Ascorbic acid and the brain: rationale for the use against cognitive decline. Nutrients 6(4):1752–1781. Published 2014 Apr 24. https://doi.org/10.3390/nu6041752

Hockenbery DM, Oltvai ZN, Yin XM, Milliman CL, Korsmeyer SJ (1993) Bcl-2 functions in an antioxidant pathway to prevent apoptosis. Cell 75(2):241–251. https://doi.org/10.1016/0092-8674(93)80066-n

Hubbard WB, Joseph B, Spry M, Vekaria HJ, Saatman KE, Sullivan PG (2019) Acute mitochondrial impairment underlies prolonged cellular dysfunction after repeated mild traumatic brain injuries. J Neurotrauma 36(8):1252–1263. https://doi.org/10.1089/neu.2018.5990

Irato P, Santovito G (2021) Enzymatic and non-enzymatic molecules with antioxidant function. Antioxidants (Basel, Switzerland) 10(4):579. https://doi.org/10.3390/antiox10040579

Jamjoom AAB, Rhodes J, Andrews PJD, Grant SGN (2021) The synapse in traumatic brain injury. Brain 144(1):18–31. https://doi.org/10.1093/brain/awaa321

Jasey N, Ward I (2019) Neuroplasticity in brain injury: maximizing recovery. Curr Phys Med Rehabil Rep 7:333–340. https://doi.org/10.1007/s40141-019-00242-7

Kahl R, Kampkötter A, Wätjen W, Chovolou Y (2004) Antioxidant enzymes and apoptosis. Drug Metab Rev 36(3–4):747–762. https://doi.org/10.1081/DMR-200033488

Kane DJ, Sarafian TA, Anton R, Hahn H, Gralla EB, Valentine JS, Ord T, Bredesen DE (1993). Bcl-2 inhibition of neural death: decreased generation of reactive oxygen species Science (New York, N.Y.), 262(5137), 1274–1277. https://doi.org/10.1126/science.8235659

Kannan K, Jain SK (2000) Oxidative stress and apoptosis. Pathophysiology 7(3):153–163. https://doi.org/10.1016/s0928-4680(00)00053-5

Karve IP, Taylor JM, Crack PJ (2016) The contribution of astrocytes and microglia to traumatic brain injury. Br J Pharmacol 173(4):692–702. https://doi.org/10.1111/bph.13125

Keaney J, Campbell M (2015) The dynamic blood-brain barrier. FEBS J 282(21):4067–4079. https://doi.org/10.1111/IJMS.13412

Khalili H, Ahl R, Cao Y et al (2017) Early selenium treatment for traumatic brain injury: does it improve survival and functional outcome? Injury 48(9):1922–1926. https://doi.org/10.1016/j.injury.2017.07.005

Khalili H, Abdollahifard S, Niakan A, Aryaie M (2022) The effect of vitamins C and E on clinical outcomes of patients with severe traumatic brain injury: a propensity score matching study. Surg Neurol Int 13:548. Published 2022 Nov 25. https://doi.org/10.25259/SNI_932_2022

Kim K, Priefer R (2020) Evaluation of current post-concussion protocols. Biomed Pharmacother 129:110406. https://doi.org/10.1016/j.biopha.2020.110406

Klein JA, Ackerman SL (2003) Oxidative stress, cell cycle, and neurodegeneration. J Clin Invest 111(6):785–793. https://doi.org/10.1172/JCI18182

Kou Z, VandeVord PJ (2014) Traumatic white matter injury and glial activation: from basic science to clinics. Glia 62(11):1831–1855. https://doi.org/10.1002/glia.22690

Lawless S, Bergold PJ (2022) Better together? Treating traumatic brain injury with minocycline plus N-acetylcysteine. Neural Regen Res 17(12):2589–2592. https://doi.org/10.4103/1673-5374.336136

Li W, Cao F, Takase H, Arai K, Lo EH, Lok J (2022) Blood-brain barrier mechanisms in stroke and trauma. Handb Exp Pharmacol 273:267–293. https://doi.org/10.1007/164_2020_426

Licastro F, Hrelia S, Porcellini E, Malaguti M, Di Stefano C, Angeloni C, Carbone I, Simoncini L, Piperno R (2016) Peripheral inflammatory markers and antioxidant response during the post-acute and chronic phase after severe traumatic brain injury. Front Neurol 7:189. https://doi.org/10.3389/fneur.2016.00189

Liu T, Han Y, Zhou T, Zhang R, Chen H, Chen S, Zhao H (2019) Mechanisms of ROS-induced mitochondria-dependent apoptosis underlying liquid storage of goat spermatozoa. Aging 11(18):7880–7898. https://doi.org/10.18632/aging.102295

Lobo V, Patil A, Phatak A, Chandra N (2010) Free radicals, antioxidants and functional foods: impact on human health. Pharmacogn Rev 4(8):118–126. https://doi.org/10.4103/0973-7847.70902

Lü JM, Lin PH, Yao Q, Chen C (2010) Chemical and molecular mechanisms of antioxidants: experimental approaches and model systems. J Cell Mol Med 14(4):840–860. https://doi.org/10.1111/j.1582-4934.2009.00897.x

Maas AI, Marmarou A, Murray GD, Steyerberg EW (2004) Clinical trials in traumatic brain injury: current problems and future solutions. Acta Neurochir Suppl 89:113–118. https://doi.org/10.1007/978-3-7091-0603-7_16

Manach C, Scalbert A, Morand C, Rémésy C, Jiménez L (2004) Polyphenols: food sources and bioavailability. Am J Clin Nutr 79:727–747

Mary CPV, Vijayakumar S, Shankar R (2018) Metal chelating ability and antioxidant properties of curcumin-metal complexes – a DFT approach. J Mol Graph Model 79:1–14. https://doi.org/10.1016/j.jmgm.2017.10.022

Mason HD, Johnson AM, Mihelson NA, Mastorakos P, McGavern DB (2021) Glia limitans superficialis oxidation and breakdown promote cortical cell death after repetitive head injury. JCI Insight 6(19):e149229. Published 2021 Oct 8. https://doi.org/10.1172/jci.insight.149229

McGeown JP, Hume PA, Theadom A, Quarrie KL, Borotkanics R (2021) Nutritional interventions to improve neurophysiological impairments following traumatic brain injury: a systematic review. J Neurosci Res 99(2):573–603. https://doi.org/10.1002/jnr.24746

McKeegan K, Mason SA, Trewin AJ et al (2021) Reactive oxygen species in exercise and insulin resistance: working towards personalized antioxidant treatment. Redox Biol 44:102005. https://doi.org/10.1016/j.redox.2021.102005

Mendes Arent A, de Souza LF, Walz R, Dafre AL (2014) Perspectives on molecular biomarkers of oxidative stress and antioxidant strategies in traumatic brain injury. Biomed Res Int 2014:723060. https://doi.org/10.1155/2014/723060

Niki E (1987) Interaction of ascorbate and alpha-tocopherol. Ann N Y Acad Sci 498:186–199. https://doi.org/10.1111/j.1749-6632.1987.tb23761.x

Niki E, Noguchi N (2021) Antioxidant action of vitamin E *in vivo* as assessed from its reaction products with multiple biological oxidants. Free Radic Res 55(4):352–363. https://doi.org/10.1080/10715762.2020.1866181

Park GJ, Ro YS, Yoon H, et al (2022) Serum vitamin E level and functional prognosis after traumatic brain injury with intracranial injury: a multicenter prospective study. Front Neurol 13:1008717. Published 2022 Oct 19. https://doi.org/10.3389/fneur.2022.1008717

Pinto M, Benfeito S, Fernandes C, Borges F (2020) Antioxidant therapy, oxidative stress, and blood-brain barrier: the road of dietary antioxidants. In: Oxidative stress and dietary antioxidants in neurological diseases. Academic, pp 125–141

Pisoschi AM, Pop A (2015) The role of antioxidants in the chemistry of oxidative stress: a review. Eur J Med Chem 97:55–74. https://doi.org/10.1016/j.ejmech.2015.04.040. Epub 2015 Apr 22

Plantier D, Luauté J, SOFMER group (2016) Drugs for behavior disorders after traumatic brain injury: systematic review and expert consensus leading to French recommendations for good practice. Ann Phys Rehabil Med 59(1):42–57. https://doi.org/10.1016/j.rehab.2015.10.003

Poljsak B, Šuput D, Milisav I (2013) Achieving the balance between ROS and antioxidants: when to use the synthetic antioxidants. Oxidative Med Cell Longev 2013:956792. https://doi.org/10.1155/2013/956792

Redza-Dutordoir M, Averill-Bates DA (2016) Activation of apoptosis signalling pathways by reactive oxygen species. Biochim Biophys Acta (BBA) – Mol Cell Res 1863(12):2977–2992. https://doi.org/10.1016/j.bbamcr.2016.09.012

Reiter RJ, Mayo JC, Tan DX, Sainz RM, Alatorre-Jimenez M, Qin L (2016) Melatonin as an antioxidant: under promises but over delivers. J Pineal Res 61(3):253–278. https://doi.org/10.1111/jpi.12360

Richer AC (2017) Functional medicine approach to traumatic brain injury. Med Acupunct 29(4):206–214. https://doi.org/10.1089/acu.2017.1217

Romeu-Mejia R, Giza CC, Goldman JT (2019) Concussion pathophysiology and injury biomechanics. Curr Rev Musculoskelet Med 12(2):105–116. https://doi.org/10.1007/s12178-019-09536-8

Ronaldson PT, Davis TP (2015) Targeting transporters: promoting blood-brain barrier repair in response to oxidative stress injury. Brain Res 1623:39–52. https://doi.org/10.1016/j.brainres.2015.03.018

Salehi B, Martorell M, Arbiser JL, Sureda A, Martins N, Maurya PK, Sharifi-Rad M, Kumar P, Sharifi-Rad J (2018) Antioxidants: positive or negative actors? Biomol Ther 8(4):124. https://doi.org/10.3390/biom8040124

Salehi B, Berkay Yılmaz Y, Antika G et al (2019) Insights on the use of α-lipoic acid for therapeutic purposes. Biomol Ther 9(8):356. Published 2019 Aug 9. https://doi.org/10.3390/biom9080356

Salucci S, Burattini S, Curzi D, Buontempo F, Martelli AM, Zappia G, Falcieri E, Battistelli M (2014) Antioxidants in the prevention of UVB-induced keratynocyte apoptosis. J Photochem Photobiol B Biol 141:1–9. https://doi.org/10.1016/j.jphotobiol.2014.09.004

Scheff SW, Ansari MA (2017) Natural compounds as a therapeutic intervention following traumatic brain injury: the role of phytochemicals. J Neurotrauma 34(8):1491–1510. https://doi.org/10.1089/neu.2016.4718

Schimmel SJ, Acosta S, Lozano D (2017) Neuroinflammation in traumatic brain injury: a chronic response to an acute injury. Brain Circ 3(3):135–142. https://doi.org/10.4103/bc.bc_18_17

Sedighi O, Zargari M, Varshi G (2014) Effect of selenium supplementation on glutathione peroxidase enzyme activity in patients with chronic kidney disease: a randomized clinical trial. Nephrourology Monthly 6(3):e17945. https://doi.org/10.5812/numonthly.17945

Shen Q, Hiebert JB, Hartwell J, Thimmesch AR, Pierce JD (2016) Systematic review of traumatic brain injury and the impact of antioxidant therapy on clinical outcomes. Worldviews Evid-Based Nurs 13(5):380–389. https://doi.org/10.1111/wvn.12167

Shi K, Zhang J, Dong JF, Shi FD (2019) Dissemination of brain inflammation in traumatic brain injury. Cell Mol Immunol 16(6):523–530. https://doi.org/10.1038/41423-019-0213-5

Sies H (2015) Oxidative stress: a concept in redox biology and medicine. Redox Biol 4:180–183. https://doi.org/10.1016/j.redox.2015.01.002

Sophie Su YR, Veeravagu A, Grant G (2016) Neuroplasticity after traumatic brain injury. In: Laskowitz D, Grant G (eds) Translational research in traumatic brain injury. CRC Press/Taylor and Francis Group, Boca Raton. Chapter 8. Available from: https://www.ncbi.nlm.nih.gov/books/NBK326735/

Sotler R, Poljšak B, Dahmane R et al (2019) Prooxidant activities of antioxidants and their impact on health. Acta Clin Croat 58(4):726–736. https://doi.org/10.20471/acc.2019.58.04.20

Stocchetti N, Zanier ER (2016) Chronic impact of traumatic brain injury on outcome and quality of life: a narrative review. Crit Care 20(1):148. Published 2016 Jun 21:148. https://doi.org/10.1186/s13054-016-1318-1

Surveillance Report of Traumatic Brain Injury-related Emergency Department Visits, Hospitalizations, and Deaths (2014) In: Centers for Disease Control and Prevention, U.S. Department of Health and Human Services. Available at: https://www.cdc.gov/traumaticbraininjury/get_the_facts.html

Teng H, Chen L (2019) Polyphenols and bioavailability: an update. Crit Rev Food Sci Nutr 59(13):2040–2051. https://doi.org/10.1080/10408398.2018.1437023

Thapa K, Khan H, Singh TG, Kaur A (2021) Traumatic brain injury: mechanistic insight on pathophysiology and potential therapeutic targets. J Mol Neurosci 71(9):1725–1742. https://doi.org/10.1007/s12031-021-01841-7

Tyuryaeva I, Lyublinskaya O (2023) Expected and unexpected effects of pharmacological antioxidants. Int J Mol Sci 24(11):9303. Published 2023 May 26. https://doi.org/10.3390/ijms24119303

Visser K, van der Horn HJ, Bourgonje AR et al (2022) Acute serum free thiols: a potentially modifiable biomarker of oxidative stress following traumatic brain injury. J Neurol 269(11):5883–5892. https://doi.org/10.1007/s00415-022-11240-6

Wang C, Youle RJ (2009) The role of mitochondria in apoptosis*. Annu Rev Genet 43:9118. https://doi.org/10.1146/annurev-genet-102108-134850

Wang HC, Lin YJ, Shih FY, Chang HW, Su YJ, Cheng BC, Su CM, Tsai NW, Chang YT, Kwan AL, Lu CH (2016) The role of serial oxidative stress levels in acute traumatic brain injury and as predictors of outcome. World Neurosurg 87:463–470. https://doi.org/10.1016/j.wneu.2015.10.010

Wang J, Jiang C, Zhang K et al (2019) Melatonin receptor activation provides cerebral protection after traumatic brain injury by mitigating oxidative stress and inflammation via the Nrf2 signaling pathway. Free Radic Biol Med 131:345–355. https://doi.org/10.1016/j.freeradbiomed.2018.12.014

Watson WH, Cai J, Jones DP (2000) Diet and apoptosis. Annu Rev Nutr 20:485–505. https://doi.org/10.1146/annurev.nutr.20.1.485

Wu AG, Yong YY, Pan YR, et al (2022) Targeting Nrf2-mediated oxidative stress response in traumatic brain injury: therapeutic perspectives of phytochemicals. Oxid Med Cell Longev 2022:1015791. Published 2022 Apr 4. https://doi.org/10.1155/2022/1015791

Zeiler FA, Iturria-Medina Y, Thelin EP et al (2021) Integrative neuroinformatics for precision prognostication and personalized therapeutics in moderate and severe traumatic brain injury. Front Neurol 12:729184. Published 2021 Sep 7. https://doi.org/10.3389/fneur.2021.729184

Zhang S, Xue R, Hu R (2020) The neuroprotective effect and action mechanism of polyphenols in diabetes mellitus-related cognitive dysfunction. Eur J Nutr 59(4):1295–1311. https://doi.org/10.1007/s00394-019-02078-2

Zhang Y, Yu W, Zhang L, Wang M, Chang W (2022) The interaction of polyphenols and the gut microbiota in neurodegenerative diseases. Nutrients 14(24):5373. https://doi.org/10.3390/nu14245373

Zhou Y, Chen Q, Wang Y et al (2020) Persistent neurovascular unit dysfunction: pathophysiological substrate and trigger for late-onset neurodegeneration after traumatic brain injury. Front Neurosci 14:581. Published 2020 Jun 9. https://doi.org/10.3389/fnins.2020.00581

Zingg JM (2019) Vitamin E: regulatory role on signal transduction. IUBMB Life 71(4):456–478. https://doi.org/10.1002/iub.1986

Part II
Dietary Interventions for TBI: Supplements and Nutraceuticals

Chapter 7
Role of Branched-Chain Amino Acids in Traumatic Brain Injury

Kholoud Elsamman

Abstract Worldwide, traumatic brain injuries (TBIs) cause a great deal of suffering and death, making them an important issue in public health. The pathophysiology of traumatic brain injury (TBI) is a complex network of events that can lead to neurologic deficits and changes in neuronal function, such as excitotoxicity, oxidative stress, inflammation, and mitochondrial malfunction. Consequently, interest in creating new therapeutic approaches to mitigate TBI's adverse effects and enhance rehabilitation is on the rise. The potential neuroprotective effects of branched-chain amino acids (BCAAs) in various neurological illnesses have been highlighted in recent publications, which has prompted additional investigation into their usage in traumatic brain injury. Amino acids that are essential for protein synthesis, energy metabolism, and neurotransmitter regulation are the BCAAs, which consist of valine, isoleucine, and leucine. Molecular processes such as excitotoxicity, oxidative stress, and inflammation complicate the pathophysiology of traumatic brain injury (TBI). As substrates for protein synthesis and modulators of signalling pathways, BCAAs exhibit complicated neuroprotective properties. Based on experimental evidence, BCAAs have the potential to improve cognitive outcomes after a traumatic brain injury (TBI) by reducing neuroinflammation, increasing cellular resilience, and promoting synaptic plasticity.

Keywords Traumatic brain injury (TBI) · Branched-chain amino acids (BCAAs) · Neuroprotection · Cognitive recovery · Nutritional intervention · Neurological disorders · Neuroprotective effects

K. Elsamman (✉)
Biochemistry and Nutrition Department, Faculty of Women for Science, Ain Shams University, Cairo, Egypt
e-mail: 29911050200347@std.women.asu.edu.eg

W. Mohamed (ed.), *Nutrition and Traumatic Brain Injury (TBI)*, Nutritional Neurosciences, https://doi.org/10.1007/978-981-97-6341-2_7

7.1 Introduction

Traumatic brain injury (TBI) is prevalent among American veterans, athletes, and members of the armed forces (Dewan et al. 2018) frequently compounded by persistent post-concussive effects in the weeks or months after injury (Hoge et al. 2008; Rutherford et al. 1977). Potentially the most prevalent and incapacitating of these post-concussion symptoms are sleep-wake disorders (Sandsmark et al. 2017).

A 50–70% prevalence of sleep problems after a TBI (Mathias and Alvaro 2012), which may endure beyond 20–25 years following the injury (Elliott et al. 2018, 2020). This is correlated with profound functional and economic consequences (Altevogt and Colten 2006).

Sleep problems have been proven to exacerbate post-concussive symptoms (Nguyen et al. 2017), influence a person's capacity to manage these symptoms (Lew et al. 2009), aggravate neuropsychiatric symptoms following a TBI (Rao et al. 2014), impede complete adherence to rehabilitation therapy programmes (Worthington and Melia 2006), can impede or entirely postpone reintegration into the labour force, community, and familial sphere (Corrigan et al. 2014). Thus, sleep disturbances caused by a traumatic brain injury may exacerbate the problem by potentially diminishing the effectiveness of rehabilitative therapies that aim to alleviate symptoms that negatively impact quality of life (Worthington and Melia 2006).

During the subacute and chronic phases of TBI recovery, cognitive problems, whether they occur alone or as a result of sleep disruptions, are very common, affecting 60–80% of patients (Scholten et al. 2012; Vasterling et al. 2006). Sleep disruptions caused by traumatic brain injury (TBI) worsen both the impaired cognitive function and the lowered quality of life that are already present in TBI patients (Boakye et al. 2016). Functional results and quality of life could be enhanced by enhancing sleep and cognitive function in traumatic brain injury (TBI) (Wiseman-Hakes et al. 2013). When it comes to sleep, conventional treatment interventions for traumatic brain injury patients do have their limitations. Unfortunately, there are currently no specific therapies that are based on mechanisms to treat sleep-wake disorders in people with traumatic brain injury. A rodent model of traumatic brain injury (TBI) has shown that branched-chain amino acid (BCAA) supplementation improves sleep and cognitive deficits (Elliott et al. 2022).

The benefits of important, branching-chain amino acids are felt all across the body. Past research has investigated the potential benefits of using BCAA supplements for a variety of neurological and psychiatric diseases, as well as for exercise physiology and hepatic pathology (Neinast et al. 2019). Research evaluating BCAA use for more than 2 years has shown no negative effects, and many athletes have been using these supplements for years without any problems. So, it's safe to take BCAAs for a long period (Fedewa et al. 2019). It is thought that BCAA has anti-nociceptive and anti-catabolic properties (Wiseman-Hakes et al. 2013; Manner et al. 1996). The primary purpose of BCAAs in the CNS is to serve as building blocks for the synthesis of de novo glutamate and, by extension, γ-aminobutyric acid (GABA), the principal excitatory and inhibitory neurotransmitters, respectively (Fernstrom 2005). Plasma concentrations of dietary BCAAs increase fast upon consumption and are directly proportional to their dosage (or content in solid food) since most of these amino acids do not undergo liver metabolism but instead enter the systemic circulation unrestrictedly (Matthews 2005). A transporter located in the central nervous system allows BCAAs in plasma to cross the blood-brain barrier (Smith et al. 1987). Other amino acids can also be transported over the blood-brain barrier by this transporter in relation to concentration gradients. Thus, central nervous system BCAA concentration increases in response to elevated plasma BCAA concentration, although the levels of amino acids that compete with it (such as phenylalanine, tyrosine, and tryptophan) drop (Hawkins et al. 2006; Robertson et al. 1988). These antagonistic amino acids serve as building blocks for further neurotransmitters (i.e., monoamines and catecholamines). Accordingly, alterations in the general biochemical composition of the central nervous system occur when BCAAs are consumed alone, as opposed to mixed with other amino acids (as in a protein shake or a mixed meal). By restoring glutamate inside pre-synaptic nerve terminals contacting orexin neurons in the hypothalamus, prior preclinical investigations utilising a mouse model of TBI have demonstrated that BCAA supplementation improves sleep-wake disorders (Elliott et al. 2018; Lim et al. 2013). Supplementing with BCAAs enhances hippocampal-dependent memory tasks and brings the hippocampus's excitation-to-inhibition ratio back into balance, according to previous research. Last but not least, this preclinical study has investigated the best way to administer BCAAs to mice, including the best dosage, duration, and delivery method (Elkind et al. 2015). Previous research has also demonstrated that hippocampus-dependent memory tasks are improved and the excitation-to-inhibition balance in the hippocampus is restored when BCAA supplementation is used. At last, this preclinical study has measured the effects of BCAAs in mice and determined the best time, way, and dose to administer them (Manner et al. 1996). The effects of BCAAs on inflammatory biomarkers in the blood as well as subjective and objective assessments of chronic pain, both of which are increased in people with traumatic brain injury, were simultaneously investigated (Elliott et al. 2018, Balba et al. 2018; Peltz et al. 2020).

7.2 Categories of TBI

Closed head, piercing, and explosive blast are the three primary types of traumatic brain injuries (TBIs) based on the specific physical mechanisms of injury. A traumatic brain injury (TBI) could cause symptoms to appear anywhere from immediately after the event to months or even years later. Extended periods of unconsciousness, nausea, headaches, trouble speaking, seizures, memory loss, and abnormal behaviours, including hostility and anxiety, are among these symptoms (Andriessen et al. 2010). Different physical mechanisms generate the three main categories of traumatic brain injuries (TBIs): closed head, piercing, and explosive explosion. Traumatic brain injury (TBI) symptoms might include altered behaviour (such as aggression or anxiety), amnesia, seizures, nausea, aphasia, seizures, and a prolonged coma. Though they may not manifest for weeks, months, or even years following a traumatic brain injury, these signs and symptoms usually show up within minutes (Black et al. 2002).

Primary blast-induced injury results in internal damage from the shock wave; secondary injuries cause penetration; tertiary injuries cause physical harm from the blast wave; and quaternary injuries result from external trauma from the blast wave (other than the first three classes) (Risdall and Menon 2011). Hyperemia, contusion, cerebral oedema, neuronal cell death, axonal injury, compromised blood-brain barrier (BBB), vasospasm, pseudoaneurysm development, hyperdeformation of the brain caused by blast kinetic energy, and hypercoagulability are among the many white and grey matter injuries that might result (Cernak and Noble-Haeusslein 2010). Along with the aforementioned symptoms, research has revealed that a significant proportion of traumatic brain injury survivors also experience post-traumatic stress disorder (PTSD) (Risdall and Menon 2011).

7.3 Physiology of Traumatic Brain Injury

Research has demonstrated a significant prevalence of post-traumatic stress disorder in survivors of traumatic brain injuries (TBI), in addition to the clinical characteristics already stated.

7.3.1 Primary Brain Injuries

There are two main categories of brain injuries that can result from various mechanical assaults to the brain: focal and diffuse brain injuries. Patients with moderate-to-severe traumatic brain injuries sometimes experience both kinds of injuries simultaneously, according to research (Skandsen et al. 2010). Nearly 70% of traumatic brain injuries (TBIs) are caused by diffuse axonal damage (DAI). Brain lesions, skull fractures, or localised contusions at the site of the injury are the symptoms of traumatic brain injuries (TBIs) caused by cuts, concussions, or compressions. Due to the clustering of dead neurons and glial cells at the coup with reduced blood flow, hematomas, epidural, subdural, and intracerebral haemorrhages can develop at certain brain layers. Tissues on either side of the coup can develop contusions due to secondary impact when the brain rebounds and meets the skull (contrecoup)(Schmidt et al. 2004). Brain damage can cause hemiparesis, behavioural abnormalities, and cognitive impairments. Diffuse brain damage differs from focal injury in that it primarily results from non-contact forces of fast acceleration and deceleration that injure cerebral brain tissues by shearing and stretching. Brain oedema and ischaemic brain injury are caused by neural axons, oligodendrocytes, and blood vasculature being damaged by the intense tensile stresses (Smith et al. 2003). Diffuse traumatic brain injury is characterised by widespread axon damage, mainly in deep white matter and subcortical areas, including the corpus callosum and brain stem, leading to impaired axonal transport and cytoskeleton degradation.

The fact that these axonal damages might last for months after a traumatic brain injury (TBI) suggests a link to the secondary pathology of cerebral oedema and delayed haemorrhage. Neuronal degeneration and axonal damage both have a role in determining the severity of traumatic brain injury. Interestingly, classic diffuse brain injury symptoms are present in explosive blast TBI, even though this type of brain injury is caused by shock waves rather than inertial forces (Saatman et al. 2008).

7.3.2 Secondary Brain Injuries

As a result of the biochemical, physiological, and post-injury processes, secondary damages are frequently either delayed or prolonged. Depending on the severity, these damages could persist for a few hours or even years. Neuroinflammation, axon degeneration, reactive oxygen species (ROS), mitochondrial dysfunction, lipid peroxidation, excitotoxicity, and apoptotic cell death are some of the processes that lead to secondary injuries (Ray et al. 2002).

7.3.2.1 Excitotoxicity

Traumatic brain injury (TBI) causes the blood-brain barrier (BBB) to dissolve, which leads to an overabundance of neurotransmitters and the inability of glutamate transporters, which are normally responsible for re-up taking glutamate, to do their job (Chamoun et al. 2010). Ionotropic and metabotropic glutamate receptors are activated when glutamate and its metabolites bind to them. A family of receptors known as ionotropic glutamate receptors includes the *N*-methyl-D-aspartate (NMDA) and α-amino-3-hydroxy-5-methyl-4-isoxazolepropionic acid receptors (AMPA) types, which open the membrane to the depolarising effects of salt, potassium, and calcium ions (Chamoun et al. 2010). In TBI conditions, these receptors are overexpressed because of the disruption of ion homeostasis caused by the overproduction of glutamate, which allows extracellular Ca^{2+} and Na^{+} ions into the cell (Meldrum 2000). The presence of GluN2B in the synaptic cytosol lends credence to the idea that it mediates the excitotoxic response (Wyllie et al. 2013). When there is an increase in intracellular Ca^{2+} levels, many downstream signalling molecules are activated, including protein phosphatases, $Ca2^{+}$/calmodulin-dependent protein kinase II, protein kinase C, mitogen-activated protein kinases (MAPK), and many more. When the cytosolic Ca^{2+} level is too high, apoptotic proteins such as calcineurin, caspases, and calpain are activated, leading to cell death (Folkerts et al. 2007; Weber 2012). Accumulated reactive oxygen species also hinder mitochondrial activity (ROS). Oxidative stress kills cells caused by excitatory neurotransmitters, and this process is known as excitotoxicity (Chamoun et al. 2010).

In the immediate aftermath of a head injury, the shear and stretch forces activate NMDA receptors, leading to glutamate-independent excitotoxicity (LaPlaca and Thibault 1998). Regarding the subunits and signalling cascades that regulate NMDA receptors in reaction to mechanical stimuli, one study also proved that NMDA receptors are mechanosensitive. The findings demonstrated that the GluN2B subunit mediates the mechanosensitive response (Singh et al. 2012).

7.3.2.2 Mitochondrial Dysfunction

One characteristic of traumatic brain injury is damage to the mitochondria, which changes metabolic and physiological processes and eventually causes cell death. Overloading mitochondria with Ca^{2+} can cause reactive oxygen species (ROS) generation and membrane depolarisation in the absence of ATP synthesis (Xiong et al. 1999). The generation of reactive oxygen species (ROS), depolarisation of the mitochondrial membrane, and inhibition of ATP synthesis are all outcomes of an invasion of excess ions into mitochondria (Singh et al. 2006). Calcium regulation and metabolic function become dysfunctional due to oxidative phosphorylation and the subsequent impairment of the electron transport chain. The mitochondrial permeability transition pore (mPTP) becomes active when the cell is under stress. The adenine nucleotide translocator protein undergoes a structural alteration when bound to cyclophilin D due to mitochondrial malfunction. This causes the mPTP to open and the inner membrane permeability to rise (Naga et al. 2007). Additional mitochondrial proteins that are crucial for cell death upon cell death include apoptosis-inducing factor (AIF) and cytochrome c (Tsujimoto and Shimizu 2007).

7.3.2.3 Oxidative Stress

Reactive nitrogen species (RNS) and reactive oxygen species (ROS) levels might rise because of oxidative stress and secondary cell death. Damage to the mitochondrial membranes occurs as a result of lipid peroxidation, which disrupts mitochondrial activity (Xiong et al. 1999; Shohami and Kohen 2011). In response to damaged cells following a TBI, the electron transport chain (ETC) increases ROS production. In contrast, Ca^{2+} accumulation after TBI helps nitric oxide synthases produce NO (Deng et al. 2007). Peroxynitrite (PN), a byproduct of the reaction between nitrogen oxide (NO) and the free radical superoxide, is another mediator of oxidative damage. By combining with polyunsaturated fatty acids to produce lipoperoxyl radicals, these reactive oxygen species (ROS) can harm cell membranes and have a negative impact on DNA and proteins (Ansari et al. 2008). Also, chronic excitotoxicity has been linked to the aberrant buildup of Ca^{2+} ions within cells. In particular, immunosuppression and brain plasticity are brought about by the continuous generation of reactive oxygen species (ROS) and lipid peroxidation, which have a negative impact on cerebral blood flow (Shohami and Kohen 2011).

7.3.2.4 Neuroinflammation

One further thing that might cause oxidative damage is peroxynitrite (PN), which is generated when superoxide and nitrogen oxide (NO) react. Lipoperoxyl radicals, formed when reactive oxygen species (ROS) combine with polyunsaturated fatty acids, damage cell membranes in addition to DNA and proteins (Werner and Engelhard 2007).

Brain tissue and cerebrospinal fluid analysis in traumatic brain injury patients (Frugier et al. 2010), and rodent models (Semple et al. 2010) found that inflammatory mediators like IL-6, IL-1β, and TNF-α were released by polymorphonuclear leukocytes and cytokines 24 h after the shock. Oedema and neurological abnormalities result from enhanced BBB permeability, as seen by the extended-release of cytokines. As a member of the Fas family, TNF-α has significant interactions with Fas ligand and triggers caspases to cause programmed cell death (Bye et al. 2007). Traumatised areas can attract leukocytes to the injury site through the presence of chemokines such as IL-8 (CXCL8), MIP-α, and MCP-1. In addition, ICAM-1 and VCAM-1 are endothelial and leukocyte cell adhesion molecules that engage with the endothelium to help bring immune cells and leukocytes to the site of injury (Frugier et al. 2010). It has been discovered in traumatic brain injury survivors years after injury that macrophages activate microglial cells, leading to an increase in the release of astrocytes, a symptom of chronic neuroinflammation (Bye et al. 2007; Johnson et al. 2013). At both the cellular and behavioural levels, researchers have examined the function of GSK-3β in the physiological model of mild traumatic brain injury (mTBI). In addition to controlling cell death, glycogen synthase kinase-3 (GSK-3) beta has been implicated in depressed behaviour. The impact of lithium or L803-mts, which are GSK-3β inhibitors, on ameliorating depression caused by mTBI was assessed by Shapira et al. (2007).

7.3.2.5 Axonal Degeneration

The axonal cytoskeletal network, which is made up of microtubules and neurofilaments, is destroyed in diffuse axonal injury (DAI) when neurons are suddenly mechanically injured (Andriessen et al. 2010). Characteristics of acute axonal injury from trauma caused by chronic calcium-mediated proteolysis include axonal transport damage, myelin sheath breakdown, and axonal transport protein accumulation (Su and Bell 2016). Axonal enlargement and cell and oligodendrocyte death are long-term effects of an overabundance of axonal transport proteins (Fünfschilling et al. 2012). The pyramidal tracts and corpus callosum of the brain stem were the most common sites of retractions. Axonal markers β-amyloid precursor protein (β-APP) and neurofilament (NF) were used to identify DAI characteristics 1 day after a traumatic brain injury (TBI). Scientists have found that the cingulum, cortex, and hippocampus contain them (Ng and Lee 2019).

7.3.2.6 Apoptotic Cell Death

Detector axons DAI was identified 1 day after a traumatic brain injury using neurofilament (NF) and β-amyloid precursor protein (β-APP); the corpus callosum and pyramidal tracts of the brain stem exhibited retraction bulbs, which were particularly characteristic (TBI). Cingulum, cortex, and hippocampus contain evidence of their existence (Smith et al. 1997). Calpain and caspases are proteases located downstream and are stimulated via many molecular pathways, including ERK, JAK/STAT, and p38 MAPK (Shim et al. 2007; Tan et al. 2006). Principal pathways through which cell death is mediated are the extrinsic pathways (EP) and the intrinsic pathways (IP). As opposed to IP, which regulates caspases 8 and 9 to activate caspase 3, EP integrates the interactions of TNF-Fas with the cell receptors associated with each receptor. IP requires depolarisation of cytochrome c emissions within mitochondria (Schmitz et al. 2000; Žvirblis et al. 2012). Caspase-independent apoptosis induces nuclear release of mitochondrial proteins including AIF, Second mitochondria-derived activator of caspase/direct inhibitor of apoptosis-binding protein with low pI (Smac/DIABLO), endonuclease G, and polymerase-1 in the context of traumatic brain damage (TBI). Following the activation of signalling molecules upstream, harm is inflicted onto glial and neuronal cells (Hasegawa et al. 2011).

7.4 Metabolic Disorders Related Lead to Cognitive Dysfunction in TBI

Neuropeptide Y/agouti-related peptide neurons and pro-opiomelanocortin neurons are two kinds of metabolism-regulating neurons found in the anterior cingulate cortex (ARC) of the hypothalamus (Li et al. 2021). Damage to the blood-brain barrier (BBB) and subsequent inflammatory infiltration after a brain injury produces hypothalamic neuronal damage and a cascade of neurometabolic dysfunctions (Mele et al. 2021). Also, the mitochondrial pathway, oxidative stress, and the inflammatory response are just a few of the ways that non-neuronal cells (such as microglia and astrocytes) might influence aberrant neuronal cell metabolism (Zhang et al. 2021). Glucose, lipid, protein, and trace element metabolic abnormalities are among the primary pathophysiological alterations following traumatic brain injury (TBI). More and more research is linking cerebral metabolic abnormalities to a range of cognitive dysfunctions following traumatic brain injury (Lai et al. 2022).

Traumatic brain injury (TBI) induces a complex milieu of metabolic alterations due to the brain's regulatory role in metabolic activity. These alterations encompass hormonal fluctuations, aberrant cellular metabolism, and a robust systemic and cerebral inflammatory response that contends for free substrate required for damaged cell metabolism. The severity of motor dysfunction and injury exhibit an inverse correlation with the intensity of this hypermetabolic state (Fruin et al. 1986). Hyperglycemia, protein loss, and increased energy expenditure are the outcomes of systemic catabolism, which happens in reaction to these changes. All these factors contribute to the seriousness of traumatic brain damage. The catabolic reaction and the harmful effects of chronic hypermetabolism can be lessened with the help of the right nutrients. An injured brain triggers a chain reaction of hormones that impact metabolic function as a defence mechanism against stress. The hypothalamic-pituitary axis secretes several hormones, including cortisol, growth hormone, prolactin, and vasopressin. Glucagon and catecholamines are also produced in excess. In addition to regulating blood pressure and cardiac output, catecholamines have several detrimental effects, such as a rise in oxygen consumption, glycogenolysis, hyperglycemia, proteolysis, and muscular atrophy (and hence brain perfusion). An increase in basal metabolism and cellular energy demand characterise TBI-induced metabolic dysfunction, which can lead to a shift in cellular supply-demand coupling toward configurations that promote energy failure. Traumatic brain injury (TBI) manifests itself in the acute ischaemia phase with the production of reactive oxygen species, hyperglycemia, and intracellular lactate (Li et al. 1999; Diaz-Parejo et al. 2003).

Disruption of cell-mediated immunity and atrophy of muscular tissue are two potential outcomes of proteolysis. These catabolic effects may be mitigated using exogenous substrates supplied by parenteral or enteral nutrition (PN) or enteral nutrition (EN), which may lessen the requirement to release endogenous substrate storage. Secondary injuries and the outcome of traumatic brain injuries (TBI) may be influenced by changes in the concentrations of acute phase reactants, electrolytes, and amino acids. Functions of the immune system, *N*-methyl-D-aspartate (NMDA) receptors, and substrate metabolism all rely on zinc as a cofactor. As a result of hepatic sequestration and enhanced renal clearance, blood zinc concentrations are reduced in TBI. Zinc supplementation shows promise in enhancing protein metabolism and neurological outcomes 1 month following traumatic brain injury (Young et al. 1996). Magnesium may possess neuroprotective characteristics due to its impact on cellular energy production, calcium influx, and the NMDA receptor (McKee et al. 2005). We remain uncertain as to whether magnesium supplements will be beneficial for them. It can be hypothesised that decreases in insulin-like growth factor binding protein-3 (IGFBP-3) subsequent to traumatic brain injury (TBI) result in enhanced elimination of IGFBP-1 activity, which is a growth hormone mediator and insulin-like growth factor-1 (IGF-1). It appears that exogenous IGF-1 supplementation improves protein preservation and decreases blood glucose levels in patients with traumatic brain damage (Hatton et al. 1997).

The availability of transportable protein precursors into the brain and the catecholamine stress response are both influenced by changes in amino acid concentrations. Glutamine may enhance brain glutamate production, despite its usual association with reduced bacterial translocation. When glutamate binds to the NMDA receptor, it triggers a cascade of events that culminate in cell death (Nishizawa 2001; Hatton 2001). Current understanding of the risks and benefits of L-arginine use is incomplete since nitric oxide concentrations may vary based on the phase of damage. For a more comprehensive understanding of the potential therapeutic implications and precise mechanisms of action of electrolytes and amino acids such as glutamine and magnesium in the treatment of traumatic brain injury (TBI), additional study is necessary (Cherian et al. 2004).

7.4.1 Disorders of Glucose Metabolism

Although most of the glucose is utilised as energy by the brain through oxidative phosphorylation, it also plays an important part in several secondary processes such as the pentose phosphate route, the folate cycle, and the hexosamine biosynthesis pathway. Glucose energy metabolism is one of the minor metabolic pathways altered by traumatic brain injury (TBI) (Zhang et al. 2021; Chitturi et al. 2019). Upon initial brain injury, it is commonly believed that a brief hypermetabolic state is followed by a long-lasting hypometabolic state (Tong et al. 2019; Park et al. 2013). In contrast, prior research found that glucose metabolism remained unchanged within 6–18 h following traumatic brain injury (TBI), indicating that glucose oxidation was not directly involved in neuronal death during the initial phases of injury (Jayakumar et al. 2016). Decreased expression of glycolysis-related kinases and glucose transporters is associated with impaired glucose metabolism in chronic traumatic brain injury (Huynh et al. 2020). Patients with traumatic brain injury (TBI)-induced cognitive impairment are likely to have impaired glucose metabolism in the brain, even though the timing and location of this effect vary depending on the kind of brain injury (Komura et al. 2019). Research into ways to enhance this hypometabolic state may, therefore, 1 day lead to better cognitive performance. Hypometabolism, it turns out, might be an adaptive mechanism for reestablishing brain function by diverting glucose away from the main oxidative energy routes and toward the secondary ones (Kumar 2019). The upregulation of S-adenosyl methionine levels during the subacute phase of traumatic brain injury impacts one-carbon/folate metabolism, which in turn impacts protein translation, nucleotide, amino acid, and lipid metabolism (Chitturi et al. 2019).

Recent research has demonstrated that folic acid supplementation can enhance cognitive performance in individuals suffering from mild cognitive impairment (Bai et al. 2021). Hence, finding out if aberrant folate metabolism impacts cognitive performance after TBI is of the utmost importance. Recent research has shown that abnormalities in glucose aerobic metabolism and mitochondrial biogenesis can be reversed by targeting the Sirtuin1 (SIRT1) signalling system (Huang et al. 2021a).

7.4.2 Disorders of Lipid Metabolism

The brain's plasma membrane, myelin sheath, and lipid rafts are all made up of lipids. Disruptions to lipid metabolism can have a major effect on mental capacity (Bowman et al. 2019). Alterations to fatty acids and other lipid byproducts in the brain (Chitturi et al. 2019; Anthonymuthu et al. 2017; Yi et al. 2016) and phospholipids (Lamade et al. 2020; Novgorodov et al. 2019) contribute to the development of traumatic brain injury. Subacute phase of traumatic brain injury is characterised by an elevation in acylcarnitine levels and a reduction in L-carnitine, palmitic acid (PA), and caprylic acid (Chitturi et al. 2019). L-Carnitine is essential for optimal mitochondrial activity in addition to facilitating functional recovery and restoring mitochondrial function in the aftermath of brain injury (Scafidi et al. 2011). Elevated levels of arachidonic acid are associated with cognitive impairment subsequent to traumatic brain injury, while reduced levels of linoleic acid and PA are detected. When the PA-mediated insulin pathway is excessively engaged, sensitivity to amino-3-hydroxy-5-methyl-4-isoxazole-propionic acid is diminished and hippocampal neuronal activity is compromised, according to another study (Loehfelm et al. 2020). Another ketogenic molecule found in the brain is caprylic acid. Notably, research in patients with mild cognitive impairment has shown that ketogenic drinks containing caprylic acid can increase brain energy and certain cognitive indicators (Fortier et al. 2019).

Fatty acid oxidation was enhanced in the subacute phase of TBI (Chitturi et al. 2019). Oxidised free fatty acid levels may rise in patients with TBI, according to another research. Proinflammatory lipid mediators, mostly caused by enzymatic lipid peroxidation, reached their peak 1 h following TBI. Lipidomic oxidation following traumatic brain injury (TBI) appears to have been a dynamic and intricate process since its distribution changed to an anti-inflammatory response within 24 h. By boosting the release of proinflammatory cytokines through toll-like receptor 4, inducing apoptosis through caspase-8 and caspase-3, and stimulating neuronal death, oxidised free fatty acids increase inflammation (Bowman et al. 2019).

Some research suggests that phospholipid metabolism may amplify the effects of traumatic brain injury on neuronal tissue (Lamade et al. 2020; Novgorodov et al. 2019; Palafox-Sánchez et al. 2021). A possible impact on membrane repair and synaptic integrity may be indicated by the marked decrease in levels of choline following TBI. Choline is the primary lipid component of phospholipids (Chitturi et al. 2019). Moreover, traumatic brain injury (TBI) can activate acid sphingomyelinase, which damages mitochondria; concurrently, it activates the NOD-like receptor protein 3 inflammasome, which causes cognitive impairment (Novgorodov et al. 2019). Phospholipase A2 activity affects the integrity of lysosomal membranes, inhibits autophagy, and facilitates neuronal cell death following TBI (Sarkar et al. 2020). By taking advantage of the compromised BBB, activated phospholipid enzymes produce hepatic lysophosphatidylcholine (LPC), a lipid inflammatory mediator, which is then delivered to the brain. Consequently, problems with lipid metabolism can hasten the ageing process, which includes cognitive decline, demyelination, and cell death (Palafox-Sánchez et al. 2021).

7.4.3 Neurotransmitter Disorders

The central nervous system neurotransmitters glutamate (Glu) and γ-aminobutyric acid (GABA) collaborate to keep various brain circuits' excitement and inhibition levels balanced. But when this equilibrium is upset, brain dysfunction follows. Acute traumatic brain injury is commonly believed to cause a temporary rise in extracellular Glu levels, while chronic traumatic brain injury is characterised by aberrant GABA function and blocked Glu signalling (McGuire et al. 2019). Oedema and cell death can result from increased intracellular calcium levels and sodium outflow caused by elevated extracellular Glu levels (Luo et al. 2019). Additionally, learning impairment results from dysregulation of Glu and GABA, which speeds up the delayed inhibition of long-term potentiation (LTP) following traumatic brain injury (TBI) (Zhang et al. 2019). These findings show that restoring cognitive function after traumatic brain injury (TBI) requires a delicate equilibrium between glutamate (Glu) and GABA.

An increasing body of data suggests that dopamine (DA), an exceptional signalling molecule located in the central brain system, is significantly involved in regulating cognitive processes (D'Amelio et al. 2018). DA release and DA transporter expression experience transient increases during the acute phase of traumatic brain injury (TBI), followed by sustained decreases in both variables subsequently (McGuire et al. 2019). Recent research has shown that a component of neurotransmitter exocytosis, the soluble *N*-ethylmaleimide-sensitive factor attachment protein receptor complex, is crucial for normal DA neurotransmission in striatal synapses following traumatic brain injury (TBI) (Carlson and Dixon 2018). Regarding norepinephrine (NE), persistent traumatic brain injury (TBI) is associated with decreased NE signalling (McGuire et al. 2019). The NE reuptake inhibitor atomoxetine failed to alleviate chronic cognitive impairment following traumatic brain injury (Dougall ct al. 2015).

The central cholinergic system relies on the neurotransmitter acetylcholine (Ach). In addition to its critical functions in learning and memory, its primary function is to keep the mind alert. Acute traumatic brain injury (TBI) is characterised by an increase in the transitory release of Ach, which is subsequently followed by a decrease in both Ach release and the expression of the Ach receptor. While changes in acetylcholinesterase activity have been linked to persistent cholinergic dysfunction (McGuire et al. 2019). No statistically significant improvement in cognitive impairment following traumatic brain injury was shown when acetylcholinesterase inhibitors were used in clinical trials (Zafonte et al. 2020). Because of its importance in many areas of cognition, scientists have recently focused on nicotinic Ach receptors (nAChRs). After traumatic brain injury (TBI), the density of α7 nAChRs decreases in most brain areas, ranging from 16% to 47%. The hippocampus is known to have high levels of these receptors (Hoffmeister et al. 2011). An additional study has shown that α7 nAChRs can be positively allosterically modulated to increase learning, memory, and LTP in the hippocampus following traumatic brain injury (TBI) (Titus et al. 2019).

To sum up, metabolic illnesses and traumatic brain injuries may share common pathways for cognitive impairment, including an inflammatory response, insulin resistance, and disruption to the blood-brain barrier (BBB). At its foundation, internal environment regulation rests on the interplay between the inflammatory response and metabolic regulation; impairment in either area can trigger or worsen a cascade of metabolic problems, both central and peripheral (Zhang et al. 2021; Hotamisligil 2006). The areas of interest are insulin resistance and the inflammatory response. The inflammatory and immunological response signalling pathways rely on insulin activity for nearly all of their functionality (Hotamisligil 2006). Thus, this inflammatory response associated with metabolism could represent a significant step forward in our understanding of cognitive impairment caused by aberrant central metabolism following traumatic brain injury.

7.4.4 Ferroptosis

More and more research is focusing on how iron metabolism contributes to the pathogenesis of traumatic brain injury (TBI). The proper functioning of the brain depends on the presence of iron, a trace element. Sphingomyelin and other neurotransmitter production, along with the mitochondrial energy route, are all impacted by it. Red blood cells flood the brain parenchyma due to intracerebral haemorrhage, broken blood vessel damage, and increased capillary permeability; these cells then burst and degrade, releasing non-transferrin-bound iron (NTBI), which is particularly prevalent following traumatic brain injury (TBI) (Tang et al. 2020). As a result of an increase in transcription factor activator protein-1 (AP-1), heme-by-heme oxygenase 1 is able to synthesise a substantial amount of iron after brain injury (Cheng et al. 2016). Also, AP-1-deficient mice had better cognitive function (Wang et al. 2018). The increased divalent metal transporter-1 is responsible for the preferential transport of iron (particularly NTBI) into cells during the inflammatory response (Wang et al. 2018). Along with this, glycolysis increases and oxidative respiration decreases. An excess of lipid reactive oxygen species is produced by intracellular iron overload via the Fenton reaction, and this process is accelerated by traumatic brain injury (TBI) due to downregulated glutathione peroxidase 4 (GPX4). To ferroptosis, both paths lead (Tang et al. 2020). In addition, cytoplasmic iron deposition and iron toxicity are caused by cells being weakened by endoplasmic reticulum stress (Lucke-Wold et al. 2017). Oligodendrocytes show high levels of iron buildup and are most severely impacted in white matter specimens from patients with chronic traumatic encephalopathy (Chancellor et al. 2021). The process of myelin formation and regeneration is carried out by oligodendrocytes. Axonal survival and central nervous system signalling are both impacted by myelin sheath loss (Liu et al. 2019), suggesting that the buildup of iron can lead to axonal rupture and oligodendrocyte necrosis. One notable treatment for traumatic brain injury (TBI) is the intraventricular injection of the ferroptosis inhibitor ferrostatin-1, which can lessen the intensity of lesions and ameliorate certain signs of cognitive impairment (Xie et al. 2019). It is anticipated that cognitive dysfunction following traumatic brain injury will become a new focus for improving cognitive function as the amount of research on the subject increases.

Research has shown that cognitive impairment following traumatic brain injury is linked to ferroptosis. Consequently, our goal is to find a way to alleviate cognitive dysfunction following a traumatic brain injury (TBI) by preventing ferroptosis. One inhibitor of ferroptosis, ferrostatin-1, has the potential to prevent neuronal degeneration and iron deposition while simultaneously enhancing motor and cognitive abilities in the long run (Xie et al. 2019; Chen et al. 2021). Similarly, ruxolitinib and other non-iron chelators (Chen et al. 2021), minocycline (Zhang et al. 2020), and polydatin (Huang et al. 2021b), have shown promise in preventing ferroptosis following traumatic brain injury.

Metabolic reprogramming and ferroptosis are intimately connected, according to recent studies. The immunological response in the central nervous system can be influenced by cellular metabolic reprogramming, which modulates microglial polarisation. Following traumatic brain injury (TBI), the iron ion could play a pivotal role in altering cellular metabolism (Kapralov et al. 2020). Scientists have also discovered that copper ions play a role in the demise of cells. By interacting with enzymes involved in the tricarboxylic acid cycle, copper ions can trigger a protein-toxic stress response, ultimately leading to cell death (Tsvetkov et al. 2022). After a traumatic brain injury (TBI), copper metabolism abnormalities may not go away (Portbury et al. 2016). Delay in injury development after traumatic brain injury may be caused in part by copper-induced cell death.

7.5 Therapeutic Intervention

Treatment plans for traumatic brain injuries primarily focus on stabilising the injury site and preventing further harm, as these injuries typically involve immediate physical damages and irreversible necrotic cell death. A variety of risk factors can lead to the progressive development of secondary injuries, as indicated earlier. This opens a door for therapeutic intervention into processes like excitotoxicity, oxidative stress, apoptotic cell death, persistent inflammatory response, and damage epicentre-induced neuronal and glial cell loss (Ray et al. 2002). To create more effective treatment methods that target several phases, a lot of study has gone into understanding the processes of secondary brain injuries. The next subject will cover the treatment of TBIs. Traumatic brain injury (TBI) is a prevalent neurological disease that can result in long-term or short-term brain damage. Also, it kills a lot of people, both young and old. Therefore, therapy is necessary. Medical treatment for traumatic brain injury mostly addresses symptoms. Maintaining and weaning patients off of sedatives and analgesics used to treat intracranial hypertension and convulsive status epilepticus are the primary objectives (Geeraerts et al. 2018). Understanding the mechanics of secondary brain injuries has been the focus of much research, with the goal of creating more effective treatment options that address many phases. After this, we will talk about how to deal with TBIs. Whether the damage is permanent or temporary, traumatic brain injury (TBI) is a prevalent neurological condition. Additionally, it is a leading killer of both adults and young people. Therefore, we need to treat you. There is a heavy emphasis on symptom management in TBI medical treatment. Patients undergoing treatment for intracranial hypertension or convulsive status epilepticus should primarily aim to maintain and eventually stop using sedation and analgesics (Delage et al. 2021). Despite the fact that many pharmacological therapies have shown promise in pre-clinical studies, no phase III trials have yielded positive results (Hiskens 2022). One type of regulated cell death (RCD) is ferroptosis, which is different from other known types. It's defined by an excess of reactive oxygen species (ROS) and iron-dependent lipid peroxide (Zhao et al. 2021). The importance of ferroptosis in the treatment of TBI has recently come to light. Choline and its metabolites play an important role in predicting the outcome of traumatic brain injuries (Javaid et al. 2021). An magnetic resonance imaging (MRI) choline peak after a craniocerebral transplant is associated with myelin degradation, as shown by Esele et al. (Eisele et al. 2020). Moreover, choline/creatine levels in the occipital grey matter were shown to be elevated in chronic stages of traumatic brain injury, according to Friedman et al. This finding offered magnetic resonance imaging (MRI) evidence of cellular damage (Friedman et al. 1998). A significant change that transpires subsequent to a traumatic brain injury is a considerable elevation in the concentration of free choline within the specifically afflicted region of the brain and its environs (Scremin and Jenden 1991). Choline depletion during the chronic phase and elevated myoinositol levels that persist for months after damage may be the result of diffuse glial growth (Ashwal et al. 2004). Another possible reason for the observed increase in choline levels during the chronic phase of traumatic brain injury is the hyperosmolarity condition of white matter, which causes the detection of elevated choline levels (Lin et al. 2012). It is well-known that brain injuries cause the blood-brain barrier (BBB) to become less effective, which in turn causes neuronal tissues to die off and changes how neuroprotective medications work. Thus, preserving or protecting the BBB may be one of the most promising ways to avoid or reduce the likelihood of neuronal inflammation, secondary brain injury, and immediate neurodegeneration after traumatic brain injury (TBI) (Sivandzade et al. 2020). The severity of traumatic brain injury (TBI) effects is influenced by both mechanical factors and patient-specific variables, such as age (Newell et al. 2015). The sexual orientation of the patient is another important consideration (Mollayeva et al. 2018).

7.5.1 Methods of Treatment

7.5.1.1 Injury Assessment

A thorough evaluation and familiarity with the injury are prerequisites for effective treatment. This is the most important part of the treatment. When treating traumatic brain injury, the doctor must first determine the extent of the damage. They must be informed about the patient's prognosis about a full recovery or the necessity of months of rehabilitation. Medications, surgeries, and physical therapy are the mainstays of care for TBI patients.

7.5.1.2 Medications

To treat traumatic brain injury (TBI), diuretics can be prescribed (Jacobson 1987). Its usual indication is the reduction of intracranial pressure, which occurs after a traumatic brain injury. Mannitol is another ingredient in the medicine (Patil and Gupta 2019). The medication is furosemide. Traumatic brain injury (TBI) treatment options also include anticonvulsants. Their usual usage is to forestall more damage caused by seizures. Levetiracetam coma is caused by the medicine carbamazepine. When there is pressure on blood vessels, they are most often used. Medications that alleviate pain are also known as analgesics. When treating traumatic brain injury, they are helpful in lowering pain. Additional medications used to treat traumatic brain injury include propofol, pentobarbital, aspirin, and many more (Capizzi et al. 2020).

7.5.1.3 Surgery

As a kind of treatment for traumatic brain injury (TBI), surgery involves removing the injured tissue, such as the skull, among other organs. For certain traumatic brain injuries, it might be essential. In such situations, there are two distinct kinds of harm. When it comes to removing the affected brain tissue, some doctors favour open brain surgery. On the other hand, closed brain surgery allows certain doctors to remove diseased or injured brain tissue without cutting into surrounding nerves. It is crucial to do a thorough study of your treatment options prior to surgery in these circumstances because the brain can be damaged during the procedure (Jha and Ghewade 2022).

7.5.1.4 Physical Therapy

Treatment for traumatic brain injury also makes use of therapies. They are just as crucial as the conventional medical and surgical approaches to treating traumatic brain injuries. Rehabilitation is the primary goal of most treatments and simpler therapies. People with traumatic brain injuries also have access to alternative treatments. Physical therapists, osteopaths, and chiropractors are all qualified to treat patients (Kane et al. 2019). After a traumatic brain injury, you may experience muscle spasms and contractions; these can be alleviated with their help. Another way they can help reduce inflammation is by increasing circulation. Traumatic brain injury (TBI) and spinal fractures are two more conditions that osteopaths can assist with (Dang et al. 2017).

7.5.1.5 Exercises

Symptoms of traumatic brain damage can be managed with exercise, but it will not cure the condition itself. Theoretically, the activity does not increase cerebral blood flow, which would otherwise lead to brain inflammation. Proper and regular exercise reduces inflammation in the brain by increasing oxygen supply to the area. Even if exercise does not immediately improve blood flow to the brain, it is nevertheless recommended to exercise consistently to decrease inflammation. On the other hand, it has an indirect effect via increasing brain's oxygen supply (Perry et al. 2020).

7.6 Importance of Adequate Nutrition in TBI Recovery

There has been minimal success with pharmaceutical medications that aim to treat traumatic brain injuries (TBIs) (Scrimgeour and Condlin 2014). The function of nutritional assistance in reducing traumatic brain injury (TBI) sequelae is an increasingly popular topic of research into the pathophysiology of TBI. The human brain uses 20% of our resting energy even though it only makes up 2% of our entire mass, therefore getting enough nutrition is crucial (Forbes et al. 2022). When critically sick people don't get enough to eat, it can lead to endocrine dysfunction, organ failure, compromised immunity, and even death. To avoid malnutrition and its consequences, nutritional therapy is used. In neurological disorders such as traumatic brain injury (TBI), sepsis, stroke, and coronavirus illness (2019), nutritional deficiencies are caused by changes in metabolism and gastrointestinal dysfunction (COVID-19). In addition, functional outcomes are frequently poor when these nutritional deficiencies are present (Pardo et al. 2020; Wischmeyer 2018). Patients in critical care require nutritional treatment optimisation to optimise both short- and long-term recovery. Furthermore, research showing a higher incidence of malnutrition in TBI patients further highlights the significance of proper nutritional assistance following a TBI (Dijkink et al. 2020; Caliri et al. 2019; Ford and Ouma 2021; Charrueau et al. 2009). It is as important to treat TBI with nutritional therapy, but the timing and route are different (Lee-anne et al. 2014).

There are two main types of traumatic brain injuries: primary and secondary. Brain injuries caused by mechanical stress primarily manifest as CNS damage (CNS). Tissue distortion, axonal shearing, and BBB failure are hallmarks of this condition. Brain oedema, elevated inflammatory cytokines, excitotoxicity, ischaemia, reactive oxygen species, and immunosuppression are hallmarks of traumatic brain injury (TBI) secondary injury, which develops in reaction to the initial primary insult (Ng and Lee 2019). The cumulative effect of these impairments on cellular metabolism is an increase in energy demand in traumatic brain injury (TBI) patients due to a hypermetabolic and hypercatabolic condition (Lee-anne et al. 2014; Kurtz and Rocha 2020). An imbalance in electrolytes and nitrogen occurs when calorie expenditure rises, which in turn causes an imbalance in lean body mass consumption, an imbalance in electrolytes, an increase in infection risk, an increase in hospital admissions, and an increase in morbidity and death (Kurtz and Rocha 2020; Twyman 1997; Pillsbury et al. 2011).

7.6.1 *Protein Needs and the Amino Acids Metabolism in Traumatic Brain Injury*

Finding the ideal amounts of glucose, protein, and lipids required for neuronal survival is of the utmost importance when thinking about nutritional needs following TBI (Cook et al. 2008). Brain injury patients are at risk for both hyperglycemia and hypoglycaemia (Shi et al. 2016). It is believed that hyperglycemia, which is more prevalent after a TBI, is caused by heightened insulin resistance and impaired stress metabolism (Sarkar et al. 2020; Shaughness et al. 2020). Enhanced post-TBI prognoses are associated with this elevation in blood glucose levels (Shi et al. 2016). Despite the fact that research indicates that insulin therapy improves outcomes in patients who have suffered a traumatic brain injury (TBI), intensive insulin therapy worsens outcomes, especially in patients with severe TBI, when compared to regular insulin therapy (Singer et al. 2019). Optimal blood glucose levels after a TBI are 6–9 mmol/L (108–162 mg/dL), according to recommendations (Béchir et al. 2010). Patients with severe traumatic brain injuries who go an entire week without eating run the risk of losing 10–15% of their lean body mass (Singer et al. 2011). Protein catabolism is a significant contributor to this depletion; it peaks between 8 and 14 days after injury and is greatly increased during the acute phase of the disorder. Protein catabolism is correlated with the severity of traumatic brain damage (Young et al. 1992). An estimated 0.22–0.28 g/kg/day of nitrogen is released following a traumatic brain injury, according to estimates (Clifton et al. 1985; Young et al. 1985).

As opposed to attempting to preserve muscle mass and limit nitrogen loss, intensive protein supplementation is incapable of reversing the catabolic state of TBI (Young et al. 1987; de Aguilar-Nascimento and Kudsk 2007). However, current guidelines suggest that individuals with traumatic brain injuries (TBIs) begin consuming 1.5–2.0 g/kg/day of protein as soon as possible. To prevent azotemia, patients with renal impairment in particular should have ample access to free water while on these high-protein diets (Cook et al. 2008). Positive outcomes have been observed in clinical trials utilising the anabolic hormone insulin growth factor 1 (IGF-1) to reinstate a positive protein balance in patients undergoing growth hormone supplementation subsequent to a traumatic brain injury (TBI) (Hatton et al. 1997). In patients undergoing growth hormone therapy following traumatic brain injury (TBI), clinical trials employing the anabolic hormone insulin growth factor 1 (IGF-1) to restore a positive protein balance have yielded encouraging results (Ruokonen and Takala 2002).

7.6.2 Nutrition Assessment: Protein

Proteins are broken down at an abnormally high rate in individuals with traumatic brain injuries (TBI) due to hypercatabolism, which is triggered by inflammatory mediators and catecholamines. There seems to be a correlation between the severity of the injury and the peak protein catabolism, which occurs 8–14 days following the injury. Nitrogen elimination in the urine of traumatic brain injury patients varies between 0.2 and 0.28 g/kg/day (Young et al. 1987). It appears that supplementing with an excessive amount of calories to prevent protein loss is not beneficial (Bivins et al. 1986). It has not been proven that aggressive protein supplementation continuously maintains muscle mass or balances nitrogen excretion following TBI (Young et al. 1987; Clifton et al. 1986). Research on other severely sick individuals experiencing catabolism has also demonstrated that consuming large amounts of protein does not result in the recovery of lost muscle mass (Hart et al. 2002). Studies have shown an increase in mortality in critically sick patients receiving growth hormone, which makes the use of anabolic medicines like IGF-1 and growth hormone contentious (Takala et al. 1999). The only treatment that has restored protein balance in traumatic brain injury patients is insulin-like growth factor-1 (Hatton et al. 1997). In any case, to compensate for the increased catabolism, current guidelines propose that patients with acute TBI get protein at a rate of 1.5–2 g/kg/day (Hatton and Ziegler 1998). Conventional intact protein formulae are well-tolerated by most individuals with traumatic brain injuries. For patients with traumatic brain injuries (TBIs), there is a dearth of information on the pros and cons of specialist goods. Patients with brain injuries had a lower infection rate and a shorter intensive care unit stay after receiving an enteral formula with glutamine and probiotics, according to one research (Hatton and Ziegler 1998). Standard intact protein formulae are well-tolerated by the majority of TBI patients. There is a dearth of information on the pros and cons of specialist goods for patients with traumatic brain injuries. Research on individuals with brain injuries found that an enteral formulation of glutamine and probiotics reduced infection rates and reduced intensive care unit stays (Fuhrman et al. 2004; Gabay and Kushner 1999; Lim et al. 2005).

7.7 Overview in Branched-Chain Amino Acids (BBCAs)

A number of diseases and conditions have been investigated in relation to the essential amino acids valine, leucine, and isoleucine, including cancer, cirrhosis of the liver, renal failure, sepsis, trauma, burns, and burns. It is believed that via enhancing anabolic pathways, BCAA supplementation can alleviate cachexia, reduce exercise exhaustion, speed wound healing, prevent or treat hepatic encephalopathy, and stimulate insulin production (Bifari and Nisoli 2017).

7.7.1 *Metabolism*

7.7.1.1 BCAA Catabolism

The absence of significant hepatic activity of branched-chain amino acid aminotransferase (BCAT), the initial enzyme in the BCAA catabolism pathway, distinguishes the early phase of BCAA catabolism from the metabolism of other amino acids. This means that extrahepatic tissues have easy access to the BCAAs because their concentration in the bloodstream increases quickly following protein consumption. The dietary formulae based on BCAAs, particularly those aimed at the muscles and brain, have a distinct edge over others due to this phenomenon (Fig. 7.1).

Skeletal muscle is the primary target of BCAA catabolism due to the high activity of BCAT. The BCAT reaction generates glutamate and other comparable branched-chain keto acids (BCKAs), including α-ketoisocaproate (KIC), α-keto-β-methylvalerate (KMV), ketoisoleucine, and α-ketoisovalerate. The transfer of the BCAA amino group to α-KG is reversible (KIV, ketovaline). The next step is for glutamate to either assist pyruvate in becoming alanine (ALA) or convert ammonia to glutamine (GLN). In the bloodstream, muscles release GLN, ALA, and a significant amount of BCKA. A complex of enzymes called branched-chain α-keto acid

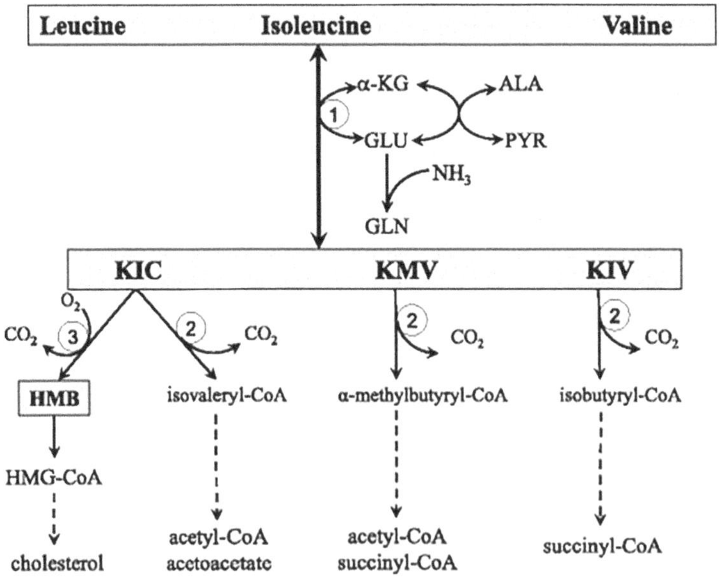

Fig. 7.1 BCAA catabolism. Main pathways of BCAA catabolism. *ALA* alanine, *GLU* glutamate, *GLN* glutamine, *HMB* β-hydroxy-β-methylbutyrate, *HMG-CoA* 3-hydroxy-3-methylglutaryl-CoA, *KIC* α-ketoisocaproate (ketoleucine), *KIV* α-ketoisovalerate (ketovaline), *KMV* α-keto-β-methylvalerate (ketoisoleucine), *α-KG* α-ketoglutarate. (1) branched-chain-amino-acid aminotransferase (BCAT); (2) branched-chain α-keto acid dehydrogenase (BCKD); (3) KIC dioxygenase (Holeček 2018)

dehydrogenase (BCKD) is located on the inner surface of the mitochondrial membrane. It is responsible for the second enzyme in the breakdown of BCAAs, the irreversible decarboxylation of BCKA to its corresponding branched-chain acyl-CoA esters. BCKD is controlled by the phosphorylation-dephosphorylation pathway.

In contrast to how phosphorylation by a specific kinase makes the enzyme inactive, dephosphorylation by a specific phosphatase activates it. Changes in kinase activity may play a crucial role. The liver, kidneys, and heart have intermediate BCKD activity, while brain, adipose tissue, and muscles have low levels. Muscles, which account for 35–40% of the body's overall mass, ought to account for a significant portion of the BCAAs used by the body when the weights of specific tissues are considered. Many tissues work together to regulate BCAA breakdown; the two most important are the liver and the muscle. The enzyme's activity status is affected by numerous factors, including cytokines, hormones, nutrition, and different metabolites (Harper et al. 1984). The infusion of endotoxin or tumour necrosis factor-alpha (TNF-α) is prompted by the significant increase in BCKD activity in muscles (Holeček 1996; Holecek et al. 1997).

Following the BCKD reaction, the metabolism of BCAAs divides into two separate pathways. There is a catabolic pathway that breaks down the ketogenic KIC into acetyl-CoA and acetoacetate; a similar pathway that breaks down the glucogenic KIV into succinyl-CoA; and a pathway that breaks down the ketogenic KMV into acetyl-CoA and succinyl-CoA. (KMV is both glycogenic and ketogenic). One distinct result of KIC catabolism is the reaction that KIC dioxygenase catalyses and produces HMB, or β-hydroxy-β-methyl butyrate.

7.7.1.2 Amination of BCKA and BCAA and BCKA Interorgan Cycling

The direction of the reversible and almost balanced BCAT reaction should shift in response to changes in BCAA and BCKA concentrations as well as the availability of nitrogen donors and acceptors. Though other tissues, like enterocytes, can convert BCKA to BCAA, the bulk of BCAA absorption and BCKA release often happens in the muscles. Results from investigations demonstrating the presence of tagged BCAA in proteins using labelled BCKA offer direct proof of BCKA amination to BCAAs (Swain et al. 1990). The amination process of the BCKA mostly utilises nitrogen that is supplied by GLN, glutamate, and ALA (Harper et al. 1984) (Fig. 7.2).

Muscles secrete large quantities of GLN and ALA into the circulation, which activates the synthesis of BCAAs from the BCKA, according to the study. Sepsis, trauma, or surgery are a few of the many causes of muscle atrophy, and this is one of their symptoms (Holecek et al. 2003; Holeček and Mičuda 2017; Holeček et al. 1998). Intravenous administration of KIC to endotoxin-treated rats' isolated livers dramatically increased leucine release (Holeček et al. 1998). Liver perfusion with medium containing 0.5 mM GLN resulted in greater synthesis of BCAA from BCKA compared to perfusion with medium without GLN (Holecek et al. 2003). During a brief period of hunger, the three BCAAs uniquely increased in blood plasma, and this rise may be due to amination of the BCKA. Muscles release ALA,

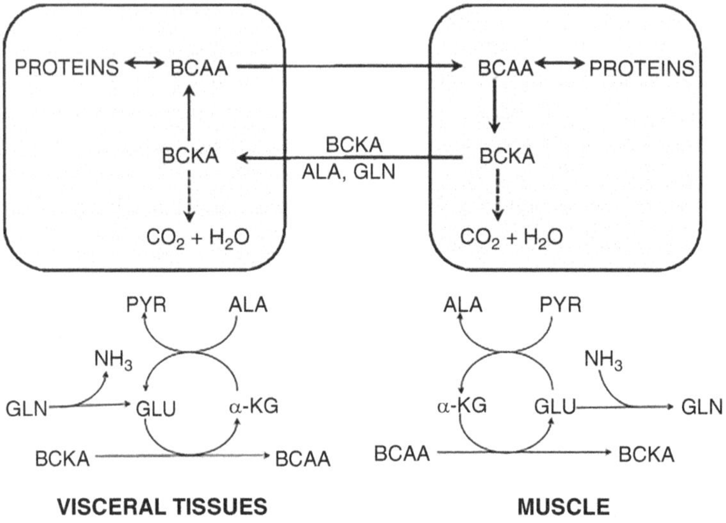

Fig. 7.2 Amination of the BCKA and interorgan cycling of the BCAA and BCKA. The schemes of the BCAT reactions (BCAA deamination and BCKA amination) and supposed cycling of the BCAA and BCKA among organs, which may in various conditions attenuate the loss of essential BCAA. ALA, alanine; BCAA, branched-chain amino acids; BCKA, branched-chain keto acids. GLU, glutamate; GLN, glutamine; PYR, pyruvate; α-KG, α-ketoglutarate (Holeček 2018)

GLN, and BCKA more quickly, and the liver increases gluconeogenesis (Holeček and Mičuda 2017). In a number of physiological and pathological states, the aforementioned results suggest that an interorgan cycle exists that reduces the loss of vital BCAA (Holeček 2001).

7.7.2 The Functions of BCAAs in Metabolism and the Body

The transfer of AAs into the cell is essential for metabolism. A unique carrier-mediated mechanism also carries AAs to the cell, much as glucose does. In the cell, each AA's unique keto acids are produced by the transamination process of BCAAs. After entering the Tricarboxylic acid cycle (TCA) cycle, the keto acids proceed through oxidative decarboxylation to form acyl-CoA derivatives. While the liver does contain the enzyme (activity) that is responsible for BCAA metabolism, it is more abundant in skeletal muscle, the heart, and the kidneys (Skeie et al. 1990) (Fig. 7.3).

Not only do BCAAs serve as a precursor for protein synthesis in muscle, but they also offer a non-specific carbon source for oxidation, which is used to produce energy. Glucose is not as energy efficient as BCAAs. As an example, compared to glucose, leucine oxidation in muscle results in a greater amount of energy available as ATP. In response to energy demands, such as those caused by famine, the

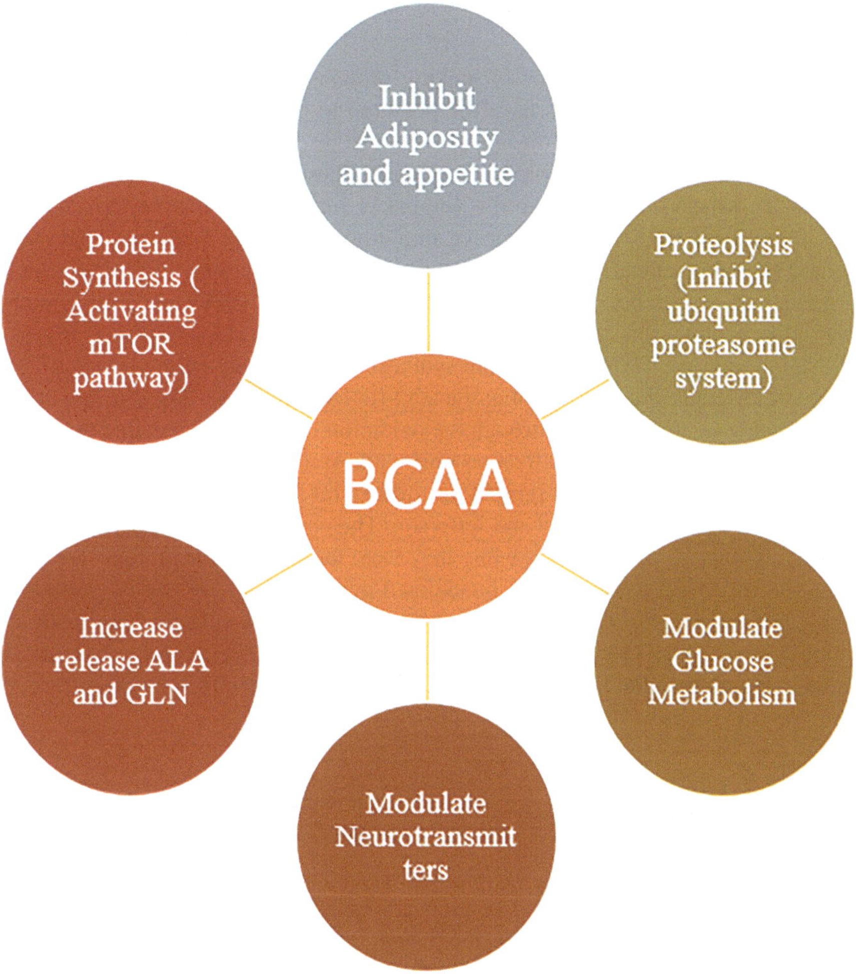

Fig. 7.3 BCAAs functions

oxidation of BCAAs rises under several physiological circumstances. By reducing protein breakdown and increasing protein synthesis, BCAAs control protein turnover in muscles (Kainulainen et al. 2013; De Bandt and Cynober 2006; Buse and Reid 1975). The loss of protein, particularly muscle protein, can occur as a result of a number of metabolic and clinical problems; taking BCAA supplements can help with these issues and others like cancer, liver disease, famine, surgery, trauma, sepsis, and liver disease (Kainulainen et al. 2013; De Bandt and Cynober 2006; Buse and Reid 1975; Manelli et al. 1984; O'Connell 2013). Under those medical circumstances, alanine, glutamine, and AAs are in high demand, and SK muscle can produce them by oxidising BCAAs. Therefore, when those medical situations arise, the body's increased demands for alanine and glutamine can be met by BCAAs (Holecek 2005).

7.7.2.1 Synthesis and Turnover

One of the most important parts of making proteins is starting the translation of messenger RNA. Amino acids have the potential to promote protein synthesis by improving this critical stage of the process (Kimball and Jefferson 2001). Deacylated tRNA impacts the beginning phase of protein synthesis, which in turn decreases or inhibits the process. By reducing this inhibitory impact, aminoacyl-tRNA derivatives formed by BCAAs may improve initiation of protein synthesis. One such method BCAAs could affect protein synthesis is by the oxidation of BCAAs, which forms an active metabolic intermediate (Li and Jefferson 1978). Because it interferes with the initiation step, deacylated tRNA decreases or stops protein synthesis. In order to improve the start of protein synthesis by reducing that inhibitory effect, BCAAs may cause the creation of aminoacyl-tRNA derivatives. The creation of an active metabolic intermediate through the oxidation of BCAAs is another potential mechanism by which BCAAs affect protein synthesis (Buse et al. 1979). The effects of BCAA supplementation on protein turnover in rat SK muscle were investigated in a separate investigation by Li and Jefferson. The juvenile rats were perfused with various combinations of BCAAs while they were fasting. Muscle protein synthesis increased by 25–50% and protein breakdown decreased by 30% in the perfused-supplemented group compared to the control group. Removing BCAAs from an amino acid mixture had no discernible impact on protein turnover, as was also noted for the remaining amino acids (Li and Jefferson 1978). Therefore, BCAAs influence both protein synthesis and turnover.

7.7.2.2 BCAAs and Signalling

Both protein synthesis and protein turnover are affected by BCAAs (Brown et al. 1994; Hay and Sonenberg 2004). The translational beginning of protein synthesis is triggered by BCAAs, especially leucine, which are strong activators of mTOR signalling (Buse et al. 1979). Ijichi found that various combinations of BCAAs caused albumin secretion in cultured rat hepatocytes. The researchers discovered that a combination of BCAAs increases albumin production in a dose-dependent way, and those BCAAs were the main effectors. Leucine promotes albumin formation; however, the powerful mTOR inhibitor rapamycin cuts this production in half (Ijichi et al. 2003).

Additionally, it was noted that leucine is the sole amino acid that can activate the mTOR, elF4E-BP1, and p70 S6 kinase downstream translational effector proteins. Furthermore, rapamycin entirely blocks the action of those proteins. From what we can tell, BCAAs and leucine seem to stimulate albumin synthesis in rat hepatocytes via mTOR signalling. Male rats who were famished were given either saline or each BCAA individually in the study conducted by Anthony. In addition, rats that had been deprived were given an intravenous injection of rapamycin prior to being given leucine. Research in SK muscle revealed that, in comparison to controls, only leucine significantly enhanced protein synthesis. Leucine also improved elF4E-BP1 and p70 S6 kinase activity and phosphorylation. Protein synthesis was reduced by rapamycin in both leucine-treated and starving rats. According to all of these studies, BCAAs and leucine, in particular, stimulate protein synthesis in skeletal muscle and liver through the same mTOR pathway (Anthony et al. 2000).

7.7.2.3 Glucose Metabolism

Many people with cirrhosis also end up with diabetes mellitus because of their impaired glucose metabolism. The liver and skeletal muscles are the primary sites for glycogen synthesis and storage. Maintaining a healthy blood sugar level relies heavily on glycogen storage. The glycogen stores in the liver and skeletal muscle of cirrhotic patients are significantly lower than those of healthy persons (Nishitani et al. 2004). In addition to improving glycogen synthesis, BCAAs aid in glucose uptake by the liver and skeletal muscle. Nishitani et al. examined the impact of leucine on glucose uptake by collecting soleus muscles from rats in good health and exposing them to the amino acid in an insulin-free environment. A similar stimulatory effect was observed in α-ketoisocaproic acid, a metabolic byproduct of leucine, and they also discovered that leucine enhances glucose absorption. Researchers found that leucine's stimulatory action is completely abolished when phosphatidylinositol 3-kinase (PI3-kinase) or protein kinase C (PKC) are inhibited using specific inhibitors. Nonetheless, there was no improvement after rapamycin treatment. According to these findings, leucine promotes glucose transport in skeletal muscle not via the mTOR route but rather via the insulin-independent PKC and PI3-kinase pathways (Nishitani et al. 2002). In a separate investigation, Peyrollier et al. subjected myoblast L6 cells to AA deprivation for approximately 1 h before exposing them to leucine. The researchers found that PI3-kinase and p70 S6 kinase activities were significantly upregulated. Additionally, they discovered that leucine renders glycogen synthase kinase-3 (GSK-3) inactive. An enzyme called GSK-3 is responsible for blocking glycogen synthase (GS) activity and halting glycogen production. These findings point to the possibility that activating GS with BCAA supplementation improves glycogen production (Peyrollier et al. 2000).

7.7.2.4 Energy Metabolism

When people are more active, they are less likely to be overweight. In contrast, metabolic illnesses are among the many consequences linked to obesity. A person's chance of developing certain diseases can be reduced with regular physical activity. We would do well to familiarise ourselves with the mechanisms that control and coordinate energy metabolism. Due to an increase in fatty acid oxidation and a decrease in the risk of obesity, it appears that oxidising BCAAs is good for our metabolic health. Blood BCAA levels are higher in people and animals who are overweight, and lower in those who are slim and active. Taking a BCAA-enriched protein supplement improves health and fitness in both humans and animals in a number of ways, including enhanced exercise performance, improved glucose tolerance and regulation, and more balanced protein composition and profile (Kainulainen et al. 2013).

To examine the impact of isoleucine supplementation on obesity, hyperglycemia, and fat homeostasis, Nishimura et al. administered isoleucine to mice who were already on a high-fat diet. They found that the group given isoleucine supplements had less weight gain and lower lipid mass than the control group. The group that took isoleucine supplements also had reduced triglyceride levels and hyperinsulinemia severity in their hepatic and skeletal muscle (Nishimura et al. 2010). Participants in the 3-day, double-blind trial took a supplement containing either BCAAs or a placebo. They were tested for lipid oxidation and exercise capacity. Research has shown that taking BCAA supplements can help you get through workout exhaustion by increasing lipid oxidation. Additionally, the volunteer's glucose metabolism was enhanced (Gualano et al. 2011). Obesity and BCAA consumption are inversely related, according to research by Qin et al. Obesity and overweight status were uncommon among middle-aged healthy persons from the United States, China, Japan, and the United Kingdom who drank more BCAAs. It is clear from all of this research that BCAAs play an important role in energy metabolism and in lowering obesity risks (Qin et al. 2011).

7.7.2.5 Effects on Protein Metabolism

Beyond their role as protein synthesis substrates, BCAAs also stimulate protein synthesis while inhibiting proteolysis. The BCAAs—and notably leucine and its metabolites—are responsible for the effects. Phosphorylation of ribosomal proteins and translation initiation factors are two mechanisms by which leucine promotes protein synthesis via the mTOR signalling pathway (Nair and Short 2005). The stimulating impact of leucine on insulin secretion is in addition to its function in protein-anabolic effects (Floyd et al. 1966). Primarily, BCKAs and HMB mediate the inhibitory impact of BCAAs on proteolysis. It has been demonstrated in vitro that BCKAs inhibit muscle proteolysis (Tischler et al. 1982). Compared to leucine, KIC infusions were more effective in keeping nitrogen balance in patients following major abdominal surgery and those who were fasting (Sapir et al. 1983; Mitch et al. 1981). Muscles benefit from HMB in both healthy and diseased situations because it reduces ubiquitin-proteasome proteolytic pathway activity (Holeček 2017).

7.7.2.6 Effects on Neurotransmission

Several neurotransmitters, including dopamine, norepinephrine, and 5-hydroxytryptamine, may be impacted by the competition between aromatic amino acids (AAA; phenylalanine, tyrosine, and tryptophan) and BCAAs since they share an entrance pathway into the brain (serotonin). Thus, an increase in BCAAs in the blood may lead to alterations in neurotransmitter levels, which in turn may influence brain function and behaviour. Patients with liver cirrhosis should be prescribed BCAAs since hepatic encephalopathy is a condition in which an imbalance between BCAAs and AAAs plays a role in its pathogenesis (Fischer et al. 1975). Supplementing with BCAAs may reduce serotonin synthesis, which is thought to be the culprit behind workout weariness. In addition, astrocytes detoxify ammonia to GLN and BCAA transamination in the brain is involved in glutamate and gamma-aminobutyric acid synthesis. Research suggests that leucine may reduce body fat by lowering hunger levels (Pedroso et al. 2015).

It's possible that BCAAs have significant functions in the brain as well. Branched-chain amino acids (BCAAs) have the potential to affect the synthesis of many neurotransmitters directly or indirectly, including serotonin, dopamine, norepinephrine, and others, as well as brain protein synthesis and energy production. Following a meal, the majority of BCAAs in the diet enter the bloodstream unmetabolised by the liver. The protein that carries aromatic amino acids into the brain is identical to that which carries BCAAs, including tryptophan (Trp), tyrosine (Tyr), and phenylalanine (Phe). Neurotransmitter serotonin has its precursor in Trp, while catecholamines have their predecessors in Tyr and Phe (dopamine, norepinephrine, and epinephrine). An increase in BCAA plasma concentration leads to an increase in BCAA brain absorption and a decrease in aromatic AA absorption. That might cause these linked neurotransmitters to be synthesised less frequently (Fernstrom 2005). The role of catecholamines in reducing blood pressure is significant. The blood pressure of hypertensive rats decreased significantly after receiving an injection of Tyr; however, this effect is blocked when an equal volume of valine is administered (Sved et al. 1979). Muscle and plasma BCAA depletion is a natural occurrence in people who engage in vigorous physical activity, such as athletes. Furthermore, it is possible that an increase in brain Trp absorption and serotonin release can result from a decrease in muscle and plasma BCAAs. Central tiredness is caused by serotonin. Taking BCAAs as a supplement might help athletes perform better and keep their bodies in good shape (Fernstrom 2005).

7.7.2.7 Effects on Glucose Metabolism

Plasma glucose levels are closely related to BCAAs. It has been extensively shown that BCAAs trigger insulin secretion and upregulate glucose transporters (Zhang et al. 2017; Nishitani et al. 2005). On the other hand, several studies have shown that insulin signalling suppression may result from an excessive consumption of amino acids (Um et al. 2006). Recent studies have shown that BCAAs activate mTOR, which can lead to insulin resistance, and that different BCAAs may have different impacts on glucose uptake (White et al. 2016). Different reports indicate either an increase in glucose consumption or the emergence of insulin resistance; further study is needed to understand these contradictory claims.

7.7.2.8 The Impacts Mediated by ALA and GLN

Skeletal muscle BCAA breakdown rate is very sensitive to BCAA supply (Manchester 1965). Because of this phenomenon, the main impacts of eating a diet rich in BCAAs are an increase in BCKA, ALA, and GLN levels in the blood and activation of BCAA catabolism (Holecek et al. 2016). Consequently, ALA and GLN modulate a number of actions associated with BCAA intake. The immune system, the creation of glutathione, the maintenance of acid-base balance by the kidneys, and the expression of heat shock proteins all depend on the availability of glutathione, and the principal gluconeogenic amino acid is ALA.

7.8 States of Hypermetabolism with Systemic Inflammatory Response Syndrome

Levels of BCAAs have been found to fluctuate between increased, unaltered, and decreased in several hypermetabolic diseases, including sepsis, burn injury, trauma, and cancer. Systemic inflammatory response syndrome (SIRS) is present in all these disorders and is marked by various neuro-humoral abnormalities, such as increased production of cytokines, activation of the sympathetic nervous system, and cortisol production. Extreme loss of lean body mass is the consequence of several metabolic changes brought about by these events, such as insulin resistance and increased breakdown of myofibrillar proteins. Death may result from organ failure affecting several systems if the hypermetabolic condition continues. Here, BCAAs play a crucial role as an energy substrate for muscles (Holeček 1996; Holecek et al. 1997; Nawabi et al. 1990). The production of GLN rises in tandem with the oxidation of BCAAs; this molecule is then secreted from muscles and used, ideally, by the immune system. When GLN consumption outpaces synthesis, GLN levels in blood and tissues drop (Fürst et al. 1987). Declining GLN supplies may impose rate limitations on critical immune cell activities such as phagocytosis and antibody production. There is some evidence that lower GLN levels promote the uptake of BCAAs by muscles (Holecek and Sispera 2014).

Inflammatory signals, according to studies, reduce BCAA absorption in the intestines, block BCAA transport in the blood to the muscles, and enhance transfer to the liver (Hasselgren et al. 1988; Gardiner and Barbul 1993). It is likely that the BCAA synthesis in visceral tissues is triggered by BCKA. The addition of KIC to the perfusion medium caused a significant rise in leucine release by the isolated liver of animals that had been treated with endotoxin (Holeček et al. 1998).

Although BCAA oxidation is strikingly triggered, there are uneven changes in BCAA levels due to various influences of individual metabolic abnormalities in SIRS. The levels of BCAAs may be elevated if insulin resistance occurs, muscle protein synthesis is reduced, or protein breakdown is increased. Protein synthesis in visceral tissues and increased ALA and GLN formation in muscles are linked to the activation of BCAA catabolism, which in turn decreases BCAA levels. Hence, changes in BCAA levels display no discernible pattern.

7.8.1 *Use of BCAAs in the Treatment of Burns, Trauma, and Sepsis*

The increased oxidation of BCAAs, which could reduce their availability in tissues, and their protein-anabolic characteristics are the rationales for taking BCAA supplements in SIRS. One possible explanation for BCAAs' beneficial effects is that they serve as a building block for GLN, an important molecule that helps keep the digestive and immune systems functioning normally while also influencing protein balance in a positive way. The effects of BCAAs in trauma, burn, or sepsis have been studied using a variety of solutions containing varying concentrations and ratios of specific BCAAs. Some studies have shown that BCAA can help with negative nitrogen balance (Jimenez Jimenez et al. 1991; Oki and Cuddy 1989). There is no universal agreement among scientists on how BCAA-enriched formulations affect protein homeostasis, hospital stays, and deaths, and some researchers' findings have been underwhelming (Mattick et al. 2013). The impact of BCAA-enriched formulations on protein balance, duration of hospital stay, and mortality has not been well-conserved, and other researchers' findings have been underwhelming (Lang and Frost 2005). Recent research has shown that BCAA may help those suffering from traumatic brain injury. Research in rodents has shown that BCAAs can improve cognitive impairment caused by injuries (Cole et al. 2010), BCAAs improve cognitive recovery in patients with severe traumatic brain injuries, according to scientific trials (Jeter et al. 2013; Aquilani et al. 2005).

7.9 Potentially Life-Saving Branched-Chain Amino Acid Therapy for Traumatic Brain Injury

A potentially crucial component of any treatment that achieves clinical efficacy may be branched-chain amino acids (BCAAs). It appears that BCAAs are the sole amino acids whose concentration is considerably changed in the affected brain areas after a TBI. Adding BCAAs to hippocampus slices in a controlled environment normalised electrophysiological parameters, and giving these amino acids to the water supply improved cognitive function 7 days following injury (Hawkins et al. 2006).

Nonetheless, the ways in which these consequences manifest have remained unknown. By studying how BCAA treatment affects calcium homeostasis following brain injury, this lab has now identified a probable reason for this improvement. Following an injury, BCAAs seem to improve calcium diastasis. By modulating calcium levels, BCAAs will exert fine control over the transport of neurotransmitters and their release from synapses, electrophysiological function, and, by extension, cognitive capacities. Because they change the activity of enzymes that are dependent on calcium and those that are not, BCAAs will improve mitochondrial and cytosolic bioenergetic performance (Hoge et al. 2008). Based on the results so far, BCAAs show great potential as a treatment for TBI. There are two major ways in which BCAAs affect neurotransmitter synthesis in the brain. They aid in the production of both excitatory glutamate and inhibitory gamma-aminobutyric acid by donating nitrogen (GABA)(Yudkoff et al. 2005). Along with tryptophan, tyrosine, and phenylalanine, they compete for transport over the blood-brain barrier (BBB), which is necessary for the production of serotonin (precursors for catecholamines) (Fernstrom 2005). Consumption of BCAAs leads to a surge in plasma concentrations and an increase in BCAA absorption to the brain, while the uptake of tryptophan, tyrosine, and phenylalanine is reduced. Reduced levels of these aromatic amino acids have an immediate impact on serotonin and catecholamine production and release. Neurological disorders including manic-depressive episodes, motor dysfunction, amyotrophic lateral sclerosis, and spinocerebral degeneration have been investigated as potential treatments with oral BCAAs. Traumatic brain injury (TBI) cells are damaged by excitotoxicity, which occurs when neurotransmitters like glutamate are stimulated to an excess (TBI). It is not known how significant a function BCAAs play in synthesis plays, though, because they also contribute to inhibitory neurotransmitter production (Fernstrom 2005).

7.9.1 Memory and Learning Circuit in the Brain

Cognitive impairment in the medial temporal lobe, particularly in the hippocampus formation, a region of the limbic system, has been linked to traumatic brain injury (TBI), according to anatomical and stereological counting research. The temporal horn of the lateral ventricle is bounded by the dentate gyrus, which contains the granule cell, the subiculum, and the hippocampus—the three subregions of this structure. Composed of four distinct regions, the hippocampus is also referred to as the cornu ammonis (CA) (CA1, CA2, CA3, and CA4). The pyramidal cell is an element in this design (Mckee and Daneshvar 2015).

Research in anatomy and stereology has linked traumatic brain injury (TBI) cognitive impairment to targeted lesions in the limbic system's hippocampus formation, which is in the medial temporal lobe. This structure comprises the dentate gyrus, which contains granule cells, the hippocampus, and the subiculum. It serves to confine the floor of the temporal horn of the lateral ventricle. The hippocampus, also known as the cornu ammonis (CA), is subdivided into four further portions (CA1, CA2, CA3, and CA4). The assemblage comprises pyramidal cells (Lynch 2004).

During the performance of complicated cognitive tasks, LTP—a mechanism associated with synaptic plasticity—strikes to strengthen the link between two neurons, namely between pre-synaptic CA3 and post-synaptic CA1 neurons. Glutamate, a neurotransmitter, is essential for reinforcing and solidifying this process. In the CA1 neurons' post-synaptic membranes, you can find two receptors: one that allows sodium ions to pass through and another that allows calcium ions to pass through more strongly. The former is called the AMPAr while the latter is called the *N*-methyl-D-aspartate receptor. The latter is inhibited by an ion of magnesium (Mg^{2+}). They are both Glu-sensitive (Lynch 2004).

Long-term potentiation (LTP) is a synaptic plasticity-related mechanism that, during the performance of complicated cognitive tasks, seeks to strengthen the connection between two neurons, namely between pre-synaptic CA3 and post-synaptic CA1 neurons. To solidify and reinforce this process, the neurotransmitter glutamate (Glu) is essential. The CA1 neurons' post-synaptic membranes include two receptors: one that allows sodium ions to pass through and the other that allows calcium ions to pass through to a higher extent. These receptors are called AMPAr and NMDAr, respectively. An ion of magnesium (Mg^{2+}) blocks the second one. Glu affects both greatly (Lüscher and Malenka 2012).

There are two ways in which an increase in post-synaptic Ca2+ concentration might strengthen a neural connection. The short-term increase in AMPAr availability may amplify the electrostatic repulsion effect on NMDAr Mg^{2+}. Over time, there is an upregulation of transcription and protein synthesis of AMPAr and other brain growth factors that help CA3 and CA1 neurons strengthen their connection by forming new synapses (Lynch 2004).

The enhancement of the neural connection is facilitated by a rise in post-synaptic Ca^{2+} concentration in two distinct ways. There may be a strengthening of the electrostatic repulsion effect on NMDAr Mg^{2+} in the near future due to the increased availability of AMPAr. In the long run, there's an uptick in the production of proteins like AMPAr and brain growth factors, which help strengthen the link between CA3 and CA1 neurons by forming new synapses (Shoji et al. 2014).

7.9.2 Excitatory and Inhibitory Neurotransmitters (Glutamate and GABA Production)

The amino acid glutamine (Glu) has an amine group in addition to two carboxyl groups (–COOH) (–NH_2). Particularly in astrocytes and neuronal cells, it carries out several tasks at the level of the brain. It helps create molecules with a lot of energy, such as adenosine triphosphate (ATP). The Krebs cycle provides most of the ATP that brain cells produce. To produce alpha-ketoglutarate, the cytoplasmic enzyme aspartate aminotransferase facilitates the reversible transamination of Glu (alpha-KG). This is crucial for the Krebs cycle, which is a process that produces ATP molecules, to work properly (Nissen et al. 2015). Glu encourages post-synaptic excitatory potentials in neurons, which depolarise the cell and lead to the formation of an amyloid precursor protein (APP), which improves the efficacy of synaptic networks via the long-term potentiation (LTP) phenomenon. The enzyme Glu-decarboxylase converts glutamate to the neurotransmitter GABA in neurons that are primarily inhibitory. Thus, Glu is thought to be an Neurotransmitters (NT) precursor of the main inhibitory kind (Lynch 2004). On a molecular level, Glu augments synaptic network efficiency via LTP by promoting post-synaptic excitatory potentials, which in turn cause depolarisation and the subsequent formation of an amyloid precursor protein (APP). In neurons that are mainly inhibitory, the enzyme Glu-decarboxylase is responsible for converting Glu into GABA. Thus, it is believed that Glu is an NT precursor that plays a significant inhibitory role (Shen 2013). Therefore, it can be inferred that the Gln-Glu cycle does not adequately guarantee steady levels of Glu and, by extension, GABA, in the brain. There are known to be other routes that facilitate the synthesis of these NTs.

7.9.3 Brain Metabolism and BCAAs

It follows that the Gln-Glu cycle is ineffective in keeping brain Glu and, by extension, GABA, levels consistent. The existence of other mechanisms that facilitate the synthesis of these NTs is well-established (Portier et al. 2008). The results of the memory test at the conclusion of the competition supported the theory stated above, even though there were no differences in the pre-trial testing. This was since the group that had supplemented with BCAAs performed better. The BCAAs play a crucial role in protein synthesis, energy production, and NT formation when they cross the BBB (Fernstrom 2005). Nitrogen from BCAAs makes up about half of the cerebral Glu, and the release of Glu from the synaptic cleft is fuelled by the breakdown of these AAs to the tune of 40% (Fernstrom 2005). Two main mechanisms underlie astrocyte BCAA catabolism, both of which contribute to the synthesis of new Glu. The astrocyte mitochondria are the site of the initial event. To create branched-chain alpha-keto acids, the mitochondrial branched-chain aminotransferase enzyme reversibly converts BCAAs (BKAAs). Glu is formed by combining the amine group with alpha-KG, a ketonic acid that is produced during the Krebs cycle. The second step is the transformation of BKAAs into the Krebs cycle substrates succinyl-Co, acetoacetate, and acetyl-COA. This cycle yields alpha-KG, carbon molecules for Glu's COOH group construction, and the ATP needed for two reactions: the Gln-Glu cycle and the astrocyte's conversion of Glu to Gln for usage by neural cells later on (Hull et al. 2012).

7.9.4 TBI Metabolic Alterations and BCAAs

The application of lateral fluid percussion injury, a methodology sanctioned by the Institutional Animal Care and Use Committee, enables scientists to investigate the cognitive deficits and molecular and biochemical alterations induced by mild-to-moderate traumatic brain injury (TBI) in animal models. Particular alterations occur in the hippocampus of the medial temporal lobe as a result of traumatic brain damage. Following traumatic brain injury (TBI), ATP generation is diminished due to a reduction in aerobic glycolysis. An ionic carrier that has been identified as malfunctioning is the Na^{+}/K^{+}ATPase pump. As a consequence, the intricate equilibrium of ions both inside and outside of cells is disrupted, hence endangering the viability of neurons. A 40% reduction in neurons may be detectable in the ipsilateral dentate gyrus, CA1, and hilar region within the initial month following a traumatic brain injury, as determined by stereological quantification techniques (TBI)(Witgen et al. 2005).

Lateral fluid percussion injury, a protocol-approved technique by the Institutional Animal Care and Use Committee, emulates the cognitive impairments associated with mild-to-moderate traumatic brain injury (TBI) in animal models. This permits scientists to investigate the molecular and biochemical alterations that transpire as a consequence of TBI. Specialised changes in the hippocampus, which is located in the medial temporal lobe, are the result of traumatic brain damage. As a result of reduced aerobic glycolysis caused by a traumatic brain injury (TBI), ATP production decreases. It is discovered that many ionic carriers, including the Na^{+}/K^{+}ATPase pump, are exhibiting dysfunction. This causes an imbalance between internal and extracellular ions, putting neurons at risk of lysis. In the ipsilateral dentate gyrus, CA1, and hilar area, stereological quantification methods may detect a 40% loss of neurons in the initial month following a traumatic brain injury (TBI) (Jeter et al. 2013). Alterations in excitability patterns of the hippocampus can be attributed to neuronal demise, ionic imbalances, and compromised NT synthesis. In contrast to CA1 and CA3, which have decreased excitability, the dentate gyrus has increased it. This results in an impairment of long-term potentiation (LTP), a synaptic plasticity phenomenon that impacts the cognitive capacities of the patient (Guerriero et al. 2015).

Modified protein synthesis is just one of many reactions that rely on high-energy chemicals. As a result, the brain is unable to properly rebuild injured structures or establish new synapses (Jeter et al. 2012). Rats that underwent local field potential analysis showed a decrease in ipsilateral hippocampus BCAAs. The three primary BCAAs are reduced in individuals with a Glasgow Coma Scale (GCS) of 15 to 3, according to chromatography and mass spectrometry methods. For low GCS scores, the decrease is most pronounced (Jeter et al. 2013).

Using the Fear Conditioning Test, researchers looked at the effects of BCAA supplementation on rats following traumatic brain injury. Evidence suggests that after 5–10 days of treatment, this supplement raises percent F to levels similar to those before the trauma. It is necessary to take BCAAs continuously for at least

10 days, as their effectiveness has been demonstrated to diminish after only 5 days of consumption. It should be mentioned that in order to achieve a more advanced enhancement of cognitive abilities, it was essential to supplement with 100 mg of BCAAs (Elkind et al. 2015).

Individuals with moderate-to-severe traumatic brain injuries (TBI) are evaluated using the Disability Rating Scale (DRS), which ranges from 0 (normal) to 29 (extreme vegetative state). It measures both the severity and course of neurocognitive impairment, a form of disability caused by traumatic brain injury. The DRS was utilised by Aquilani et al. to confirm that taking BCAA supplements improved cognitive function. After receiving 19.6 g of BCAAs intravenously over a 5-h period every day, TBI victims with a DRS of 20 showed improved scores and had multiple therapeutic effects (Aquilani et al. 2005). It was demonstrated later in 2008 that people whose initial DRS scores were between 22 and 29 experienced an even more dramatic drop in score (68.2% of the traumatised patients had been considered as minimally conscious) (Aquilani et al. 2008). The results of these most recent trials point to the possibility that BCAA supplementation can enhance TBI patients' cognitive abilities. However, a single study indicated that severe traumatic brain injury in patients with intravenous BCAA supplementation led to elevated intracranial pressure and reduced jugular vein oxygen saturation (Vuille-Dit-Bille et al. 2012).

7.9.4.1 Dosage

Leucine, isoleucine, and valine each have estimated average requirements (EARs) of 34, 15, and 19 mg/kg/day, correspondingly. In the US, men consume about three times as much BCAAs as advised and women consume about twice as much, based on the required consumption for a 70 kg individual (Trumbo et al. 2002). Proteins or amino acid combinations enriched with BCAAs or BCAAs alone have been used to treat metabolic illnesses such as chronic liver disease, encephalopathy, sepsis, and others. The goal is to increase low circulation levels of BCAAs while decreasing brain uptake of aromatic amino acids. Even though there is little evidence from peer-reviewed studies to back up the claims that BCAA supplements increase muscle mass, decrease post-workout pain, and lessen central fatigue, these products are being sold to healthy persons (Wagenmakers 1999).

Adding more BCAAs and other particular amino acids to rations for short-term, high-intensity military operations was something that an earlier Institute of Medicine (IOM) committee looked into. The committee's findings did not support the recommendation to include certain amino acids in such rations. The public's intake of BCAAs may be lower than that of athletes and military people who strive to improve their physical performance, due to claims about the fitness benefits of BCAAs (Medicine et al. 2005).

Lieberman et al. (2010) discovered that 23% of combat arms soldiers and 47% of Special Forces personnel were supplementing with protein or amino acids. There were no trials that revealed any adverse events linked with BCAAs in regular diets or with infused supplement doses up to 9.75 g, according to a 2005 IOM study. However, due to a lack of dose-response data, no Tolerable Upper Intake Level (UL) was determined (Lieberman et al. 2010). No negative effects have been documented in studies involving the acute or chronic oral administration of BCAAs, even at a high single dose of 60 g (Fernstrom 2005).

Cognitive and neurological disorders, phenylketonuria, hepatic cirrhosis, and physical performance were the endpoints that this research sought to examine. Nevertheless, it's not easy to tell if any of the results that weren't considered negative would be a problem in a military context. For instance, the committee determined that taking too many BCAAs could be harmful after reviewing a study that found no negative effects. The study found that healthy humans experienced minor increases in spatial recognition memory latency after taking oral doses of 10, 30, and 60 g of BCAAs, but no changes in visual information processing or pattern recognition. The scientists postulated that a decrease in the ratio of tyrosine to phenylalanine relative to BCAAs was responsible for the longer delay since this would lead to less tyrosine and phenylalanine transported across the blood-brain barrier and, ultimately, less dopamine production (Gijsman et al. 2002).

7.10 Conclusions and Recommendations

There is substantial evidence that consuming essential amino acids, including leucine, improves protein synthesis and increases lean body composition, therefore they are necessary. Dietary BCAAs may help treat or lessen TBI, but there isn't enough proof to recommend them. Seizures and other neurological issues are possible outcomes of traumatic brain injury (TBI). Combining a ketogenic diet with BCAAs was successful in a pilot trial; this finding has the potential to revolutionise the field. No evidence has been found that BCAAs reduce mortality or morbidity. Nutritional intake and status prior to injury and nutritional intake during therapy can be better assessed with the use of BCAAs in dietary intake studies for traumatic brain injury treatment facilities. For simple reasons, BCAAs are a favourite dietary supplement among service members. Analysing the collected data could uncover potential correlations between the victim's pre-injury nutritional consumption and status and their post-injury healing progress.

References

Altevogt BM, Colten HR (2006) Sleep disorders and sleep deprivation: an unmet public health problem. Public Health Rep

Andriessen TM, Jacobs B, Vos PE (2010) Clinical characteristics and pathophysiological mechanisms of focal and diffuse traumatic brain injury. J Cell Mol Med 14:2381–2392

Ansari MA, Roberts KN, Scheff SW (2008) Oxidative stress and modification of synaptic proteins in hippocampus after traumatic brain injury. Free Radic Biol Med 45:443–452

Anthony JC, Yoshizawa F, Anthony TG, Vary TC, Jefferson LS, Kimball SR (2000) Leucine stimulates translation initiation in skeletal muscle of postabsorptive rats via a rapamycin-sensitive pathway. J Nutr 130:2413–2419

Anthonymuthu TS, Kenny EM, Amoscato AA, Lewis J, Kochanek PM, Kagan VE et al (2017) Global assessment of oxidized free fatty acids in brain reveals an enzymatic predominance to oxidative signaling after trauma. Biochim Biophys Acta Mol basis Dis 1863:2601–2613

Aquilani R, Iadarola P, Contardi A, Boselli M, Verri M, Pastoris O et al (2005) Branched-chain amino acids enhance the cognitive recovery of patients with severe traumatic brain injury. Arch Phys Med Rehabil 86:1729–1735

Aquilani R, Boselli M, Boschi F, Viglio S, Iadarola P, Dossena M et al (2008) Branched-chain amino acids may improve recovery from a vegetative or minimally conscious state in patients with traumatic brain injury: a pilot study. Arch Phys Med Rehabil 89:1642–1647

Ashwal S, Holshouser B, Tong K, Serna T, Osterdock R, Gross M et al (2004) Proton spectroscopy detected myoinositol in children with traumatic brain injury. Pediatr Res 56:630–638

Bai D, Fan J, Li M, Dong C, Gao Y, Fu M et al (2021) Effects of folic acid combined with DHA supplementation on cognitive function and amyloid-β-related biomarkers in older adults with mild cognitive impairment by a randomized, double blind, placebo-controlled trial. J Alzheimers Dis 81:155–167

Balba NM, Elliott JE, Weymann KB, Opel RA, Duke JW, Oken BS et al (2018) Increased sleep disturbances and pain in veterans with comorbid traumatic brain injury and posttraumatic stress disorder. J Clin Sleep Med 14:1865–1878

Béchir M, Meierhans R, Brandi G, Sommerfeld J, Fasshauer M, Cottini S et al (2010) Insulin differentially influences brain glucose and lactate in traumatic brain injured patients. Minerva Anestesiol 76:896–904

Bifari F, Nisoli E (2017) Branched-chain amino acids differently modulate catabolic and anabolic states in mammals: a pharmacological point of view. Br J Pharmacol 174:1366–1377

Bivins BA, Twyman DL, Young AB (1986) Failure of nonprotein calories to mediate protein conservation in brain-injured patients. J Trauma 26:980–986

Black KL, Hanks RA, Wood DL, Zafonte RD, Cullen N, Cifu DX et al (2002) Blunt versus penetrating violent traumatic brain injury: frequency and factors associated with secondary conditions and complications. J Head Trauma Rehabil 17:489–496

Boakye PA, Olechowski C, Rashiq S, Verrier MJ, Kerr B, Witmans M et al (2016) A critical review of neurobiological factors involved in the interactions between chronic pain, depression, and sleep disruption. Clin J Pain 32:327–336

Bowman CE, Scafidi J, Scafidi S (2019) Metabolic perturbations after pediatric TBI: It's not just about glucose. Exp Neurol 316:74–84

Brown EJ, Albers MW, Bum Shin T, Ichikawa K, Keith CT, Lane WS et al (1994) A mammalian protein targeted by G1-arresting rapamycin–receptor complex. Nature 369:756–758

Buse MG, Reid SS (1975) Leucine. A possible regulator of protein turnover in muscle. J Clin Invest 56:1250–1261

Buse M, Atwell R, Mancusi V (1979) In vitro effect of branched chain amino acids on the ribosomal cycle in muscles of fasted rats. Horm Metab Res 11:289–292

Bye N, Habgood MD, Callaway JK, Malakooti N, Potter A, Kossmann T et al (2007) Transient neuroprotection by minocycline following traumatic brain injury is associated with attenuated microglial activation but no changes in cell apoptosis or neutrophil infiltration. Exp Neurol 204:220–233

Caliri S, Andaloro A, Corallo F, Donato A, Marino S, Mantarro C et al (2019) Recovery of malnutrition in a patient with severe brain injury outcomes: a case report. Medicine:98, e16755

Capizzi A, Woo J, Verduzco-Gutierrez M (2020) Traumatic brain injury: an overview of epidemiology, pathophysiology, and medical management. Med Clin 104:213–238

Carlson SW, Dixon CE (2018) Lithium improves dopamine neurotransmission and increases dopaminergic protein abundance in the striatum after traumatic brain injury. J Neurotrauma 35:2827–2836

Cernak I, Noble-Haeusslein LJ (2010) Traumatic brain injury: an overview of pathobiology with emphasis on military populations. J Cereb Blood Flow Metab 30:255–266

Chamoun R, Suki D, Gopinath SP, Goodman JC, Robertson C (2010) Role of extracellular glutamate measured by cerebral microdialysis in severe traumatic brain injury. J Neurosurg 113:564–570

Chancellor KB, Chancellor SE, Duke-Cohan JE, Huber BR, Stein TD, Alvarez VE et al (2021) Altered oligodendroglia and astroglia in chronic traumatic encephalopathy. Acta Neuropathol 142:295–221

Charrueau C, Belabed L, Besson V, Chaumeil J-C, Cynober L, Moinard C (2009) Metabolic response and nutritional support in traumatic brain injury: evidence for resistance to renutrition. J Neurotrauma 26:1911–1920

Chen X, Gao C, Cheng Z, Chen G, Rui T, Luo C et al (2021) Ruxolitinib exerts neuroprotection via repressing ferroptosis in a mouse model of traumatic brain injury. Exp Neurol 342:113762

Cheng T, Wang W, Li Q, Han X, Xing J, Qi C et al (2016) Cerebroprotection of flavanol (−)-epicatechin after traumatic brain injury via Nrf2-dependent and-independent pathways. Free Radic Biol Med 92:15–28

Cherian L, Hlatky R, Robertson CS (2004) Nitric oxide in traumatic brain injury. Brain Pathol 14:195–201

Chitturi J, Li Y, Santhakumar V, Kannurpatti SS (2019) Consolidated biochemical profile of subacute stage traumatic brain injury in early development. Front Neurosci 13:431

Clifton GL, Robertson CS, Contant CF (1985) Enteral hyperalimentation in head injury. J Neurosurg 62:186–193

Clifton GL, Robertson CS, Choi SC (1986) Assessment of nutritional requirements of head-injured patients. J Neurosurg 64:895–901

Cole JT, Mitala CM, Kundu S, Verma A, Elkind JA, Nissim I et al (2010) Dietary branched chain amino acids ameliorate injury-induced cognitive impairment. Proc Natl Acad Sci USA 107:366–371

Cook AM, Peppard A, Magnuson B (2008) Nutrition considerations in traumatic brain injury. Nutr Clin Pract 23:608–620

Corrigan JD, Cuthbert JP, Harrison-Felix C, Whiteneck GG, Bell JM, Miller AC et al (2014) US population estimates of health and social outcomes 5 years after rehabilitation for traumatic brain injury. J Head Trauma Rehabil 29:1–9

D'Amelio M, Puglisi-Allegra S, Mercuri N (2018) The role of dopaminergic midbrain in Alzheimer's disease: translating basic science into clinical practice. Pharmacol Res 130:414–419

Dang B, Chen W, He W, Chen G (2017) Rehabilitation treatment and progress of traumatic brain injury dysfunction. Neural Plast 2017:1582182

de Aguilar-Nascimento JE, Kudsk KA (2007) Clinical costs of feeding tube placement. JPEN J Parenter Enteral Nutr 31:269–273

De Bandt J-P, Cynober L (2006) Therapeutic use of branched-chain amino acids in burn, trauma, and sepsis. J Nutr 136:308–313

Delage C, Taib T, Mamma C, Lerouet D, Besson VC (2021) Traumatic brain injury: an age-dependent view of post-traumatic neuroinflammation and its treatment. Pharmaceutics 13:1624

Deng Y, Thompson BM, Gao X, Hall ED (2007) Temporal relationship of peroxynitrite-induced oxidative damage, calpain-mediated cytoskeletal degradation and neurodegeneration after traumatic brain injury. Exp Neurol 205:154–165

Dewan MC, Rattani A, Gupta S, Baticulon RE, Hung Y-C, Punchak M et al (2018) Estimating the global incidence of traumatic brain injury. J Neurosurg 130:1080–1097

Diaz-Parejo P, Ståhl N, Xu W, Reinstrup P, Ungerstedt U, Nordström C-H (2003) Cerebral energy metabolism during transient hyperglycemia in patients with severe brain trauma. J Intensive Care Med 29:544–550

Dijkink S, Meier K, Krijnen P, Yeh DD, Velmahos GC, Schipper IB (2020) Malnutrition and its effects in severely injured trauma patients. Eur J Trauma Emerg Surg 46:993–1004
Dougall D, Poole N, Agrawal N (2015) Pharmacotherapy for chronic cognitive impairment in traumatic brain injury. Cochrane Database Syst Rev 12:CD009221
Eisele A, Hill-Strathy M, Michels L, Rauen K (2020) Magnetic resonance spectroscopy following mild traumatic brain injury: a systematic review and meta-analysis on the potential to detect posttraumatic neurodegeneration. J Neurodegener Dis 20:2–11
Elkind JA, Lim MM, Johnson BN, Palmer CP, Putnam BJ, Kirschen MP et al (2015) Efficacy, dosage, and duration of action of branched chain amino acid therapy for traumatic brain injury. Front Neurol 6:73
Elliott JE, De Luche SE, Churchill MJ, Moore C, Cohen AS, Meshul CK et al (2018) Dietary therapy restores glutamatergic input to orexin/hypocretin neurons after traumatic brain injury in mice. Sleep 41:212
Elliott JE, Opel RA, Pleshakov D, Rachakonda T, Chau AQ, Weymann KB et al (2020) Posttraumatic stress disorder increases the odds of REM sleep behavior disorder and other parasomnias in veterans with and without comorbid traumatic brain injury. Sleep 43:237
Elliott JE, Keil AT, Mithani S, Gill JM, O'Neil ME, Cohen AS et al (2022) Dietary supplementation with branched chain amino acids to improve sleep in veterans with traumatic brain injury: a randomized double-blind placebo-controlled pilot and feasibility trial. Front Syst Neurosci 16:854874
Fedewa MV, Spencer SO, Williams TD, Becker ZE, Fuqua CA (2019) Effect of branched-chain amino acid supplementation on muscle soreness following exercise: a meta-analysis. Int J Vitam Nutr Res 89:a000543
Fernstrom JD (2005) Branched-chain amino acids and brain function. J Nutr 135:1539–1546
Fischer JE, Funovics JM, Aguirre A, James JH, Keane JM, Wesdorp RI et al (1975) The role of plasma amino acids in hepatic encephalopathy. Surgery 78:276–290
Floyd J, Fajans SS, Conn JW, Knopf RF, Rull J (1966) Stimulation of insulin secretion by amino acids. J Clin Invest 45:1487–1502
Folkerts MM, Parks EA, Dedman JR, Kaetzel MA, Lyeth BG, Berman RF (2007) Phosphorylation of calcium calmodulin—dependent protein kinase II following lateral fluid percussion brain injury in rats. J Neurotrauma 24:638–650
Forbes SC, Cordingley DM, Cornish SM, Gualano B, Roschel H, Ostojic SM et al (2022) Effects of creatine supplementation on brain function and health. Nutrients 14:921
Ford LM, Ouma JR (2021) A descriptive study of malnutrition in traumatic brain injury patients. Acta Neuropathol 10:107
Fortier M, Castellano C-A, Croteau E, Langlois F, Bocti C, St-Pierre V et al (2019) A ketogenic drink improves brain energy and some measures of cognition in mild cognitive impairment. Alzheimers Dement 15:625–634
Friedman SD, Brooks WM, Jung RE, Hart BL, Yeo RA (1998) Proton MR spectroscopic findings correspond to neuropsychological function in traumatic brain injury. AJNR Am J Neuroradiol 19:1879–1885
Frugier T, Morganti-Kossmann MC, O'Reilly D, McLean CA (2010) In situ detection of inflammatory mediators in post mortem human brain tissue after traumatic injury. J Neurotrauma 27:497–507
Fruin AH, Taylon C, Pettis MS (1986) Caloric requirements in patients with severe head injuries. Surg Neurol 25:25–28
Fuhrman MP, Charney P, Mueller CM (2004) Hepatic proteins and nutrition assessment. J Am Diet Assoc 104:1258–1264
Fünfschilling U, Supplie LM, Mahad D, Boretius S, Saab AS, Edgar J et al (2012) Glycolytic oligodendrocytes maintain myelin and long-term axonal integrity. Nature 485:517–521
Fürst P, Albers S, Stehle P (1987) Stress-induced intracellular glutamine depletion. The potential use of glutamine-containing peptides in parenteral nutrition. Beitr Infusther Wien Klin Wochenschr 17:117–136

Gabay C, Kushner I (1999) Acute-phase proteins and other systemic responses to inflammation. N Engl J Med 340:448–454

Gardiner K, Barbul A (1993) Intestinal amino acid absorption during sepsis. JPEN J Parenter Enteral Nutr 17:277–283

Geeraerts T, Velly L, Abdennour L, Asehnoune K, Audibert G, Bouzat P et al (2018) Management of severe traumatic brain injury (first 24 hours). Anaesth Crit Care Pain Med 37:171–186

Gijsman HJ, Scarnà A, Harmer CJ, McTavish SF, Odontiadis J, Cowen PJ et al (2002) A dose-finding study on the effects of branch chain amino acids on surrogate markers of brain dopamine function. J Psychopharmacol 160:192–197

Gualano A, Bozza T, Lopes De Campos P, Roschel H, Dos Santos CA, Luiz Marquezi M et al (2011) Branched-chain amino acids supplementation enhances exercise capacity and lipid oxidation during endurance exercise after muscle glycogen depletion. J Sports Med Phys Fitness 51:82–88

Guerriero RM, Giza CC, Rotenberg A (2015) Glutamate and GABA imbalance following traumatic brain injury. Curr Neurol Neurosci Rep 15:1–11

Harper A, Miller R, Block K (1984) Branched-chain amino acid metabolism. Annu Rev Nutr 4:409–454

Hart DW, Wolf SE, Herndon DN, Chinkes DL, Lal SO, Obeng MK et al (2002) Energy expenditure and caloric balance after burn: increased feeding leads to fat rather than lean mass accretion. Ann Surg 235:152–161

Hasegawa Y, Suzuki H, Sozen T, Altay O, Zhang JH (2011) Apoptotic mechanisms for neuronal cells in early brain injury after subarachnoid hemorrhage. In: Early brain injury or cerebral vasospasm: Volume 1: Pathophysiology, pp 43–48

Hasselgren P-O, Pedersen P, Sax HC, Warner BW, Fischer JE (1988) Current concepts of protein turnover and amino acid transport in liver and skeletal muscle during sepsis. Arch Surg 123:992–999

Hatton J (2001) Pharmacological treatment of traumatic brain injury: a review of agents in development. CNS Drugs 15:553–581

Hatton J, Ziegler T (1998) Nutritional support of the neurosurgical patient. J Neurol Neurosurg Psychiatry 60:381–396

Hatton J, Rapp RP, Kudsk KA, Brown RO, Luer MS, Bukar JG et al (1997) Intravenous insulin-like growth factor-I (IGF-I) in moderate-to-severe head injury: a phase II safety and efficacy trial. J Neurosurg 86:779–786

Hawkins RA, O'Kane RL, Simpson IA, Vina JR (2006) Structure of the blood–brain barrier and its role in the transport of amino acids. J Nutr 136:218–226

Hay N, Sonenberg N (2004) Upstream and downstream of mTOR. Genes Dev 18:1926–1945

Hiskens MI (2022) Targets of neuroprotection and review of pharmacological interventions in traumatic brain injury. J Pharmacol Exp Ther 382:149–166

Hoffmeister P-G, Donat CK, Schuhmann MU, Voigt C, Walter B, Nieber K et al (2011) Traumatic brain injury elicits similar alterations in α7 nicotinic receptor density in two different experimental models. NeuroMolecular Med 13:44–53

Hoge CW, McGurk D, Thomas JL, Cox AL, Engel CC, Castro CA (2008) Mild traumatic brain injury in US soldiers returning from Iraq. N Engl J Med 358:453–463

Holeček M (1996) Leucine metabolism in fasted and tumor necrosis factor-treated rats. Clin Nutr 15:91–93

Holeček M (2001) The BCAA–BCKA cycle: its relation to alanine and glutamine synthesis and protein balance. Nutr J 17:70

Holecek M (2005) Glutamine and branched-chain amino acids--practical importance of their metabolic relations. Cas Lek Cesk 144:9–12

Holeček M (2017) Beta-hydroxy-beta-methylbutyrate supplementation and skeletal muscle in healthy and muscle-wasting conditions. J Cachexia Sarcopenia Muscle 8:529–541

Holeček M (2018) Branched-chain amino acids in health and disease: metabolism, alterations in blood plasma, and as supplements. J Nutr Metab 15:1–12

Holeček M, Mičuda S (2017) Amino acid concentrations and protein metabolism of two types of rat skeletal muscle in postprandial state and after brief starvation. Physiol Res 66:959–967

Holecek M, Sispera L (2014) Glutamine deficiency in extracellular fluid exerts adverse effects on protein and amino acid metabolism in skeletal muscle of healthy, laparotomized, and septic rats. J Amino Acids 46:1377–1384

Holecek M, Sprongl L, Skopec F, Andrys C, Pecka M (1997) Leucine metabolism in TNF-α- and endotoxin-treated rats: contribution of hepatic tissue. Am J Physiol Endocrinol Metab 273:1052–1058

Holeček M, Šprongl L, Tichý M, Pecka M (1998) Leucine metabolism in rat liver after a bolus injection of endotoxin. Metabolism 47:681–685

Holecek M, Rysava R, Safranek R, Kadlcikova J, Sprongl L (2003) Acute effects of decreased glutamine supply on protein and amino acid metabolism in hepatic tissue: a study using isolated perfused rat liver. Metabolism 52:1062–1067

Holecek M, Siman P, Vodenicarovova M, Kandar R (2016) Alterations in protein and amino acid metabolism in rats fed a branched-chain amino acid-or leucine-enriched diet during postprandial and postabsorptive states. Nutr Metab 13:1–12

Hotamisligil GS (2006) Inflammation and metabolic disorders. Nature 444:860–867

Huang Q, Su H, Qi B, Wang Y, Yan K, Wang X et al (2021a) A SIRT1 activator, ginsenoside Rc, promotes energy metabolism in cardiomyocytes and neurons. J Am Chem Soc 143:1416–1427

Huang L, He S, Cai Q, Li F, Wang S, Tao K et al (2021b) Polydatin alleviates traumatic brain injury: role of inhibiting ferroptosis. Biochem Biophys Res Commun 556:149–155

Hull J, Hindy ME, Kehoe PG, Chalmers K, Love S, Conway ME (2012) Distribution of the branched chain aminotransferase proteins in the human brain and their role in glutamate regulation. J Neurochem 123:997–1009

Huynh LM, Burns MP, Taub DD, Blackman MR, Zhou J (2020) Chronic neurobehavioral impairments and decreased hippocampal expression of genes important for brain glucose utilization in a mouse model of mild TBI. Front Endocrinol 11:556380

Ijichi C, Matsumura T, Tsuji T, Eto Y (2003) Branched-chain amino acids promote albumin synthesis in rat primary hepatocytes through the mTOR signal transduction system. Biochem Biophys Res Commun 303:59–64

Jacobson HR (1987) Diuretics: mechanisms of action and uses. Hosp Pract 22:129–156

Javaid S, Farooq T, Rehman Z, Afzal A, Ashraf W, Rasool MF et al (2021) Dynamics of choline-containing phospholipids in traumatic brain injury and associated comorbidities. Int J Mol Sci 22:11313

Jayakumar A, Bak L, Rama Rao K, Waagepetersen H, Schousboe A, Norenberg M (2016) Neuronal cell death induced by mechanical percussion trauma in cultured neurons is not preceded by alterations in glucose, lactate and glutamine metabolism. Neurochem Res 41:307–315

Jeter CB, Hergenroeder GW, Ward NH III, Moore AN, Dash PK (2012) Human traumatic brain injury alters circulating L-arginine and its metabolite levels: possible link to cerebral blood flow, extracellular matrix remodeling, and energy status. J Neurotrauma 29:119–127

Jeter CB, Hergenroeder GW, Ward NH III, Moore AN, Dash PK (2013) Human mild traumatic brain injury decreases circulating branched-chain amino acids and their metabolite levels. J Neurotrauma 30:671–679

Jha S, Ghewade P (2022) Management and treatment of traumatic brain injuries. Cureus 14:10

Jimenez Jimenez F, Ortiz Leyba C, Morales Mendez S, Barros Perez M, Munoz GJ (1991) Prospective study on the efficacy of branched-chain amino acids in septic patients. JPEN J Parenter Enteral Nutr 15:252–261

Johnson VE, Stewart JE, Begbie FD, Trojanowski JQ, Smith DH, Stewart W (2013) Inflammation and white matter degeneration persist for years after a single traumatic brain injury. Brain 136:28–42

Kainulainen H, Hulmi JJ, Kujala UM (2013) Potential role of branched-chain amino acid catabolism in regulating fat oxidation. Exerc Sport Sci Rev 41:194–200

Kane AW, Diaz DS, Moore C (eds) (2019) Physical therapy management of adults with mild traumatic brain injury. Seminars in speech and language. Thieme Medical Publishers

Kapralov AA, Yang Q, Dar HH, Tyurina YY, Anthonymuthu TS, Kim R et al (2020) Redox lipid reprogramming commands susceptibility of macrophages and microglia to ferroptotic death. Nat Chem Biol 16:278–290

Kimball SR, Jefferson LS (2001) Regulation of protein synthesis by branched-chain amino acids. Curr Opin Clin Nutr Metab Care 4:39–43

Komura A, Kawasaki T, Yamada Y, Uzuyama S, Asano Y, Shinoda J (2019) Cerebral glucose metabolism in patients with chronic mental and cognitive sequelae after a single blunt mild traumatic brain injury without visible brain lesions. J Neurotrauma 36:641–649

Kumar A (2019) Sugar as a therapeutic target for the cognitive restoration following traumatic brain injury. Curr Opin Neurol 32:815–821

Kurtz P, Rocha EE (2020) Nutrition therapy, glucose control, and brain metabolism in traumatic brain injury: a multimodal monitoring approach. Front Neurosci 14:190

Lai J-Q, Shi Y-C, Lin S, Chen X-R (2022) Metabolic disorders on cognitive dysfunction after traumatic brain injury. Trends Endocrinol Metab 33:451

Lamade AM, Anthonymuthu TS, Hier ZE, Gao Y, Kagan VE, Bayır H (2020) Mitochondrial damage & lipid signaling in traumatic brain injury. Exp Neurol 329:113307

Lang CH, Frost RA (2005) Endotoxin disrupts the leucine-signaling pathway involving phosphorylation of mTOR, 4E-BP1, and S6K1 in skeletal muscle. J Cell Physiol 203:144–155

LaPlaca MC, Thibault LE (1998) Dynamic mechanical deformation of neurons triggers an acute calcium response and cell injury involving the N-methyl-D-aspartate glutamate receptor. J Neurosci Res 52:220–229

Lee-anne SC, Lithander FE, Gruen RL, Williams LT (2014) Nutrition therapy in the optimisation of health outcomes in adult patients with moderate to severe traumatic brain injury: findings from a scoping review. Injury 45:1834–1841

Lew HL, Otis JD, Tun C, Kerns RD, Clark ME, Cifu DX (2009) Prevalence of chronic pain, posttraumatic stress disorder, and persistent postconcussive symptoms in OIF/OEF veterans: polytrauma clinical triad. J Rehabil Res Dev 46:697–702

Li JB, Jefferson LS (1978) Influence of amino acid availability on protein turnover in perfused skeletal muscle. Biochim Biophys Acta Mol basis Dis 544:351–359

Li P-A, Liu G-J, He QP, Floyd RA, Siesjö BK (1999) Production of hydroxyl free radical by brain tissues in hyperglycemic rats subjected to transient forebrain ischemia. Free Radic Biol Med 27:1033–1040

Li C, Xu J-J, Hu H-T, Shi C-Y, Yu C-J, Sheng J-Z et al (2021) Amylin receptor insensitivity impairs hypothalamic POMC neuron differentiation in the male offspring of maternal high-fat diet-fed mice. Mol Metab 44:101135

Lieberman HR, Stavinoha TB, McGraw SM, White A, Hadden LS, Marriott BP (2010) Use of dietary supplements among active-duty US Army soldiers. Am J Clin 92:985–995

Lim SH, Lee JS, Chae SH, Ahn BS, Chang DJ, Shin CS (2005) Prealbumin is not sensitive indicator of nutrition and prognosis in critical ill patients. Yonsei Med J 46:21–26

Lim MM, Elkind J, Xiong G, Galante R, Zhu J, Zhang L et al (2013) Dietary therapy mitigates persistent wake deficits caused by mild traumatic brain injury. Sci Transl Med 5:215

Lin A, Liao H, Merugumala S, Prabhu S, Meehan W, Ross B (2012) Metabolic imaging of mild traumatic brain injury. Brain Imaging Behav 6:208–223

Liu Z, Yan M, Liang Y, Liu M, Zhang K, Shao D et al (2019) Nucleoporin Seh1 interacts with Olig2/Brd7 to promote oligodendrocyte differentiation and myelination. Neuron 102:587–601

Loehfelm A, Elder MK, Boucsein A, Jones PP, Williams JM, Tups A (2020) Docosahexaenoic acid prevents palmitate-induced insulin-dependent impairments of neuronal health. FASEB J 34:4635–4652

Lucke-Wold BP, Logsdon AF, Turner RC, Huber JD, Rosen CL (2017) Endoplasmic reticulum stress modulation as a target for ameliorating effects of blast induced traumatic brain injury. J Neurotrauma 34:62–70

Luo M-l, Pan L, Wang L, Wang H-y, Li S, Long Z-y et al (2019) Transplantation of NSCs promotes the recovery of cognitive functions by regulating neurotransmitters in rats with traumatic brain injury. Neurochem Res 44:2765–2775
Lüscher C, Malenka RC (2012) NMDA receptor-dependent long-term potentiation and long-term depression (LTP/LTD). Cold Spring Harb Perspect Biol 4:5710
Lynch MA (2004) Long-term potentiation and memory. Physiol Rev 84:87–136
Manchester K (1965) Oxidation of amino acids by isolated rat diaphragm and the influence of insulin. Biochim Biophys Acta Mol basis Dis 100:295–298
Manelli J, Garabedian M, Ounis N, Houvenaeghel M, Ottomani A, Bimar J (eds) (1984) Effects on muscular and general proteolysis in burn patients of a solution enriched with branched amino acids. Anesthesie Reanim
Manner T, Katz D, Askanazi J (1996) The antinociceptive effects of branched-chain amino acids: evidence for their ability to potentiate morphine analgesia. Pharmacol Biochem Behav 53:449–454
Mathias J, Alvaro P (2012) Prevalence of sleep disturbances, disorders, and problems following traumatic brain injury: a meta-analysis. Sleep Med 13:898–905
Matthews DE (2005) Observations of branched-chain amino acid administration in humans. J Nutr 135:1580–1584
Mattick JS, Kamisoglu K, Ierapetritou MG, Androulakis IP, Berthiaume F (2013) Branched-chain amino acid supplementation: impact on signaling and relevance to critical illness. Wiley Interdiscip Rev Syst Biol Med 5:449–460
McGuire JL, Ngwenya LB, McCullumsmith RE (2019) Neurotransmitter changes after traumatic brain injury: an update for new treatment strategies. Mol Psychiatry 24:995–1012
Mckee AC, Daneshvar DH (2015) The neuropathology of traumatic brain injury. Handb Clin Neurol 127:45–66
McKee JA, Brewer RP, Macy GE, Phillips-Bute B, Campbell KA, Borel CO et al (2005) Analysis of the brain bioavailability of peripherally administered magnesium sulfate: a study in humans with acute brain injury undergoing prolonged induced hypermagnesemia. J Crit Care Med 33:661–666
Medicine IO, Food BN, Research CMN (2005) Nutrient composition of rations for short-term, high-intensity combat operations. National Academies Press
Meldrum BS (2000) Glutamate as a neurotransmitter in the brain: review of physiology and pathology. J Nutr 130:1007–1015
Mele C, Pingue V, Caputo M, Zavattaro M, Pagano L, Prodam F et al (2021) Neuroinflammation and hypothalamo-pituitary dysfunction: focus of traumatic brain injury. Int J Mol Sci 22:2686
Mitch WE, Walser M, Sapir DG (1981) Nitrogen sparing induced by leucine compared with that induced by its keto analogue, alpha-ketoisocaproate, in fasting obese man. J Clin Invest 67:553–562
Mollayeva T, Mollayeva S, Colantonio A (2018) Traumatic brain injury: sex, gender and intersecting vulnerabilities. Nat Rev Neurol 14:711–722
Naga KK, Sullivan PG, Geddes JW (2007) High cyclophilin D content of synaptic mitochondria results in increased vulnerability to permeability transition. J Neurosci 27:7469–7475
Nair KS, Short KR (2005) Hormonal and signaling role of branched-chain amino acids. J Nutr 135:1547–1552
Nawabi MD, Block KP, Chakrabarti MC, Buse MG (1990) Administration of endotoxin, tumor necrosis factor, or interleukin 1 to rats activates skeletal muscle branched-chain alpha-keto acid dehydrogenase. J Clin Invest 85:256–263
Neinast M, Murashige D, Arany Z (2019) Branched chain amino acids. Annu Rev Physiol 81:139–164
Newell E, Shellington DK, Simon DW, Bell MJ, Kochanek PM, Feldman K et al (2015) Cerebrospinal fluid markers of macrophage and lymphocyte activation after traumatic brain injury in children. Pediatr Crit Care Med 16:549–557

Ng SY, Lee AYW (2019) Traumatic brain injuries: pathophysiology and potential therapeutic targets. Front Cell Neurosci 13:528
Nguyen S, McKay A, Wong D, Rajaratnam SM, Spitz G, Williams G et al (2017) Cognitive behavior therapy to treat sleep disturbance and fatigue after traumatic brain injury: a pilot randomized controlled trial. Arch Phys Med Rehabil 98:1508–1517
Nishimura J, Masaki T, Arakawa M, Seike M, Yoshimatsu H (2010) Isoleucine prevents the accumulation of tissue triglycerides and upregulates the expression of PPARα and uncoupling protein in diet-induced obese mice. J Nutr 140:496–500
Nishitani S, Matsumura T, Fujitani S, Sonaka I, Miura Y, Yagasaki K (2002) Leucine promotes glucose uptake in skeletal muscles of rats. Biochem Biophys Res Commun 299:693–696
Nishitani S, Ijichi C, Takehana K, Fujitani S, Sonaka I (2004) Pharmacological activities of branched-chain amino acids: specificity of tissue and signal transduction. Biochem Biophys Res Commun 313:387–389
Nishitani S, Takehana K, Fujitani S, Sonaka I (2005) Branched-chain amino acids improve glucose metabolism in rats with liver cirrhosis. Am J Physiol Endocrinol Metab 288:1292–1300
Nishizawa Y (2001) Glutamate release and neuronal damage in ischemia. Life Sci 69:369–381
Nissen JD, Pajęcka K, Stridh MH, Skytt DM, Waagepetersen HS (2015) Dysfunctional TCA-cycle metabolism in glutamate dehydrogenase deficient astrocytes. Glia 63:2313–2326
Novgorodov SA, Voltin JR, Wang W, Tomlinson S, Riley CL, Gudz TI (2019) Acid sphingomyelinase deficiency protects mitochondria and improves function recovery after brain injury. J Lipid Res 60:609–623
O'Connell TM (2013) The complex role of branched chain amino acids in diabetes and cancer. Metabolites 3:931–945
Oki JC, Cuddy PG (1989) Branched-chain amino acid support of stressed patients. DICp 23:399–410
Palafox-Sánchez V, Ying Z, Royes LFF, Gomez-Pinilla F (2021) The interaction between brain and liver regulates lipid metabolism in the TBI pathology. Biochim Biophys Acta Mol basis Dis 1867:166078
Pardo E, Constantin J-M, Bonnet F, Verdonk F (2020) Nutritional support for critically ill patients with COVID-19: new strategy for a new disease? Anaesth Crit Care Pain Med 39:738–739
Park SH, Kim JH, Choi KH, Jang YJ, Bae SS, Choi BT et al (2013) Hypercholesterolemia accelerates amyloid β-induced cognitive deficits. Int J Mol Sci 31:577–582
Patil H, Gupta R (2019) A comparative study of bolus dose of hypertonic saline, mannitol, and mannitol plus glycerol combination in patients with severe traumatic brain injury. World Neurosurg 125:221–228
Pedroso JA, Zampieri TT, Donato J Jr (2015) Reviewing the effects of L-leucine supplementation in the regulation of food intake, energy balance, and glucose homeostasis. Nutrients 7:3914–3937
Peltz CB, Kenney K, Gill J, Diaz-Arrastia R, Gardner RC, Yaffe K (2020) Blood biomarkers of traumatic brain injury and cognitive impairment in older veterans. J Neurol 95:1126–1133
Perry SA, Coetzer R, Saville CW (2020) The effectiveness of physical exercise as an intervention to reduce depressive symptoms following traumatic brain injury: a meta-analysis and systematic review. Neuropsychol Rehabil 30:564–578
Peyrollier K, Hajduch E, Blair AS, Hyde R, Hundal HS (2000) L-leucine availability regulates phosphatidylinositol 3-kinase, p70 S6 kinase and glycogen synthase kinase-3 activity in L6 muscle cells: evidence for the involvement of the mammalian target of rapamycin (mTOR) pathway in the L-leucine-induced up-regulation of system a amino acid transport. Biochem J 350:361–368
Pillsbury L, Oria M, Erdman J (2011) Nutrition and traumatic brain injury: improving acute and subacute health outcomes in military personnel. National Academic Press
Portbury SD, Hare DJ, Sgambelloni C, Finkelstein DI, Adlard PA (2016) A time-course analysis of changes in cerebral metal levels following a controlled cortical impact. Metallomics 8:193–200

Portier H, Chatard J, Filaire E, Jaunet-Devienne M, Robert A, Guezennec C (2008) Effects of branched-chain amino acids supplementation on physiological and psychological performance during an offshore sailing race. Eur J Appl Physiol 104:787–794
Qin LQ, Xun P, Bujnowski D, Daviglus ML, Van Horn L, Stamler J et al (2011) Higher branched-chain amino acid intake is associated with a lower prevalence of being overweight or obese in middle-aged East Asian and Western adults. J Nutr 141:249–254
Rao V, McCann U, Han D, Bergey A, Smith MT (2014) Does acute TBI-related sleep disturbance predict subsequent neuropsychiatric disturbances? Brain Inj 28:20–26
Ray S, Dixon C, Banik N (2002) Molecular mechanisms in the pathogenesis of traumatic brain injury. Histol Histopathol 17:4
Risdall JE, Menon DK (2011) Traumatic brain injury. Philos Trans R Soc Lond Ser B Biol Sci 366:241–250
Robertson CS, Clifton GL, Grossman RG, Ching-Nan OU, Goodman JC, Borum P et al (1988) Alterations in cerebral availability of metabolic substrates after severe head injury. J Trauma Acute Care Surg 28:1523–1532
Ruokonen E, Takala J (2002) Dangers of growth hormone therapy in critically ill patients. Curr Opin Clin Nutr Metab Care 5:199–209
Rutherford W, Merrett J, Mcdonali J (1977) Sequelae of concussion caused by minor head injuries. Lancet 309:1–4
Saatman KE, Duhaime A-C, Bullock R, Maas AI, Valadka A, Manley GT (2008) Classification of traumatic brain injury for targeted therapies. J Neurotrauma 25:719–738
Sandsmark DK, Elliott JE, Lim MM (2017) Sleep-wake disturbances after traumatic brain injury: synthesis of human and animal studies. Sleep 40:44
Sapir D, Walser M, Moyer E, Rosenshein N, Stewart P, Moreadith C et al (1983) Effects of α-ketoisocaproate and of leucine on nitrogen metabolism in postoperative patients. Lancet 321:1010–1014
Sarkar C, Jones JW, Hegdekar N, Thayer JA, Kumar A, Faden AI et al (2020) PLA2G4A/cPLA2-mediated lysosomal membrane damage leads to inhibition of autophagy and neurodegeneration after brain trauma. Autophagy 16:466–485
Scafidi S, Racz J, Hazelton J, McKenna MC, Fiskum G (2011) Neuroprotection by acetyl-L-carnitine after traumatic injury to the immature rat brain. Dev Neurosci 32:480–487
Schmidt OI, Infanger M, Heyde CE, Ertel W, Stahel PF (2004) The role of neuroinflammation in traumatic brain injury. Eur J Trauma 30:135–149
Schmitz I, Kirchhoff S, Krammer PH (2000) Regulation of death receptor-mediated apoptosis pathways. Int J Biochem Cell Biol 32:1123–1136
Scholten JD, Sayer NA, Vanderploeg RD, Bidelspach DE, Cifu DX (2012) Analysis of US Veterans Health Administration comprehensive evaluations for traumatic brain injury in Operation Enduring Freedom and Operation Iraqi Freedom veterans. Brain Inj 26:1177–1184
Scremin OU, Jenden D (1991) Time-dependent changes in cerebral choline and acetylcholine induced by transient global ischemia in rats. Stroke J 22:643–647
Scrimgeour AG, Condlin ML (2014) Nutritional treatment for traumatic brain injury. J Neurotrauma 31:989–999
Semple BD, Bye N, Rancan M, Ziebell JM, Morganti-Kossmann MC (2010) Role of CCL2 (MCP-1) in traumatic brain injury (TBI): evidence from severe TBI patients and CCL2−/− mice. J Cereb Blood Flow Metab 30:769–782
Shapira M, Licht A, Milman A, Pick CG, Shohami E, Eldar-Finkelman H (2007) Role of glycogen synthase kinase-3β in early depressive behavior induced by mild traumatic brain injury. Mol Cell Neurosci 34:571–577
Shaughness M, Acs D, Brabazon F, Hockenbury N, Byrnes KR (2020) Role of insulin in neurotrauma and neurodegeneration: a review. Front Neurosci 14:547175
Shen J (2013) Modeling the glutamate–glutamine neurotransmitter cycle. Front Neuroenerg 5:1
Shi J, Dong B, Mao Y, Guan W, Cao J, Zhu R et al (2016) Traumatic brain injury and hyperglycemia, a potentially modifiable risk factor. Oncotarget 7:71052

Shim H-Y, Park J-H, Paik H-D, Nah S-Y, Kim DS, Han YS (2007) Genistein-induced apoptosis of human breast cancer MCF-7 cells involves calpain–caspase and apoptosis signaling kinase 1–p38 mitogen-activated protein kinase activation cascades. Anti-Cancer Drugs 18:649–657

Shohami E, Kohen R (2011) The role of reactive oxygen species in the pathogenesis of traumatic brain injury. In: Oxidative stress and free radical damage in neurology, pp 99–118

Shoji H, Takao K, Hattori S, Miyakawa T (2014) Contextual and cued fear conditioning test using a video analyzing system in mice. J Vis Exp 85:50871

Singer P, Anbar R, Cohen J, Shapiro H, Shalita-Chesner M, Lev S et al (2011) The tight calorie control study (TICACOS): a prospective, randomized, controlled pilot study of nutritional support in critically ill patients. J Intensive Care Med 37:601–609

Singer P, Blaser AR, Berger MM, Alhazzani W, Calder PC, Casaer MP et al (2019) ESPEN guideline on clinical nutrition in the intensive care unit. Clin Nutr 38:48–79

Singh IN, Sullivan PG, Deng Y, Mbye LH, Hall ED (2006) Time course of post-traumatic mitochondrial oxidative damage and dysfunction in a mouse model of focal traumatic brain injury: implications for neuroprotective therapy. J Cereb Blood Flow Metab 26:1407–1418

Singh P, Doshi S, Spaethling JM, Hockenberry AJ, Patel TP, Geddes-Klein DM et al (2012) N-methyl-D-aspartate receptor mechanosensitivity is governed by C terminus of NR2B subunit. J Biol Chem 287:4348–4359

Sivandzade F, Alqahtani F, Cucullo L (2020) Traumatic brain injury and blood–brain barrier (BBB): underlying pathophysiological mechanisms and the influence of cigarette smoking as a premorbid condition. Int J Mol Sci 21:2721

Skandsen T, Kvistad KA, Solheim O, Strand IH, Folvik M, Vik A (2010) Prevalence and impact of diffuse axonal injury in patients with moderate and severe head injury: a cohort study of early magnetic resonance imaging findings and 1-year outcome. J Neurosurg 113:556–563

Skeie B, Kvetan V, Gil KM, Rothkopf MM, Newsholme EA, Askanazi J (1990) Branch-chain amino acids: their metabolism and clinical utility. J Crit Care Med 18:549–571

Smith QR, Momma S, Aoyagi M, Rapoport SI (1987) Kinetics of neutral amino acid transport across the blood-brain barrier. J Neurochem 49:1651–1658

Smith DH, Chen X-H, Pierce JE, Wolf JA, Trojanowski JQ, Graham DI et al (1997) Progressive atrophy and neuron death for one year following brain trauma in the rat. J Neurotrauma 14:715–727

Smith DH, Meaney DF, Shull WH (2003) Diffuse axonal injury in head trauma. J Head Trauma Rehabil 18:307–316

Su E, Bell M (2016) Diffuse axonal injury. Transl Res 57:41

Sved AF, Fernstrom JD, Wurtman RJ (1979) Tyrosine administration reduces blood pressure and enhances brain norepinephrine release in spontaneously hypertensive rats. Proc Natl Acad Sci USA 76:3511–3514

Swain LM, Shiota T, Walser M (1990) Utilization for protein synthesis of leucine and valine compared with their keto analogues. Am J Clin 51:411–415

Takala J, Ruokonen E, Webster NR, Nielsen MS, Zandstra DF, Vundelinckx G et al (1999) Increased mortality associated with growth hormone treatment in critically ill adults. N Engl J Med 341:785–792

Tan Y, Dourdin N, Wu C, De Veyra T, Elce JS, Greer PA (2006) Ubiquitous calpains promote caspase-12 and JNK activation during endoplasmic reticulum stress-induced apoptosis. J Biol Chem 281:16016–16024

Tang S, Gao P, Chen H, Zhou X, Ou Y, He Y (2020) The role of iron, its metabolism and ferroptosis in traumatic brain injury. Front Cell Neurosci 14:590789

Tischler ME, Desautels M, Goldberg AL (1982) Does leucine, leucyl-tRNA, or some metabolite of leucine regulate protein synthesis and degradation in skeletal and cardiac muscle? J Biol Chem 257:1613–1621

Titus DJ, Johnstone T, Johnson NH, London SH, Chapalamadugu M, Hogenkamp D et al (2019) Positive allosteric modulation of the α7 nicotinic acetylcholine receptor as a treatment for cognitive deficits after traumatic brain injury. PLoS One 14:223180
Tong X-K, Trigiani LJ, Hamel E (2019) High cholesterol triggers white matter alterations and cognitive deficits in a mouse model of cerebrovascular disease: benefits of simvastatin. Cell Death Dis 10:89
Trumbo P, Schlicker S, Yates AA, Poos M (2002) Dietary reference intakes for energy, carbohydrate, fiber, fat, fatty acids, cholesterol, protein and amino acids (Commentary). J Am Diet Assoc 102:1621–1631
Tsujimoto Y, Shimizu S (2007) Role of the mitochondrial membrane permeability transition in cell death. Apoptosis 12:835–840
Tsvetkov P, Coy S, Petrova B, Dreishpoon M, Verma A, Abdusamad M et al (2022) Copper induces cell death by targeting lipoylated TCA cycle proteins. Science 375:1254–1261
Twyman D (1997) Nutritional management of the critically ill neurologic patient. Crit Care Clin 13:39–49
Um SH, D'Alessio D, Thomas G (2006) Nutrient overload, insulin resistance, and ribosomal protein S6 kinase 1, S6K1. Cell Metab 3:393–402
Vasterling JJ, Proctor SP, Amoroso P, Kane R, Heeren T, White RF (2006) Neuropsychological outcomes of army personnel following deployment to the Iraq war. JAMA 296:519–529
Vuille-Dit-Bille RN, Ha-Huy R, Stover JF (2012) Changes in plasma phenylalanine, isoleucine, leucine, and valine are associated with significant changes in intracranial pressure and jugular venous oxygen saturation in patients with severe traumatic brain injury. J Amino Acids 43:1287–1296
Wagenmakers AJ (1999) Amino acid supplements to improve athletic performance. Curr Opin Clin Nutr Metab Care 2:539–544
Wang X-H, Liu Q, Shao Z-T (2018) Deletion of JDP2 improves neurological outcomes of traumatic brain injury (TBI) in mice: inactivation of Caspase-3. Biochem Biophys Res Commun 504:805–811
Weber JT (2012) Altered calcium signaling following traumatic brain injury. Front Pharmacol 3:60
Werner C, Engelhard K (2007) Pathophysiology of traumatic brain injury. Br J Anaesth 99:4–9
White PJ, Lapworth AL, An J, Wang L, McGarrah RW, Stevens RD et al (2016) Branched-chain amino acid restriction in Zucker-fatty rats improves muscle insulin sensitivity by enhancing efficiency of fatty acid oxidation and acyl-glycine export. Mol Metab 5:538–551
Wischmeyer PE (2018) Nutrition therapy in sepsis. Crit Care Clin 34:107–125
Wiseman-Hakes C, Murray B, Moineddin R, Rochon E, Cullen N, Gargaro J et al (2013) Evaluating the impact of treatment for sleep/wake disorders on recovery of cognition and communication in adults with chronic TBI. Brain Inj 27:1364–1376
Witgen B, Lifshitz J, Smith M, Schwarzbach E, Liang S-L, Grady M et al (2005) Regional hippocampal alteration associated with cognitive deficit following experimental brain injury: a systems, network and cellular evaluation. Neurosci J 133:1–15
Worthington AD, Melia Y (2006) Rehabilitation is compromised by arousal and sleep disorders: results of a survey of rehabilitation centres. Brain Inj 20:327–332
Wyllie D, Livesey M, Hardingham G (2013) Influence of GluN2 subunit identity on NMDA receptor function. Curr Neuropharmacol 74:4–17
Xie BS, Wang YQ, Lin Y, Mao Q, Feng JF, Gao GY et al (2019) Inhibition of ferroptosis attenuates tissue damage and improves long-term outcomes after traumatic brain injury in mice. CNS Neurosci Ther 25:465–475
Xiong Y, Peterson P, Lee C (1999) Effect of N-acetylcysteine on mitochondrial function following traumatic brain injury in rats. J Neurotrauma 16:1067–1082
Yi L, Shi S, Wang Y, Huang W, Xia Z-a, Xing Z et al (2016) Serum metabolic profiling reveals altered metabolic pathways in patients with post-traumatic cognitive impairments. Sci Rep 6:21320
Young B, Ott L, Norton J, Tibbs P, Rapp R, McClain C et al (1985) Metabolic and nutritional sequelae in the non-steroid treated head injury patient. J Neurosurg 17:784–791

Young B, Ott L, Twyman D, Norton J, Rapp R, Tibbs P et al (1987) The effect of nutritional support on outcome from severe head injury. J Neurosurg 67:668–676
Young B, Ott L, Yingling B, McClain C (1992) Nutrition and brain injury. J Neurotrauma 9:375–383
Young B, Ott L, Kasarskis E, Rapp R, Moles K, Dempsey RJ et al (1996) Zinc supplementation is associated with improved neurologic recovery rate and visceral protein levels of patients with severe closed head injury. J Neurotrauma 13:25–34
Yudkoff M, Daikhin Y, Nissim I, Horyn O, Luhovyy B, Lazarow A et al (2005) Brain amino acid requirements and toxicity: the example of leucine. J Nutr 135:1531–1538
Zafonte RD, Fregni F, Bergin MJ, Goldstein R, Boudreau N, Monge I et al (2020) Huperzine A for the treatment of cognitive, mood, and functional deficits after moderate and severe TBI (HUP-TBI): results of a phase II randomized controlled pilot study: implications for understanding the placebo effect. Brain Inj 34:34–41
Zhang S, Zeng X, Ren M, Mao X, Qiao S (2017) Novel metabolic and physiological functions of branched chain amino acids: a review. J Anim Sci Biotechnol 8:1–12
Zhang D, Xiao M, Wang L, Jia W (2019) Blood-based glutamate scavengers reverse traumatic brain injury-induced synaptic plasticity disruption by decreasing glutamate level in hippocampus interstitial fluid, but not cerebral spinal fluid, in vivo. Neurotox Res 35:360–372
Zhang L, Xiao H, Yu X, Deng Y (2020) Minocycline attenuates neurological impairment and regulates iron metabolism in a rat model of traumatic brain injury. Arch Biochem Biophys 682:108302
Zhang S, Lachance BB, Mattson MP, Jia X (2021) Glucose metabolic crosstalk and regulation in brain function and diseases. Prog Neurobiol 204:102089
Zhao Y, Huang Z, Peng H (2021) Molecular mechanisms of ferroptosis and its roles in hematologic malignancies. Front Oncol 11:743006
Žvirblis T, Ivanauskaitė V, Tavorienė I, Gerbutavičius R, Juozaitytė E, Malciūtė L et al (2012) Increasing relative survival in chronic myeloid leukemia (CML): a population-based study in Lithuania. In: Haematologica: 17th congress of the European Hematology Association: Amsterdam, The Netherlands, June 14–17, 2012: abstract book/European Hematology Association Pavia, Italy: Ferrata Storti Foundation, 2012, vol 97, suppl 1

Chapter 8
Application of Choline for Treatment of Traumatic Brain Injury

Fatima Rachidi, Hajar Ouaki, Rabii Hilali, Abdelmonaim Azzouz, Youssef Aoulad El Hadj Ali, Dia Eddine Oudghiri, Ibtissam Chakir, Rabia Ajana, Abdelaati El Khiat, and Lahcen Tamegart

F. Rachidi
Laboratory of Biostatistics, Clinical Research and Epidemiology (LBRCE), Faculty of Medicine and Pharmacy, Mohammed V University, Rabat, Morocco

Biology Department, Higher School of Teachers (ENS), Mohammed V University in Rabat, Rabat, Morocco

H. Ouaki
Laboratory of Biostatistics, Clinical Research and Epidemiology (LBRCE), Faculty of Medicine and Pharmacy, Mohammed V University, Rabat, Morocco

R. Hilali · R. Ajana · L. Tamegart (✉)
Biology Ecology and Health Laboratory, Department of Biology, Faculty of Sciences, Abdelmalek Essaadi University, Tetouan, Morocco
e-mail: l.tamegart@uae.ac.ma

A. Azzouz · Y. A. E. H. Ali
Department of Chemistry, Faculty of Sciences, University of Abdelmalek Essaâdi, Tetouan, Morocco

D. E. Oudghiri
Biology Ecology and Health Laboratory, Department of Biology, Faculty of Science, Abdelmalek Essaadi University, Tetouan, Morocco

Biology Higher Institute of Nursing Professions and Health Techniques (ISPITS), Tetouan, Morocco

I. Chakir
Biology Ecology and Health Laboratory, Department of Biology, Faculty of Science, Abdelmalek Essaadi University, Tetouan, Morocco

Multidisciplinary Faculty, LPRD, Sultan Moulay Slimane University, Beni-Mellal, Morocco

A. El Khiat
Biology Ecology and Health Laboratory, Department of Biology, Faculty of Science, Abdelmalek Essaadi University, Tetouan, Morocco

Higher Institute of Nursing Professions and Health Techniques in Ouarzazate Ministry of Health, Ouarzazate, Morocco

W. Mohamed (ed.), *Nutrition and Traumatic Brain Injury (TBI)*, Nutritional Neurosciences, https://doi.org/10.1007/978-981-97-6341-2_8

Abstract The primary and secondary injuries are the phases of the pathophysiology of traumatic brain injury (TBI). Primary injury happens when the brain is subjected to an external force that directly affects the vascular system, glial cells, and neuronal tissue according to their physical properties. The development of biological processes brought on by the first damage leads to secondary injuries. These may include cellular malfunction, resulting in apoptosis, ischemia, glutamate toxicity, neuroinflammation, edema, and increased blood–brain barrier (BBB) permeability. Choline acts as a modulator of the inflammatory response, a neuroprotective agent.

In addition, nanomaterials associated with choline and their strong loading capacity could play an important role in the treatment of TBI.

Keywords Traumatic brain injury · Trauma · Choline · Nanoparticles

8.1 Introduction

Traumatic brain injury (TBI) can differ in severity, location, and patient features. It can also be diffuse or focal, piercing or non-penetrating. Additionally, there are not many basic preventive strategies because TBI is frequently associated with accidents. Secondary production of tissue damage and inflammation is linked to a large portion of the acute and chronic harm that follows a TBI (Andriessen et al. 2010).

Furthermore, oxidative factors, such as reactive oxygen species (ROS) and reactive nitrogen species (RNS) that are formed in excess after a TBI, consequently contribute to secondary brain damage (Davis and Vemuganti 2022).

Choline exerts its impact on the regulation of ion channels through its binding to the sigma 1 receptor (Sig1-R) rather than direct actions on ion channels. Specifically, choline acts as an agonist for Sig1-R expressed on the endoplasmic reticulum membrane, enhancing Ca^{2+} signals elicited by inositol 1,4,5-triphosphate receptors (IP3Rs) (Su et al. 2016; Brailoiu et al. 2019). Furthermore, choline absorption, originating from the breakdown of acetylcholine at the synaptic level toward Sig1-R, facilitated by high-affinity transporters, serves as a paracrine activity messenger at cholinergic synapses, resulting in potentiation of Ca^{2+} signals (Haga 2014; Brailoiu et al. 2019).

Furthermore, a study conducted by Wehrwein et al. (2004) emphasizes the neuroprotective effect of acetylcholine against glutamate-induced neurotoxicity through the activation of α7 nicotinic acetylcholine receptors (α7nAChRs). These discoveries suggest that choline and its derivatives exert a neuroprotective effect against neuronal alterations induced by excitotoxicity.

8.2 Pathophysiology of TBI

TBI can be divided into three categories including: the initial phase with many symptoms such as cell death secondary to excitotoxicity and shearing of neurons, and blood vessels caused by mechanical damage; the intermediate phase is characterized by neuroinflammation; and the late phase with an obvious increase in susceptibility to seizures and epilepsy (Algattas and Huang 2013).

Early phase: Normal energy processes are delicately regulated, and interference with this complex pathway results in a reduction in the use of glucose, the buildup of lactic acid, a decrease in ATP and the activity of ion pumps that depend on ATP, depolarization brought on by Ca^{2+}, excitotoxicity, and cell death. The ischemic cascade starts sequentially when normal blood flow is disrupted. The blood flow in the brain is incredibly tightly regulated, and any disruption to this pattern modifies the brain's typical metabolic environment. The brain's reduced oxygen metabolism and blood flow encourage the metabolic shift from the typical aerobic process to an anaerobic one. Anaerobic respiration is indicated by the accumulation of lactate in tissue that is oxygen-deficient (Glenn et al. 2003; Cunningham et al. 2005). Not only may cerebral blood flow and oxygen levels impact metabolic processes, but they may also impact the brain's capacity to absorb glucose (Selwyn et al. 2013). Significant fluctuations in glucose levels have been linked to worse long-term outcomes in later stages of TBI pathophysiology, indicating a more complex metabolic relationship (Matsushima et al. 2012). A deficit in cerebral energy production is inescapably caused by dysregulated cerebral metabolism and the preferential breakdown of lactate rather than glucose. Consequently, ATP-dependent ion channels and proteins fail as a result of decreases in ATP (Werner and Engelhard 2007).

TBI can directly harm the cerebral vasculature. Blood–brain barrier (BBB) rupture, edema, hemorrhage, and irregular blood flow are common early and immediate phenomena seen in humans and animal models with TBI (Salehi et al. 2017). Based on where in the brain they occur, there are three different forms of hemorrhage following traumatic brain injury (TBI): subdural hematomas (between the dura and arachnoid mater), subarachnoid hemorrhage (between the arachnoid and pia mater), and epidural hematomas (between the dura and skull). Increased intracranial hypertension, vasospasms, and compression of brain structures are the outcomes of bleeding. Hemosiderin and ferritin buildup is known to be harmful to brain tissue and can cause an inflammatory response. Delay in microbleeds after traumatic brain injury is linked to white matter damage.

Brain edema is an excessive buildup of fluid in the brain. Cerebral edema develops at the site of a TBI and melds with the surrounding tissue. There are two types of edema that have been observed: vasogenic and cytotoxic. Dysregulation of the sodium and potassium pumps on the cell membrane causes cytotoxic edema, which causes water to build up in the cells. Retention of ions and water can cause apoptosis and have a negative impact on cell function. In cases of cytotoxic edema, the BBB is unaffected. The breakdown of the BBB leads to vascular edema, which is characterized by an accumulation of water in the extracellular space. Elevated intracranial pressure, tissue edema, altered blood flow, and compression of brain structures are all linked to vascular edema (Algattas and Huang 2013; Thapa et al. 2021).

Vasospasms abnormal constriction of blood vessels have the potential to negatively impact subarachnoid hemorrhage, hypertension, and functional recovery following traumatic brain injury. Moreover, vasospasms have the potential to cause ischemic injury and infarction by reducing cerebral perfusion to specific brain regions. Ectotoxicity-mediated cell death, encompassing necrosis and apoptosis, is ultimately brought about by ischemia, decreased cerebral blood flow (CBF), and modified metabolic processes (Bullock et al. 1998; Raghupathi et al. 2000). The BBB plays a vital role in preserving cellular homeostasis by controlling the flow of substances between the blood and the central nervous system (CNS). Tightly packed endothelial cells, astrocytes, neurons, and pericytes make up this selective barrier (forming the expanded neurovascular unit (eNVU)). After TBI, either the contusion or elevated intracranial pressure causes disruptions to the endothelial tight junction proteins that seal the blood–brain barrier. By secreting cytokines and proteases, neutrophils also worsen the disruption of the blood–brain barrier. Toxic waste substances like iron, reactive oxygen species, and blood poisoning are released into the local injury microenvironment as a result of the BBB disruption. These compounds have the ability to worsen inflammation and raise the risk of hypertension and edema (Salehi et al. 2017). Although there are many different hemostasis pathways that can be affected by TBI, they all eventually result in inflammation, hyperfibrinolysis, endothelial dysfunction, and abnormal platelets.

Intermediate phase: One important factor that leads to secondary brain injury is microglial activation. Damaged cells release molecules known as danger-associated molecular pattern factors, which can cause microglia to become hyperactive. Consequently, elevated levels of inflammatory cytokines and cytotoxic factors are produced by hyperactive microglia, leading to additional neurodegeneration. Furthermore, when these factors are released continuously, the blood–brain barrier is disrupted, which allows inflammatory factors and peripheral immune cells to enter the brain and intensify the inflammatory response (Salehi et al. 2017; Thapa et al. 2021).

Another point is that glutamate, the primary excitatory amino acid, is released via presynaptic vesicles or seeps out of damaged membranes after TBI. Ca^{2+}-mediated release and reduced glial glutamate uptake cause glutamate levels to rise (Hinzman et al. 2012). Subsequently, the process by which excess glutamate binds to the NMDA receptor causes a huge influx of Ca^{2+} and Na^{+}, which in turn activates several enzymes that cause cellular damage. Additionally, astrocytes are susceptible to cell death caused by excitotoxicity (Floyd et al. 2005). Similarly, nitric oxide (NO) production is somewhat dependent on Ca^{2+}, and excessive NO production after TBI can cause serious oxidative damage. While NO plays a direct role in the neuroinflammatory cascade, it is interesting to note that glutamate also indirectly stimulates inflammatory responses. On top of that, increased quantities of reactive oxygen species (ROS), which cause more harm, are produced when Ca^{2+} enters mitochondria (Maciel et al. 2001). In fact, increased edema, astrocytosis, neutrophil infiltration, and pro- and anti-inflammatory cytokines are observed after TBI. Tumor necrosis factor alpha (TNFα), interleukin-1β (IL-1β), and interleukin-6 (IL-6) are the main pro-inflammatory cytokines secreted. Interleukin-10 (IL-10) and transforming growth factor beta (TGFβ) are the anti-inflammatory cytokines (C.-H. Yu et al. 2011). The control of ionic and water flow between glia and extracellular fluid is closely related to edematous alterations that occur during the inflammatory phase of TBI. Aquaporin-4 (AQP4) is an AQP that mediates water homeostasis and seems to be the main AQP in charge of edema production after TBI (Satoh et al. 2007).

Late phase: A role for infiltration and inflammation in epileptogenesis, as well as in TBI, affects GABAergic hippocampal interneurons of the immunoreactive parvalbumin, calretinin, or neuropeptide Y classes (Zhang et al. 2011).

8.3 Neuroprotective Properties of Choline

8.3.1 Role of Choline in Ion Channel Regulation

As a precursor to acetylcholine (ACh), choline can indirectly influence the modulation of ion channels through nicotinic acetylcholine receptors (nAChR). Acetylcholine, formed in specific neurons from choline acetylated by the enzyme choline O-acetyltransferase (CAT), in conjunction with acetyl coenzyme A (Derbyshire and Obeid 2020), is released into the synaptic cleft and binds to nAChR, causing a receptor conformational change that leads to the opening of ion channels (Wu et al. 2015).

8.3.2 Anti-Inflammatory Effects of Choline

Choline has demonstrated anti-inflammatory effects through the cholinergic anti-inflammatory pathway (CAP), an interaction between the autonomic nervous system and the immune system (Lv et al. 2021). This pathway involves vagus nerve stimulation, leading to the release of norepinephrine (NA) by the splenic sympathetic nerve. NA, binding to β_2 adrenergic receptors (β_2ARs) on CD4$^+$T cells, triggers the release of acetylcholine. Once released, acetylcholine binds to nicotinic acetylcholine receptors of type 7 (α7nAChR), primarily expressed on immune cells (macrophages, monocytes, and lymphocytes), and modulates the inflammatory response by inhibiting the release of pro-inflammatory cytokines (W. Wang et al. 2021).

As previously demonstrated, the cholinergic anti-inflammatory pathway plays a protective role in various inflammatory diseases, such as cardiovascular diseases, inflammatory bowel diseases, and autoimmune rheumatic diseases (Lv et al. 2021; Wang et al. 2021). In an experimental study using a murine model of nonalcoholic steatohepatitis (NASH), activation of α7nAChR in Kupffer cells reduced hepatic inflammatory symptoms by negatively affecting the nuclear factor kappa-light-chain-enhancer of activated B cells (NFκB) pathway involved in inflammation and suppressing the release of inflammatory cytokines (Nishio et al. 2017).

The finding of a study conducted by Pavlov et al. in Pavlov et al. 2009 using a murine model of endotoxemia suggests that administration of the acetylcholinesterase inhibitor, which prevents the enzymatic degradation of acetylcholine, leads to a decrease in serum levels of tumor necrosis factor (TNF) and interleukin-6 (IL-6). Furthermore, in another series of experiments, choline treatment was investigated, and the results demonstrated that this molecule attenuates the inflammatory response by reducing TNF production through modulation of the NFκB pathway (Ilcol et al. 2005; Parrish et al. 2008).

8.3.3 Influence of Choline on Synaptic Plasticity and Neuronal Recovery

A study conducted by Hurtado et al. in Hurtado et al. 2007 demonstrated that treatment with cytidine diphosphate-choline (CDP-choline) enhances neuronal plasticity and contributes to the recovery of sensorimotor functions after experimental stroke. Consistently, the work of Gutiérrez-Fernández et al. in Gutiérrez-Fernández et al. 2012 highlighted the ability of CDP-choline to improve functional recovery while increasing levels of cell proliferation, vasculogenesis, synaptophysin, and reducing lesion volume and cell death. Another experimental study on a murine model shows that treatment with α-glycerophosphocholine (α-GPC) improves memory deficits and protects against disruptions in neurotransmitter expression in the brain induced by double stress (H. J. Yu et al. 2020).

Furthermore, it has been demonstrated that choline supplementation restores the structure of hippocampal neuron dendrites in iron-deficient rats and improves learning, memory, and synaptic plasticity (Bastian et al. 2021). Other studies indicate that postnatal choline intake modulates neuronal plasticity and prevents motor coordination deficits (Bastian et al. 2021; Derbyshire and Maes 2023).

8.4 Underlying Molecular Mechanisms

8.4.1 Impact of Choline on Excitotoxicity

Excitotoxicity is a pathological process characterized by neuronal damage resulting from the abnormal release of excitatory neurotransmitters such as glutamate, known as the primary cause of neuronal death in various conditions such as neurodegenerative diseases and traumatic brain injuries (Lewerenz and Maher 2015). An experimental study on cerebellar granule cells has demonstrated that CDP-choline prevents glutamate-induced neuronal death. This prevention depends on both the dose and duration of treatment. These observations suggest that CDP-choline acts by inhibiting apoptotic pathways, thus providing protection against the deleterious effects of glutamate (Mir et al. 2003). Another study, utilizing pharmacological approaches, indicates that CDP-choline administration appears to have a protective effect by attenuating changes associated with programmed neuronal death through its inhibitory effect on the activation of caspase-mediated cell death pathways (Matyja et al. 2008).

8.4.2 *Effects of Choline on Neuroinflammatory Processes*

The impact of choline on the neuroinflammatory process has been the subject of extensive research. In a promising study conducted by Pollak et al. in Pollak et al. 2005, it was revealed that the administration of acetylcholinesterase inhibitors decreases the production of IL1-β both in the hippocampus and in the blood, in the context of Alzheimer's disease. These results highlight the potential of acetylcholinesterase inhibitors to modulate the inflammatory response associated with neurodegenerative diseases. Concurrently, findings from a study on trophoblastic cells showed that choline deficiency leads to a significant increase in pro-inflammatory cytokines (Jiang et al. 2014). Additionally, supplementation with choline during the prenatal period significantly attenuates the expression of TNF-α and IL-1β (King et al. 2019; Kwan et al. 2017).

Choline appears to act as a modulator of the inflammatory response in the hippocampus, exerting a protective effect against neuroinflammation and cognitive impairment induced by lipopolysaccharide (LPS) in a murine model. This protective effect is mediated through the activation of α7 nicotinic acetylcholine receptors (7nAChR), which counteract p38 mitogenactivated protein kinases (p38 MAPK) phosphorylation and attenuate the expression of IL-1β (J. Yu et al. 2017).

Inflammation has become a central aspect of understanding the underlying mechanisms of Alzheimer's disease. Experimental studies have demonstrated that choline inhibits the activation of the nucleotide-binding and oligomerization domain-like receptor protein 3 (NLRP3) inflammasome complex (Wang et al. 2019). These observations highlight the various mechanisms through which choline acts in the neuroinflammatory process.

8.5 Choline in Humans with Traumatic Brain Injury

Choline is an essential nutrient that enters the synthesis of structural cell membrane phospholipids, other signaling molecules, and a precursor for acetylcholine synthesis (Zeisel and Niculescu 2006). Choline treatment can be useful to enhance recovery following brain injury in rats (Guseva et al. 2008). In humans, serum choline levels decrease after some stressful situations like surgery, childbirth, and traumatic head injury. These changes could be altered cholinergic neurotransmission as well as membrane phospholipids synthesis (Ulus et al. 1998). Memory impairment and cognitive dysfunction after head injury are not only attributable to decreased concentrations of choline but also associated with a loss of cholinergic neurotransmission (Murdoch et al. 1998).

8.6 Preclinical and Clinical Studies with Choline in Brain Injury

Choline is an essential nutrient that is crucial for brain health (Zeisel and da Costa 2009; Holt et al. 2012), particularly as a precursor of acetylcholine. The neurotransmitter is crucial for memory, learning, and muscle coordination (Blusztajn et al. 2017). The involvement of it in brain plasticity and neuroprotection is leading to the exploration of its therapeutic potential in brain damage (J. Yu et al. 2019; Borges et al. 2015).

Preclinical and clinical studies have shown that choline can have beneficial effects in the treatment of brain lesions (Holt et al. 2012) (Eisele et al. 2020).

8.6.1 Findings from Observational Studies

Observational studies contribute to understanding the role of choline in brain damage (Javaid et al. 2021). They analyze the association between choline exposure (via diet, dietary supplements, or biological status) and various aspects of brain damage, such as symptom severity, disease progression, and response to treatment (Eisele et al. 2020; Institute of Medicine (US) Committee on Nutrition, Trauma, and the Brain 2011).

8.6.2 Examples of Observational Studies

- Rotterdam cohort study: This study followed over 10,000 elderly people for several years and found that those with high blood choline status had a lower risk of cognitive decline and dementia (Gámiz and Gallo 2021).
- Pediatric head injury study: This study examined the association between choline levels in cerebrospinal fluid and symptom severity in children with head injury. Results showed that higher levels of choline were associated with better cognitive function and reduced neuronal inflammation (Guseva et al. 2008).

- Alzheimer's disease study: An analysis of data from several studies has suggested that dietary choline intake may be associated with a lower risk of Alzheimer's disease (Yuan et al. 2022).

Observational studies cannot establish cause and effect, but they can identify potential associations that require further investigation (Geerts et al. 2018). The results of observational studies may be influenced by various factors, such as age, gender, lifestyle, and general health (Institute of Medicine (US) Committee on Nutrition, Trauma, and the Brain 2011; Yuan et al. 2022; Secades 2021).

It is important to combine the results of observational studies with data from preclinical and clinical studies to obtain a more complete understanding of the role of choline in brain injury (Riley and Renshaw 2018).

8.6.3 Clinical Trials Evaluating the Efficacy of Choline in TBI

Traumatic brain injury (TBI) is a common brain injury with potentially devastating consequences. Research into choline, an essential nutrient playing a crucial role in brain health, is exploring its therapeutic potential in TBI (Table 8.1) (J. Yu et al. 2019; Borges et al. 2015; Joyce et al. 2022).

Clinical trials investigating the efficacy of choline in TBI offer promising prospects for the development of new therapies. Choline may play an important role in

Table 8.1 Clinical trials that evaluate the effectiveness of choline in TBI

Clinical trials	Objective	Population	Intervention	Results
Choline for mild-to-moderate TBI (CITRIC) (Amyot et al. 2015)	Assessing the effect of choline on cognitive function and quality of life after mild-to-moderate TBI	Adults with mild-to-moderate TBI	Choline supplementation or placebo for 12 weeks	Significant improvements in cognitive function were observed in the choline group.
Citicoline for cognitive function and quality of life in TBI (CHOLINE)	Evaluating the effect of citicoline (a form of choline) on cognitive function and quality of life after a TBI	Adults with a TBI	Supplementation with citicoline or placebo for 24 weeks	Significant improvements in cognitive function and quality of life were observed in the citicoline group.
Study on choline and neuroimaging in TBI	Assessing the effect of choline on brain structure and function after TBI	Adults with a TBI	Choline supplementation for 12 weeks	Structural and functional changes in the brain were observed in the choline group, suggesting possible neuroprotection.

repairing neuronal damage, improving cognitive function, and protecting against the adverse effects of TBI (J. Yu et al. 2019; Joyce et al. 2022).

Further research is needed to confirm its potential and determine its optimal clinical applications.

8.7 Therapeutic Implications

8.7.1 Potential Use of Choline as a Preventive Treatment

8.7.1.1 Combination of Choline with Nanoparticles for the Treatment of TBI

Traumatic brain injury (TBI) is a brain injury that is caused by a traumatic incident, such as a blow or jolt to the head or a penetrating head injury, which alters the normal functioning of the brain (Halim et al. 2021). About 69 million individuals are estimated to suffer TBI from all causes each year (Halim et al. 2021). Hematoma, subarachnoid hemorrhage, contusion, and diffuse axonal injury are the common pathological consequences of TBI (Halim et al. 2021). Brain damage is linked to a cholinergic system spatial imbalance, which seriously damages the recovery of the damaged neurons in the brain (Squitti et al. 2023) [4]. Thus, preservation of the cholinergic system is of utmost relevance. Recently, some studies have demonstrated that nanomaterials and nanoparticles (e.g., silica nanoparticles, silver nanoparticles, and carbon materials) can cross the blood–brain barrier (BBB) (Tang et al. 2008; Zarouki et al. 2024). In addition, it has been reported that some kinds of nanomaterials (i.e., black phosphorus, graphene oxide, MXenes, and layered double hydroxide) can significantly reduce neuroinflammation, which is one of the important features of ABI (Wang et al. 2024) (Li et al. 2023). Graphene nanomaterials [e.g., graphene quantum dots (GQDs), graphene oxide (GO), reduced graphene oxide (rGO), and functionalized GO)] stand out from the various allotropic forms of carbon materials for their high surface area, resistance, lightweight, chemical stability, enhanced cell adhesion, proliferation, low toxicity, high biocompatibility differentiation, and application in tissue repair (Zarouki et al. 2024; Ghosal et al. 2021; Oukhrib et al. 2023). For instance, Pradhan et al. designed an injectable hydrogel with acetylcholine-functionalized GO and poly(acrylic acid) (choline functionalized graphene oxide (CFGO)) to repair focal brain injury (Pradhan et al. 2019). Not only their hydrogels were able to support neuronal cell growth and differentiation, but they also stabilized the actin cytoskeleton. As choline is involved in enhancing neural recovery in TBI treatment, they injected their hydrogels in mice with parietal cortex brain injuries. They found scaffolds were able to restore cortical loss in just 7 days of treatment (Pradhan et al. 2019). In this work, they reported the effectiveness of GO nanosheets as a delivery vehicle of RNA interference (RNAi) in the brain-injured mice model (Joo et al. 2016). The in vivo mouse model that is being used in this study involves a penetrating brain injury on the right hemisphere, which generates localized inflammation and immune deficiencies that can impede systemic delivery of diagnostic and therapeutic agents. Moreover, the major challenge of the application of RNAi-based therapy is the extracellular nucleolytic degradation and the inefficient delivery into the cytoplasm of target cells. In this study, they coated the mesoporous silicon nanoparticles (pSiNPs) with GO to slow the release of the silicon- RNA (siRNA) payload, and as a result, a larger dose of siRNA can be delivered to the injured site (Joo et al. 2016).

8.8 Conclusion

The pathophysiology of TBI shows complex pathological processes, a series of nervous system disorders, and the structural destruction and dysfunction such as changes in cerebral blood flow, angiogenesis, and vasculogenesis and metabolic dysfunctions. Application of choline and its derivatives have shown promising results in modulating neuronal plasticity and restoring neurons in various contexts.

References

Algattas H, Huang JH (2013) Traumatic brain injury pathophysiology and treatments: early, intermediate, and late phases post-injury. Int J Mol Sci 15(1):309–341. https://doi.org/10.3390/ijms15010309

Amyot F, Arciniegas DB, Brazaitis MP, Curley KC, Diaz-Arrastia R, Gandjbakhche A, Herscovitch P, Hinds SR, Manley GT, Pacifico A, Razumovsky A, Riley J, Salzer W, Shih R, Smirniotopoulos JG, Stocker D (2015) A review of the effectiveness of neuroimaging modalities for the detection of traumatic brain injury. J Neurotrauma 32(22):1693–1721. https://doi.org/10.1089/neu.2013.3306

Andriessen TMJC, Jacobs B, Vos PE (2010) Clinical characteristics and pathophysiological mechanisms of focal and diffuse traumatic brain injury. J Cell Mol Med 14(10):2381–2392. https://doi.org/10.1111/j.1582-4934.2010.01164.x

Bastian TW, von Hohenberg WC, Kaus OR, Lanier LM, Georgieff MK (2021) Choline supplementation partially restores dendrite structural complexity in developing iron-deficient mouse hippocampal neurons. J Nutr 152(3):747–757. https://doi.org/10.1093/jn/nxab429

Blusztajn JK, Slack BE, Mellott TJ (2017) Neuroprotective actions of dietary choline. Nutrients 9(8):815. https://doi.org/10.3390/nu9080815

Borges AA, El-Batah PN, Yamashita LF, Santana A d S, Lopes AC, Freymuller-Haapalainen E, Coimbra CG, Sinigaglia-Coimbra R (2015) Neuroprotective effect of oral choline administration after global brain ischemia in rats. Nutr Neurosci 18(6):265–274. https://doi.org/10.1179/1476830514Y.0000000125

Brailoiu E, Chakraborty S, Brailoiu GC, Zhao P, Barr JL, Ilies MA, Unterwald EM, Abood ME, Taylor CW (2019) Choline is an intracellular messenger linking extracellular stimuli to IP3-evoked Ca2+ signals through sigma-1 receptors. Cell Rep 26(2):330–337.e4. https://doi.org/10.1016/j.celrep.2018.12.051

Bullock R, Zauner A, Woodward JJ, Myseros J, Choi SC, Ward JD, Marmarou A, Young HF (1998) Factors affecting excitatory amino acid release following severe human head injury. J Neurosurg 89(4):507–518. https://doi.org/10.3171/jns.1998.89.4.0507

Cunningham AS, Salvador R, Coles JP, Chatfield DA, Bradley PG, Johnston AJ, Steiner LA, Fryer TD, Aigbirhio FI, Smielewski P, Williams GB, Carpenter TA, Gillard JH, Pickard JD, Menon DK (2005) Physiological thresholds for irreversible tissue damage in contusional regions following traumatic brain injury. Brain J Neurol 128(Pt 8):1931–1942. https://doi.org/10.1093/brain/awh536

Davis CK, Vemuganti R (2022) Antioxidant therapies in traumatic brain injury. Neurochem Int 152:105255. https://doi.org/10.1016/j.neuint.2021.105255

Derbyshire E, Maes M (2023) The role of choline in neurodevelopmental disorders—a narrative review focusing on ASC, ADHD and dyslexia. Nutrients 15(13):13. https://doi.org/10.3390/nu15132876

Derbyshire E, Obeid R (2020) Choline, neurological development and brain function: a systematic review focusing on the first 1000 days. Nutrients 12(6):1731. https://doi.org/10.3390/nu12061731

Eisele A, Hill-Strathy M, Michels L, Rauen K (2020) Magnetic resonance spectroscopy following mild traumatic brain injury: a systematic review and meta-analysis on the potential to detect posttraumatic neurodegeneration. Neurodegener Dis 20(1):2–11. https://doi.org/10.1159/000508098

Floyd CL, Gorin FA, Lyeth BG (2005) Mechanical strain injury increases intracellular sodium and reverses Na+/Ca2+ exchange in cortical astrocytes. Glia 51(1):35–46. https://doi.org/10.1002/glia.20183

Gámiz F, Gallo M (2021) A systematic review of the dietary choline impact on cognition from a psychobiological approach: insights from animal studies. Nutrients 13(6):1966. https://doi.org/10.3390/nu13061966

Geerts H, Gieschke R, Peck R (2018) Use of quantitative clinical pharmacology to improve early clinical development success in neurodegenerative diseases. Expert Rev Clin Pharmacol 11(8):789–795. https://doi.org/10.1080/17512433.2018.1501555

Ghosal K, Mondal P, Bera S, Ghosh S (2021) Graphene family nanomaterials-opportunities and challenges in tissue engineering applications. FlatChem 30:100315

Glenn TC, Kelly DF, Boscardin WJ, McArthur DL, Vespa P, Oertel M, Hovda DA, Bergsneider M, Hillered L, Martin NA (2003) Energy dysfunction as a predictor of outcome after moderate or severe head injury: indices of oxygen, glucose, and lactate metabolism. J Cereb Blood Flow Metab 23(10):1239–1250. https://doi.org/10.1097/01.WCB.0000089833.23606.7F

Guseva MV, Hopkins DM, Scheff SW, Pauly JR (2008) Dietary choline supplementation improves behavioral, histological, and neurochemical outcomes in a rat model of traumatic brain injury. J Neurotrauma 25(8):975–983. https://doi.org/10.1089/neu.2008.0516

Gutiérrez-Fernández M, Rodríguez-Frutos B, Fuentes B, Vallejo-Cremades MT, Álvarez-Grech J, Expósito-Alcaide M, Díez-Tejedor E (2012) CDP-choline treatment induces brain plasticity markers expression in experimental animal stroke. Neurochem Int 60(3):310–317. https://doi.org/10.1016/j.neuint.2011.12.015

Haga T (2014) Molecular properties of the high-affinity choline transporter CHT1. J Biochem 156(4):181–194. https://doi.org/10.1093/jb/mvu047

Halim A, Qu K-Y, Zhang X-F, Huang N-P (2021) Recent advances in the application of two-dimensional nanomaterials for neural tissue engineering and regeneration. ACS Biomater Sci Eng 7(8):3503–3529. https://doi.org/10.1021/acsbiomaterials.1c00490

Hinzman JM, Thomas TC, Quintero JE, Gerhardt GA, Lifshitz J (2012) Disruptions in the regulation of extracellular glutamate by neurons and glia in the rat striatum two days after diffuse brain injury. J Neurotrauma 29(6):1197–1208. https://doi.org/10.1089/neu.2011.2261

Holt R, Uriu-Adams J, Keen C (2012) Present knowledge in nutrition, 10th edn. Wiley, pp 521–539. https://doi.org/10.1002/9781119946045.ch34

Hurtado O, Cárdenas A, Pradillo JM, Morales JR, Ortego F, Sobrino T, Castillo J, Moro MA, Lizasoain I (2007) A chronic treatment with CDP-choline improves functional recovery and increases neuronal plasticity after experimental stroke. Neurobiol Dis 26(1):105–111. https://doi.org/10.1016/j.nbd.2006.12.005

Ilcol YO, Yilmaz Z, Ulus IH (2005) Endotoxin alters serum-free choline and phospholipid-bound choline concentrations, and choline administration attenuates endotoxin-induced organ injury in dogs. Shock 24(3):288–293. https://doi.org/10.1097/01.shk.0000174018.02688.4b

Institute of Medicine (US) Committee on Nutrition, Trauma, and the Brain (2011) Nutrition and traumatic brain injury: improving acute and subacute health outcomes in military personnel (Erdman J, Oria M, Pillsbury L, Éds.). National Academies Press. http://www.ncbi.nlm.nih.gov/books/NBK209314/

Javaid S, Farooq T, Rehman Z, Afzal A, Ashraf W, Rasool MF, Alqahtani F, Alsanea S, Alasmari F, Alanazi MM, Alharbi M, Imran I (2021) Dynamics of choline-containing phospholipids in traumatic brain injury and associated comorbidities. Int J Mol Sci 22(21):21. https://doi.org/10.3390/ijms222111313

Jiang X, Jones S, Andrew BY, Ganti A, Malysheva OV, Giallourou N, Brannon PM, Roberson MS, Caudill MA (2014) Choline inadequacy impairs trophoblast function and vascularization in cultured human placental trophoblasts. J Cell Physiol 229(8):1016–1027. https://doi.org/10.1002/jcp.24526

Joo J, Kwon EJ, Kang J, Skalak M, Anglin EJ, Mann AP, Ruoslahti E, Bhatia SN, Sailor MJ (2016) Porous silicon-graphene oxide core-shell nanoparticles for targeted delivery of siRNA to the injured brain. Nanoscale Horizons 1(5):407–414. https://doi.org/10.1039/C6NH00082G

Joyce JM, La PL, Walker R, Harris AD (2022) Magnetic resonance spectroscopy of traumatic brain injury and subconcussive hits: a systematic review and meta-analysis. J Neurotrauma 39(21-22):1455–1476. https://doi.org/10.1089/neu.2022.0125

King JH, Kwan ST(C), Yan J, Jiang X, Fomin VG, Levine SP, Wei E, Roberson MS, Caudill MA (2019) Maternal choline supplementation modulates placental markers of inflammation, angiogenesis, and apoptosis in a mouse model of placental insufficiency. Nutrients 11(2):374. https://doi.org/10.3390/nu11020374

Kwan ST, King JH, Yan J, Jiang X, Wei E, Fomin VG, Roberson MS, Caudill MA (2017) Maternal choline supplementation during murine pregnancy modulates placental markers of inflammation, apoptosis and vascularization in a fetal sex-dependent manner. Placenta 53:57–65. https://doi.org/10.1016/j.placenta.2017.03.019

Lewerenz J, Maher P (2015) Chronic glutamate toxicity in neurodegenerative diseases—what is the evidence? Front Neurosci 9:469. https://www.frontiersin.org/journals/neuroscience/articles/10.3389/fnins.2015.00469

Li K, Ji Q, Liang H, Hua Z, Hang X, Zeng L, Han H (2023) Biomedical application of 2D nanomaterials in neuroscience. J Nanobiotechnol 21(1):181. https://doi.org/10.1186/s12951-023-01920-4

Lv J, Ji X, Li Z, Hao H (2021) The role of the cholinergic anti-inflammatory pathway in autoimmune rheumatic diseases. Scand J Immunol 94(4):e13092. https://doi.org/10.1111/sji.13092

Maciel EN, Vercesi AE, Castilho RF (2001) Oxidative stress in Ca(2+)-induced membrane permeability transition in brain mitochondria. J Neurochem 79(6):1237–1245. https://doi.org/10.1046/j.1471-4159.2001.00670.x

Matsushima K, Peng M, Velasco C, Schaefer E, Diaz-Arrastia R, Frankel H (2012) Glucose variability negatively impacts long-term functional outcome in patients with traumatic brain injury. J Crit Care 27(2):125–131. https://doi.org/10.1016/j.jcrc.2011.08.012

Matyja E, Taraszewska A, Nagańska E, Grieb P, Rafałowska J (2008) CDP-choline protects motoneurons against apoptotic changes in a model of chronic glutamate excitotoxicity in vitro. Folia Neuropathol 46:139–148

Mir C, Clotet J, Aledo R, Durany N, Argemí J, Lozano R, Cervós-Navarro J, Casals N (2003) CDP-choline prevents glutamate-mediated cell death in cerebellar granule neurons. J Mol Neurosci 20(1):53–60. https://doi.org/10.1385/JMN:20:1:53

Murdoch I, Perry EK, Court JA, Graham DI, Dewar D (1998) Cortical cholinergic dysfunction after human head injury. J Neurotrauma 15(5):295–305. https://doi.org/10.1089/neu.1998.15.295

Nishio T, Taura K, Iwaisako K, Koyama Y, Tanabe K, Yamamoto G, Okuda Y, Ikeno Y, Yoshino K, Kasai Y, Okuno M, Seo S, Sakurai T, Asagiri M, Hatano E, Uemoto S (2017) Hepatic vagus nerve regulates Kupffer cell activation via α7 nicotinic acetylcholine receptor in nonalcoholic steatohepatitis. J Gastroenterol 52(8):965–976. https://doi.org/10.1007/s00535-016-1304-z

Oukhrib M, Tamegart L, Assafi A, Hejji L, Azzouz A, Villarejo LP, Haida M, Mohamed C, Gamrani H (2023) Effects of graphene oxide nanoparticles administration against reserpine-induced neurobehavioral damage and oxidative stress in an animal model of Parkinson's disease. Food Chem Toxicol 178:113904. https://doi.org/10.1016/j.fct.2023.113904

Parrish WR, Rosas-Ballina M, Gallowitsch-Puerta M, Ochani M, Ochani K, Yang L-H, Hudson L, Lin X, Patel N, Johnson SM, Chavan S, Goldstein RS, Czura CJ, Miller EJ, Al-Abed Y, Tracey KJ, Pavlov VA (2008) Modulation of TNF release by choline requires α7 subunit nicotinic acetylcholine receptor-mediated signaling. Mol Med 14(9-10):567–574. https://doi.org/10.2119/2008-00079.Parrish

Pavlov VA, Parrish WR, Rosas-Ballina M, Ochani M, Puerta M, Ochani K, Chavan S, Al-Abed Y, Tracey KJ (2009) Brain acetylcholinesterase activity controls systemic cytokine levels through the cholinergic anti-inflammatory pathway. Brain Behav Immun 23(1):41–45. https://doi.org/10.1016/j.bbi.2008.06.011

Pollak Y, Gilboa A, Ben-Menachem O, Ben-Hur T, Soreq H, Yirmiya R (2005) Acetylcholinesterase inhibitors reduce brain and blood interleukin-1β production. Ann Neurol 57(5):741–745. https://doi.org/10.1002/ana.20454

Pradhan K, Das G, Khan J, Gupta V, Barman S, Adak A, Ghosh S (2019) Neuro-regenerative choline-functionalized injectable graphene oxide hydrogel repairs focal brain injury. ACS Chem Neurosci 10(3):1535–1543. https://doi.org/10.1021/acschemneuro.8b00514

Raghupathi R, Graham DI, McIntosh TK (2000) Apoptosis after traumatic brain injury. J Neurotrauma 17(10):927–938. https://doi.org/10.1089/neu.2000.17.927

Riley CA, Renshaw PF (2018) Brain choline in major depression: a review of the literature. Psychiatry Res Neuroimag 271:142–153. https://doi.org/10.1016/j.pscychresns.2017.11.009

Salehi A, Zhang JH, Obenaus A (2017) Response of the cerebral vasculature following traumatic brain injury. J Cereb Blood Flow Metab 37(7):2320–2339. https://doi.org/10.1177/0271678X17701460

Satoh J, Tabunoki H, Yamamura T, Arima K, Konno H (2007) Human astrocytes express aquaporin-1 and aquaporin-4 in vitro and in vivo. Neuropathology 27(3):245–256. https://doi.org/10.1111/j.1440-1789.2007.00774.x

Secades JJ (2021) Role of citicoline in the management of traumatic brain injury. Pharmaceuticals 14(5):410. https://doi.org/10.3390/ph14050410

Selwyn R, Hockenbury N, Jaiswal S, Mathur S, Armstrong RC, Byrnes KR (2013) Mild traumatic brain injury results in depressed cerebral glucose uptake: an (18)FDG PET study. J Neurotrauma 30(23):1943–1953. https://doi.org/10.1089/neu.2013.2928

Squitti R, Reale G, Tondolo V, Crescenti D, Bellini S, Moci M, Caliandro P, Padua L, Rongioletti M (2023) Imbalance of essential metals in traumatic brain injury and its possible link with disorders of consciousness. Int J Mol Sci 24(7):6867. https://doi.org/10.3390/ijms24076867

Su T-P, Su T-C, Nakamura Y, Tsai S-Y (2016) Sigma-1 receptor as a pluripotent modulator in the living system. Trends Pharmacol Sci 37(4):262–278. https://doi.org/10.1016/j.tips.2016.01.003

Tang J, Xiong L, Wang S, Wang J, Liu L, Li J, Wan Z, Xi T (2008) Influence of silver nanoparticles on neurons and blood-brain barrier via subcutaneous injection in rats. Appl Surf Sci 255(2):502–504. https://doi.org/10.1016/j.apsusc.2008.06.058

Thapa K, Khan H, Singh TG, Kaur A (2021) Traumatic brain injury: mechanistic insight on pathophysiology and potential therapeutic targets. J Mol Neurosci 71(9):1725–1742. https://doi.org/10.1007/s12031-021-01841-7

Ulus IH, Ozyurt G, Korfali E (1998) Decreased serum choline concentrations in humans after surgery, childbirth, and traumatic head injury. Neurochem Res 23(5):727–732. https://doi.org/10.1023/a:1022455325657

Wang Y, Guan X, Chen X, Cai Y, Ma Y, Ma J, Zhang Q, Dai L, Fan X, Bai Y (2019) Choline supplementation ameliorates behavioral deficits and alzheimer's disease-like pathology in

transgenic APP/PS1 mice. Mol Nutr Food Res 63(18):1801407. https://doi.org/10.1002/mnfr.201801407
Wang W, Xu H, Lin H, Molnar M, Ren H (2021) The role of the cholinergic anti-inflammatory pathway in septic cardiomyopathy. Int Immunopharmacol 90:107160. https://doi.org/10.1016/j.intimp.2020.107160
Wang Y, Zhang X, Yue H (2024) Two-dimensional nanomaterials induced nano-bio interfacial effects and biomedical applications in cancer treatment. J Nanobiotechnol 22(1):67. https://doi.org/10.1186/s12951-024-02319-5
Wehrwein E, Thompson SA, Coulibaly SF, Linn DM, Linn CL (2004) Acetylcholine protection of adult pig retinal ganglion cells from glutamate-induced excitotoxicity. Invest Ophthalmol Vis Sci 45(5):1531–1543. https://doi.org/10.1167/iovs.03-0406
Werner C, Engelhard K (2007) Pathophysiology of traumatic brain injury. Br J Anaesth 99(1):4–9. https://doi.org/10.1093/bja/aem131
Wu Z, Cheng H, Jiang Y, Melcher K, Xu HE (2015) Ion channels gated by acetylcholine and serotonin: structures, biology, and drug discovery. Acta Pharmacol Sin 36(8):Article 8. https://doi.org/10.1038/aps.2015.66
Yu C-H, Yhee J-Y, Kim J-H, Im K-S, Kim N-H, Jung D-I, Lee H-C, Chon S-K, Sur J-H (2011) Pro- and anti-inflammatory cytokine expression and histopathological characteristics in canine brain with traumatic brain injury. J Vet Sci 12(3):299–301. https://doi.org/10.4142/jvs.2011.12.3.299
Yu J, Zhao W-X, Du C-Y, Zhang N, Zhang W-D, Jin S-Y, Wang H, Feng Z-G (2017) Choline improves lipopolysaccharide-induced central nervous system inflammatory response and cognitive dysfunction in mice. J South Med Univ 37(5):600–606. https://doi.org/10.3969/j.issn.1673-4254.2017.05.06
Yu J, Zhu H, Taheri S, Monday WL, Perry S, Kindy MS (2019) Reduced neuroinflammation and improved functional recovery after traumatic brain injury by prophylactic diet supplementation in mice. Nutrients 11(2):299. https://doi.org/10.3390/nu11020299
Yu HJ, Kim MJ, Park JM, Park SY, Park SN, Yang DW (2020) The effect of choline alphoscerate on non spatial memory and neurogenesis in a rat model of dual stress. bioRxiv, p 154310. https://doi.org/10.1101/2020.06.16.154310
Yuan J, Liu X, Liu C, Ang AF, Massaro J, Devine SA, Auerbach SH, Blusztajn JK, Au R, Jacques PF (2022) Is dietary choline intake related to dementia and Alzheimer's disease risks? Results from the Framingham heart study. Am J Clin Nutr 116(5):1201–1207. https://doi.org/10.1093/ajcn/nqac193
Zarouki MA, Tamegart L, Hejji L, El Hadj Ali YA, Ayadi AE, Villarejo LP, Mennane Z, Souhail B, Azzouz A (2024) Graphene quantum dots based on cannabis seeds for efficient wound healing in a mouse incisional wound model: link with stress and neurobehavioral effect. Int J Pharm 649:123658. https://doi.org/10.1016/j.ijpharm.2023.123658
Zeisel SH, da Costa K-A (2009) Choline: an essential nutrient for public health. Nutr Rev 67(11):615–623. https://doi.org/10.1111/j.1753-4887.2009.00246.x
Zeisel SH, Niculescu MD (2006) Perinatal choline influences brain structure and function. Nutr Rev 64(4):197–203. https://doi.org/10.1111/j.1753-4887.2006.tb00202.x
Zhang B-L, Chen X, Tan T, Yang Z, Carlos D, Jiang R-C, Zhang J-N (2011) Traumatic brain injury impairs synaptic plasticity in hippocampus in rats. Chin Med J 124(5):740–745

Chapter 9
Creatine and TBI

Adejoke Elizabeth Memudu, Gideon S. Alex, and Amadi Ogonda Ihunwo

Abstract Traumatic brain injury (TBI) is a global health issue due to its pathophysiological aftermath that results in long-term cognitive and psychological impairments. Nutritional or dietary interventions have emerged as a vital measure to combat neuropathology and recovery of TBI. An example of this dietary intervention is creatine supplementation. This chapter explores TBI pathophysiology while delving into the multifaceted mechanism through which creatine ameliorates TBI-mediated neurological deficits in a translational approach which bridges the gap between experimental animal research and clinical application. The biochemistry of creatine synthesis, its roles in neuroprotection, and the mechanisms underlying its cellular energy (ATP) restoration, reduction of oxidative stress-mediated brain damage, and maintenance of mitochondrial homeostasis were discussed. We also reviewed preclinical studies demonstrating the efficacy of creatine in animal models of TBI as the foundation for its potential therapeutic benefits which led us to delve into clinical trials that assess the impact of creatine supplementation on TBI in humans, focusing on cognitive function and physical recovery to validate creatine's efficacy and safety in TBI patients. Translation neuroscience report on creatine and TBI helps demonstrate the integration of bench research with bedside practice directed to provide a detailed understanding of creatine's role in TBI recovery and patient outcomes.

A. E. Memudu (✉)
Anatomy Department, Neuroscience Unit, Faculty of Basic Medical Sciences, Edo State University Uzairue, Auchi, Edo State, Nigeria
e-mail: adejoke.elizabeth@edouniversity.edu.ng

G. S. Alex
University of Port Harcourt, Port-Harcourt, Rivers State, Nigeria

A. O. Ihunwo
School of Anatomical Sciences, Faculty of Health Sciences, University of the Witwatersrand, Johannesburg, South Africa
e-mail: amadi.ihunwo@wits.ac.za

W. Mohamed (ed.), *Nutrition and Traumatic Brain Injury (TBI)*, Nutritional Neurosciences, https://doi.org/10.1007/978-981-97-6341-2_9

Keywords Creatine · Nutrition · Translational neuroscience · TBI · Mitochondrial dysfunction · Neuroprotection · Oxidative stress · Neuroregeneration

9.1 Introduction

Traumatic brain injury (TBI) is the leading cause of brain-injury-related death and disability (Van Gils et al. 2020) with incidence and prevalence being 27.16 million and 48.99 million, respectively, from a report in 2019 (Guan et al. 2023). Traumatic brain injury ranks among the most common reasons for admission to accident and emergency rooms and contributes to 30% of deaths globally (Huseyin Fatih et al. 2021; Ochems et al. 2021). Traumatic brain injury (TBI) remains one of the leading causes of morbidity and mortality globally. Hallmark neuropathological changes include lesions, contusions, bleeding, edema, axonal damage or Wallerian degeneration of axons, mitochondrial dysfunction, excitotoxicity, oxidative stress, and apoptotic cell death of neurons and glia (Ng and Lee 2019; Slavoaca et al. 2020). This results in a wide spectrum of neurological dysfunctions such as anxiety, depression, post-traumatic stress disorder (PTSD), and post-concussive symptoms. This condition is attributed to loss of neuronal integrity or density in the prefrontal cortex and maladaptive neuroplasticity (Robayo et al. 2023). Creatine kinase (CK) isoenzyme, creatine kinase BB (CKBB), is mainly found in the central nervous system (CNS) where it is used as a plasma biomarker for traumatic brain injury (Forouzan et al. 2023). TBI patients experience mitochondrial dysfunction, neuropsychological burden, and deficits in cognitive performance due to malperformance of brain creatine levels, diminished brain adenosine triphosphate (ATP) levels, glutamate toxicity, and oxidative stress (Newman et al. 2023). Oxidative stress is the fundamental cause of many vascular and neurodegenerative diseases, and creatine has been reported to have antioxidant function (Senoner and Dichtl 2019). An increased brain creatine level is potentially beneficial in the treatment of neurodegenerative diseases such as Alzheimer's, Huntington's, and Parkinson's diseases (Roschel et al. 2021), which are characterized by progressive and irreversible neuronal dysfunction, mitigating the ability to perform cognitive and motor function (Harmon et al. 2021). Over the past 20 years, the use of dietary supplements has greatly increased, and it has been recommended by health-care professionals to their patients (Bredahl et al. 2021) for injury prevention and improving recovery (Papadopoulou 2020). Nutritional supplements containing amino acids, antioxidants, creatine, and omega-3 have significant implications for preventing injury, improving tissue healing, and enhancing the swift recovery process (Turnagöl et al. 2021). Creatine (N-(aminoiminomethyl)-N-methyl glycine), commonly known as an amino acid, is a naturally occurring nitrogen-containing compound with an integral role in cellular metabolism (Bonilla et al. 2021). Creatine is a commonly used ergogenic dietary supplement (Hall et al. 2021). Creatine and phosphagens are required to maintain energy balance, especially in high-energy-demanding tissues such as CNS and muscle tissues (Hall et al. 2021); they also help improve neuron functions and replenishe adenosine triphosphate without needing oxygen (Marques and Wyse 2019; Chang and Leem 2023). Creatine is a key player in energy metabolism and cellular energy production (Balestrino and Adriano 2019; Kreider and Stout 2021). Creatine occurs in various forms such as monohydrate, malate, hydrochloride, orotate, and kre-alkalyn (Antonio et al. 2021; Abod et al. 2023). Creatine supplementation such as creatine monohydrate is generally safe (Balestrino and Adriano 2019). Daily consumption of creatine is naturally obtained from a diet such as red meat and fish (Clarke et al. 2020) or dietary supplements (Marques and Wyse 2019), while the other portion is provided through endogenous synthesis (Kreider and Stout 2021). Creatine is synthesized endogenously in the liver, pancreas, kidney, and spleen with some quantity manufactured in the brain using amino acids, arginine and glycine, catalyzed by arginine glycine amidinotransferase (AGAT) to form guanidinoacetate (GAA) which is then methylated by the enzyme guanidinoacetate N-methyltransferase (GAMT) with S-adenosyl methionine (SAMe) to form creatine; 95% of creatine formed endogenously is stored in muscle tissue, while the remaining lies in the tissues, like the heart, brain, and testes (Forbes et al. 2021). The availability of substrate and arginine transferase are the main factors regulating the creatine pathway (Arazi et al. 2021). Creatine has robust evidence as a supplement beneficial to brain health relating to cognitive processing and recovery from traumatic injury in neurological conditions such as mild traumatic brain injury, aging, Alzheimer's disease, and depression (Roschel et al. 2021; Kreider and Stout 2021; Forbes et al. 2022a) has been reported and contributed research done to unreveal its mechanism of action. Creatine supplementation has not been extensively researched for traumatic brain injury (TBI); however, studies suggest its potential as a neuroprotective agent and a potential treatment for brain-injury complications in individuals with TBI (Newman et al. 2023). This chapter discusses TBI pathology and explores the frontiers involving the supplementation of creatinine following traumatic brain injury.

9.2 The Role of Creatine in Brain Function

9.2.1 Definition and Natural Occurrence in the Body

Creatine (N-(aminoiminomethyl)-N-methyl glycine) is a natural form of a nitrogen-containing compound that plays an integral role in cellular energy metabolism (Bonilla et al. 2021). Creatine is described as a non-proteinogenic amino acid (amino acid derivative) present in various forms of animal-based foods such as human breast milk, red meat, poultry, and fish as well as artificial diets such as infant formulas and dietary supplements containing artificially prepared and purified creatine such as creatine monohydrate (Ostojic 2021a; Kreider et al. 2022). The amount of creatine across natural food sources varies from 0.2 g/kg in breast milk to 11.0 g/kg in raw and dried fish; however, plant-based foods do not contain any creatine (Wu 2020). Creatine synthesis provides about half of the daily need for normal tissue function, while the other is obtained through dietary intake (Brosnan and Brosnan 2016; Kerksick et al. 2018). Creatine has demonstrated therapeutic properties on muscle, kidney, bone, and brain function (Forbes et al. 2022a). The endogenous production of creatine is from amino acids arginine and glycine, whereby L-arginine:glycine forms guanidinoacetate (GAA) (Kreider et al. 2022) through the enzymatic reaction of arginine glycine amidinotransferase (AGAT) to form guanidinoacetate (GAA) (Borchel et al. 2019). Then GAA is methylated by adding a CH3 group through the enzymatic action of guanidinoacetate N-methyltransferase (GAMT) with S-adenosyl methionine (SAMe) to form creatine in the kidney, liver, pancreas, and some areas within the brain containing amidinotransferase (AGAT) (Brosnan and Brosnan 2016; Kreider et al. 2022). However, most GAA formed in the kidney is converted by GMAT to creatine in the liver (Brosnan and Brosnan 2016). In vertebrates, creatine is converted into phosphocreatine for rapid ATP generation by creatine kinase (CK),; hence, an increased CK activity correlates with creatine content and energy demands in tissue (Han et al. 2020; Forouzan et al. 2023). Creatine supplementation is in the form of creatine monohydrate, which is an effective and bioavailable form of creatine for increasing plasma creatine levels, tissue creatine content (i.e., brain and muscle), and performance outcome measures (Kreider et al. 2022).

9.2.2 *Creatine Biochemistry and Synthesis*

Creatine (N-aminoiminomethyl-N-methyl glycine) is a natural form of a nitrogen-containing compound made up of amino acids and is categorized within the family of guanidine phosphagens (Borchel et al. 2019). Creatine synthesis is required for cellular metabolism (Brosnan et al. 2011). Creatine is converted to creatinine at a rate of 1.7% per day and excreted in urine (da Silva et al. 2014). Creatine is endogenously synthesized in the kidney, liver, and pancreas from the amino acids: glycine, arginine, and methionine (Jagim and Kerksick 2021; Forbes et al. 2022a). The endogenous synthesis of creatine requires three amino acids such as glycine, methionine and arginine and three enzymes such as L-arginine-glycine amidinotransferase (AGAT), methionine adenosyltransferase (MAT), and guanidinoacetate methyltransferase (GAMT) (Brosnan et al. 2011; Kreider et al. 2022). The enzymatic processes involve L-Arginine glycine aminotransferase (AGAT) that converts L-arginine and glycine into GAA and L-ornithine within the mitochondrial intermembrane space, whereby N-guanidinoacetate methyltransferase (GAMT) transfers a methyl group from S-adenosylmethionine (SAM) to GAA to produce creatine (da Silva et al. 2014; Barcelos et al. 2016). During synthesis, glycine is consumed, but the addition of methyl and amidino groups to glycine is obtained from the methionine and arginine, respectively, whereby methyl groups are provided by S-adenosylmethionine (SAM) or by de novo methylneogenesis (Brosnan et al. 2011). The liver has high GAMT activity, and the kidney has high AGAT activity, while the pancreas has both AGAT and GAMT activities but an increased AGAT activity and lower GAMT activity in the pancreas; with pancreatic acini having an increased activities of both AGAT and GAMT geared toward synthesis of GAA and creatine from amino acids (da Silva et al. 2014), Creatine supplementation led to a decrease in AGAT activity in the pancreas (da Silva et al. 2014). Creatine permeates tissues through a specific creatine plasma membrane transporter (SLC6A8) found in the kidney, brain, and liver (da Silva et al. 2014; Barcelos et al. 2016). Close to 95% of Creatine in the body is kept in the muscle, with the remaining 5% stored in the heart, brain, and testes (Kreider and Stout 2021).

9.2.3 Transport of Creatine Through Nervous Tissue

Nutritional or dietary creatine is distributed by the blood to tissues. They permeate the brain through the blood–brain barrier (BBB) by a specific sodium- and chloride-dependent membrane protein transporter called the creatine transporter (CRT1, CT1, CreaT, or SLC6A8) (Forbes et al. 2022a; Samborska et al. 2022). These specific transporters are expressed in the microcapillaries of the BBB, neurons, and oligodendrocytes but not in perivascular astrocytes (Farr et al. 2022). In the brain, both AGAT and GAMT are expressed in astrocytes, and released Cr is taken up by neurons expressing SLC6A8, a specific Cr transporter (Tropak et al. 2023; Farr et al. 2022). This explains the balance between endogenous production in the brain and dietary uptake of intracellular Cr content to benefit brain function by increasing bioenergetics and strengthening neuroprotective functions (Chang and Leem 2023).

9.2.4 General Functions of Creatine

Creatine is critical in providing energy and maintaining energy balance for cellular functions, especially in high energy-demanding tissues such as nervous tissue's energetically demanding cognitive functions in learning and memory (Forbes et al. 2022a) in addition to muscle tissues (Hall et al. 2021); it also helps improve neuron functions and replenishes adenosine triphosphate without needing oxygen (Balestrino and Adriano 2019; Marques and Wyse 2019; Chang and Leem 2023). Furthermore, creatine has anti-oxidant, anti-apoptotic, and anti-excitotoxic properties (Marques and Wyse 2019; Chang and Leem 2023). Creatine maintains balance in continual creatine utilization requirement for cellular energy flow (Kazak and Cohen 2020). This is done through the endogenous synthesis of creatine to regulate AMPK signaling in the brain, increase brain creatine kinase expression in cortical neurons, prevent axonal damage, enhance myelination, and provide energy for maturation of oligodendrocytes (Rosko et al. 2023). Creatine has immunomodulatory and neuromodulator (Marques and Wyse 2019) functions. Creatine helps in mitochondrial transport and cellular energetics that promotes regeneration and functional restoration after muscle and nerve injuries (Han et al. 2020). Creatine supplementation helps upregulate mTORC1 signaling resulting in protein synthesis in aging rodent brains (Hodges et al. 2018; Mao et al. 2021).

9.2.5 Creatine as an Energy Source for Brain Cells/Role of Creatine in Cellular Energy Metabolism

Creatine and phosphagens are vital sources of energy production endogenously through the creatine kinase (CK) and the phosphate creatine system (Hall et al. 2021). In cellular bioenergetic, creatine binds with inorganic phosphate (Pi) in the cell to form phosphocreatine (PCr), a phosphate source of energy to resynthesize a high-energy adenosine triphosphate (ATP) (Kreider and Stout 2021). This process is termed the creatine phosphate shuttle or phosphocreatine shuttle (Hall et al. 2021). The creatine phosphate shuttle is critical in translocating energy in the form of ATP formed in the mitochondria through oxidative phosphorylation into the cytosol and to the cells needing ATP for energy metabolism (Balestrino and Adriano 2019). Creatine obtained via endogenous synthesis or through dietary means permeates enters the brain tissue into the cytosol via the BBB via creatine transporter (CRT1 or SLC6A8) expressed on microcapillaries of the BBB, neurons, and oligodendrocytes (Balestrino and Adriano 2019; Farr et al. 2022; Samborska et al. 2022; Tropak et al. 2023), within the cytosol, creatine along with its associated cytosolic and glycolytic creatine kinase helps maintain glycolytic ATP levels, the cytosolic ATP/ADP ratio, and cytosolic ATP-consumption (Balestrino and Adriano 2019) through the creatine phosphate shuttle that mediates cellular bioenergetic whereby free energy is being produced from the enzymatic degradation of ATP into ADP and inorganic phosphate (Pi) by creatine kinase to replenish ATP expended for cellular metabolism and neuronal activity (Balestrino and Adriano 2019). The enzymatic breakdown of phosphocreatine molecules into inorganic phosphates and creatine using the enzyme creatine kinase yields about 10.3 kcals of free energy available to resynthesize ADP ± Pi into ATP (Balestrino and Adriano 2019; Bonilla et al. 2021). This explains creatine bioenergetics in the neurons or nerve cells through the phosphocreatine shuttle for transport of intracellular energy molecules when mitochondrial creatine kinase cleaves with phosphate to form phosphocreatine molecules that release energy in the form of adenosine diphosphate (ADP) (Hall et al. 2021; Kreider and Stout 2021), ADP is then recycled to regenerate adenosine triphosphate (ATP) through the enzymatic reaction of intracellular creatine kinase that removes a phosphate group from phosphocreatine molecules and phosphorylates ADP to form new ATP energy to fuel cellular activities such as muscular contraction and action potential during neuronal activity (Hall et al. 2021; Borchel et al. 2019). This explains the balance between endogenous production in the brain and dietary uptake of intracellular Cr content to benefit brain function by increasing bioenergetics and strengthening neuroprotective functions (Chang and Leem 2023).

9.2.6 Emerging Research on Creatine and Brain Health

Brain tissue requires high energy consumption (Grimm et al. 2016). The disruption of brain energy metabolism is linked to a number of metabolic, neurological, and neurodegenerative disorders (Dwir et al. 2019). This is because brain tissue is vulnerable to oxidative stress triggered by hypoxia, mitochondrial damage, reactive oxygen species (ROS), and energy deficit (Grimm et al. 2016; Dwir et al. 2019). Oxidative stress is caused by the excessive production of ROS due to abnormal mitochondrial and cytosolic energy metabolism (Tarafdar and Pula 2018). Creatine's direct effect as an antioxidant helps scavenge ROS, thereby attenuating the deleterious effects of oxidative stress in the brain as seen in cognitive impairment (Chang and Leem 2023). Creatine is a widely used ergogenic dietary supplement among athletes because of its ability to boost recovery after an injury and prevent injury in athletes (Hall et al. 2021). Over the past two decades, scientific studies have confirmed other beneficial effects of creatine aside from its ergogenic role, including a role as therapeutics in neuromuscular and cardiometabolic diseases, post-traumatic rehabilitation, and aging (Snow et al. 2020; Kreider and Stout 2021). Several preclinical and clinical studies have tested the potential of creatine to treat degenerative disorders (Marques and Wyse 2019). However, although in vitro and in vivo animal models are highly encouraging, most clinical trials fail to reproduce positive results suggesting the prophylactic use as a neuroprotective agent (Marques and Wyse 2019). However, continued research is required to prove its positive outcome in clinical trials. Studies show that short- and long-term supplementation is safe and well-tolerated in healthy individuals and in a number of patient populations (Hall et al. 2021). Creatine maintains a balance between creatine utilization and substitution essential in maintaining normal cellular energy flow (Kazak and Cohen 2020). Creatine has a robust evidence of its role as a supplement beneficial to brain health relating to cognitive processing and recovery from traumatic injury evident in neurological conditions such as mild traumatic brain injury, aging, Alzheimer's disease, and depression (Roschel et al. 2021; Kreider and Stout 2021; Forbes et al. 2022a). Creatine improved prolonged cognitive performance in healthy participants (Van Cutsem et al. 2020). Neuroregeneration is a high energy-demanding process, regenerating axons face an energy crisis due to injury-induced mitochondrial dysfunction (Han et al. 2020). A study conducted to assess the role of creatine in boosting corticospinal tract regeneration in Snph−/− mice deduced that creatine supplementation reversed spinal cord injury in Snph−/− mice through enhancing mitochondrial transport of creatine to induce the production of energy required for axonal mitochondrial repair function and energy production metabolism that promotes axonal sprouting and regeneration after spinal cord injury (Han et al. 2020). Creatine administration in lipopolysaccharide-induced neurocognitive impairment in Wistar helps upregulate the mammalian target of rapamycin (mTOR) signaling, synapsin, and postsynaptic density protein within the dentate gyrus, resulting in an improved learning and memory (Mao et al. 2021). A single dose of creatinine supplementation reversed corticosterone-induced depressive-like behavior by activating the

PI3K/Akt/mTORC1 pathway and increasing PSD95 levels in the hippocampus (Escudero-Martinez et al. 2019). Investigative studies on the efficacy of creatine monohydrate supplementation for cognitive functions are gradually increasing in human clinical studies (Dolan et al. 2019; Roschel et al. 2021; Forbes et al. 2022a). An increased intracellular creatine content helps increase bioenergetics and neuroprotection (Chang and Leem 2023). Dietary supplements like creatine have helped reduce the risk of stroke in the aging population by reducing excitotoxicity mediated by ischemic brain injury (Escudero-Martinez et al. 2019).

9.3 TBI /Understanding Traumatic Brain Injury

9.3.1 Prevalence of TBI, Types, and Clinical Manifestation

Traumatic brain injury (TBI) is due to traumatic assault on brain integrity from an external mechanical force. This can lead to a temporary or permanent impairment (Capizzi et al. 2020). A demography report on TBI conducted in 2019 showed that TBI had an incidence case of 27.16 million and 48.99 million prevalent cases (Guan et al. 2023). The overall burden of traumatic brain injury (TBI) is three times greater in low- to middle-income nations than in high-income nations (Abio et al. 2021). Traffic accidents were estimated to be the cause of 56% of TBIs in lower-income nations and 25% in high-income ones (Abio et al. 2021; Maas et al. 2022; Clark et al. 2022). Traumatic brain injury (TBI) remains one of the leading causes of morbidity and mortality globally (Ng and Lee 2019). Traumatic brain injury can be classified into primary and secondary injury: primary traumatic injury occurs at the moment of the initial trauma and is characterized by contusions, concussions, laceration, skull fracture, and diffuse neuronal damage, while secondary traumatic injury is characterized by meningitis, cerebral edema, cerebral ischemia, increased intracranial pressure, increased or decreased blood pressure, and distortion of neurotransmitters (Capizzi et al. 2020; Mao 2023). Traumatic brain injury can occur in three forms based on its clinical manifestation such as mild, moderate, and severe (Rodríguez-Triviño et al. 2019). Mild TBI is the most prevalent form of TBI characterized by a variety of diffuse, nonspecific symptoms such as headache, nausea, fatigue, dizziness, hypersomnolence, attention deficit disorder, photosensitivity, photosensitivity, irritability, and depersonalization, while moderate-to-severe traumatic brain injury (TBI) is the leading cause of injury-related death and disability (Rodríguez-Triviño et al. 2019; Van Gils et al. 2020). TBI has a long-term impact, especially for those with moderate-to-severe injuries. Patients of all ages frequently express a heterogeneous array of symptoms that can progress and activate numerous pathophysiological pathways that lead to long-term impairments in motor and cognitive function through excitotoxicity, apoptosis, inflammatory events, seizures, demyelination, white matter disease, and neurodegeneration (Haarbauer-Krupa et al. 2021). The global prevalence, health impact, physical challenges, cognitive impairment, and emotional consequences of traumatic brain injury require effective and multifaceted treatment strategies to reduce the burden of the disorder (Li et al. 2021a). TBI causes mitochondrial dysfunction, neuropsychological burden, and deficits in cognitive performance as a result of abysmal brain creatine levels, low brain adenosine triphosphate (ATP) levels, glutamate toxicity, and oxidative stress (Newman et al. 2023). Creatine kinase BB (CKBB), which is an isoenzyme of creatine kinase is mainly found in the central nervous system and hence used as a diagnostic biomarker for traumatic brain injury (Forouzan et al. 2023). Traumatic brain injury (TBI) can trigger various forms of neurological disorder like anxiety, depression, seizures, epilepsy, stroke, post-traumatic stress disorder (PTSD), and post-concussive symptoms as a result of poor energy metabolism attributed to loss of neuronal integrity in the prefrontal cortex inclusive of abnormal neuroplasticity (Fordington and Manford 2020; Galovic et al. 2021; Robayo et al. 2023).

9.3.2 Pathophysiology of TBI

Hypoxic ischemia in TBI causes energy failure attributed to the reduction of cerebral blood flow, which causes marked oxygen and glucose deprivation this affects the normal osmotic gradients within the neuronal cells resulting in depolarization and excessive release of glutamate, which causes excitotoxicity-mediated intracellular influx of calcium that triggers apoptosis and necrotic pathways, induced mitochondrial dysfunction resulting in the depletion of ATP production and the overproduction of reactive oxidative species (ROS), neuronal cell death and cognitive impairment (Lu et al. 2015; Greco et al. 2020). Ischemia activates pro-inflammatory responses such as microglia, dendritic cells, macrophages, and lymphocytes and release of pro-inflammatory cytokines such as tumor necrosis factor α (TNF-α) and several interleukins (e.g., IL-1β, IL-6), resulting in the breakdown of the blood-brain barrier (BBB); this allows the infiltration of immune cells into the brain parenchyma resulting in edema and tissue degeneration (Greco et al. 2020). A significant pathological feature of TBI is glial activation which results in the development of chronic neuroinflammation (Mira et al. 2020; Lira et al. 2021; Yang et al. 2019). The hallmark of TBIs includes low energy supply due to low cerebral energy and cerebral blood flow (Vedung et al. 2022) resulting in secondary CNS damage such as Wallerian degeneration of axons, oxidative stress, and apoptotic cell death of neurons and astrogliosis (Ng and Lee 2019). TBI declines brain creatine content (Dolan et al. 2019; Forbes et al. 2022a). Hence, TBI patients experience mitochondrial dysfunction, neuropsychological stress, and deficits in cognitive performance due to low brain creatine and ATP along with glutamate toxicity and oxidative stress (Newman et al. 2023). TBI results in the release of excitatory amino acid glutamate which hyperactivates the N-methyl-d-aspartate (NMDA) receptor, leading to an excitotoxic reaction in neurons such as an increased cellular calcium (Ca^{2+}), mitochondrial dysfunction resulting in neuronal cell death and demyelination (Masse et al. 2021; Forbes et al. 2022a).

9.3.3 Consequences of TBI and Need for Effective TBI Treatments

Traumatic brain injury (TBI) results in substantial neurological disabilities and mental distress arising from its being a leading cause of mortality (Pavlovic et al. 2019). TBI patients or survivors are prone to various mental health problems such as sensorimotor, behavioral, and neuropsychological deficits (Ng and Lee 2019). TBI is one of the leading causes of disability (Ng and Lee 2019; Pavlovic et al. 2019; Capizzi et al. 2020). Recent studies have also shown that at the time of diagnosis, children and adults with mild TBIs experience difficulties with visual smooth pursuit, saccades, vestibular-ocular reflex, convergence, and gait (Denby et al. 2020). Other effects of this injury includes dementia, stroke, post-traumatic epilepsy, chronic traumatic encephalopathy, psychiatric illness, motor disorders, and death (LoBue and Cullum 2019). Mild traumatic brain injury (MTBI) result is in various forms of neurological, cognitive, and behavioral symptoms and can lead to increased intracranial pressure as well as postconcussion syndrome in the long term (Forouzan et al. 2023). TBI pathophysiology can trigger other health conditions such as diabetic peripheral neuropathy that can lead to impaired mobility (Li et al. 2021b), hence the need for an effective therapy for precise treatment of the broad spectrum of health challenges aggravated by TBI (Li et al. 2021b). A nutritional diet effective in injury prevention and improving recovery (Papadopoulou 2020) can be used to avert TBI pathophysiology and improve the quality of life of TBI patients or survivors. Although some reports on creatine said it is largely ineffective to tackle neurodegenerative diseases (Ostojic 2021b), other studies on creatine have shown its neuroprotective ability to alleviate the consequences of traumatic brain injury (Newman et al. 2023). One consequence of TBI to delve into in this chapter is spontaneous epileptic seizures (Thijs et al. 2019; Balestrino and Adriano 2019) which occur due to a short sudden excessive discharge of electrical activity through abnormal action potential in neurons resulting in loss of consciousness, abnormal movements, and aberrant perceptions of events following the seizure (Falco-Walter 2020). TBI-medicated seizures are postulated to arise due to excessive electrical activity which culminates in increased synthesis of ROS, resulting in mitochondrial dysfunction leading to poor cellular energy supply for cells, triggering more ROS generation and abnormal calcium homeostasis, generating free radicals from the mitochondrial due to mitochondrial injury or assault (Yang et al. 2020).

9.4 Creatine as a Potential Treatment for TBI

Creatine has promising efficacy in protecting the neurons and supplying energy demand during oxidative-stress-mediated cognitive deficit following brain injury (Newman et al. 2023). Creatine supplementation is effective in endogenous creatine deficiency due to inborn errors in CRT genes responsible for creatine synthesis in creatine transporter syndrome or disease (Balestrino and Adriano 2019). Furthermore, creatine is gaining interest in clinical trials whereby creatine is administered in neurological conditions attributed to energy depletion, oxidative stress, and mitochondrial dysfunction (Alraddadi et al. 2023). A number of preclinical (animal model) and human clinical trial studies using the exogenous, commercially available form of creatine known as creatine monohydrate (CM) given orally have been conducted to evaluate its efficacy and mechanism of absorption from the intestinal mucosa into the blood through creatine transporters (CRT1) into various tissues (Jomura et al. 2022). Creatine supplementation through various studies has the potential to modulate various neurological dysfunctions following TBI (Alraddadi et al. 2023) and creatine efficiency to improve cognitive function and prevent loss of cell loss neurons such as GABAergic neurons associated with TBI after a week of discontinuation of creatine supplementation (Gerbatin et al. 2019). Creatine supplementation increases creatine content in the brain which results in the alleviation of TBI clinical manifestations such as concussion and cognitive impairment by facilitating protein translation for neuron healing post injury through an increase in neural transcription factors, glial cell activity, insulin-like growth factor 1, and protein kinases downstream in the mammalian target of rapamycin (mTOR) (Candow et al. 2019, 2021). Through investigative reports from clinical and animal models of TBI, we delve into the reported potential mechanism through which creatine averts neurological deficits in TBI.

9.4.1 Mechanisms by Which Creatine May Benefit TBI Patients

9.4.1.1 Neuroprotective Effects of Creatine

Creatine has emerged as a promising neuroprotective agent in the adult, developing brain (Tran et al. 2021) and in a number of animal models of neurodegenerative diseases such as Alzheimer's disease (AD), Parkinson's disease (PD), TBI, and seizures, among others (Forbes et al. 2022a; Alraddadi et al. 2023). The neuroprotective mechanism is via its potent ability to provide cellular energy production (Kreider and Stout 2021) and act as an antioxidant (Ostojic 2021a) to fight against physiologic mechanism for the progression of oxidative tissue damage in TBI (Newman et al. 2023). The proposed neuroprotective therapeutic intervention of creatine supplementation in TBI (Forbes et al. 2022a) is through an elevation of brain creatine level which helps provide energy, reduce mitochondrial dysfunction and hyperproduction of ROS (Dolan et al. 2019) that alters neuropathophysiology and neurometabolic activities in TBI (Ainsley Dean et al. 2017) leading to progressive neuron recovery following TBIs (Dolan et al. 2019; Candow et al. 2023). Animal models provide promising evidence that creatine supplementation prior to inducing TBIs may be neuroprotective (Sullivan et al. 2000). Sullivan's experiment on mice given 3 mg/g per day of creatine for 3–5 days before a controlled cortical contusion had a reduction in cortical damage as compared to the control this demonstrates the neuroprotective potential of creatine. Also, Sullivan further conducted a study to evaluate creatinine neuroregeneration potential by administration of 1% of dietary creatine for 1-month post-experimental TBI. This study shows that dietary creatine ameliorates cortical damage hence creatine neuroprotective and neuroregenerative effects of creatine against ischemic and oxidative insults in TBI is proposed to be due to the maintenance of mitochondrial homeostasis and the ability of the phosphocreatine shuttle to provide energy (ATP) buffer thereby reducing free radical production and intramitochondrial calcium (Ca^{2+}) generation due to a decline in energy supply (Masse et al. 2021; Forbes et al. 2022a).

9.4.1.2 Anti-Inflammatory Effects of Creatine

Creatine as a dietary supplement has anti-inflammatory properties by reducing inflammation (Candow et al. 2019; Bredahl et al. 2021; Ostojic 2021b), resulting in a decline in oxidative DNA damage and lipid peroxidation culminating in a significant increase in muscle performance or neural activity (Rahimi 2011; Candow et al. 2019). Creatine transporter for creatine uptake is immunoregulatory (Di Biase et al. 2019) whereby it regulates the activation of macrophages (Ji et al. 2019) or modulates CD8 T cell response to antigen exposure (Di Biase et al. 2019). Creatine dietary interventions reduce pro-inflammatory cascade reactions triggered by oxidative stress (Escudero-Martinez et al. 2019; Cella et al. 2020).

9.4.1.3 Antioxidant Role of Creatine

Oxidative stress occurs due to an imbalance between reactive oxygen species (ROS) and antioxidant enzyme status or mechanisms whereby ROS overpowers antioxidants resulting in the degradation of biogenic compounds such as deoxyribonucleic acid, lipids, and proteins, leading to oxidative tissue damage in active cells like neurons or muscle cells (Arazi et al. 2021). Oxidative stress is associated with neurodegenerative diseases such as Alzheimer's disease, Parkinson's disease, Huntington's disease (Singh et al. 2019), and TBI (Dennis et al. 2022). The consumption of antioxidants in the form of dietary supplements has promising effects to avert or ameliorate oxidative stress-related diseases (Losada-Barreiro et al. 2022). The potential therapeutic role of creatine in health has been proven as a result of its antioxidant properties (Kreider and Stout 2021; Balestrino and Adriano 2019; Clarke et al. 2021; Tsvetkova et al. 2023). The use of antioxidant dietary supplements such as creatine has been reported to have a positive impact on the antioxidant system this is linked to creatine's amino acids, arginine, methionine, and glycine having potent antioxidant roles in tissues by reducing oxidative stress (Quan et al. 2020; Chaouachi et al. 2022) in metabolic diseases (Singh et al. 2019). One mechanism postulated to explain the antioxidant properties of creatine is linked to arginine's ability to oxidatively disintegrate nitric oxide (NO) into stable inorganic nitrogen oxides such as nitrite (NO2) and nitrate (NO3) excreted and detectable in plasma and urine (Clarke et al. 2021; Quan et al. 2020). According to Arazi et al. (2021), the exact mechanism of creatine as an antioxidant is not known; however, creatine has demonstrated its ability to increase the activity of antioxidant enzymes to eliminate ROS and reactive nitrogen species (RNS); this is done through an indirect mechanism to increase cell energy status or directly by maintaining mitochondrial integrity. Creatine enters the mitochondria to couple with ATP and the adenine nucleotide translocator, then diffuses back into the cytosol via the phosphate creatine shuttle to meet cellular energy needs; this coupling by creatine helps reduce the formation of reactive oxygen species (ROS) and therefore acts as a direct and/or indirect antioxidant (Borchel et al. 2019; Marques and Wyse 2019; Hall et al. 2021).

9.4.1.4 Potential Neuroregenerative Properties of Creatine in TBI

As TBI neuropathology progresses it could trigger nerve injury, apoptosis, and necrotic pathways resulting in neuronal cell death and cognitive impairment through mitochondrial dysfunction, depletion in ATP synthesis, and hyperproduction of reactive oxidative species (ROS) (Greco et al. 2020). However, creatine as a bioenergetic amino acid can mediate nerve and neuron regeneration. This was deduced by Han et al. (2020) using the Snph deletion model in mice where creatine administration in Snph−/− mice boosts corticospinal tract regeneration (Han et al. 2020) through an enhanced mitochondrial transport and cellular bioenergetics to recover axonal mitochondrial integrity, increased energy metabolism that promotes axonal sprouting and regeneration of injured corticospinal tract, removal of damaged mitochondria and synaptic connection following CNS injury to restore cognitive functions (Han et al. 2020). Studies have investigated the protective and regenerative properties of dietary creatine on brain health this shows that creatine improves cognition and enhances hippocampal-dependent spatial memory (Smolarek et al. 2020; Gualano et al. 2021). Creatine supplementation reverses nerve or neuron injury in a cellular target manner which stimulates neuron/nerve regeneration and function after CNS injuries through AMPK signaling in the brain that increases brain creatine kinase expression in cortical neurons, prevents axonal damage, enhances myelination, and provides energy for maturation of oligodendrocytes (Rosko et al. 2023). Another evidence on creatine and nerve injury proposed that creatine regulates AMPK signaling in the brain required for endogenous release of creatine through the creatine/phosphocreatine system this results in buffering cellular energy to restore cellular ATP without depending on oxygen (Hall et al. 2021; Borchel et al. 2019; Marques and Wyse 2019).

9.4.1.5 Creatine Helps Improve Neurocognition in TBI

Creatine through its neuroprotective and neuroregenerative ability has resulted in improved cognition through enhanced hippocampal-dependent spatial memory (Smolarek et al. 2020; Gualano et al. 2021) in neural rehabilitation studies. Ion reduces hypoxic brain injury by sustaining intracellular ATP levels, reducing oxidative and metabolic stress responses during oxygen deprivation (Tran et al. 2022). Creatine supplementation increases intracellular energy consumption which improves cognitive performance and reverses metabolic alterations and neuron damage linked with neural and cognitive deterioration (Gordji-Nejad et al. 2024) this demonstrates creatine's neuroprotective role (Tran et al. 2021). Creatine, phosphate creatine, and creatine kinase activities in neurodegenerative conditions decrease (Cunnane et al. 2020); however, creatine supplementation mediates energy shuttle within the neuron milieu, resulting in neuroregeneration and enhanced cognition in the brain (Marques and Wyse 2019). There are increasing reports from preclinical and clinical trials on the neuroprotective role of creatine supplementation efficacy on preclinical neurodegenerative diseases, including Huntington's disease (HD), Alzheimer's disease (AD), stroke, seizures, Parkinson's disease (PD), and amyotrophic lateral sclerosis (ALS) through beneficial mechanism of action attributed to mitochondrial function, oxidative stress, and energy production (Alraddadi et al. 2023).

9.4.1.6 Creatine Acts as a Neurotransmitter

It has been postulated that creatine acts as a possible neurotransmitter in the central nervous system (Bian et al. 2022); this is associated with astrocytes that contain high levels of an enzyme called GAMT, which is involved in the production of creatine. Astrocytic creatine serves as a neuromodulatory agent a study conducted by Bian et al. 2022 reported that AGAT and SLC6A8 are found in different cells in the brain, and upon stimulation of the brain tissue, creatine was released from stimulated coronal brain slices, but creatine release reduced in brain slices lacking the gene that codes for an enzyme called AGAT and the SLC6A8 creatine transporter (Alessandrì et al. 2020; Bian et al. 2022).

9.4.2 Studies on the Neuroprotective Effects of Creatine in TBI Animal Model and Clinical Trials

Creatine supplementation increases brain creatine content to improve cognition and memory in human and animal studies, thereby alleviating some symptoms of TBI such as concussion, depression, anxiety, and neuron injury (Candow et al. 2023). Creatine supplementation helps to alter brain pathophysiology and neurometabolic events induced by TBIs in animal model (Ainsley et al. 2017; Forbes et al. 2022b) by providing energy and increasing creatine content (Vedung et al. 2022), thereby reversing the TBI-mediated decline in brain creatine content (Forbes et al. 2022b). Exogenous creatine supplementation has been found to have neuroprotective and regeneration roles which enhance neuron recovery in TBIs (Dolan et al. 2019). Mice given 3 mg/g/day of creatine supplement showed a decline in cortical neuron damage. Additionally, a 1% dietary creatine supplement for a month post-TBI in an experimental animal model also resulted in a decline in cortical neuron damage (Sullivan et al. 2000). Creatine helps to prevent concussion and TBI-mediated release of glutamate, which mediates an excitotoxic influx of calcium (Ca^{2+}), resulting in mitochondrial dysfunction and loss of neural function (Masse et al. 2021). Creatine (300 mg/kg) helps to protect against TBI-mediated oxidative. Beyond animal studies, a growing number of human clinical studies have investigated creatine monohydrate supplementation on cognitive function and TBI (Dolan et al. 2019; Roschel et al. 2021). In an open-label randomized controlled trial for creatine supplementation (0.4 g of creatine/kg/day) in patients with TBI for 6 months, the results showed that creatine supplementation reduced the duration of post-traumatic amnesia, reduced post-traumatic headaches (Leung 2020), dizziness, and fatigue, and improved disability, recovery, self-care, communication, locomotion, sociability, personality, and behavior, as well as neurophysical and cognitive function (Borchio et al. 2020; Antonio et al. 2021; Chaychi et al. 2022). Traumatic brain injury (TBI) often results in behavioral disabilities, including post-traumatic epilepsy (PTE). A study using an animal model of TBI with fluid percussion injury and tonic-clonic seizures induced by a sub-convulsant dose of PTZ (35 mg/kg, i.p) found that administering 300 mg/kg of creatine for 4 weeks results in significant outcomes. The results showed that creatine supplementation helped to reduce seizure intensity and epileptiform discharges through a creatine-mediated decline in neuronal cell loss, number of parvalbumin positive (PARV+) cells in CA3 region of the hippocampus, and prevention of the decline in GAD67 levels and GAD activity in the hippocampus. These findings suggest a neuroprotective role of creatine supplementation that helped in controlling the GABAergic function against the hyperactivity of glutamate post-TBI (Gerbatin et al. 2019). This creatine-mediated neuroprotective effect persists for up to 1 week even after creatine supplementation has been discontinued (Gerbatin et al. 2019). Recurrent subconcussive injuries are one type of traumatic brain injury that can lead to neurologic damage. Traumatic brain injury (TBI) causes dysregulation of ions (Na^+, K^+, and Ca^{2+}), neuroinflammation, cellular and axonal destruction, which are exacerbated by oxidative stress and mitochondrial dysfunction caused by the influx of calcium (Ca^{2+}) (Smith-Ryan et al. 2020; Fesharaki-Zadeh 2022). Currently, there are 12 research studies examining the effects of creatine supplementation on brain creatine or PCr concentrations, as thoroughly reviewed by Dolan et al. (2019). Nine of these studies demonstrated a statistically significant increase in brain creatine, averaging between 5% and 10%. Changes in PCr/Pi and ATP can be mediated by creatine, which also enhances cognitive function in fatigue-induced cognitive decline (Mohan MarugaRaja et al. 2022; Gordji-Nejad et al. 2024). The administration of creatine has been shown to lessen the effects of brain trauma. In rodents, a combination of omega fatty acid and creatine therapy has proven significant in repairing axonal depletion, reducing cellular death, and decreasing inflammation. Omega-3 fatty acids and creatine monohydrate supplements may be beneficial for post-TBI symptoms, as they can help to restore adenosine triphosphate storage, alleviate the detrimental impact on energy levels, and enhance cognitive performance, thereby reducing TBI symptoms (Dolan et al. 2019). GABAergic neurons linked to traumatic brain injury showed reduced cell loss with creatine supplementation (Gerbatin et al. 2019). The ability of creatine to decrease cell loss, including the quantity of parvalbumin positive (PARV+) cells in the CA3 region of the hippocampus, has been linked to its anticonvulsant action (Alraddadi at al. 2023). Similarly, supplementing with creatine also appeared to guard against the decline in GAD activity, GAD67 levels, and specific [3H] flunitrazepam binding in the hippocampal regions (Gerbatin et al. 2019; Royes at al. 2023). Creatine ameliorates the severity of seizures, anxiety, and depressive-like behaviors associated with TBI in animal model (Okwuofu et al. 2021). Evidence-based studies demonstrate that supplementing with creatine is generally well-tolerated and effective, particularly when taken at the recommended dosages (i.e., 3–5 g/day or 0.1 g/kg of body mass/day) (Antonio et al. 2021). There is no risk or negative impact associated with creatine supplementation following traumatic brain injury, according to data from numerous studies (Bødker and Marcussen 2023; Antonio et al. 2021). However, further extensive study is needed to examine the adverse effects of varying dosages of creatine supplementation.

9.5 Conclusion

Nutritional intervention such as creatine supplementation in traumatic brain injury (TBI) is a promising therapeutics to improve patient recovery based on creatine neuroprotective properties to inhibit the progression of TBI neuropathological detrimental effects through energy restoration, reduction of oxidative stress, and stabilization of mitochondrial function which results in clinical research providing promising evidence of creatine supplementation role in neuroregeneration, improved cognitive function, and physical recovery in neurorehabilitation of TBI patients. Hence, the exploration of the synergistic effects of creatine with other nutritional interventions could offer a more comprehensive approach to TBI management, thereby resulting in a transition from bench to bedside in the application of creatine for TBI patients a better strategy to support TBI recovery.

References

Abio A, Bovet P, Valentin B, Bärnighausen T, Shaikh MA, Posti JP, Lowery Wilson M (2021) Changes in mortality related to traumatic brain injuries in the Seychelles from 1989 to 2018. Front Neurol 12:720434

Abod L, Ilnicka N, Matyja D, Sadlik M, Zuziak P (2023) Supplementation of creatine and its role in brain function. J Educ Health Sport 15(1):175–185

Ainsley Dean PJ, Arikan G, Opitz B, Sterr A (2017) Potential for use of creatine supplementation following mild traumatic brain injury. Concussion 2:CNC34. https://doi.org/10.2217/cnc-2016-0016

Alessandrì MG, Strigini F, Cioni G, Battini R (2020) Increased creatine demand during pregnancy in Arginine: Glycine Amidino-Transferase deficiency: a case report. BMC Pregnancy Childbirth 20(1):506. https://doi.org/10.1186/s12884-020-03192-4

Alraddadi EA, Khojah AM, Alamri FF, Kecheck HK, Altaf WF, Khouqeer Y (2023) Potential role of creatine as an anticonvulsant agent: evidence from preclinical studies. Front Neurosci 17:1201971. https://doi.org/10.3389/fnins.2023.1201971

Antonio J, Candow DG, Forbes SC, Gualano B, Jagim AR, Kreider RB, Rawson ES, Smith-Ryan AE, VanDusseldorp TA, Willoughby DS, Ziegenfuss TN (2021) Common questions and misconceptions about creatine supplementation: what does the scientific evidence really show? J Int Soc Sports Nutr 18(1):13. https://doi.org/10.1186/s12970-021-00412-w

Arazi H, Eghbali E, Suzuki K (2021) Creatine supplementation, physical exercise and oxidative stress markers: a review of the mechanisms and effectiveness. Nutrients 13(3):1–17. Article 869. https://doi.org/10.3390/nu13030869

Balestrino M, Adriano E (2019) Beyond sports: efficacy and safety of creatine supplementation in pathological or paraphysiological conditions of brain and muscle. Med Res Rev 39(6):2427–2459. https://doi.org/10.1002/med.21590

Barcelos RP, Stefanello ST, Mauriz JL, Gonzalez-Gallego J, Soares FA (2016) Creatine and the liver: metabolism and possible interactions. Mini Rev Med Chem 16(1):12–18. https://doi.org/10.2174/1389557515666150722102613

Bian X, Zhu J, Jia X, Liang W, Li Z, Yu S, Rao Y (2022) Evidence suggesting creatine as a new central neurotransmitter: presence in synaptic vesicles, release upon stimulation, effects on cortical neurons and uptake into synaptosomes and synaptic vesicles. bioRxiv 2022-12

Bødker RL, Marcussen M (2023) Pilot study protocol of a randomized controlled trial for the potential effects of creatine monohydrate on persistent post-concussive symptoms. Front Neurol 14:1209548

Bonilla DA, Kreider RB, Stout JR, Forero DA, Kerksick CM, Roberts MD, Rawson ES (2021) Metabolic basis of creatine in health and disease: a bioinformatics-assisted review. Nutrients 13(4):1238. https://doi.org/10.3390/nu13041238

Borchel A, Verleih M, Kühn C, Rebl A, Goldammer T (2019) Evolutionary expression differences of creatine synthesis-related genes: implications for skeletal muscle metabolism in fish. Sci Rep 9(1):5429. https://doi.org/10.1038/s41598-019-41907-6

Borchio L, Machek SB, Machado M (2020) Supplemental creatine monohydrate loading improves cognitive function in experienced mountain bikers. J Sports Med Phys Fitness 60(8):1168–1170. https://doi.org/10.23736/S0022-4707.20.10589-9

Bredahl EC, Eckerson JM, Tracy SM, McDonald TL, Drescher KM (2021) The role of creatine in the development and activation of immune responses. Nutrients 13(3):751. https://doi.org/10.3390/nu13030751

Brosnan ME, Brosnan JT (2016) The role of dietary creatine. Amino Acids 48:1785–1791. https://doi.org/10.1007/s00726-016-2188-1

Brosnan JT, da Silva RP, Brosnan ME (2011) The metabolic burden of creatine synthesis. Amino Acids 40(5):1325–1331. https://doi.org/10.1007/s00726-011-0853-y

Candow DG, Forbes SC, Chilibeck PD, Cornish SM, Antonio J, Kreider RB (2019) Effectiveness of creatine supplementation on aging muscle and bone: focus on falls prevention and inflammation. J Clin Med 8(4):488. https://doi.org/10.3390/jcm8040488

Candow DG, Forbes SC, Kirk B, Duque G (2021) Current evidence and possible future applications of creatine supplementation for older adults. Nutrients 13:745. https://doi.org/10.3390/nu13030745

Candow DG, Forbes SC, Ostojic SM, Prokopidis K, Stock MS, Harmon KK, Faulkner P (2023) "Heads up" for creatine supplementation and its potential applications for brain health and function. Sports Med 53(Suppl 1):49–65. https://doi.org/10.1007/s40279-023-01870-9

Capizzi A, Woo J, Verduzco-Gutierrez M (2020) Traumatic brain injury: an overview of epidemiology, pathophysiology, and medical management. Med Clin North Am 104(2):213–238. https://doi.org/10.1016/j.mcna.2019.11.001

Cella PS, Marinello PC, Borges FH, Ribeiro DF, Chimin P, Testa MTJ, Guirro PB, Duarte JA, Cecchini R, Guarnier FA, Deminice R (2020) Creatine supplementation in Walker-256 tumor-bearing rats prevents skeletal muscle atrophy by attenuating systemic inflammation and protein degradation signaling. Eur J Nutr 59:661–669

Chang H, Leem YH (2023) The potential role of creatine supplementation in neurodegenerative diseases. Phys Act Nutr 27(4):48–54. https://doi.org/10.20463/pan.2023.0037

Chaouachi M, Gautier S, Carnot Y, Guillemot P, Pincemail J, Moison Y, Collin T, Groussard C, Vincent S (2022) Spirulina supplementation prevents exercise-induced lipid peroxidation, inflammation and skeletal muscle damage in elite rugby players. J Hum Nutr Diet 35(6):1151–1163. https://doi.org/10.1111/jhn.13014

Chaychi S, Valera E, Tartaglia MC (2022) Sex and gender differences in mild traumatic brain injury/concussion. Int Rev Neurobiol 164:349–375. https://doi.org/10.1016/bs.irn.2022.07.004

Clark D, Joannides A, Adeleye AO, Bajamal AH, Bashford T, Biluts H, Budohoski K, Ercole A, Trivedi R, Devi BI, Servadei F, Hutchinson P, Global Neurotrauma Outcomes Study collaborative (2022) Casemix, management, and mortality of patients rreseceiving emergency neurosurgery for traumatic brain injury in the Global Neurotrauma Outcomes Study: a prospective observational cohort study. Lancet Neurol 21(5):438–449. https://doi.org/10.1016/S1474-4422(22)00037-0. Epub 2022 Mar 16. PMID: 35305318

Clarke H, Kim DH, Meza CA, Ormsbee MJ, Hickner RC (2020) The evolving applications of creatine supplementation: could creatine improve vascular health? Nutrients 12(9):2834. https://doi.org/10.3390/nu12092834

Clarke H, Hickner RC, Ormsbee MJ (2021) The potential role of creatine in vascular health. Nutrients 13(3):857

Cunnane SC et al (2020) Brain energy rescue: an emerging therapeutic concept for neurodegenerative disorders of ageing. Nat Rev Drug Discov 19(9):609–633. https://doi.org/10.1038/s41573-020-0072-x

da Silva RP, Clow K, Brosnan JT, Brosnan ME (2014) Synthesis of guanidinoacetate and creatine from amino acids by rat pancreas. Br J Nutr 111(4):571–577. https://doi.org/10.1017/S0007114513003012

Denby E, Murphy D, Busuttil W, Sakel M, Wilkinson D (2020) Neuropsychiatric outcomes in UK military veterans with mild traumatic brain injury and vestibular dysfunction. J Head Trauma Rehabil 35:57–65

Dennis D, Memudu AE, Abel S, Adebayo G (2022) Honey-induced elevation of superoxide dismutase, declined lipid peroxidation and attenuates neurodegeneration in weight drop model of traumatic brain injury in adult wistar rats. Acta Sci Anat 1(5):10–19

Di Biase S, Ma X, Wang X, Yu J, Wang Y, Smith DJ et al (2019) Creatine uptake regulates CD8 T cell antitumor immunity. J Exp Med 216:2869–2882. https://doi.org/10.1084/jem.20182044

Dolan E, Gualano B, Rawson ES (2019) Beyond muscle: the effects of creatine supplementation on brain creatine, cognitive processing, and traumatic brain injury. Eur J Sport Sci 19(1):1–14. https://doi.org/10.1080/17461391.2018.1500644

Dwir D, Giangreco B, Xin L, Tenenbaum L, Cabungcal J-H, Steullet P et al (2019) MMP9/RAGE pathway overactivation mediates redox dysregulation and neuroinflammation, leading to inhibitory/excitatory imbalance: a reverse translation study in schizophrenia patients. Mol Psychiatry 25:2889. https://doi.org/10.1038/s41380-019-0393-5

Escudero-Martinez I, Mancha F, Vega-Salvatierra A, Ayuso MI, Ocete RF, Algaba P, Lopez-Rueda A, Pinero P, Fajardo E, Fernandez-Engo JR et al (2019) Mediterranean diet and physical activity protect from silent brain infarcts in a cohort of patients with atrial fibrillation. J Stroke 21:353–355. https://doi.org/10.5853/jos.2019.01949

Falco-Walter J (2020) Epilepsy-definition, classification, pathophysiology, and epidemiology. Semin Neurol 40:617–623. https://doi.org/10.1055/s-0040-1718719

Farr CV, El-Kasaby A, Erdem FA, Sucic S, Freissmuth M, Sandtner W (2022) Cooperative binding of substrate and ions drives forward cycling of the human creatine transporter-1. Front Physiol 13:919439. https://doi.org/10.3389/fphys.2022.919439

Fesharaki-Zadeh A (2022) Oxidative stress in traumatic brain injury. Int J Mol Sci 23(21):13000. https://doi.org/10.3390/ijms232113000

Forbes SC, Candow DG, Ostojic SM, Roberts MD, Chilibeck PD (2021) Meta-analysis examining the importance of creatine ingestion strategies on lean tissue mass and strength in older adults. Nutrients 13:1912. https://doi.org/10.3390/nu13061912

Forbes SC, Cordingley DM, Cornish SM, Gualano B, Roschel H, Ostojic SM, Rawson ES, Roy BD, Prokopidis K, Giannos P, Candow DG (2022a) Effects of creatine supplementation on brain function and health. Nutrients 14(5):921. https://doi.org/10.3390/nu14050921

Forbes SC, Candow D, Ferreira L, Souza-Junior T (2022b) Effects of creatine supplementation on properties of muscle, bone, and brain function in older adults: a narrative review. J Diet Suppl 19(3):318–335. https://doi.org/10.1080/19390211.2021.1877232

Fordington S, Manford M (2020) A review of seizures and epilepsy following traumatic brain injury. J Neurol 267:3105–3111. https://doi.org/10.1007/s00415-020-09926-w

Forouzan A, Fahimi MA, Bastan ASI, Delirrooyfard A (2023) Diagnostic competence of creatine kinase BB, in mild traumatic brain injury and its prognostic value. Adv Biomed Res 12:84. https://doi.org/10.4103/abr.abr_122_21

Galovic M, Ferreira-Atuesta C, Abraira L, Dohler N, Sinka L, Brigo F et al (2021) Seizures and epilepsy after stroke: epidemiology biomarkers and management. Drugs Aging 38:285–299. https://doi.org/10.1007/s40266-021-00837-7

Gerbatin RR, Silva LFA, Hoffmann MS, Della-Pace ID, do Nascimento PS, Kegler A, de Zorzi VN, Cunha JM, Botelho P, Neto JBT, Furian AF, Oliveira MS, Fighera MR, Royes LFF (2019) Delayed creatine supplementation counteracts reduction of GABAergic function and protects against seizures susceptibility after traumatic brain injury in rats. Prog Neuro-Psychopharmacol Biol Psychiatry 92:328–338. https://doi.org/10.1016/j.pnpbp.2019.02.004

Gordji-Nejad A, Matusch A, Kleedörfer S, Jayeshkumar Patel H, Drzezga A, Elmenhorst D, Binkofski F, Bauer A (2024) Single dose creatine improves cognitive performance and induces changes in cerebral high energy phosphates during sleep deprivation. Sci Rep 14(1):4937. https://doi.org/10.1038/s41598-024-54249-9

Greco P, Nencini G, Piva I, Scioscia M, Volta CA, Spadaro S, Neri M, Bonaccorsi G, Greco F, Cocco I et al (2020) Pathophysiology of hypoxic-ischemic encephalopathy: a review of the past and a view on the future. Acta Neurol Belg 120:277–288. https://doi.org/10.1007/s13760-020-01308-3

Grimm A, Friedland K, Eckert A (2016) Mitochondrial dysfunction: the missing link between aging and sporadic Alzheimer's disease. Biogerontology 17:281–296

Gualano B, Ostojic SM, Rawson ES (2021) Creatine supplementation and brain health. Nutrients 13(2):586. https://doi.org/10.3390/nu13020586

Guan B, Anderson DB, Chen L, Feng S, Zhou H (2023) Global, regional and national burden of traumatic brain injury and spinal cord injury, 1990–2019: a systematic analysis for the Global Burden of Disease Study 2019. BMJ Open 13(10):e075049

Haarbauer-Krupa J, Pugh MJ, Prager EM, Harmon N, Wolfe J, Yaffe K (2021) Epidemiology of chronic effects of traumatic brain injury. J Neurotrauma 38(23):3235–3247

Hall M, Manetta E, Tupper K (2021) Creatine supplementation: an update. Curr Sports Med Rep 20(7):338–344. https://doi.org/10.1249/JSR.0000000000000863

Han Q, Xie Y, Ordaz JD, Huh AJ, Huang N, Wu W, Liu N, Chamberlain KA, Sheng ZH, Xu XM (2020) Restoring cellular energetics promotes axonal regeneration and functional recovery after spinal cord injury. Cell Metab 31(3):623–641.e8. https://doi.org/10.1016/j.cmet.2020.02.002

Harmon KK, Stout JR, Fukuda DH, Pabian PS, Rawson ES, Stock MS (2021) The application of creatine supplementation in medical rehabilitation. Nutrients 13(6):1825. https://doi.org/10.3390/nu13061825

Hodges SL, Reynolds CD, Smith GD, Jefferson TS, Nolan SO, Lugo JN (2018) Molecular interplay between hyperactive mammalian target of rapamycin signaling and Alzheimer's disease neuropathology in the NS-Pten knockout mouse model. Neuroreport 29(13):1109–1113

Huseyin Fatih G, Dolanbay T, Simsek AT, Aras M (2021) Evaluation of blood urea, creatinine, and glucose levels as biochemical indicators of the type and severity of traumatic brain injury. Turk Neurosurg 31(3):333–338

Jagim AR, Kerksick CM (2021) Creatine supplementation in children and adolescents. Nutrients 13(2):664. https://doi.org/10.3390/nu13020664

Ji L, Zhao X, Zhang B, Kang L, Song W, Zhao B et al (2019) Slc6a8-mediated creatine uptake and accumulation reprogram macrophage polarization *via* regulating cytokine responses. Immunity 51:272.e7–284.e7. https://doi.org/10.1016/j.immuni.2019.06.007

Jomura R, Akanuma SI, Tachikawa M, Hosoya KI (2022) SLC6A and SLC16A family of transporters: contribution to transport of creatine and creatine precursors in creatine biosynthesis and distribution. Biochim Biophys Acta Biomembr 1864:183840. https://doi.org/10.1016/j.bbamem.2021.183840

Kazak L, Cohen P (2020) Creatine metabolism: energy homeostasis, immunity and cancer biology. Nat Rev Endocrinol 16(8):421–436. https://doi.org/10.1038/s41574-020-0365-5

Kerksick CM, Wilborn CD, Roberts MD, Smith-Ryan A, Kleiner SM, Jäger R, Collins R, Cooke M, Davis JN, Galvan E, Greenwood M, Lowery LM, Wildman R, Antonio J, Kreider RB (2018) ISSN exercise & sports nutrition review update: research & recommendations. J Int Soc Sports Nutr 15(1):38. https://doi.org/10.1186/s12970-018-0242-y

Kreider RB, Stout JR (2021) Creatine in health and disease. Nutrients 13(2):447. https://doi.org/10.3390/nu13020447

Kreider RB, Jäger R, Purpura M (2022) Bioavailability, efficacy, safety, and regulatory status of creatine and related compounds: a critical review. Nutrients 14(5):1035. https://doi.org/10.3390/nu14051035

Leung A (2020) Addressing chronic persistent headaches after MTBI as a neuropathic pain state. J Headache Pain 21(1):77. https://doi.org/10.1186/s10194-020-01133-2

Li LM, Dilley MD, Carson A, Twelftree J, Hutchinson PJ, Belli A, Betteridge S, Cooper PN, Griffin CM, Jenkins PO, Liu C, Sharp DJ, Sylvester R, Wilson MH, Turner MS, Greenwood R (2021a) Management of traumatic brain injury (TBI): a clinical neuroscience-led pathway for the NHS. Clin Med (Lond) 21(2):e198–e205. https://doi.org/10.7861/clinmed.2020-0336. PMID: 33762387; PMCID: PMC8002793

Li S, Bianconi S, van der Veen JW, Dang Do A, Stolinski J, Cecil KM et al (2021b) Oxidative phosphorylation in creatine transporter deficiency. NMR Biomed 34:e4419. https://doi.org/10.1002/nbm.4419

Lira M, Zamorano P, Cerpa W (2021) Exo70 intracellular redistribution after repeated mild traumatic brain injury. Biol Res 54:5. https://doi.org/10.1186/s40659-021-00329-3

LoBue C, Cullum CM (2019) POINT/COUNTER-POINT—beyond the headlines: the actual evidence that traumatic brain injury is a risk factor for later-in-life dementia. Arch Clin Neuropsychol 35:123–127

Losada-Barreiro S, Sezgin-Bayindir Z, Paiva-Martins F, Bravo-Díaz C (2022) Biochemistry of antioxidants: mechanisms and pharmaceutical applications. Biomedicines 10(12):3051. https://doi.org/10.3390/biomedicines10123051

Lu Y, Tucker D, Dong Y, Zhao N, Zhuo X, Zhang Q (2015) Role of mitochondria in neonatal hypoxic-ischemic brain injury. J Neurosci Rehabil 2:1–14

Maas AIR, Menon DK, Manley GT, Abrams M, Åkerlund C, Andelic N, Aries M, Bashford T, Bell MJ, Bodien YG, Brett BL, Büki A, Chesnut RM, InTBIR Participants and Investigators (2022) Traumatic brain injury: progress and challenges in prevention, clinical care, and research. Lancet Neurol 21(11):1004–1060. https://doi.org/10.1016/S1474-4422(22)00309-X. Erratum in: Lancet Neurol. 2022 Oct 7

Mao G (2023) Traumatic brain injury (TBI). MSD Manual

Mao X, Kelty TJ, Kerr NR, Childs TE, Roberts MD, Booth FW (2021) Creatine supplementation upregulates mTORC1 signaling and markers of synaptic plasticity in the dentate gyrus while ameliorating LPS-induced cognitive impairment in female rats. Nutrients 13:2758. https://doi.org/10.3390/nu13082758

Marques EP, Wyse ATS (2019) Creatine as a neuroprotector: an actor that can play many parts. Neurotox Res 36(2):411–423. https://doi.org/10.1007/s12640-019-00053-7

Masse I, Moquin L, Bouchard C, Gratton A, De Beaumont L (2021) Efficacy of prophylactic versus therapeutic administration of the NMDA receptor antagonist MK-801 on the acute neurochemical response to a concussion in a rat model combining force and rotation. J Neurosurg 136:1–10

Mira RG, Lira M, Quintanilla RA, Cerpa W (2020) Alcohol consumption during adolescence alters the hippocampal response to traumatic brain injury. Biochem Biophys Res Commun 528:514–519. https://doi.org/10.1016/j.bbrc.2020.05.160

Mohan MarugaRaja MK, Devarajan A, Dhote VV (2022) Dietary supplementation for traumatic brain injury. In: Diagnosis and treatment of traumatic brain injury, pp 485–494

Newman JM, Pekari TB, Van Wyck DW (2023) Neuroprotection and therapeutic implications of creatine supplementation for brain injury complications. Med J Per 23-4/5/6:31–38

Ng SY, Lee AYW (2019) Traumatic brain injuries: pathophysiology and potential therapeutic targets. Front Cell Neurosci 13:528. https://doi.org/10.3389/fncel.2019.00528

Ochems D, van Rein E, Niemeijer M et al (2021) Incidence, causes and consequences of moderate and severe traumatic brain injury as determined by Abbreviated Injury Score in The Netherlands. Sci Rep 11:19985. https://doi.org/10.1038/s41598-021-99484-6

Okwuofu EO, Ogundepo GE, Akhigbemen AM, Abiola AL, Ozolua RI, Igbe I et al (2021) Creatine attenuates seizure severity, anxiety and depressive-like behaviors in pentylenetetrazole kindled mice. Metab Brain Dis 36:571–579. https://doi.org/10.1007/s11011-021-00684-w

Ostojic SM (2021a) Creatine as a food supplement for the general population. J Funct Foods 83:104568. https://doi.org/10.1016/j.jff.2021.104568

Ostojic SM (2021b) Modulation of CT1 function: from klotho protein to ammonia and beyond. Front Nutr 8:660021. https://doi.org/10.3389/fnut.2021.660021

Papadopoulou SK (2020) Rehabilitation nutrition for injury recovery of athletes: the role of macronutrient intake. Nutrients 12:2449. https://doi.org/10.3390/nu12082449

Pavlovic D, Pekic S, Stojanovic M, Popovic V (2019) Traumatic brain injury: neuropathological, neurocognitive and neurobehavioral sequelae. Pituitary 22(3):270–282. https://doi.org/10.1007/s11102-019-00957 9

Quan Y, Xin Y, Tian G, Zhou J, Liu X (2020) Mitochondrial ROS-modulated mtDNA: a potential target for cardiac aging. Oxidative Med Cell Longev 2020:9423593. https://doi.org/10.1155/2020/9423593

Rahimi R (2011) Creatine supplementation decreases oxidative DNA damage and lipid peroxidation induced by a single bout of resistance exercise. J Strength Cond Res 25(12):3448–3455. https://doi.org/10.1519/JSC.0b013e3182162f2b

Robayo LE, Govind V, Salan T, Cherup NP, Sheriff S, Maudsley AA, Widerström-Noga E (2023) Neurometabolite alterations in traumatic brain injury and associations with chronic pain. Front Neurosci 17:1125128. https://doi.org/10.3389/fnins.2023.1125128

Rodríguez-Triviño CY, Torres Castro I, Dueñas Z (2019) Hypochloremia in patients with severe traumatic brain injury: a possible risk factor for increased mortality. World Neurosurg 124:e783–e788

Roschel H, Gualano B, Ostojic SM, Rawson ES (2021) Creatine supplementation and brain health. Nutrients 13(2):586. https://doi.org/10.3390/nu13020586

Rosko LM, Gentile T, Smith VN, Manavi Z, Melchor GS, Hu J, Shults NV, Albanese C, Lee Y, Rodriguez O, Huang JK (2023) Cerebral creatine deficiency affects the timing of oligodendrocyte myelination. J Neurosci 43(7):1143–1153. https://doi.org/10.1523/JNEUROSCI.2120-21.2022

Royes LFF, Busanello GL, Godinho D, Nascimento AS, Lima GC, dos Navegantes Marçal JDS, Neto JBT, Sosthenes MCKS, Gomez-Pinilla F, Krishna G, Furian AF (2023) Creatine maintains mitochondrial integrity and protects against dysfunction in molecular systems involved in early epileptiform activity and cognitive impairment in young rats submitted to traumatic brain injury. Mol Neurobiol 1:275

Samborska B, Roy DG, Rahbani JF, Hussain MF, Ma EH, Jones RG, Kazak L (2022) Creatine transport and creatine kinase activity is required for $CD8^{+}$ T cell immunity. Cell Rep 38(9):110446. https://doi.org/10.1016/j.celrep.2022.110446

Senoner T, Dichtl W (2019) Oxidative stress in cardiovascular diseases: still a therapeutic target? Nutrients 11:2090

Singh A, Kukreti R, Saso L, Kukreti S (2019) Oxidative stress: a key modulator in neurodegenerative diseases. Molecules 24(8):1583. https://doi.org/10.3390/molecules24081583

Slavoaca D, Muresanu D, Birle C (2020) Biomarkers in traumatic brain injury: new concepts. Neurol Sci 41:2033–2044. https://doi.org/10.1007/s10072-019-04238-y

Smith-Ryan AE, Hirsch KR, Saylor HE, Gould LM, Blue MN (2020) Nutritional considerations and strategies to facilitate injury recovery and rehabilitation. J Athl Train 55(9):918–930

Smolarek AC, McAnulty SR, Ferreira LH, Cordeiro GR, Alessi A, Rebesco DB et al (2020) Effect of 16 weeks of strength training and creatine supplementation on strength and cognition in older adults: a pilot study. J Exerc Physiol 23(4):88–94

Snow WM, Cadonic C, Cortes-Perez C, Adlimoghaddam A, Roy Chowdhury SK, Thomson E et al (2020) Sex-specific effects of chronic creatine supplementation on hippocampal-mediated spatial cognition in the 3xTg mouse model of Alzheimer's disease. Nutrients 12(11):3589

Sullivan P, Geiger J, Mattson M, Scheff S (2000) Dietary supplement creatine protects against traumatic brain injury. Ann Neurol 48:723–729. https://doi.org/10.1002/1531-8249(200011)48:5<723::AID-ANA5>3.0.CO;2-W

Tarafdar A, Pula G (2018) The role of NADPH oxidases and oxidative stress in neurodegenerative disorders. Int J Mol Sci 19(12):3824

Thijs RD, Surges R, O'Brien TJ, Sander JW (2019) Epilepsy in adults. Lancet 393:689–701. https://doi.org/10.1016/S0140-6736(18)32596-0

Tran NT, Kelly SB, Snow RJ, Walker DW, Ellery SJ, Galinsky R (2021) Assessing creatine supplementation for neuroprotection against perinatal hypoxic-ischaemic encephalopathy: a systematic review of perinatal and adult pre-clinical studies. Cells 10(11):2902. https://doi.org/10.3390/cells10112902

Tran NT, Kowalski GM, Muccini AM, Nitsos I, Hale N, Snow RJ, Walker DW, Ellery SJ (2022) Creatine supplementation reduces the cerebral oxidative and metabolic stress responses to acute in utero hypoxia in the late-gestation fetal sheep. J Physiol 600(13):3193–3210. https://doi.org/10.1113/JP282840

Tropak MB, Tkachyova I, Gu R, Lee A, Schulze A (2023) Evidence of an intracellular creatine-sensing mechanism that modulates creatine biosynthesis via AGAT expression in human HAP1 cells. Sci Rep 13(1):22392. https://doi.org/10.1038/s41598-023-49860-1

Tsvetkova D, Kostadinova I, Landzhov B, Vezenkov L, Marinov L, Ivanova I (2023) Application of ABTS method for assessment of radicalbinding effect of creatine monohydrate. J Adv Pharm Educ Res 13(2):92–98. https://doi.org/10.51847/rxgOGbIUkJ

Turnagöl HH, Koşar ŞN, Güzel Y, Aktitiz S, Atakan MM (2021) Nutritional considerations for injury prevention and recovery in combat sports. Nutrients 14(1):53. https://doi.org/10.3390/nu14010053

Van Cutsem J, Roelands B, Pluym B, Tassignon B, Verschueren JO, DE Pauw K, Meeusen R (2020) Can creatine combat the mental fatigue-associated decrease in visuomotor skills? Med Sci Sports Exerc 52(1):120–130. https://doi.org/10.1249/MSS.0000000000002122

Van Gils A, Stone J, Welch K, Davidson LR, Kerslake D, Caesar D, McWhirter L, Carson A (2020) Management of mild traumatic brain injury. Pract Neurol 20(3):213–221

Vedung F, Fahlström M, Wall A, Antoni G, Lubberink M, Johansson J et al (2022) Chronic cerebral blood flow alterations in traumatic brain injury and sports-related concussions. Brain Inj 36:948–960. https://doi.org/10.1080/02699052.2022.2109746

Wu G (2020) Important roles of dietary taurine, creatine, carnosine, anserine and 4-hydroxyproline in human nutrition and health. Amino Acids 52(3):329–360

Yang Y, Ye Y, Kong C, Su X, Zhang X, Bai W et al (2019) MiR-124 enriched exosomes promoted the M2 polarization of microglia and enhanced hippocampus neurogenesis after traumatic brain injury by inhibiting TLR4 pathway. Neurochem Res 44:811–828. https://doi.org/10.1007/s11064-018-02714-z

Yang N, Guan QW, Chen FH, Xia QX, Yin XX, Zhou HH et al (2020) Antioxidants targeting mitochondrial oxidative stress: promising Neuroprotectants for epilepsy. Oxidative Med Cell Longev 2020:1–14. https://doi.org/10.1155/2020/6687185

Chapter 10
Magnesium and Traumatic Brain Injury

Abdelaati El Khiat, Sana El Foutat, Bilal El-Mansoury, Driss Ait Ali, Youssef Ait Hamdan, Sakhr Dadda, Kamal Kaoutar, Younsse Korchyou, Lahcen Tamegart, Mohamed El Koutbi, and Brahim El Houate

A. El Khiat (✉)
Biological and Health Sciences Team, Higher Institute of Nursing Professions and Health Techniques, Ministry of Health, Ouarzazate, Morocco

Laboratory of Clinical and Experimental Neurosciences and Environment, Faculty of Medicine and Pharmacy, Cadi Ayyad University, Marrakech, Morocco

Biological Engineering Laboratory, Faculty of Sciences and Techniques (FST), Sultan Moulay Slimane University, Beni Mellal, Morocco

Laboratory of Anthropogenic, Biotechnology and Health, Team Physiopathology Nutritional, Neurosciences and Toxicology, Faculty of Sciences, Chouaib Doukkali University, El Jadida, Morocco

Department of Biology, Faculty of Science, Abdelmalek Essaadi University, Tetouan, Morocco

Interdisciplinary Laboratory in Bio-Resources, Environment and Materials, Higher Normal School, Cadi Ayyad University, Marrakech, Morocco

S. El Foutat
Metabolics platform, Biochemistry Laboratory, Faculty of Medicine, Cadi Ayyad University, Marrakech, Morocco

B. El-Mansoury
Laboratory of Anthropogenic, Biotechnology and Health, Team Physiopathology Nutritional, Neurosciences and Toxicology, Faculty of Sciences, Chouaib Doukkali University, El Jadida, Morocco

D. A. Ali · K. Kaoutar · Y. Korchyou · M. El Koutbi · B. El Houate
Biological and Health Sciences Team, Higher Institute of Nursing Professions and Health Techniques, Ministry of Health, Ouarzazate, Morocco

Y. A. Hamdan
Interdisciplinary Laboratory in Bio-Resources, Environment and Materials, Higher Normal School, Cadi Ayyad University, Marrakech, Morocco

S. Dadda
Department of Biology, Faculty of Sciences and Techniques, University Moulay Ismail, Errachidia, Morocco

W. Mohamed (ed.), *Nutrition and Traumatic Brain Injury (TBI)*, Nutritional Neurosciences, https://doi.org/10.1007/978-981-97-6341-2_10

Abstract One of the extremes that has been associated with traumatic brain injuries (TBIs) is the general public health burden with the most significant implication in terms of serious morbidity and mortality. In spite of the fact that medicine is constantly improving, the problem of TBI treatment still is the most complicated and is very poorly solved in our health-care system. With increasing interest in recent years about the protective effects of magnesium in TBIs against neurodegeneration, this potential is certainly noteworthy. This part of the chapter is meant to provide an in-depth look at magnesium in the course of TBI pathophysiology and management. It explores the pathways by which magnesium is able to provide this neuroprotection and features such properties as anti-inflammation and anti-excitotoxicity. Along with this, the chapter exhibits the results from preclinical and clinical trials that buffer the neuronal damage instigated by TBI, and thus, a better outcome is observed. Other than that, the chapter emphasizes the outcome of the preclinical and clinical studies which demonstrated the ability of magnesium to mitigate the neuronal damages resulting from TBI and majorly lead to an improved prognosis. Not only does it concentrate on issues that surround the translation of preclinical findings to clinical context regarding the administration of magnesium, including the most appropriate dose and time of administration, but also it highlights the limitations. In general, this chapter reveals the multiple diagnostic points of magnesium as a possible remedy for TBI and points out the necessity of the next studies to clarify its exact mechanisms of action and augment the clinical efficacy.

Keywords Traumatic brain injuries · Magnesium · Neuroprotective · Excitotoxicity · Oxidative stress · Neuroinflammation · Therapeutic implications · Pathophysiology · Functional outcomes

10.1 Introduction

Patients with traumatic brain injury (TBI) experience brain damage due to sudden trauma, affecting individuals of all ages and resulting in significant morbidity and mortality (Ayt Hamdan et al. 2024). The aftermath of TBI in afflicted individuals leads to permanent disability, causing enduring financial, medical, emotional, familial, and social repercussions.

L. Tamegart
Laboratory of Clinical and Experimental Neurosciences and Environment, Faculty of Medicine and Pharmacy, Cadi Ayyad University, Marrakech, Morocco

Department of Biology, Faculty of Science, Abdelmalek Essaadi University, Tetouan, Morocco

After damage, studies on animals revealed a magnesium (Mg) deficit in the brain tissues and human blood, respectively. In animal models of brain injuries, magnesium supplementation has been shown to be an effective means of mitigating neurobehavioural and lesional changes. On the contrary, two prospective clinical researches which explored magnesium as a neuroprotective agent in patients with TBI provided the opposite results (Dhandapani et al. 2008). The treatment outcome in brain injury patients is compromised by the secondary insults and other existing factors, the adverse impact of which may be unknown by the Mg clinical studies assessing the patient group with TBI. The kinetic and dynamics pharmacological researches of magnesium administered intravenously in normal rats have demonstrated that this metal can cross the blood-brain barrier (Hallak et al. 1992). Nevertheless, there were human case studies of patients with brain injuries, which showed that intracerebral injection of magnesium did not consistently cause an increase in the concentration of this element in the cerebrospinal fluid (CSF) (Thurnau et al. 1987). The central nervous system (CNS) may control magnesium levels in both the brain and cerebrospinal fluid. This suggests that the blood-brain barrier (BBB) might be less permeable to magnesium if it is administered in a manner other than peripherally. The possibility of this is a critical factor that can affect the merging use of magnesium in TBI patients.

The construction of new brain pathways that follow TBI is diffusely spread through multiple biochemical pathways. But, on the other hand, employing one agent may mean that the safest dose does not work or the patient experiences the adverse side effects at the therapeutic dose and upon repeated action. The cardinal pathways of neuronal damage in TBI need to be targeted and controlled by using multiple agents working together to synergize the overall therapeutic intervention in neuroprotective therapy for TBI. Magnesium is among the drugs, further physiological options such as hyperoxia and hypothermia (described later in this chapter) are under investigation for TBI treatment (the rest of the drugs are mentioned earlier) (Vink and Nimmo 2009).

The tested hypothesis is that magnesium's effect on the cells' bioavailability when administered adjunctively with pharmaceutical agents and physiological interventions at their safest and lowest level of doses could produce a clinically effective neuroprotection protocol for decreased TBI-related complications. This approach is aimed at maximizing therapeutic advantages by overcoming the limitations of single-agent therapies in shortening or abolishing treatment courses to prevent or combat relapses, all while simultaneously minimizing potential adverse effects.

10.2 Pathophysiology of TBI

In the early post-injury period, primary brain injury results in tissue distortion and disintegration as a result of the original damage forces. The ability to mediate the cellular and molecular alterations that follow the initial insult is crucial for determining the clinical results.

Secondary injuries from TBI affect cell function and spread through mechanisms such as excitotoxicity, depolarization, free radical production, BBB disruption, altered calcium homeostasis, neuroinflammation, ischemic injury, edema, and intracranial hypertension (Mark W. Greve and Zink 2009; Zink et al. 2010; Dixon 2017).

10.2.1 Neuroinflammation

After a TBI, the brain displays the typical signs of inflammation (Louveau et al. 2015). Following TBI, oligodendrocyte cell counts are thought to be reduced, glial reactivity is maintained, and pro-inflammatory cytokines are elevated (Lotocki et al. 2011; Mouzon et al. 2014; Yang et al. 2013).

Complex penumbral inflammation is activated in the brain after the traumatic episode involving the activation of CNS immune cells along with the infiltration of peripheral immune cells which cross the specialized BBB in some cases. Taken together, these processes are the results of the synergistic effect of the previously mentioned cytokines including chemokines, adhesion molecules, reactive oxygen and nitrogen species, complement factors, and other mediators. Along with the process of inflammation, cerebral edema develops, and as a result, victims experience edema of the brain and increased intracranial pressure (ICP)—which are generally hazardous after brain injury (McGinn and Povlishock 2016).

The inflammation is initiated by the liberation of intracellular and intraneural components, increased production of free radicals, and matrix interference as well as the primary tissue damage caused by the mechanical injury to the brain (P. Mathew et al. 1994; Rock and Kono 2008). As a result, there is a local activation of CNS microglia (Kreutzberg 1996; Nimmerjahn et al. 2005; Streit 2000). These patrol cells, that go on living images of the debris, also set their own secret youth of a wide variety of cytokines, proteases, as well as reactive free radicals, which result in the inflammation of the tissues (Benveniste 1998; Gehrmann et al. 1995; Nakajima and Kohsaka 2004). These cytokines (and chemokines) prompt inflammation by advocating excitotoxicity, attracting peripheral immune cells, and intensifying cerebrovascular permeability (McGinn and Povlishock 2016).

10.2.2 Altered Blood-Brain Barrier, Cerebral Edema, and Impaired Regulation of Cerebral Blood Flow

It was longed not learnt to exist brain vascular dysfunction in modern investigation of anatomy of a human and animal brain tissue, until early 1990. This finding raised the possibility that neurotrauma-induced BBB malfunction could be the cause of this dysfunction. In fact, it is currently accepted as a central feature common in both traumatic (with the TBI) and nontraumatic (such as ischemic stroke or intracerebral hemorrhage) forms of brain injuries. Dysfunction (vegetative subsystem) may cause damage and also impact its evolution process. There are lots of pathological processes in which the brain intima has a role irrespective of whether it is dysfunctional or functional. Procedures such as loss and/or displacement of junction proteins, increased permeability, production of pro-inflammatory mediators, and/or altered BBB transporter activity (Abbott et al. 2006) accompany these phenomena (Abbott et al. 2006). Research indicates that brain damage can alter the BBB's function, affecting injury progression, treatment response, and neuronal healing (Zink et al. 2010). A physical barrier is formed following brain damage, and this barrier is closely linked to various parameters, such as vascular endothelial growth factor, reactive oxygen species (ROS), matrix metalloproteinases (MMPs), and pro-inflammatory cytokines (Zink et al. 2010).

Cerebral injury at the initial stage of TBI is often characterized by too many cellular damage and insufficient cerebral blood flow (CBF). The initial days after trauma are also related to vascular regression of the blood vessels which are situated in the neighboring area (Hayward et al. 2011; Park et al. 2009). This "ischemia-like" pattern leads to the accumulation of lactic acid due to anaerobic glycolysis, increased membrane permeability, and consecutive edema formation (Werner and Engelhard 2007). Besides, there is a drop in CBF that occurs immediately after a traumatic injury (within hours), and depending on the degree of the lesion, this reduction may last for days (DeWitt and Prough 2003; Golding et al. 1999; McQuire et al. 1998; Pasco et al. 2007). Increasing nitric oxide expression following damage may be the cause of this decreased blood flow, leading to vasodilation rather than an increase in vasoconstriction brought on by pressure (Villalba et al. 2014).

Cerebral edema in TBI patients is a significant and difficult treatment issue, particularly in moderate-to-severe TBI. Cerebral edema is traditionally classified into two types: vasogenic edema and cytotoxic edema. TBI is likely to cause cytotoxic and vasogenic edema. Increased permeability of the BBB to both low- and high-molecular-weight indicators and a build-up of osmotically active, plasma-derived molecules, like plasma proteins, in the interstitial fluid of the brain are the hallmarks of vasogenic edema (Zink et al. 2010). Following TBI, the combined action of cytotoxic edema caused by the swelling of millions of brain cells can increase cerebral volume, which in turn can raise tissue pressure and ICP. The cumulative impact of cytotoxic edema in the wounded brain is largely determined by the role played by astrocytes and endothelial cells (Zink et al. 2010).

10.2.3 Excitotoxicity and Oxidative Stress

Excessive excitatory amino acid transmission has been proven to be a key factor that acts as a trigger for the disease's pathophysiological processes following traumatic brain injury. After traumatic brain injury (TBI), the saturating concept of amino acid neurotransmitter—primarily glutamate—occurs abundantly (Werner and Engelhard 2007). In fact, it has been found that extracellular glutamate increases promptly and then returns to the normal level in the minutes after a brain injury; the length of the abovementioned increasing process and its duration are associated with the degree of the damage (Chamoun et al. 2010; Faden et al. 1989; Katayama et al. 1990; Koura et al. 1998). Ionotropic and metabotropic glutamate receptors are overstimulated as a result of the increased extracellular glutamate availability, which impacts neurons and astrocytes (Werner and Engelhard 2007). In fact, it has been found that extracellular glutamate increases promptly and then returns to the normal level in the minutes after a brain injury; the length of the abovementioned increasing process and its duration are associated with the degree of the damage (Cohen et al. 2007). According to experimental researches, the neurons in the neocortex become hyperexcited after a couple of days of the traumatic brain injury, which contributes to brain dysfunction (McGinn and Povlishock 2016). On the contrary, other post-traumatic brain changes of the hippocampus circuits, similar to those observed in the hippocampal domains, have also been reported using slice techniques. Several electrophysiologic studies with the use of the slice preparations revealed that traumatic brain injury may affect both excitatory and inhibitory synaptic transmissions, thereby leading to the dysfunction of hippocampal circuit (Reeves et al. 1997).

Reactive oxygen species (ROS) are known to cause oxidative stress after traumatic brain injury (TBI). When the body's natural antioxidant defenses, including glutathione peroxidase, catalase, and superoxide dismutase, are depleted, an excess of reactive oxygen species (ROS) is created. This results in DNA breakage, protein oxidation, and peroxidation of vascular and cellular components (Werner and Engelhard 2007). Oxidative stress induces inflammation and early or late apoptotic programs even though these pathways are enough to contribute to rapid cell death (Chong et al. 2005).

10.3 Neuroprotective Properties of Magnesium

10.3.1 Role of Magnesium in Ion Channel Regulation

Ion channels are a type of proteins which provide passageways for ions to travel through the cellular membranes by means of action potentials, which have a crucial role in the transmitter function of the nervous and muscular systems. As a matter of fact, magnesium has been identified to be the backbone of the process. On the other hand, it is a basic necessity in the metabolism of cells, furnishing more than 300 cofactors for involved enzymatic reactions. It is bound to the adenine triple of adenosine triphosphate (ATP) counterion which thus stabilizes numerous ATP-dependent processes like glucose utilization, protein synthesis, and nucleic acid synthesis (Jahnen-Dechent and Ketteler 2012). Magnesium, in addition, plays a significant role in the structural construction of nucleic acids, proteins, and mitochondria (Swaminathan 2003).

As an inside calcium blocker, it works to balance calcium within synapses of the neurons and neuromuscular. Furthermore, calcium tends to inhibit channel opens at the presynaptic junction, prohibiting acetylcholine over release and neuromuscular stimulation. In addition, magnesium functions as a depressant at the postsynaptic layer by blocking N-methyl-D-aspartate (NMDA) receptors through the voltage-dependent response (Fawcett et al. 1999). This inhibitory action on NMDA receptors is considered one of the primary mechanisms underlying magnesium's neuroprotective effects.

Seeking to understand the exact mechanism of how magnesium works as a neuroprotective, which is still unclear, many researchers share the common hypothesize that in response to this magnesium, the NMDA receptor blocking occurs that could lead to the prevention of the toxic effects of overstimulation on nerve cells; therefore, the problem is solved. NMDA receptors, which are sited at the postsynaptic in, provide the necessary reinforcement for stimulated efficacy of the connections between the see the gaps that exist in the primary connections between neurons where the repeated stimulations take effect (Bliss and Collingridge 1993; Newcomer et al. 2000).

NMDA receptors become overstimulated by excessive glutamate release in the weeks of anoxia brain, while neurons can no longer keep a normal balance. This occurrence leads to a sequence of cascade effects, which cause apoptosis, and hence, the nerve cells begin to die. There are developmentally enabled intracellular pathways such as calcium influx and by-products of activated catabolic enzymes followed by the production of free radicals that lead to apoptosis at the end (Choi 1992).

The construction of the NMDA receptor consists of four subunits with a homotetramer structure just as the hemoglobin molecule. Subunits that receive NR2B have high calcium permeability and are the ones mostly present in developing white matter during preterm fetuses (Jantzie et al. 2015; Paoletti et al. 2013). This abundance may explain the vulnerability of preterm white matter and contribute to distinct injury patterns between preterm and term hypoxia-ischemia (Jantzie et al. 2015).

Due to the current flow through the neuron cell, with the opening of magnesium channels at the NMDA receptor that act as endogenous calcium antagonists, the NMDA receptor acquires voltage-dependent blockade. By slowing down the elegant ingoing of calcium, magnesium saves the infection of another harmful pathway which leads to the death of cells. In preclinical data, there is a specific explanation in vitro and in vivo that agonists of magnesium reduce touch-and-go damage of glutamatergic and they improve conditions with oxygen glucose deprivation (Temkin et al. 2007).

Nevertheless, still debatable is whether it is the degree of injury caused by excessive activation of the NMDA receptors or not. Besides, some NMDA receptor antagonists faded in time and have been achieving limited functional improvement, but the efficiency of magnesium has been questioned (Dean et al. 2006). Trials involving a dissociative anesthetic, xenon, that also acts as an NMDA receptor antagonist but is used with cooling, have also not shown efficacy; this was possibly affected by a delay in starting therapy after birth (Azzopardi et al. 2016).

10.3.2 Anti-Inflammatory Effects of Magnesium

Both inflammation and infection are known to lead to neuron damage, and magnesium sulfate stabilizes the damage by downregulating on the inflammatory route. Magnesium has been found to have a pervasive effect on producing anti-inflammatory monocytes which secrete less pro-inflammatory cytokines like TNF-α and interleukin (IL)-6 when exposed to lipopolysaccharide (LPS) in in vitro studies (Sugimoto et al. 2012). Apart from that, preclinical results reveal that magnesium lowers the levels of pro-inflammatory cytokines (IL-6 and TNF-α) in LPS-treated pregnant rodents; and it improves the learning ability of the ones at 3 months (Lamhot et al. 2015; Tam et al. 2011).

There is a possibility that magnesium may exert its anti-inflammatory activity due to the fact that it inhibits the nuclear factor kappa B (NF-κB) signal pathway. NF-κB, being a transcription factor responding to the stimuli of inflammation, translocates into the nucleus and starts the transcription of the genes regulating inflammation. NF-κB activity, reduced by magnesium sulfate, has been shown to result from LPS sensitizing the microglia of the adult rodent brain to dislodge them from the nucleus (Gao et al. 2013; Lawrence 2009).

In prematurely delivered children, inflammation may contribute to the development of brain damage; therefore, cofactors such as chorioamnionitis, prolongation of membrane rupture, and maternal infection complicate the risk of a neurological disorder called cerebral palsy (Murphy et al. 1995). Preterm labor by intrinsic affection may be indicated considering elevated pro-inflammatory cytokines in cord blood. Or it may be of exogenous affection by bacteria. Infection of a woman also raises the cerebral palsy etiology risk in a term infant, especially when classified as a hypoxia-ischemia condition (Grether and Nelson 1997).

However, the theoretical premise that this element modifies the two processes is not always confirmed by the randomized studies related to magnesium. Segregation analysis of the fetus from the NICHD cohort exposed to antenatal magnesium to prevent cerebral palsy indicated no benefits in infants inflicted by chorioamnionitis. The exact effectiveness of magnesium during the clinical modulation of inflammatory responses has not been adequately validated through more research and trial runs (Kamyar et al. 2016).

10.4 Underlying Molecular Mechanisms

10.4.1 Impact of Magnesium on Excitotoxicity

Magnesium is essential for the nervous system, thereby ensuring perfect functionality of neurons while guarding against CNS disease. However, the excess glutamate stimulates the excitotoxicity imbalance causing intracellular magnesium depletion and subsequently disrupting the neuronal survival episode (Shindo et al. 2020).

From the physiological viewpoint, these sciences are linked to the NMDA receptor which is activated by glutamate and glycine and is thought to play a vital role in memory and synaptic plasticity. The most widespread activator of this receptor is glutamate, which pushes the calcium through the pore after displacing the magnesium ion that was blocking the way (Guo et al. 2017). Nevertheless, the precise mechanism of the Ras receptors' activation still remains unclear, as a few other factors can trigger the lifting of the magnesium block of the channels, pushing the calcium inward from outside (Maneshi et al. 2017).

Following traumatic brain injury, excessive release of certain chemicals, including NMDA receptors, is a major contributor to excitotoxicity. These receptors, which contain various types of subunits, control the flow of ions and are sensitive to glutamate molecules. Some calved particles are more sensitive to calcium and have a greater liking for magnesium (Maneshi et al. 2017). Another sub-faction of transient melastatin receptor potential (TRPM) channels typically associated with the TRPM7 subtype is acquired during ischemia (Lin and Xiong 2017). In addition to their role as calcium-selective, TRPM7 may be activated by ROS (Aarts et al. 2003). TRPM7 is particular because of their insensitivity to decreasing internal Mg^{2+} concentration, their ability to permeate through leaky channels, and their activation under acidic conditions (Chubanov et al. 2017). However, magnesium plays no less vital role in TRPM7 channel activation. In vivo, magnesium ions' levels are usually much higher in the cell than in the extracellular medium (Romani 2011). The TRPM7 channels are now known to be important in the magnesium transfer as shown by the decreased amount of magnesium in the fibroblasts. In addition, the fibroblast depleted of TRPM7 also showed diminished levels of intracellular ROS (Chen et al. 2012). Decreased intracellular magnesium levels are thought to have a consequent effect on the ROS-generating enzyme NADPH oxidase (Heyno et al. 2008). In addition, research has proved a decreased intracellular magnesium level is related to an increase in the expression of glutathione S-transferase which is crucial in destroying free radicals in the body (Wolf et al. 2008).

10.4.2 Effects of Magnesium on Neuroinflammatory Processes

Inflammation in the brain cells is one of the evidences of magnesium deficiency (Maier et al. 2021). Research corroborates the fact that magnesium deficiency signals a gene expression change involved in the inflammatory process, resulting in brain inflammation (Tsuji et al. 2021). All these processes (the decreased estrogen levels due to ovariectomy, magnesium deficiency, and brain inflammation) are characterized by increased inflammatory cytokines and microglia activation. Administration of magnesium helps mitigate any inflammation response seen (Zhang et al. 2021). With regard to ovariectomized rats, a lack of estrogen and a resulting magnesium deficiency can cause neuroinflammation marked by elevated levels of inflammatory cytokines and microglia activation. Thanks to the administration of magnesium, inflammatory responses are reduced (J. Zhang et al. 2021).

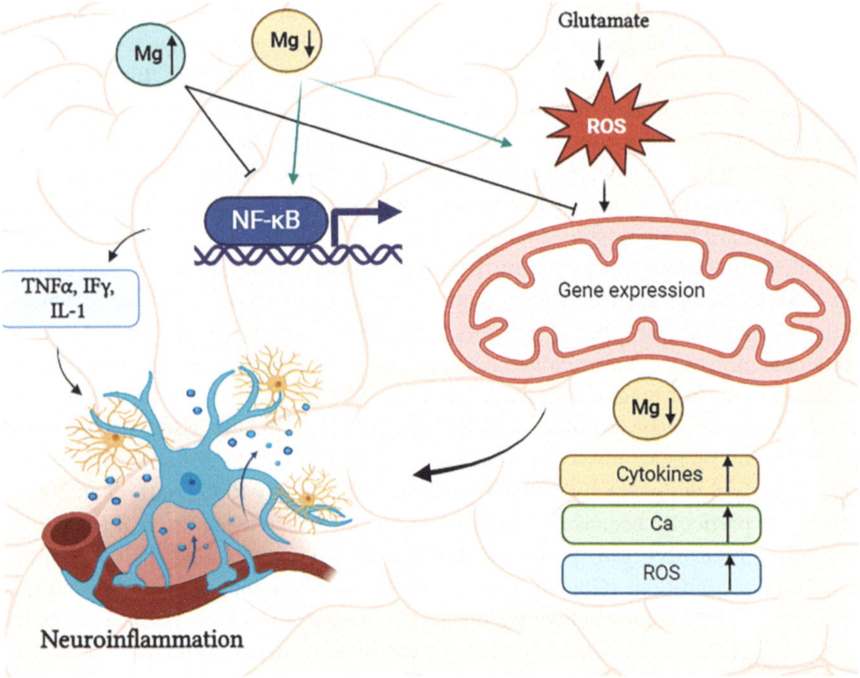

Fig. 10.1 Effects of magnesium in the brain and its relationship with neuroinflammation

Cytokines cytokines are highly involved in the physiological functions of the CNS. These cytokines have different origins with the release by various types of brain cells such as neurons, microglia, and astrocytes and include molecules like TNFα and IL-1 and interferon γ. They regulate synapsis formation and play a crucial role in learning and memory processes (Sugimoto et al. 2012). Magnesium sulfate ($MgSO_4$) treatment causes the lowering of cytokines in peripheral blood samples. Such a drop is linked to the hampering of NF-κB activation and its translocation into the nucleus caused by the rise of the levels of its constitutive inhibitor called IκBα (Bourgognon and Cavanagh 2020). The importance of magnesium administration can be related to the fact that it is the very substance that suppresses inflammation and significantly decreases the risk of cerebral palsy and severe motor dysfunction in premature babies (Marret et al. 2007). The administration of a dose of magnesium has the single effect of reducing inflammation and function when the tissues are hypoxic/ischemic (Le Dieu-Lugon et al. 2021). In Fig. 10.1, it is described how the magnesium deficit leads to the development of neuroinflammation.

10.4.3 Magnesium in Humans with Traumatic Brain Injury

Traumatic brain injuries (TBIs) are a significant cause of death and disability among the young adult population worldwide. The pathophysiology of acute traumatic brain injury consists of an initial event followed by a series of subsequent insults. While a significant portion of neurological damage occurs during the initial traumatic head injury, it is known that the next cascade of pathophysiological processes occurring in the hours or days following the injury can further worsen the damage (Arango and Bainbridge 2008).

Magnesium is known for its vitality in the human system which is important for energy metabolism, neurotransmitter function, gene transmission, and muscle contraction. It serves as a mostly important carrier for neural transmission and in neuromuscular conduction from a neurogenic perspective. It plays a role in the reduction of excitability of the nerve cells. Such excitability's excess has been linked to the genesis of a variety of neurological disorders (Fromm et al. 2004). Besides, through the action of magnesium upon vascular smooth muscle relaxation and the effect on vasodilation, more blood brings oxygen to brain cells. More and more, there is strong evidence to support the significance of magnesium in this perspective in the prevention and management of neurological diseases.

Furthermore, the involvement of magnesium in the neuronal pathology of post-traumatic brain injury is another function it plays. This neuroprotective action is achieved by blocking off the NMDA receptors without competing for their position, thereby actively controlling the amount of glutamates and calcium ions that enter the neurons. Magnesium deficiency caused by cerebral injury that would eliminate magnesium from the control loop of calcium would bring about neuron damage and cell death (Sen and Gulati 2010).

Traumatic brain injury leads to a substantial decrease in intracellular magnesium ion levels, which can potentially compromise essential cellular functions (Fromm et al. 2004). A decrease in magnesium levels following a traumatic brain injury acts as a trigger for the neurochemical cascade that occurs when excessive calcium enters neurons (Heath and Vink 1996). Several studies have established a connection between low magnesium levels and a higher likelihood of severe neurological complications following a brain injury.

By conducting experiments, components of the mechanism of antidepressants of magnesium have been elucidated, such as the monoaminergic and serotonergic systems of neurotransmitter functioning. Such multimode systems are usually exposed to disruptive infractions like those that occur as a secondary injury as a result of being part of the brain injury induced by trauma (Fromm et al. 2004).

The delivery of magnesium could parallel the calcium influx into nerve cells which subsequently drives the excitotoxicity observed in a traumatic brain injury. Research shows that magnesium in the prevention of concussions has even proven to be more beneficial. Besides, it has been shown that magnesium (Sen and Gulati 2010) contributes to alleviating traumatic brain injuries.

10.4.4 Preclinical and Clinical Studies with Magnesium in Brain Injury

In the past few years, magnesium has gained enormous attention in the area of neuroprotection where it protects the brain against damage as was evidenced in preclinical and clinical studies (Mathew and Panonnummal 2022). Among the four most abundant cations in the body, magnesium might have the biggest chance to influence the body's physiological functions, which range from energy and protein metabolism to intracellular calcium regulation and even the coagulation process.

It was demonstrated that it regulates anti-oxidative processes, among which are the reduction of excitotoxicity, inflammations, and oxidative stress. Nevertheless, these levels may sometimes be either abnormal or else are commonly associated with the prognosis of patients. Thus, carrier of enough magnesium is of all importance for normal brain functioning, as magnesium takes part in the control of many neurotransmitter substances and their receivers (Mathew and Panonnummal 2022; Nayak et al. 2018; Wang et al. 2022b).

10.5 Findings from Observational Studies

A study conducted by Han et al. (2019) among 323 patients with acute intracerebral hemorrhage (ICH) divided them into three groups based on their admission magnesium levels (T1: low (<0.84 mmol/L), T2: normal (0.84–0.91 mmol/L), and T3: high (≥0.91 mmol/L). The analysis of the study pointed out that 3 months after the patient follow-ups, T3 patients with the higher magnesium level had a lower frequency of mortality compared to those with the lower one (T1) (adjusted odds ratio [OR], 0.10; 95% confidence interval [CI], 0.02–0.54; and P-trend = 0.010). Despite this, the groups with varying surgical outcomes did not show any apparent differences (adjusted OR, 1.80; 95% CI, 0.71–4.56; and P-trend = 0.227) after considering potential confounders. So ICH patients with low serum magnesium levels have a decrease in the mortality rate, but this correlation has nothing to do with functional impairment 3 months from the day of the disease onset (Han et al. 2019).

Statistic from numerous studies shows that the incidence of hypomagnesemia among TBI patients reaches 42.7%–66%. Some statistics indicate that the incidence is even higher than 66% (Dey et al. 2021; Nayak et al. 2018; Wang et al. 2022b).

Researchers discovered that electrolyte constraints, such as hypomagnesemia, had been present in most traumatic brain injury (TBI) patients compared to what was found in their healthy counterparts (Dey et al. 2021).

In connection with that, the admission of 72 people with severe traumatic brain injury (TBI) under the trial was subjected to a prospective test. After that, they were separated into two groups: Participants were randomly assigned to the low serum magnesium or the normal serum magnesium group which is based on their original level of serum magnesium when they joined the study. On the contrary, while 81%

of patients who were poorly neurologically developed had low magnesium levels in their serum at the 6-month scale-up, only 19% of them obtained good outcomes ($P = 0.01$).

Besides, regression analysis demonstrated that there was a linear association between hypomagnesemia and poor neurological (OR = 2.1, $P = 0.04$, and 95% CI = 1.0–8.8), suggesting that hypomagnesemia was an independent prognostic marker for severe TBI patients (Nayak et al. 2018).

Another research topic has looked at patients who had suffered from traumatic brain injury and compared their serum magnesium level before hospitalization and the likelihood of their death following trauma. The risk factors related to deaths which were found to be among the TBI patients under research were determined by the method of logistic regression analysis. A restricted cubic spline (RCS) was used to investigate the link between preoperative serum magnesium levels and mortality in TBI patients. It was established that patients with abnormal magnesium levels regardless of either being too high or too low were coupled with high dying rates. Additionally, the research revealed that blood magnesium levels before the onset of TBI influence the chance of survival of the affected patient (Wang et al. 2022b).

A recent study was conducted to evaluate initial serum calcium and magnesium levels and their predictive roles in inpatient traumatic brain injury patients. Interestingly, the conclusion is that abnormal electrolyte levels upon admission of patients may be linked to unfavorable outcomes. The study enrolled 922 patients whose clinical records were taken on them, indicating the value of serum electrolytes on admission. A mortality rate of 24% was observed in the hypocalcemia group, while only 12% was discovered in the normal calcium group ($p < 0.001$). The normal magnesium level group recorded an MI rate of 23%, while the hypomagnesemia group had a 15% MI rate with p equal to 0.006.

Moreover, of the 12 cases of hypermagnesemia, 7 (58%) patients died during hospital stay. The regression model, which included various factors, showed that hypocalcemia was not a significant predictor of mortality after TB (odds ratio [OR] 0.59, 95% confidence interval [CI]: Mild hypertension: systolic blood pressure 120–139; diastolic blood pressure 80–89. However, hypermagnesemia was a significant predictor of mortality (OR 16 and 95% CI: 2.1–111) (Mekkodathil et al. 2023).

A systematic review was performed to examine the effects of electrolyte abnormalities among traumatic brain injury (TBI) patients with current evidence being taken into consideration. Fourteen articles met the inclusion criteria. The outcomes showed that the incidences of hyponatremia, hyperkalemia, hypercalcemia, hypomagnesemia, and hypermagnesemia were not linked to death in the TBI population. Nevertheless, symptoms of both hyponatremia and hypomagnesemia were the reason for low Glasgow Outcome Scale (GOS) scores at 6 months.

On one hand, no unanimous definition exists among these studies regarding what hypomagnesemia is, and there is no consensus yet on the relationship between hypomagnesemia and the outcomes of patients with TBI (Dey et al. 2021; Pin-On et al. 2018; Wang et al. 2022b).

10.6 Clinical Trials Evaluating the Efficacy of Magnesium in TBI

The discoveries in the animal experiments have shown that magnesium can be effectively administered to treat such factors that are the tertiary brain injury processes, like glutamate discharge, N-methyl-D-aspartate (NMDA) receptor activity, calcium channels, intensification of the antioxidative system, and restraint of the free radical actions (Lee et al. 2004; Wang et al. 2022b).

In a preclinical research employing male Sprague Dawley rats, scientists tested the magnesium effect on the expression of cell death and survival-related proteins after traumatic brain injury (TBI). Rats were divided into two groups: the sham group that underwent uninjured head surgery and the other rats that received moderate TBI through a fluid percussion injury model. The results indicated a prompt surge in proteins that were related to cell death, of which the p53 and the responding proteins to it peaked on the first day but decreased to the base level by the fourth day. The contradiction is that on the fourth day, the apoptosis rate decreased, and a lot of proteins that have the function of survival increased. It was found that magnesium treatment resulted in a reduction in the expression of proteins which is associated with p53 and apoptosis, however, did not influence making changes in the expression of proteins which are associated with survival connotation. In view of Lee et al. (2004), these appear to propose that the effects of magnesium on TBI-induced tissue damage seem to be working in ways of altering other molecular pathways (Lee et al. 2004).

Treating the effects of traumatic brain injury (TBI), Ameliorate et al. (2017) investigated the combined therapy of magnesium and an NK1 tachykinin receptor antagonist as opposed to treatment with either method alone. The investigation involved untreated rats with TBI, $MgSO_4$, and the NK1 antagonist n-acetyl L tryptophan or both of these. $MgSO_4$ and n-acetyl Ltryptophan attenuated the elevation of blood-brain barrier permeability and led to better functional outcomes. However, we found that the combination treatment of n-acetyl L tryptophan and $MgSO_4$ produced superior results in the field of decrement of blood-brain barrier permeability and functional outcomes compared to administering the drugs separately. These findings mean that discord of magnesium and NK1 receptor blockers can serve the purpose of treating traumatic brain injury better (Ameliorate et al. 2017).

Magnesium sulfate is evaluated with a view to determining the potentiality of healing it in the treatment of severe brain injury (diffuse axonal injury [DAI]). The present study consists of 128 adult patients with initial Glasgow Coma Scale (GCS) sub-scores ranging from 3 to 8. The patients were randomly divided into two groups: The research group was divided into the Magnesium Sulfate treatment (MST) treatment (MST) group (n = 64) and the control (C) group (n = 64). The MST group received magnesium sulfate 20 min through the vein and in a dose of 250 μmol/kg. From then on, they always received at least 750 μmol/kg per day through the next 5 days. Normal patients subjected to other ordinary magnesium sulfate managed the treatment. GCS evaluation, as well as the blood count, were performed at the admission and the third and seventh days after the injury. At leaving, the scores based on the Glasgow Coma Scale were recorded at the 3-month follow-up after discharge. After the 7-day treatment, the MST group registered a significantly lower serum neuron-specific enolase level than the control group (25.40 ± 6.66 vs. 29.58 ± 7.32; P = 0.001) and a significantly higher GCS score (8.23 ± 2.72 vs. 7.05 ± 2.6).

The mean score of GCS of the MST group had significantly decreased at the moment of discharge (3.30 ± 1.35 vs. 3.90 ± 1.10; P = 0.004) and was lower 3 months later (2.95 ± 1.48 vs. 3.66 ± 1.44; P = 0.009) than the control. The obtained outcome implies that introducing magnesium sulfate at early stages of diffuse axonal injury (DAI) results in significant improvement of the condition (Zhao et al. 2016).

Kazim and his colleagues carried out an assessment of the utilization of magnesium sulfate in the administration of traumatic brain injury in 2020. Prospective cases of 60 TBIs, selected through non-probability consecutive sampling, were included in the study. Subjects had been treated with a standard protocol of TBI along with magnesium sulfate as an added therapy. Glasgow Coma Scale (GCS) scores were scored at admission and on the third day of ward stay. The study with t-tests is conducted to compare results, and it reveals a statistically significant difference between GCS scores at admission and GCS scores on the third day with a p value of <0.0001. The mean GCS score was 7.69 ± 2.55 among patients under 30 years old and 8.29 ± 2.57 for those over 30 years old. As for this differentiation, it was not statistically significant (Ali et al. 2021). Similar findings were reported in a study conducted by Hussain Arain et al. (2021).

The early clinical trials suggest that the improvement of outcomes and reducing mortality rates in specific injuries of the brain is possible by means of magnesium supplementation, and more research could eventually lead to understanding the full range of magnesium's therapeutic potential and optimal dosage requirements. Thus, it is advised that the right fluids should be carefully selected in order to prevent or control the condition of electrolyte imbalances right from the beginning. In addition to this, physicians should pay attention to the treatment (clinical management) of patients with TBI, especially those with higher magnesium levels (Ngatuvai et al. 2023).

In a preliminary double-blind study involving 499 patients aged 14 years or older, who suffered moderate-to-severe traumatic brain injury (TBI) within 8 hours of the incident, participants were randomly assigned to one of three groups: a placebo group and two groups receiving different dosages of magnesium. Treatment was administered within 8 hours of the injury and continued for five days. The primary outcomes, assessed 6 months post-injury, included a composite of mortality, seizures, functional status, and neuropsychological evaluations.

This same research revealed that the high-dose magnesium compared to the placebo did not provide a significant difference in the primary result (mean average percentile ranking 55 vs. 52, 95% CI for difference −7 to 14, and p = 0.70). In line with expectations, the patients who received the lower dose of magnesium did worse than those who received the placebo (48% vs. 54%, 95% CI −10.5 to −2, and $p = 0.007$). In addition, the mortality rate was higher when a high dose of magnesium was used instead of the placebo. The investigation found that the body cannot tolerate continuous magnesium infusion lasting 5 days in patients with moderate or severe traumatic brain injury (TBI). Besides, it could be even worse for the future treatment of brain injury patients (Temkin et al. 2007).

A systematic evaluation concluding with a meta-analysis was performed to evaluate whether or not magnesium supplements are an effective adjuvant during the treatment of traumatic brain injuries for severe adults. The results of the review have confirmed that no apparent difference was noticed in the cause of death results for all the treatment and control groups. On the contrary, the GOS scores were also demonstrated to have no significant difference between the treatment and control groups. In conclusion, it was, unfortunately, found that the previous data are not consistent between themselves, and it was proved that there are not enough important findings to give magnesium therapy prescriptions to patients (Lyons and Blackshaw 2018).

The role of magnesium in negotiations regarding traumatic brain injury (TBI) is essentially dealing with the controversial issue. The preceding research studies imply variable results with respect to magnesium supplementation in patients diagnosed with traumatic brain injuries (TBIs) (Temkin et al. 2007). Though there are studies that show its contribution juxtaposed with those that do not display proper objectiveness, its high dosage is lethal; hence, researchers need to carry out more investigations to determine their mode of action (Sen and Gulati 2010; Temkin et al. 2007; Wang et al. 2022b). The inconsistency further supports the complexity of TBI and the need for deeper studies.

10.7 Therapeutic Implications

10.7.1 Magnesium's Possible Application as a Preventative Measure

Some medical conditions can be treated using magnesium as a preventative measure. The evidence shows that going with a magnesium supplement will control migraines, decrease the frequency of type 2 diabetes, reduce hypertension, and avoid osteoporosis. For instance, these studies show that the addition of magnesium, mainly in people with low magnesium levels, is impactful in terms of frequency and the severity of their headaches. The mechanism responsible for this is due to NMDA receptors blocking its calcium channels and protecting the cell from the influx of abnormally high calcium ions (Domitrz and Cegielska 2022; Holland et al. 2012).

Equally important, a lesser rate of diabetes type 2 is also associated with a greater magnesium intake that is widely studied by numerous researchers and authors (Domitrz and Cegielska 2022; Holland et al. 2012). Magnesium's ability to raise insulin levels, decrease glycated hemoglobin, and quicken wound healing all contributed to this outcome (Piuri et al. 2021). At the same time, magnesium supplementation may lead to small but significant reductions in blood pressure levels in individuals with hypertension (Jee et al. 2002). Still, this effect is controversial for other studies stating a lack of consistency of results regarding the benefits of magnesium supplementation for hypertension (Dickinson et al. 2006; X. Zhang et al. 2016). Furthermore, it is important to note that the amount of magnesium intake has been positively correlated with bone mineral density levels which resulted in the prevention of osteoporosis (Veronese et al. 2014).

10.7.2 Magnesium Administration in the Initial Hours after Traumatic Brain Injury

Hypomagnesaemia is common among patients with traumatic brain injury (TBI) (Mekkodathil et al. 2023; Wang et al. 2022a). In order to reduce subsequent damage processes and enhance prognosis, magnesium supplementation is being investigated as a possible therapeutic intervention in the early hours following traumatic brain injury.

Giving magnesium sulfate right away after inducing brain trauma in rats has been shown to reduce traumatic brain injury and attenuate blood-brain barrier damage in preclinical trials (Feng et al. 2004). Another similar study found that magnesium treatment reduced apoptosis and expression of p53-related proteins among rats with induced brain damage (Lee et al. 2004).

Nevertheless, in other studies like Li et al. (2015), mortality rates were unchanged when magnesium was given as an after-injury therapy (Li et al. 2015). Temkin et al. (2007) found that continuous magnesium infusions were not neuroprotective and might potentially interfere with the management of severe head injuries in patients with moderate-to-severe traumatic brain injury (Temkin et al. 2007). Therefore, although magnesium is thought to have neuroprotective properties, human clinical trials failed to demonstrate these effects.

10.7.3 Combination of Magnesium with Pharmacological Agents for the Treatment of TBI

Magnesium in combination with other drugs to treat traumatic brain damage (TBI) aims to enhance neuroprotection and improve prognosis. In rats used in preclinical trials, magnesium and MK-801 combined therapy reduced brain edema development and helped restore blood-brain barrier permeability following experimental diffuse brain damage (Temkin et al. 2007). Similar studies revealed that polyethylene glycol (PEG) increases neuroprotection following traumatic brain injury (TBI) by facilitating central penetration of magnesium while concurrently lowering the dose needs for magnesium (Busingye et al. 2016), and combining magnesium with an NK1 antagonist may be a more effective therapy for traumatic brain injury than using either medication alone (Ameliorate et al. 2017).

10.8 Conclusion

Ultimately, the comprehensive approach toward TBI, which goes beyond traditional strategies, takes into account the inherent health complications and metabolic consequences, and utilizes specialized nutrition interventions, would be very useful. The practical recommendations should be implemented because this will refine the recovery process and result in a better outcome for an individual who has a TBI. This chapter forms a meaningful toolkit for medical staff, patients, and caregivers looking for aid in dealing with the challenges of nutrition related to TBI recovery.

References

Aarts M, Iihara K, Wei W-L, Xiong Z-G, Arundine M, Cerwinski W, MacDonald JF, Tymianski M (2003) A key role for TRPM7 channels in anoxic neuronal death. Cell 115(7):863–877

Abbott NJ, Rönnbäck L, Hansson E (2006) Astrocyte–endothelial interactions at the blood–brain barrier. Nat Rev Neurosci 7(1):41–53

Ali K, Irfan M, Abbas R (2021) Effects of magnesium sulfate therapy on GCS scores in patients with severe traumatic brain injury. Pak J Neurol Surg 24(4):322–327. https://doi.org/10.36552/pjns.v24i4.497

Ameliorate JL, Ghabriel MN, Vink R (2017) Magnesium enhances the beneficial effects of NK1 antagonist administration on blood-brain barrier permeability and motor outcome after traumatic brain injury. Magnes Res 30(3):88–97. https://doi.org/10.1684/mrh.2017.0427

Arango MF, Bainbridge D (2008) Magnesium for acute traumatic brain injury. Cochrane Database Syst Rev 4:CD005400

Azzopardi D, Robertson NJ, Bainbridge A, Cady E, Charles-Edwards G, Deierl A, Fagiolo G, Franks NP, Griffiths J, Hajnal J (2016) Moderate hypothermia within 6 h of birth plus inhaled xenon versus moderate hypothermia alone after birth asphyxia (TOBY-Xe): a proof-of-concept, open-label, randomised controlled trial. Lancet Neurol 15(2):145–153

Benveniste EN (1998) Cytokine actions in the central nervous system. Cytokine Growth Factor Rev 9(3–4):259–275

Bliss TVP, Collingridge GL (1993) A synaptic model of memory: long-term potentiation in the hippocampus. Nature 361(6407):31–39

Bourgognon J-M, Cavanagh J (2020) The role of cytokines in modulating learning and memory and brain plasticity. Brain Neurosci Adv 4:2398212820979802

Busingye DS, Turner RJ, Vink R (2016) Combined magnesium/polyethylene glycol facilitates the neuroprotective effects of magnesium in traumatic brain injury at a reduced magnesium dose. CNS Neurosci Ther 22(10):854–859

Chamoun R, Suki D, Gopinath SP, Goodman JC, Robertson C (2010) Role of extracellular glutamate measured by cerebral microdialysis in severe traumatic brain injury. J Neurosurg 113(3):564–570

Chen H-C, Su L-T, González-Pagán O, Overton JD, Runnels LW (2012) A key role for Mg2+ in TRPM7's control of ROS levels during cell stress. Biochem J 445(3):441–448

Choi DW (1992) Excitotoxic cell death. J Neurobiol 23(9):1261–1276

Chong ZZ, Li F, Maiese K (2005) Oxidative stress in the brain: novel cellular targets that govern survival during neurodegenerative disease. Prog Neurobiol 75(3):207–246

Chubanov V, Ferioli S, Gudermann T (2017) Assessment of TRPM7 functions by drug-like small molecules. Cell Calcium 67:166–173

Cohen AS, Pfister BJ, Schwarzbach E, Grady MS, Goforth PB, Satin LS (2007) Injury-induced alterations in CNS electrophysiology. Prog Brain Res 161:143–169

Dean JM, George SA, Wassink G, Gunn AJ, Bennet L (2006) Suppression of post-hypoxic-ischemic EEG transients with dizocilpine is associated with partial striatal protection in the preterm fetal sheep. Neuropharmacology 50(4):491–503

DeWitt DS, Prough DS (2003) Traumatic cerebral vascular injury: the effects of concussive brain injury on the cerebral vasculature. J Neurotrauma 20(9):795–825

Dey S, Kumar R, Cureus AT et al (2021) Evaluation of electrolyte imbalance in patients with traumatic brain injury admitted in the central ICU of a tertiary care centre: a prospective observational. Cureus 13(8):e17517. https://doi.org/10.7759/cureus.17517

Dhandapani SS, Gupta A, Vivekanandhan S, Sharma BS, Mahapatra AK (2008) Randomized controlled trial of magnesium sulphate in severe closed traumatic brain injury. Indian J Neurotrauma 5(01):27–33

Dickinson HO, Nicolson D, Campbell F, Cook JV, Beyer FR, Ford GA, Mason J (2006) Magnesium supplementation for the management of primary hypertension in adults. Cochrane Database Syst Rev 3:CD004640

Dixon KJ (2017) Pathophysiology of traumatic brain injury. Phys Med Rehabil Clin N Am 28(2):215–225. https://doi.org/10.1016/j.pmr.2016.12.001

Domitrz I, Cegielska J (2022) Magnesium as an important factor in the pathogenesis and treatment of migraine—from theory to practice. Nutrients 14(5):1089

Faden AI, Demediuk P, Panter SS, Vink R (1989) The role of excitatory amino acids and NMDA receptors in traumatic brain injury. Science 244(4906):798–800

Fawcett WJ, Haxby EJ, Male DA (1999) Magnesium: physiology and pharmacology. Br J Anaesth 83(2):302–320

Feng DF, Zhu ZA, Lu YC (2004) Effects of magnesium sulfate on traumatic brain edema in rats. Chin J Traumatol 7(3):148–152

Fromm L, Heath DL, Vink R, Nimmo AJ (2004) Magnesium attenuates post-traumatic depression/anxiety following diffuse traumatic brain injury in rats. J Am Coll Nutr 23(5):529S–533S

Gao F, Ding B, Zhou L, Gao X, Guo H, Xu H (2013) Magnesium sulfate provides neuroprotection in lipopolysaccharide-activated primary microglia by inhibiting NF-κB pathway. J Surg Res 184(2):944–950

Gehrmann J, Matsumoto Y, Kreutzberg GW (1995) Microglia: intrinsic immuneffector cell of the brain. Brain Res Rev 20(3):269–287

Golding EM, Robertson CS, Bryan RM (1999) The consequences of traumatic brain injury on cerebral blood flow and autoregulation: a review. Clin Exp Hypertens 21(4):299–332

Grether JK, Nelson KB (1997) Maternal infection and cerebral palsy in infants of normal birth weight. JAMA 278(3):207–211

Greve MW, Zink BJ (2009) Pathophysiology of traumatic brain injury. Mt Sinai J Med 76:15–21. https://doi.org/10.1002/MSJ

Guo H, Camargo LM, Yeboah F, Digan ME, Niu H, Pan Y, Reiling S, Soler-Llavina G, Weihofen WA, Wang H-R (2017) A NMDA-receptor calcium influx assay sensitive to stimulation by glutamate and glycine/D-serine. Sci Rep 7(1):11608

Hallak M, Berman RF, Irtenkauf SM, Evans MI, Cotton DB (1992) Peripheral magnesium sulfate enters the brain and increases the threshold for hippocampal seizures in rats. Am J Obstet Gynecol 167(6):1605–1610
Hamdan YA, Rhazi M, El-Mansoury B, Kabdy H, Oukhrib M, Baba AA, Elhousseini A, Ahmad ZH, Oudadesse H (2024) Astrocytes functions and their involvement in brain injury: a focus on the biomaterials. Role in mitigating traumatic brain injury. In: Physiology and function of glial cells in health and disease. IGI Global, pp 145–163
Han X, You S, Huang Z, Han Q, Zhong C, Xu J, Shi R, Chen D, Zhang Y, Xiao G, Cao Y, Liu C-F (2019) Prognostic significance of serum magnesium in acute intracerebral hemorrhage patients. Curr Neurovasc Res 16(2):123–128. https://doi.org/10.2174/1567202616666190412124539
Hayward NMEA, Tuunanen PI, Immonen R, Ndode-Ekane XE, Pitkänen A, Gröhn O (2011) Magnetic resonance imaging of regional hemodynamic and cerebrovascular recovery after lateral fluid-percussion brain injury in rats. J Cereb Blood Flow Metab 31(1):166–177
Heath DL, Vink R (1996) Traumatic brain axonal injury produces sustained decline in intracellular free magnesium concentration. Brain Res 738(1):150–153
Heyno E, Klose C, Krieger-Liszkay A (2008) Origin of cadmium-induced reactive oxygen species production: mitochondrial electron transfer versus plasma membrane NADPH oxidase. New Phytol 179(3):687–699
Holland S, Silberstein SD, Freitag F, Dodick DW, Argoff C, Ashman E (2012) Evidence-based guideline update: NSAIDs and other complementary treatments for episodic migraine prevention in adults:[RETIRED] report of the quality standards Subcommittee of the American Academy of neurology and the American headache society. Neurology 78(17):1346–1353
Hussain Arain S, Saqib Ansari N, Akbar Shaikh H, Ali Narejo M, Kanwal R, Professor A (2021) Effectiveness of magnesium sulfate for the treatment of severe traumatic brain injury. Pak J Med Health Sci 15(2):347
Jahnen-Dechent W, Ketteler M (2012) Magnesium basics. Clin Kidney J 5(Suppl_1):i3–i14
Jantzie LL, Talos DM, Jackson MC, Park H-K, Graham DA, Lechpammer M, Folkerth RD, Volpe JJ, Jensen FE (2015) Developmental expression of N-methyl-D-aspartate (NMDA) receptor subunits in human white and gray matter: potential mechanism of increased vulnerability in the immature brain. Cereb Cortex 25(2):482–495
Jee SH, Miller ER, Guallar E, Singh VK, Appel LJ, Klag MJ (2002) The effect of magnesium supplementation on blood pressure: a meta-analysis of randomized clinical trials. Am J Hypertens 15(8):691–696
Kamyar M, Manuck TA, Stoddard GJ, Varner MW, Clark EAS (2016) Magnesium sulfate, chorioamnionitis, and neurodevelopment after preterm birth. BJOG Int J Obstet Gynaecol 123(7):1161–1166
Katayama Y, Becker DP, Tamura T, Hovda DA (1990) Massive increases in extracellular potassium and the indiscriminate release of glutamate following concussive brain injury. J Neurosurg 73(6):889–900
Koura SS, Doppenberg EMR, Marmarou A, Choi S, Young HF, Bullock R (1998) Relationship between excitatory amino acid release and outcome after severe human head injury. In: Intracranial pressure and neuromonitoring in brain injury: proceedings of the tenth international ICP symposium, Williamsburg, Virginia, May 25–29, 1997, pp 244–246
Kreutzberg GW (1996) Microglia: a sensor for pathological events in the CNS. Trends Neurosci 19(8):312–318
Lamhot VB, Khatib N, Ginsberg Y, Anunu R, Richter-Levin G, Weiner Z, Ross MG, Divon MY, Hallak M, Beloosesky R (2015) Magnesium sulfate prevents maternal inflammation–induced impairment of learning ability and memory in rat offspring. Am J Obstet Gynecol 213(6):851–8e1
Lawrence T (2009) The nuclear factor NF-κB pathway in inflammation. Cold Spring Harb Perspect Biol 1(6):a001651

Le Dieu-Lugon B, Dupré N, Derambure C, Janin F, Gonzalez BJ, Marret S, Arabo A, Leroux P (2021) Effect of neuroprotective magnesium sulfate treatment on brain transcription response to hypoxia ischemia in neonate mice. Int J Mol Sci 22(8):4253

Lee JS, Han YM, Yoo DS, Choi SJ, Choi BH, Kim JH, Kim YH, Huh PW, Ko YJ, Rha HK, Cho KS, Kim DS (2004) A molecular basis for the efficacy of magnesium treatment following traumatic brain injury in rats. J Neurotrauma 21(5):549–561. https://doi.org/10.1089/089771504774129883

Li W, Bai Y-A, Li Y-J, Liu K-G, Wang M-D, Xu G-Z, Shang H-L, Li Y-F (2015) Magnesium sulfate for acute traumatic brain injury. J Craniofac Surg 26(2):393–398

Lin J, Xiong Z-G (2017) TRPM7 is a unique target for therapeutic intervention of stroke. Int J Physiol Pathophysiol Pharmacol 9(6):211

Lotocki G, de Rivero Vaccari J, Alonso O, Molano JS, Nixon R, Safavi P, Dietrich WD, Bramlett HM (2011) Oligodendrocyte vulnerability following traumatic brain injury in rats. Neurosci Lett 499(3):143–148

Louveau A, Smirnov I, Keyes TJ, Eccles JD, Rouhani SJ, Peske JD, Derecki NC, Castle D, Mandell JW, Lee KS (2015) Structural and functional features of central nervous system lymphatic vessels. Nature 523(7560):337–341

Lyons MWH, Blackshaw WJ (2018) Does magnesium sulfate have a role in the management of severe traumatic brain injury in civilian and military populations? A systematic review and meta-analysis. BMJ Mil Health 164(6):442–449. https://doi.org/10.1136/JRAMC-2018-000916

Maier JA, Castiglioni S, Locatelli L, Zocchi M, Mazur A (2021) Magnesium and inflammation: advances and perspectives. Semin Cell Dev Biol 115:37–44

Maneshi MM, Maki B, Gnanasambandam R, Belin S, Popescu GK, Sachs F, Hua SZ (2017) Mechanical stress activates NMDA receptors in the absence of agonists. Sci Rep 7(1):39610

Marret S, Marpeau L, Zupan-Simunek V, Eurin D, Lévêque C, Hellot M, Bénichou J, Group PT (2007) Magnesium sulphate given before very-preterm birth to protect infant brain: the randomised controlled PREMAG trial. BJOG Int J Obstet Gynaecol 114(3):310–318

Mathew AA, Panonnummal R (2022) A mini review on the various facets effecting brain delivery of magnesium and its role in neurological disorders. Biol Trace Elem Res 201(9):4238–4253. https://doi.org/10.1007/S12011-022-03517-8

Mathew P, Graham DI, Bullock R, Maxwell W, McCulloch J, Teasdale G (1994) Focal brain injury: histological evidence of delayed inflammatory response in a new rodent model of focal cortical injury. In: Brain Edema IX: proceedings of the ninth international symposium Tokyo, May 16–19, 1993, pp 428–430

McGinn MJ, Povlishock JT (2016) Pathophysiology of traumatic brain injury. Neurosurg Clin N Am 27(4):397–407. https://doi.org/10.1016/j.nec.2016.06.002

McQuire JC, Sutcliffe JC, Coats TJ (1998) Early changes in middle cerebral artery blood flow velocity after head injury. J Neurosurg 89(4):526–532

Mekkodathil A, El-Menyar A, Hakim S, Al Jogol H, Parchani A, Peralta R, Rizoli S, Al-Thani H (2023) Initial serum levels of magnesium and calcium as predictors of mortality in traumatic brain injury patients: a retrospective study. Diagnostics 13(6):1172. https://doi.org/10.3390/DIAGNOSTICS13061172

Mouzon BC, Bachmeier C, Ferro A, Ojo J, Crynen G, Acker CM, Davies P, Mullan M, Stewart W, Crawford F (2014) Chronic neuropathological and neurobehavioral changes in a repetitive mild traumatic brain injury model. Ann Neurol 75(2):241–254

Murphy DJ, Johnson AM, Sellers S, MacKenzie IZ (1995) Case-control study of antenatal and intrapartum risk factors for cerebral palsy in very preterm singleton babies. Lancet 346(8988):1449–1454

Nakajima K, Kohsaka S (2004) Microglia: neuroprotective and neurotrophic cells in the central nervous system. Curr Drug Targets Cardiovasc Hematol Disorders 4(1):65–84

Nayak R, Attry S, Nath Ghosh S (2018) Serum magnesium as a marker of neurological outcome in severe traumatic brain injury patients. Asian J of Neurosurg 13(03):685–688. https://doi.org/10.4103/AJNS.AJNS_232_16

Newcomer JW, Farber NB, Olney JW (2000) NMDA receptor function, memory, and brain aging. Dialogues Clin Neurosci 2(3):219–232

Ngatuvai M, Martinez B, Sauder M, Beeton G, Andrade R, Maka P, Smith CP, Kornblith L, Elkbuli A (2023) Traumatic brain injury, electrolyte levels, and associated outcomes: a systematic review. In: Journal of surgical research, vol 289. Academic Press, pp 106–115. https://doi.org/10.1016/j.jss.2023.03.029

Nimmerjahn A, Kirchhoff F, Helmchen F (2005) Resting microglial cells are highly dynamic surveillants of brain parenchyma in vivo. Science 308(5726):1314–1318

Paoletti P, Bellone C, Zhou Q (2013) NMDA receptor subunit diversity: impact on receptor properties, synaptic plasticity and disease. Nat Rev Neurosci 14(6):383–400

Park E, Bell JD, Siddiq IP, Baker AJ (2009) An analysis of regional microvascular loss and recovery following two grades of fluid percussion trauma: a role for hypoxia-inducible factors in traumatic brain injury. J Cereb Blood Flow Metab 29(3):575–584

Pasco A, Lemaire L, Franconi F, Lefur Y, Noury F, Saint-André J-P, Benoit J-P, Cozzone PJ, Le Jeune J-J (2007) Perfusional deficit and the dynamics of cerebral edemas in experimental traumatic brain injury using perfusion and diffusion-weighted magnetic resonance imaging. J Neurotrauma 24(8):1321–1330

Pin-On P, Saringkarinkul A, Punjasawadwong Y, Kacha S, Wilairat D (2018) Serum electrolyte imbalance and prognostic factors of postoperative death in adult traumatic brain injury patients. Medicine 97(45):e13081. https://doi.org/10.1097/MD.0000000000013081

Piuri G, Zocchi M, Della Porta M, Ficara V, Manoni M, Zuccotti GV, Pinotti L, Maier JA, Cazzola R (2021) Magnesium in obesity, metabolic syndrome, and type 2 diabetes. Nutrients 13(2):320

Reeves TM, Lyeth BG, Phillips LL, Hamm RJ, Povlishock JT (1997) The effects of traumatic brain injury on inhibition in the hippocampus and dentate gyrus. Brain Res 757(1):119–132

Rock KL, Kono H (2008) The inflammatory response to cell death. Annu Rev Pathol Mech Dis 3:99–126

Romani AMP (2011) Cellular magnesium homeostasis. Arch Biochem Biophys 512(1):1–23

Sen AP, Gulati A (2010) Use of magnesium in traumatic brain injury. Neurotherapeutics 7(1):91–99. https://doi.org/10.1016/J.NURT.2009.10.014/METRICS

Shindo Y, Yamanaka R, Hotta K, Oka K (2020) Inhibition of mg2+ extrusion attenuates glutamate excitotoxicity in cultured rat hippocampal neurons. Nutrients 12(9):2768

Streit WJ (2000) Microglial response to brain injury: a brief synopsis. Toxicol Pathol 28(1):28–30

Sugimoto J, Romani AM, Valentin-Torres AM, Luciano AA, Ramirez Kitchen CM, Funderburg N, Mesiano S, Bernstein HB (2012) Magnesium decreases inflammatory cytokine production: a novel innate immunomodulatory mechanism. J Immunol 188(12):6338–6346

Swaminathan R (2003) Magnesium metabolism and its disorders. Clin Biochem Rev 24(2):47

Tam HBT, Dowling O, Xue X, Lewis D, Rochelson B, Metz CN (2011) Magnesium sulfate ameliorates maternal and fetal inflammation in a rat model of maternal infection. Am J Obstet Gynecol 204(4):364–3e1

Temkin NR, Anderson GD, Winn HR, Ellenbogen RG, Britz GW, Schuster J, Lucas T, Newell DW, Mansfield PN, Machamer JE, Barber J, Dikmen SS (2007) Magnesium sulfate for neuroprotection after traumatic brain injury: a randomised controlled trial. Lancet Neurol 6(1):29–38. https://doi.org/10.1016/S1474-4422(06)70630-5

Thurnau GR, Kemp DB, Jarvis A (1987) Cerebrospinal fluid levels of magnesium in patients with preeclampsia after treatment with intravenous magnesium sulfate: a preliminary report. Am J Obstet Gynecol 157(6):1435–1438

Tsuji R, Inoue H, Uehara M, Kida S (2021) Dietary magnesium deficiency induces the expression of neuroinflammation-related genes in mouse brain. Neuropsychopharmacol Rep 41(2):230–236

Veronese N, Berton L, Carraro S, Bolzetta F, De Rui M, Perissinotto E, Toffanello ED, Bano G, Pizzato S, Miotto F (2014) Effect of oral magnesium supplementation on physical performance in healthy elderly women involved in a weekly exercise program: a randomized controlled trial. Am J Clin Nutr 100(3):974–981

Villalba N, Sonkusare SK, Longden TA, Tran TL, Sackheim AM, Nelson MT, Wellman GC, Freeman K (2014) Traumatic brain injury disrupts cerebrovascular tone through endothelial inducible nitric oxide synthase expression and nitric oxide gain of function. J Am Heart Assoc 3(6):e001474

Vink R, Nimmo AJ (2009) Multifunctional drugs for head injury. Neurotherapeutics 6(1):28–42

Wang GH, Yan Y, Shen HP, Chu Z (2022a) The clinical characteristics of electrolyte disturbance in patients with moderate and severe traumatic brain injury who underwent craniotomy and its influence on prognosis. J Korean Neurosurg Soc 66(3):332–339

Wang R, He M, Xu J (2022b) Initial serum magnesium level is associated with mortality risk in traumatic brain injury patients. Nutrients 14(19):4174. https://doi.org/10.3390/NU14194174

Werner C, Engelhard K (2007) Pathophysiology of traumatic brain injury. Br J Anaesth 99(1):4–9. https://doi.org/10.1093/bja/aem131

Wolf FI, Trapani V, Simonacci M, Boninsegna A, Mazur A, Maier JAM (2008) Magnesium deficiency affects mammary epithelial cell proliferation: involvement of oxidative stress. Nutr Cancer 61(1):131–136

Yang SH, Gangidine M, Pritts TA, Goodman MD, Lentsch AB (2013) Interleukin 6 mediates neuroinflammation and motor coordination deficits after mild traumatic brain injury and brief hypoxia in mice. Shock 40(6):471

Zhang X, Li Y, Del Gobbo LC, Rosanoff A, Wang J, Zhang W, Song Y (2016) Effects of magnesium supplementation on blood pressure: a meta-analysis of randomized double-blind placebo-controlled trials. Hypertension 68(2):324–333

Zhang J, Mai C-L, Xiong Y, Lin Z-J, Jie Y-T, Mai J-Z, Liu C, Xie M-X, Zhou X, Liu X-G (2021) The causal role of magnesium deficiency in the neuroinflammation, pain hypersensitivity and memory/emotional deficits in ovariectomized and aged female mice. J Inflamm Res 14:6633

Zhao L, Wang W, Zhong J, Li YY, Cheng YZ, Su Z, Zheng W, Guan XD (2016) The effects of magnesium sulfate therapy after severe diffuse axonal injury. Ther Clin Risk Manag 12:1481–1486. https://doi.org/10.2147/TCRM.S109482

Zink BJ, Szmydynger-Chodobska J, Chodobski A (2010) Emerging concepts in the pathophysiology of traumatic brain injury. Psychiatr Clin N Am 33(4):741–756. https://doi.org/10.1016/j.psc.2010.08.005

Chapter 11
Polyunsaturated Fatty Acids and Traumatic Brain Injury

Kholoud Elsamman

Abstract Polyunsaturated fatty acids (PUFA), including precursors and long-chain PUFAs (LC-PUFAs), are necessary fatty acids. Due to the inability of mammals to create precursors, they must be obtained from nutrition. They can be converted into LC-PUFAs. It is wise to take LC-PUFA supplements, which control cell and brain function, because humans have such a low conversion rate. Long-chain polyunsaturated fatty acids (LC-PUFAs) are characterized by their carbon chain length of 20 or more carbon atoms and multiple double bonds. Key members of this group include eicosapentaenoic acid (EPA, 20:5 ω-3), docosahexaenoic acid (DHA, 22:6 ω-3), and arachidonic acid (AA, 20:4 ω-6).The composition of LC-PUFAs is significant in both n-3 and n-6 series, with n-3 LC-PUFAs, such as EPA and DHA, being derived from dietary sources like fish oil, while n-6 LC-PUFAs, such as AA, are primarily obtained from vegetable oils. The synthesis of these fatty acids involves a series of enzymatic reactions, including desaturation and elongation, facilitated by specific enzymes such as delta-5 desaturase (D5D) and delta-6 desaturase (D6D). These enzymes introduce double bonds at specific positions in the carbon chain, which is essential for the formation of bioactive lipid mediators, including eicosanoids and specialized pro-resolving mediators (SPMs) that regulate inflammation and promote tissue repair.

Keywords Traumatic brain injury (TBI) · Polyunsaturated fatty acids (PUFAs) · Neuroprotection · Cognitive recovery · Nutritional intervention · Neurological disorders · Neuroprotective effects

---

K. Elsamman (✉)
Biochemistry and Nutrition Department, Faculty of Women for Science, Ain Shams University, Cairo, Egypt
e-mail: 29911050200347@std.women.asu.edu.eg

W. Mohamed (ed.), *Nutrition and Traumatic Brain Injury (TBI)*, Nutritional Neurosciences, https://doi.org/10.1007/978-981-97-6341-2_11

11.1 Sources of PUFAs in the Brain

11.1.1 Metabolism of PUFAs

There are several double bonds in the carbon chain of polyunsaturated fatty acids (PUFAs). The n-6 PUFAs and the n-3 PUFAs are distinguished from one another by the first double bond from the methyl terminal end. A first double bond forms on the third carbon atom in n-3 PUFAs and on the sixth in n-6 PUFAs. Linoleic acid (LA) and alpha-linolenic acid (ALA) are two types of essential fatty acids that mammals are unable to produce naturally. These molecules can transform into long-chain polyunsaturated fatty acids (LC-PUFAs, with 20 carbon atoms or more) in living things by undergoing elongation, desaturation, and beta-oxidation. LC-PUFAs have more carbon and double bonds. Both metabolic routes comprise Δ6 and Δ5 desaturases and elongases that are specific to n-6 and n-3 PUFA, which is why they are competing (Simopoulos 2011). The liver's microsomes and peroxisomes play a pivotal role in LC-PUFA synthesis (Sprecher 2000).

However, the brain's enzymatic machinery can generate LC-PUFAs. Docosahexaenoic acid (DHA; 22:6 n-3) and arachidonic acid (AA; 20:4 n-6) are the principal n-3 and n-6 LC-PUFAs due to their roles as bioactive derivative precursors and their brain levels, respectively (Joffre et al. 2016a).

Essential n-3 LC-PUFA and components of bioactive substances, eicosapentaenoic acid (EPA; 20:5 n-3) is present in low amounts in the brain and undergoes fast β-oxidation (Chen and Bazinet 2015). Docosapentaenoic acid (DPA; 22:5 n-6) may substitute docosahexaenoic acid (DHA) in cell membranes if one does not obtain sufficient n-3 PUFA from diet. The main locations for LC-PUFA esterification to occur are phospholipids. In contrast to the 10 μmoles/g of tissue, the free LC-PUFAs are present albeit in extremely little amounts: 1 nmole/g (Green et al. 2010).

11.1.2 Nutritional Source

Seeds, oils, and vegetables are excellent sources of the precursors of LA and ALA. In comparison to sunflower oil, which contains 60% LA, rapeseed oil only has 10% ALA (Lewinska et al. 2015). There is some evidence that healthy people can convert these precursors into LC-PUFAs although the conversion efficiency is pitiful, coming in at less than 5% (Plourde and Cunnane 2007). Food is the primary source of LC-PUFAs. Fatty fish, such as salmon (18.7% EPA + DHA) and tuna (32.9% EPA + DHA), contain AA in far higher quantities than eggs and meats (5–8%) (Meyer et al. 2003; Strobel et al. 2012).

Lean fish, such as cod and sole, have relatively high levels of EPA and DHA. It is essential to eat LC-PUFAs to maintain an adequate amount of these fats in the cell membrane. This is why recommendations for diets rich in polyunsaturated fatty acids have been made. Two weekly meals of fish with an LA/ALA ratio of 4:5 mg each may provide 500 mg of EPA and DHA, the recommended daily allowance. This is far more than enough to meet the DHA requirements of the body and lower the danger of cardiovascular disease (Lucas et al. 2009).

Eating more ALA and less LA increases the bioavailability of n-3 LC-PUFA, according to clinical and preclinical research (Blanchard et al. 2013; Taha et al. 2014). Even after following these recommendations, the typical American diet is woefully deficient in n-3 polyunsaturated fatty acids (ALA, EPA, and DHA) and long-chain polyunsaturated fatty acids (LC-PUFAs). The proportion of monounsaturated to polyunsaturated fatty acids is 12–20 times greater in Western diets (Simopoulos 2002). N-3 polyunsaturated fatty acids (PUFAs) DHA and EPA, as well as the long-chain PUFA ALA, are still severely under-consumed by the general population. The Western diet is imbalanced because n-6 PUFA intake is 12–20 times greater than n-3 PUFA consumption (Rao et al. 2007).

The brain's DHA half-life is prolonged when n-3 PUFA levels are low, which is in contrast to a healthy diet (DeMar Jr et al. 2004). Lipids, namely triglycerides—chains of three fatty acids connected by a glycogen backbone—contribute most of our caloric intake (about 35–40%). Phospholipids are another possible form they take (in which the third position of glycerol is replaced by phosphorylated alcohol function to obtain phospholipids). Although there is an ongoing dispute, it is believed that krill oil, which includes phospholipids, and fish oil, which contains triglycerides, both enhance the bioavailability of EPA and DHA (Yurko-Mauro et al. 2015).

Cod and sole are examples of lean fish that have relatively high levels of EPA and DHA. Essential fatty acid synthesis (EFAs) requires dietary LC-PUFA consumption to maintain an adequate concentration in the membrane. Dietary recommendations based on polyunsaturated fatty acids have been proposed because of this. Even with only two fish meals a week and an LA/ALA ratio of 4:5, you can get 500 mg of EPA and DHA per day. The body's DHA requirements will be more than met, and the risk of cardiovascular disease will be decreased (Lucas et al. 2009).

Scientific research has shown that n-3 LC-PUFA bioavailability can be enhanced by reducing LA and increasing ALA in the diet (Blanchard et al. 2013; Taha et al. 2014). N-3 polyunsaturated fatty acids (ALA, DHA, and EPA) and long-chain polyunsaturated fatty acids (LC-PUFAs) are still severely underrepresented in the average American diet, even when these guidelines are applied. Western diets have a 12- to 20-fold higher n-6 to n-3 polyunsaturated fatty acid imbalance (Simopoulos 2002). Despite these recommendations, the average American diet is severely lacking in n-3 polyunsaturated fatty acids (PUFAs), such as ALA, and the long-chain PUFAs, such as DHA and EPA. In reality, n-6 PUFA consumption is 12–20 times more than n-3 PUFA consumption, which causes an imbalance in the Western diet (Rao et al. 2007).

Furthermore, when n-3 PUFA levels are low in the diet, the brain's DHA half-life is enhanced, which is contrary to a healthy diet (DeMar Jr et al. 2004). Triglycerides, which are chains of three fatty acids connected by a glycogen backbone, provide the majority of the calories that our bodies absorb, accounting for around 35–40% of total caloric intake. They can also occur in phospholipid form (in which the

3-position on the glycerol is replaced by a phosphorylated alcohol function). To enhance EPA and DHA bioavailability, it is suggested to include krill oil's phospholipids and fish oil's triglycerides; however, this is still debatable (Yurko-Mauro et al. 2015).

11.2 The Brain's PUFA Concentration

Most brain tissue is composed of polyunsaturated fatty acids (PUFAs), with 8–10% AA and 12–14% DHA (n-3 PUFA) accounting for the bulk (Chung et al. 2008; McNamara and Carlson 2006). Unlike rats, which achieve their peak between seven and twenty-first days after birth, humans attain it between 2 years of age and the beginning of the third trimester (Green et al. 1999). During these moments, there is an increase in gray matter, rapid neuronal development, and synaptogenesis (Giedd et al. 1999).

The effects on LC-PUFA concentration vary among brain regions (Joffre et al. 2016a; McNamara et al. 2009), For example, in adult C57Bl6/J mice, the concentration of amyloid beta is 10.2% in the hippocampus, 9.7% in the prefrontal cortex, 8.5% in the hypothalamus, 7.7% in the cortex, 6.5% in the cerebellum, and 9.0% in the brain stem (6.5%). Here are the top brain areas in terms of DHA content: prefrontal cortex (14.3%), hippocampus (13.7%), cerebellum (12.2%), cortex (11.9%), hypothalamus (10.1%), and brain stem (0.3%) among others (Joffre et al. 2016a).

Although the ratio of AA to DHA drops to 0.54 in the cerebellum, it ranges from 0.75 to 0.85 in the hypothalamus and hippocampus. Perhaps different approaches are to blame for the discrepancies. There are a number of pathways by which LC-PUFA reaches the brain, one of which is by incorporation into the cell membranes of the appropriate tissues. In contrast to the 12.3–15.9% levels seen in rats and mice, the human prefrontal cortex contains DHA levels ranging from 14.1% to 15.9% (Joffre et al. 2016a; Hamazaki et al. 2015; Moriguchi et al. 2001).

Seventy percent of brain cells are astrocytes, 15% are oligodendrocytes, and 15% are microglial cells (Morand et al. 1982). Information about the precise makeup of cellular fatty acids is scarce. Bourre et al. validated earlier findings by assessing the fatty acid composition of oligodendrocytes, neurons, and astrocytes in rats at 15 and 60 days of life (Morand et al. 1982; Cohen and Bernsohn 1973).

Astrocytes are glial cells that provide support to many parts of the nervous system. They store energy metabolites and play a role in synaptic transmission. An injury to the central nervous system triggers a condition called reactive astrogliosis (central nervous system [CNS]). Astrocytes malfunctions lead to abnormal changes in the brain and spinal cord. Astrocytes contain a variety of fatty acids, including 10.1–10.3% AA, 2.5–2.7% n-6 DPA, 2.4–2.7% adrenic acid, 10.6–12.1% DHA, and 0.7–1.3% n-3 LC-PUFAs (22:4 n-6). Their PUFA precursor content is low, with just 1.2–1.4% LA. They are also lacking in ALA and other similar compounds. The ratio ranges from 0.72 to 0.76 for $n = -3$ (Bourre et al. 1984).

Among the many roles played by these multiuse glial cells in brain function are the maintenance of synaptic transmission and the supply of energy metabolites. When they detect brain damage, one of their responses is reactive astrogliosis (CNS). Damage to the central nervous system (CNS) could result from malfunctioning astrocytes. Astrocytes include 10.6–12.1% DHA, 0.7–1.3% n-3 LC-PUFAs, 10.1–10.3% AA, 2.5–2.7% n-6 DPA, and 2.4–2.7% adrenic acid in their fatty acid composition (22:4 n-6). They are severely deficient in PUFA precursors, with only 1.2–1.4% LA and no ALA at all. The value of the ratio n-3/n-6 is around 0.72–0.76. (Bourre et al. 1984).

One kind of immune cell found in the brain is microglial cells. Their involvement in neuroinflammation, synaptogenesis, and the maturation and maintenance of synapses is crucial. They keep a close eye on their surroundings and, when triggered, secrete substances that promote inflammation and cell renewal. They are unique among brain cells in their fatty acid content. DHA is the primary fatty acid found in every one of these cells. Microscopic glial cells are easily identifiable by their abundant EPA and relatively modest levels of DHA and n-3 DPA (Rey et al. 2018). In the brain, microglial cells stand in for the innate immune system. Neuroinflammation, synaptogenesis, and synapse function and structure are all profoundly affected by them. In response to stimulation, they constantly assess and regulate their surroundings, and their reactions might range from pro-inflammatory to pro-regenerative. They are distinct from other brain cells in the makeup of the fatty acids they contain. All of these cells rely on DHA as their primary fatty acid. There is a significant concentration of EPA (3.7%) in microglial cells but low amounts of DHA (1%) and n-3 DPA (0.1%) (Madore et al. 2014). Brain cells called microglia play an important role in the body's innate immune response. They are essential in the processes of neuroinflammation, synaptogenesis, and synapse formation and function. They are constantly watching and controlling their surroundings, and when triggered, they create responses that encourage inflammation and regeneration. The fatty acid makeup of these brain cells is different from that of other brain cells. The primary lipid found in all these cell kinds is DHA. Although microglial cells have a little amount of DHA (1%), they have an abundance of EPA (3.7%) (Chen et al. 2009).

11.3 The Role of Diet in Determining the PUFA Content of the Brain and Brain Cells

Throughout their lifetimes, people are subject to a variety of environmental factors, including dietary choices. As an environmental feature, it has changed significantly during the past few decades. In Western societies, the amount of n-6 PUFA has been rising consistently, whereas the level of n-3 PUFA has been decreasing. The finding is that industrialized nations' n-6/n-3 PUFA ratio is 12:20, rather than the five that were indicated (Simopoulos 2011). The importance of this cannot be overstated because the composition of brain fatty acids is affected by dietary fatty acid consumption (Calder 2007). All areas of the brain are significantly affected by the dietary PUFA content (Joffre et al. 2016a; Alashmali et al. 2016). Brain DHA declines in all brain regions when n-3 PUFA precursor is not consumed throughout development and/or maturity. Most severely impacted are the areas of the brain with the highest concentrations of DHA, such as the prefrontal cortex and hippocampus, and least so in the hypothalamus, which has the lowest concentrations of DHA (Joffre et al. 2016a; Delpech et al. 2015a). One probable explanation for these variations is the regular progression of brain function (Broadhurst et al. 2002).

Metabolic alterations occur in n-3 PUFA insufficiency; for instance, DHA's half-life in the brain is prolonged to decrease loss (Rapoport et al. 2007). Moreover, the enzymes Δ6 desaturase and elongase, which are responsible for DHA synthesis, show an increase in activity in the liver (Wang et al. 2005). The brain's DHA levels rise in response to an abundance of n-3 LC-PUFAs, in contrast to times of shortage (Kitson et al. 2016). Taking a supplement containing DHA has a greater impact on improving cognitive performance than taking an ALA supplement (Lacombe et al. 2017; Rey et al. 2019). Taking a DHA supplement can help prevent age-related or n-3 PUFA deficiency–induced DHA depletion in the brain (Bascoul et al. 2016). Brain DHA levels were also greater in fat-1 mice, a genetic model of n-3 PUFA enrichment (Joffre et al. 2016a; Delpech et al. 2015b).

Diet is just one of many environmental factors that people are subject to on a daily basis. The past few decades have seen significant changes to it as an environmental feature. The n-6 PUFA level has been rising gradually in Western societies, whereas the n-3 PUFA concentration has been declining. Consequently, the n-6/n-3 PUFA ratio in developed nations is today 12:20, rather than the five that were originally proposed (Simopoulos 2011).

Because the makeup of fatty acids in the brain is affected by what people eat, this is of the utmost importance (Calder 2007). Every part of the brain is significantly affected by the amount of PUFA in the diet (Joffre et al. 2016a; Alashmali et al. 2016). All regions of the brain experience a decline in DHA when the n-3 PUFA precursor is not consumed throughout development and/or adulthood. The hypothalamus is the least impacted, in contrast to the highly affected prefrontal cortex and hippocampus, which contain the lowest concentrations of DHA (Joffre et al. 2016a; Delpech et al. 2015a). These differences may have their origins in the natural progression of brain function (Broadhurst et al. 2002).

Metabolic alterations, such as an extension of DHA's half-life in the brain to decrease loss, occur under circumstances of n-3 PUFA insufficiency (Rapoport et al. 2007). In addition, the enzymes Δ6 desaturase and elongase, which are responsible for producing DHA, show an increase in activity in the liver (Wang et al. 2005). The availability of n-3 LC-PUFAs causes a rise in brain DHA levels, in contrast to situations when they are scarce (Kitson et al. 2016). Supplementing with DHA has a greater impact on enhancing brain function compared to ALA (Chung et al. 2008; Rey et al. 2019). When you take a DHA supplement, you can prevent the loss of DHA in your brain that comes with becoming older or not getting enough n-3 PUFAs (Bascoul et al. 2016). In addition, DHA levels were greater in the brains of fat-1 mice, which serve as genetic models for n-3 PUFA enrichment (Joffre et al. 2016a; Delpech et al. 2015b).

11.4 Therapeutic PUFA Compounds

11.4.1 The Metabolic Pathway of Bioactive Polyunsaturated Fatty Acid Derivatives

By stimulating the synthesis of bioactive lipid mediators, LC-PUFAs may, in theory, regulate immune responses. One kind of lipid mediator is the eicosanoids, which include prostaglandins, leukotrienes, and thromboxanes; the other is the specialized pro-resolving mediators, or PRMs (SPMs; resolvins, protectins, and maresins). Given that eicosanoids derived from n-3 PUFAs are less powerful inflammatory than those derived from n-6 PUFAs, it is important for the membrane PUFA concentration to rise while the n-6 PUFA concentration falls. When AA and EPA both produce eicosanoids, in a co-occurrence scenario, the two compete with one another (Calder 2002). Three major mediators derived from EPA have been determined to be inactive: 3-series prostaglandin (PG), 5-series leukotriene (LT), and 3-series thromboxane (TX). DHA is also used to make 3-series PG. The binding sites on G-protein-coupled receptors are fiercely contested by the eicosanoids produced by AA and EPA. Furthermore, the EPA limits AA's competitiveness. A decrease in AA production is caused by a decrease in the amount of D5 desaturase activity that can convert dihomo-gamma-linolenic acid (dGLA) to AA (Calder 2002).

Eicosanoids derived from AA are reduced in vitro by EPA due to their ability to inhibit the enzymes cyclooxygenase-2 (COX-2) and 5-lipoxygenase (5-LOX). To initiate the inflammatory response, eicosanoids must first be synthesized. An additional step in reducing inflammation and reestablishing homeostasis is the creation of specialized pro-resolving mediators (SPMs) from bioactive lipid mediators known as n-3 long-chain polyunsaturated fatty acids (LC-PUFAs) (Obata et al. 1999). The first thing that happens when inflammation starts is the production of eicosanoids. The next stage is to create SPMs from n-3 LC-PUFAs, which are bioactive lipid mediators, in order to reduce inflammation and return to homeostasis. While EPA is the progenitor of the E-series resolvins, DHA is the progenitor of the D-series resolvins, which include neuroprotectin D1 (NPD1) and maresin 1–2 (Mar1–2) (Serhan and Chiang 2013).

These substances can alleviate inflammation and hasten recovery without interfering with immune system activity (Serhan et al. 2008). Active coordination and fine-tuning of the inflammatory response is facilitated by SPMs. They improve phagocytosis of dead cells and cell debris, decrease pro-inflammatory cytokines, and raise anti-inflammatory ones even in the absence of immune suppression. After phospholipase A2 releases them from membrane phospholipids in reaction to stimulation, CYP450, LOX, and COX-2 produce them. The brain is where enzymes are most visibly expressed (Shalini et al. 2018).

Hippocampal COX-2 production is in response to inflammatory lipopolysaccharide (Czapski et al. 2010), and COX-2 inhibition prolongs the duration of acute inflammation before it goes away. The concentrations of 15-LOX and 5-LOX are highest in the brain (Yigitkanli et al. 2017). Decreased SPM production and cognitive impairments are associated with 15-LOX deletion or inhibition (Shalini et al.

2018). Epoxides derived from n-6 that have anti-inflammatory properties are produced by CYP450 (Gilroy et al. 2016). Other cell types that express these enzymes include microglia, astrocytes, oligodendrocytes, and neurons (Farooqui et al. 2007).

Eicosanoids derived from AA are reduced in vitro by EPA due to their ability to inhibit the enzymes cyclooxygenase-2 (COX-2) and 5-lipoxygenase (5-LOX) (Obata et al. 1999). To begin the inflammatory response, eicosanoids must first be synthesized. An additional step in reducing inflammation and reestablishing homeostasis is the creation of SPMs from bioactive lipid mediators known as n-3 LC-PUFAs. NPD1 is known for its neuroprotective and anti-inflammatory properties, playing a crucial role in promoting cell survival and reducing oxidative stress in neuronal tissues, particularly in the brain and retina, while also aiding in the resolution of inflammation and protecting against neurodegeneration. In contrast, Mar1-2, derived from EPA, enhance the clearance of apoptotic cells and promote tissue repair, modulating immune responses to further facilitate the resolution of inflammation and restore homeostasis in affected tissues. Together, NPD1 and Mar1-2 exemplify the beneficial effects of n-3 LC-PUFAs in mitigating inflammation and supporting recovery from injury or stress in various tissues, particularly in the nervous system.

Without interfering with immune system function, these chemicals can decrease inflammation and accelerate healing (Serhan et al. 2008). As a whole, SPMs are responsible for coordinating and fine-tuning the inflammatory response. They elevate anti-inflammatory cytokines, reduce pro-inflammatory ones, and improve phagocytosis of dead cells and cell debris even in the absence of immune suppression. In reaction to stimuli, phospholipase A2 releases them from membrane phospholipids, while CYP450, LOX, and COX-2 synthesize them. Enzyme expression occurs in the brain (Shalini et al. 2018).

The hippocampus releases COX-2 as soon as it detects the inflammatory lipopolysaccharide (Czapski et al. 2010) A longer duration of acute inflammation until it subsides is caused by decreasing COX-2. For both 15-LOX and 5-LOX, the brain is where it is at (Yigitkanli et al. 2017). Cognitive impairments and reduced SPM generation in the brain have been linked to 15-LOX loss or inhibition (Shalini et al. 2018). Epoxides generated from n-6 that have anti-inflammatory properties are produced by CYP450 (Gilroy et al. 2016). In addition to neurons, these enzymes are expressed by astrocytes, microglia, oligodendrocytes, and astrocytes (Farooqui et al. 2007).

11.4.2 Dietary Factors Influencing SPM Concentrations

The dietary intake of these lipid byproducts can be controlled. Taking n-3 LC-PUFA supplements increases levels of SPMs produced from n-3 PUFA and decreases levels of SPMs derived from n-6 PUFA; this leads to an increase in n-3 LC-PUFA in the hippocampus, as demonstrated earlier. What biological mechanisms give birth to these bioactive lipid derivatives is now a mystery. This paragraph explains how dietary PUFA supplements change the PUFA makeup of brain cells, which could affect the PUFA lipid derivatives found in brain cells. Animals supplemented with n-3 LC-PUFA showed anti-inflammatory SPM patterns when exposed to LPS, in contrast to those without this fatty acid (Rey et al. 2019).

Enzymes involved in biosynthesis can be controlled to adjust the concentration of SPM. In contrast to the group that does not receive a diet high in n-3 LC-PUFA, the group that does has an upregulated expression of 15-LOX messenger RNA (Rao et al. 2007; Rey et al. 2018). The ability of 15-LOX to activate PPAR-γ, which protects neurons, and its ability to produce RvD1 to sustain cognitive function are among its advantageous characteristics (Sun et al. 2015). Another negative consequence of 15-LOX that causes neurotoxicity and neurodegeneration is increased oxidative stress (Yigitkanli et al. 2017). This lends credence to the idea that an n-3 PUFA–rich diet could have a significant impact on the pro- and anti-inflammatory status of brain cells and the hippocampus through dietary changes in SPM quantity and composition.

11.5 PUFAs and the Maturation of the Brain

11.5.1 PUFA Accretion in the Brain in Development

Half- to three-quarters of the membranes of brain cells are composed of lipids (Sastry 1985). Aside from long-chain polyunsaturated fatty acids (LC-PUFAs) like DHA and ARA, there may be other kinds of fatty acids as well (such as monounsaturated and saturated fatty acids). This explains why PUFAs make up more than 30% of the human prefrontal cortex, with ARA accounting for around 9% and DHA accounting for 12%. (McNamara and Carlson 2006). The 12% DHA and 10% ARA content in rat brains is similar to that of human brains (Joffre et al. 2016b). Some examples of phospholipids that are esterified with fatty acids include phosphatidylcholine (PC; 42–44%), phosphatidylethanolamine (PE; 36–40%), phosphatidylserine (PS; 11–13%), and phosphatidylinositol (PI; 2–3%) (Sastry 1985). Most polymers, including PE and PS, esterify DHA, but PC and PI esterify ARA (Bascoul et al. 2016). The most striking difference between white matter and gray matter is the amount of DHA and ARA. White matter is significantly smaller in quantity (Brenna and Diau 2007). These variations could be explained by the fact that synaptosomes and vascular cell membranes, both of which are thought to serve as ARA depots in the brain, are different (Brenna and Diau 2007; Di Miceli et al. 2020). The high cholesterol and polyunsaturated fatty acid content of myelinated axons and grey matter, respectively, may be accounted for by the lipidic makeup of the myelin sheath (Poitelon et al. 2020). Consistent with other studies, the presence of high amounts of DHA in synaptosomes shows that DHA encourages the development of new neurons and synapses (Kim and Spector 2018).

It all begins in the brain during pregnancy with "accretion," the buildup of n-3 and n-6 polyunsaturated fatty acids. Babies absorb different amounts of DHA after delivery, and the amount of DHA a mother and her newborn take in throughout pregnancy is also dose-dependent (Haggarty 2010). During infancy and early childhood, the brain's DHA storage capacity increases by a factor of 30 (Lauritzen et al. 2001). During the first 2 years of life, the brain's capacity to retain DHA increases by around 30 times (Clandinin et al. 1980; Percy et al. 1996). In contrast to EPA's poor aggregation performance and the little accumulation of precursors LA and ALA, DHA and ARA exhibit rapid aggregation during the third trimester (Clandinin et al. 1980). In contrast to the rather steady postnatal LC PUFA accretion, the exponential n-3 and n-6 PUFA accretion occurs in humans between 26 and 43 weeks of gestation (Martinez 1991).

By the time a human baby reaches the fourth week of gestation, its DHA levels can reach 40 mg/day (Kuipers et al. 2012). Higher DHA synthesis from ALA and mobilization of maternal DHA reserves from adipose tissue allow mothers to satisfy the increased DHA demand of their fetal brains (Metherel et al. 2017). During the first 3 weeks of a newborn rat's life, the accumulation of DHA in the brain grows tenfold (Sinclair and Crawford 1972).

The rate of LC-PUFA accumulation by specific cell types during brain development is poorly understood. These cell types include neurons, astrocytes, microglia, and oligodendrocytes. Knowing this would be fascinating because neuronal and glial cell division and maturation do not occur simultaneously. We also have no idea how DHA/ARA aids in development specialization. Synaptosome accumulation of DHA regulates synaptic protein activity, as mentioned before (Sidhu et al. 2011). DHA is also found in astrocytes (Champeil-Potokar et al. 2006), microglia (Rey et al. 2018), and oligodendrocyte membranes (Brand et al. 2008). Interesting new evidence suggests that DHA in glial cell membranes controls glial activity, syncytium development in astrocytes, and microglial neuroinflammation (Laye et al. 2018) or myelin formation in oligodendrocytes (Harel et al. 2018). Of note, EPA is found in microglia (Madore et al. 2020). It provides strong evidence that these cells may mediate the neuroprotective effects of EPA supplementation in psychological illnesses (Bazinet et al. 2020).

We still do not know how altered accretion affects glial cell development or what role ARA, DHA, and EPA accretion play in these cells throughout development in either full- or preterm neonates. Due to the anomalies in DHA and ARA accretion, preterm newborns are at genuine risk of LC n-3 PUFA insufficiency. During the third trimester of pregnancy, these fatty acids tend to build up (Lapillonne and Moltu 2016). Recent research has shown that ARA stunts the maturation of a crucial step in the wiring of brain networks: intraspinal microglia phagocytic function (Madore et al. 2020).

11.6 Naturally Occurring Polyunsaturated Fatty Acids in Brain Development

11.6.1 The Expression of Essential Enzymes in Brain Development

Although it seems that the brain can produce LC n-6 and n-3 PUFAs from their respective precursors, as mentioned earlier, the supply of PUFAs to the brain of a baby happens during gestation and breastfeeding. Extensive in vivo rat studies established that elongation and saturation were the primary mechanisms for LC-PUFA synthesis, with subsequent release in the maturing brain (Scott and Bazan 1989). Postnatal rats were tagged LA and ALA intracranially at 11 and 13 days postpartum by researchers to circumvent the liver's metabolism. These studies show that ARA, DHA, and LC-PUFAs may be produced in the brain after birth (Dhopeshwarkar and Subramanian 1976).

Sanders et al. showed that the brains of 21-day-old fetal rats could produce LC PUFAs. They also discovered the presence of D6D, an enzyme that converts LA and ALA (Sanders and Naismith 1980; Sanders and Rana 1987). Toward the decade's close, scientists were able to clone and characterize the fads1 genes in both humans and other animals. The fads1 gene, which encodes the D5D enzyme, is also highly expressed in the developing brain of a human embryo, according to the research (Cho et al. 1999; Aki et al. 1999; Leonard et al. 2000).

In addition, the expression of many members of the Elovl family exhibits significant species-specific diversity (Elovl1, Elovl3, Elovl4, Elovl5, and Elovl6). While Elovl7 and Elovl2 are expressed at extremely low levels in mammalian brains, Elovl2 is found in non-mammal species (such as fish). The Elovl4 protein, which is present in both fish and mammals, helps in the synthesis of very low-constant polyunsaturated fatty acids. Incorporating more EPA and DHA into your diet enhances Delta-5 desaturase (D5D) activity while decreasing Delta-6 desaturase (D6D) activity. D5D activity was regulated by dietary EPA and DHA in the presence of the rs174537 genotype (Al-Hilal et al. 2013).

11.7 Brain Developmental Effects of Polyunsaturated Fatty Acids

11.7.1 Neuronal Development and the Function of Polyunsaturated Fatty Acids (PUFAs)

Synaptogenesis and the formation of structural connections are ongoing processes during fetal development, a time when DHA levels rise and the ARA/DHA ratio falls (Dubois et al. 2014). In the perinatal period, when synaptogenesis is active and structural connection is being established, the ARA/DHA ratio falls, and DHA increases (Coti Bertrand et al. 2006; Yavin et al. 2009). How exactly LC-PUFAs help with brain growth is a mystery. Although this article mainly focuses on the most well-known developmental processes—from myelinization to synaptogenesis—the mechanisms described here also govern many other processes, such as cell migration, proliferation, differentiation, neurogenesis, and myelinization. PUFAs also control these processes. Research has shown that DHA and ARA can influence brain stem cell proliferation, differentiation, and neurogenesis (NSCs). It is during the postnatal period, in newborns, and during embryonic brain development when neuronal cells migrate when n-3 polyunsaturated fatty acid (PUFA) deficiencies are present in the diet throughout early life (Sakayori et al. 2016). In keeping with these results, other ARA metabolites, including as PGE2, have been seen to promote NSC proliferation and neuronal lineage cell differentiation (Wong et al. 2016). The binding of DHA to GPR120 and β2-AR helps in the in vitro differentiation of astrocytes (Joardar et al. 2006; Das and Das 2019). Notably, ARA metabolites contribute to gliogenesis, as dietary LA enhances ARA-derived endocannabinoids, which promote astrogliogenesis from NSCs (Shinjyo et al. 2013).

Enzymes involved in lipid metabolism have been discovered at synaptic terminals, enabling local modulation of synaptic transmission, which is relevant to the influence of long-chain polyunsaturated fatty acids on synaptogenesis (Rohrbough et al. 2004). For mature synapses to develop throughout maturation, the brain's membranes must expand, which in turn requires a high concentration of ARA and DHA (Martin and Bazan 1992). Membrane signaling and synaptogenesis are both influenced by DHA during prenatal brain development in rats (Martin 1998) by accumulating in the neuronal growth cone (Innis and de la Presa 2001) and mature synaptic membranes (Suzuki et al. 1997).

In an in vivo model of *Xenopus laevis* embryos, researchers discovered that the quantity of n-3 PUFA that mothers ingest impacts the development of neuronal branching and synaptic connection. Mothers' diets were supplemented with fish oil after the embryos were given diets that were either sufficient or lacking in n-3 PUFA. These changes were also linked to a decrease in brain-derived neurotrophic factor (BDNF), according to this model (Igarashi et al. 2015). Drosophila lacking polyunsaturated fatty acids also has problems with synaptic transmission in the visual system (Ziegler et al. 2015). There may be a connection between the formation and improvement of synapses and the esterified version of DHA, which influences the physical properties of membranes (Manni et al. 2018). Consequently, in vivo DHA supplementation increases levels of pre- and postsynaptic proteins involved in synaptic transmission and long-term potentiation, and it improves neurite formation and synaptogenesis (Cao et al. 2009).

11.7.2 Microglia and Neuroinflammation: PUFAs' Role in Brain Developmental Regulation

Recent years have seen a surge in studies investigating the role of long-chain polyunsaturated fatty acids (LC-PUFAs) in neuroinflammation and microglia activity in developing brains (Rey et al. 2018). As a result of their interactions with immune cells that have infiltrated the brain, microglia which comprise around 5–15% of brain cells and look like macrophages are well-recognized as playing a crucial role in neuroinflammatory reactions to injury or lesions (Greenhalgh et al. 2020).

The regulation of neuroinflammation is maintained even throughout brain development by n-3 and n-6 polyunsaturated fatty acids (PUFAs), either directly or through their metabolites, oxylipins, and endocannabinoids (Madore et al. 2016). The effects of inflammation and, perhaps, neuroinflammation can be influenced by single nucleotide polymorphisms (SNPs) that influence PUFA metabolism, according to scientific investigations. Research has demonstrated that when exposed to environmental triggers, peripheral blood mononuclear cells from children whose mothers have minor alleles of the trend rs174556 generate elevated levels of IL-10, IL-17, and IL-5 (Muc et al. 2015).

Despite microglia's reputation for causing neuroinflammation, mounting research has identified their function in developing brain wiring (Thion et al. 2018). The embryonic brain first develops glial cells called microglia. There has been a great deal of discussion over where microglia come from throughout their development (Askew and Gomez-Nicola 2018). Prior to fate-mapping research, it was thought that microglia originated from erythro-myeloid progenitors in the yolk sac around embryonic day (ED) rather than mesoderm or monocytic progenitors (Hoeffel and Ginhoux 2018). In the early stages of hematopoiesis, pre-macrophages are produced; by the time they reach embryonic development (ED), they have transformed into microglia and have started to encompass the whole brain (Mass et al. 2016). In the second week following birth, microglia in mice complete their maturation process, which consists of three distinct phases (Bennett et al. 2016). Inflammatory events during pregnancy or early life impact microglial developmental trajectories, which is crucial. Neurodevelopmental disorders may be more likely to occur as a result of these inflammatory events' potentially harmful effects on brain circuitry (Paolicelli and Gross 2011; Lenz and Nelson 2018). Synaptic pruning and brain architecture mediated by microglia have been the subject of much research, both in normal and abnormal states (Paolicelli and Gross 2011).

11.8 PUFA Role in TBI Treatment

An important part of the early brain development process is the presence of omega-3 polyunsaturated fatty acids (PUFAs), which are necessary for the creation of neurons and brain tissue. There is hope for the treatment of neurotrauma using docosahexaenoic acid and eicosatetraenoic acid. Significant disability and social and economic impacts can result from spinal cord injuries (SCIs) and traumatic brain injuries (TBIs). Consequently, there is a lack of solutions for both treatments that hasten neurodegeneration and those that provide quick protection for neurons. After an experimental traumatic brain injury (TBI) or spinal cord injury (SCI), neurological results are improved by immediate food exposure to omega-3 polyunsaturated fatty acids (PUFAs) or by fast infusion of these PUFAs. Factors that are at work include providing neurotrophic support, activating pathways for cell survival, and reducing neuroinflammation and oxidative stress (Risdall and Menon 2011).

Omega-3 polyunsaturated fatty acids (PUFAs) play an important function in the early phases of brain development since they are necessary for the creation of neurons and brain tissue. Results from treating neurotrauma with docosahexaenoic acid and eicosatetraenoic acid seem promising. Traumatic brain injuries (TBIs) and spinal cord injuries (SCIs) are major public health and economic concerns due to the long-term disability they can produce. This is why treatments that hasten neurodegeneration as well as those that offer quick neuroprotection are underfunded. To enhance neurological outcomes in experimental TBIs and SCIs, omega-3 polyunsaturated fatty acids (PUFAs) can be administered acutely after injury or consumed before or after damage. Efforts are underway to decrease neuroinflammation and oxidative stress, activate pathways that aid cell survival, and provide neurotrophic support (Risdall and Menon 2011).

Omega-3 polyunsaturated fatty acids (PUFAs) are integral structural constituents of human cell membranes, playing a pivotal role in maintaining cellular integrity and functionality. Their significance is particularly pronounced in the context of neurological health, where they are implicated in mitigating the harmful effects of physical trauma on the nervous system. Numerous investigations conducted in the past 10 years have shown that these naturally occurring chemicals possess a crucial quality: they have the ability to alter the response of the nervous system to sudden stress. The treatment of patients suffering from spinal cord injuries (SCIs) and traumatic brain injuries (TBIs) is a formidable task for neurologists. Usually, these conditions strike young adults. Both kinds of injuries are costly to treat and severely limit patients' mobility and independence. There is a serious lack of representation in the current therapeutic landscape for acute-phase neuroprotective therapies that aim to maintain the brain and spinal cord, as well as treatments that encourage positive neuroplasticity and regeneration, resulting in partial functional recovery. Based on the results that have been reported, omega-3 polyunsaturated fatty acids (PUFAs) are going to be big in this field of medicine shortly (Risdall and Menon 2011).

Traumatic brain injuries (TBIs) in civilians are mostly caused by falls (20–30%) and car accidents before contact with sports (60%). The long-term consequences on young adults affected by mild traumatic brain injuries (TBIs) and their recurrence in popular sports such as football, hockey, boxing, wrestling, and rugby are quite concerning. Blast waves generated by explosive devices are among the leading causes of traumatic brain injuries (TBIs) in combat (Risdall and Menon 2011). As the population ages, the demographics of those affected by neurotrauma are changing. Statistically, TBIs and SCIs are more common in adults over the age of 50, and the mortality and morbidity associated with these injuries are worse in this age group (Franceschini et al. 2011).

The prognosis for traumatic brain injuries (TBIs) is greatly affected by the care administered in the initial minutes after the damage. The current focus of treatment regimens for traumatic brain injuries includes managing intracranial pressure, preventing and treating hypotension, and ensuring appropriate breathing. At this time, there is no known neuroprotective medication. Long-term effects of traumatic brain injury (TBI) include secondary epilepsy and persistent headaches; "post-concussion syndrome" describes changes in personality, cognitive impairment, anxiety, and depression. Medical treatment for traumatic brain injury (TBI) has improved overall during the past 30 years, but the associated death and morbidity rates have not decreased significantly (Roozenbeek et al. 2013).

The prognosis for traumatic brain injuries is greatly affected by how quickly they are treated. The current treatment guidelines for traumatic brain injury include managing intracranial pressure, preventing and treating hypotension, and proper breathing. At this time, there is no known medicine that can adequately shield neurons from damage. Traumatic brain injury (TBI) comes with a host of long-term effects, including neuropsychiatric sequelae including anxiety, depression, cognitive failure, and changes in personality (sometimes known as "post-concussion syndrome"). While there has been some progress in TBI treatment over the past few decades, the associated mortality and morbidity rates have remained stagnant (Mann et al. 2013).

Blood and cerebrospinal fluid biomarkers have been the subject of a great deal of research on traumatic brain injuries and spinal cord injuries. Obtaining cerebrospinal fluid (CSF) can be a bit of a hassle in cases of severe traumatic brain injury (TBI) when ventriculostomies are prevalent. However, in most cases, it is more feasible to use blood, urine, or saliva as biomarkers for TBI and SCI (Hurlbert et al. 2013).

11.9 The Pathophysiology of TBI

The prognosis for traumatic brain injuries (TBIs) can be greatly affected by the therapy administered in the minutes immediately after the accident. Current guidelines for the management of TBI emphasize the significance of preserving a proper airway, avoiding and managing hypotension, and regulating intracranial pressure. None of the therapies have proven to be neuroprotective. Traumatic brain damage can lead to long-term effects such as secondary epilepsy and persistent headaches. Negative emotions, cognitive impairment, personality changes, and anxiety are other neuropsychiatric consequences (dubbed "post-concussion syndrome"). Although there has been significant progress in the treatment of traumatic brain injuries (TBIs) during the past few decades, the rates of morbidity and mortality have not decreased in tandem (Loane and Faden 2010; Priestley et al. 2012). Several causes contribute to ischemia, such as reduced systemic blood pressure, poor tissue oxygenation, and localized blood flow control issues (Senter and Venes 1979). The secondary injury wave is characterized by a series of biochemical events that increase cell death and tissue damage (Profyris et al. 2004). Neurons and glia are vulnerable to excitotoxicity, which results from damaged cells overflowing glutamate (Liu et al. 1999). When glutamate receptors and voltage-gated calcium channels allow calcium ions to enter the cell, fatty acid release happens. Phospholipase A2 and other calcium-dependent proteases, including calpains, are activated during this process. Following a traumatic brain injury (TBI), the intensity of the activation of phospholipase A2, as determined by the release of fatty acids into cerebrospinal fluid (CSF), is linked to the outcome 1 week later (Pilitsis et al. 2003).

Omega-6 polyunsaturated fatty acid (AA) produced from cell membranes is converted into eicosanoids that induce inflammation by the enzymes cyclooxygenase (COX-1 and COX-2) and lipoxygenase. Eicosanoids exacerbate tissue damage by tightening blood vessels, causing platelets to clump together, and producing free radicals. Plasma AA levels in rats with simulated spinal cord injuries begin to climb 1 day after damage and remain elevated for a minimum of 4 weeks thereafter. Alternatively, plasma levels of docosahexaenoic acid (DHA) increase on the first day after SCI but decrease on the third day (López-Vales et al. 2010).

As a result of mitochondrial dysfunction, ATP synthesis decreases, Na+/K+ ATPases fail, and cells enlarge due to an inflow of Na^+. Oligodendrocytic and neuronal apoptosis follow caspase activation, which follows the first cascade of necrotic cell death. Oligodendrocyte apoptosis occurs in the first 24 h after an SCI and continues along white matter pathways for the first 7 days. Following a traumatic brain injury or spinal cord injury, the body's inflammatory response serves a dual purpose. Even while inflammatory cells remove necrotic material, their prolonged presence can exacerbate subsequent damage (Beck et al. 2010; Donnelly and Popovich 2008).

Neutrophils start to congregate at the site of an injury right away, but they disperse within a few days. In the first week, microglia activation is at its highest. Macrophages from the periphery invade, and local microglia also take on a macrophage character. Human inflammation follows a pattern similar to that observed in rats (Fleming et al. 2006). There are intricate dynamics in the local production of cytokines and chemokines because of injury. In a few hours, levels of cytokines, including tumor necrosis factor-a (TNFa), interleukin (IL)-1b, and IL-6, as well as chemokines like IP-10, KC, GRO, CINC, and CCL-2/monocyte chemoattractant protein-1 (MCP-1), all increase (Frugier et al. 2010). Injuries can also set off a chain reaction of systemic inflammation. An elevation in C-reactive protein in plasma was seen in asymptomatic individuals with chronic SCI in the initial hours following rat compression injury (Hall et al. 2012; Frost et al. 2005).

Within the first few hours following a TBI, astrocytes begin to activate; within a week, a scar composed of these cells will develop. A cavity may form to encase this scar, which will assist in containing the wound over time. Although astrocyte loss can have detrimental repercussions, pro-regenerative therapies have focused on the scar (Bradbury 2002). Damage to the blood-brain barrier, further cellular deterioration, and inflammation result from eliminating reactive astrocytes (Faulkner et al. 2004; Myer et al. 2006).

Several therapeutic approaches have targeted the post-TBI processes of reactive angiogenesis and neurogenesis (Myer et al. 2006). The observable behavioral impairments following traumatic brain injury are significantly impacted by axonal damage (Felix et al. 2012). Axon injury can occur in moderate and focused traumatic brain injuries (TBIs) although it was initially described in patients with severe brain injuries (Bazarian et al. 2007). Proteoglycans significantly impact plasticity after traumatic brain injury. Treatment with chondroitinase improves plasticity after TBI by eliminating inhibitory chondroitin sulfate side chains (Harris et al. 2013).

11.9.1 Role of Polyunsaturated Fatty Acids (PUFAs) in Traumatic Brain Injury

Researchers initially found that DHA may help with traumatic brain injury (TBI) in adult rats by employing an impact accelerated damage model (TBI). The rats were administered an omega-3 fatty acid supplement 30 days following injury. Sixty percent of the supplement's overall makeup was EPA and DHA, which were present in a 2:1 ratio. The dosage for EPA was 6 mg/kg daily, whereas DHA was 24 mg/kg. Treatment results included reduced caspase-3 levels and axonal protection (Bailes and Mills 2010; Mills et al. 2011a).

One possible substitute for TBI models is oils rich in omega-3 fatty acids, often known as polyunsaturated fatty acids (PUFAs). Wu et al. found that rats were given an 8% fish oil supplement 4 weeks before mild fluid percussion TBI. The prescription was continued for an additional week following the injury. As a measure of spatial learning, the Morris water maze showed improved performance with reduced oxidative stress. The levels of synapsin I, BDNF, and CREB proteins returned to normal following therapy. Omega-3 polyunsaturated fatty acid treatment before injury also reduced oxidative stress and trauma-induced alterations in cellular energy metabolism (Wu et al. 2004).

Thirty days before the injury, rats took DHA supplements at doses of 4, 12, or 40 mg/kg/day. After the rats had been injured for 7 days, Bailes and colleagues put them through their paces in the Morris water labyrinth. Improvements in behavior, reductions in microglial activation and caspase-3 levels, and axonal dysfunction were all noted at the maximum dosage studied (Mills et al. 2011b). In the first 2 weeks following fluid percussion-induced mild traumatic brain injury (TBI), the study found that the neuroprotective effects of DHA, when coupled with exercise, were at their peak in rats. It seems like the two treatments worked together more effectively (Wu et al. 2013). Damage reduced the tissue content of DHA, but supplementation brought it back to normal levels. Subcortical shielding has also been discovered. Supplementing the diets of rats with fish oil can counteract the effects of traumatic brain injury on dopamine release in the striatum, according to a study by Shin and Dixon (Shin and Dixon 2011).

11.9.2 Effects of Omega-3 Polyunsaturated Fatty Acids and Their Cellular Receptors

Similar to other neuroprotective drugs with a favorable translational profile, such as progesterone, erythropoietin, and minocycline, omega-3 polyunsaturated fatty acids (PUFA) provide neuroprotection in models of spinal cord injury and traumatic brain injury. Supplementation with DHA and EPA after acute parenteral, systemic, or chronic oral delivery has been investigated in neurotrauma and other neurological areas. The exact processes that cause these impacts are not yet fully understood (Dyall and Michael-Titus 2008). When compared to other neuroprotective methods with a proven track record of translation, such as progesterone, erythropoietin, and minocycline, the level of neuroprotection shown with omega-3 PUFA in TBI and SCI models is comparable. Acute parenteral, systemic, and chronic oral supplementation with DHA and EPA has been demonstrated to be beneficial in neurological disorders and other related fields. Nevertheless, it is still not known how precisely this happens (Vreugdenhil et al. 1996; Heurteaux et al. 2004). Curiously, in a brain damage model involving mice, TREK-1 ablation did not result in neuroprotection (Namiranian et al. 2011).

Omega-3 polyunsaturated fatty acids (PUFA) provide neuroprotection in models of traumatic brain injury and spinal cord injury that is comparable to other neuroprotective medicines with a favorable translational profile, such as minocycline, erythropoietin, and progesterone. Acute parenteral, systemic, and chronic oral administration of DHA and EPA have been investigated in the context of neurotrauma and other neurological disorders. We still do not know how these impacts work (Dyall and Michael-Titus 2008). While other neuroprotective methods like progesterone, erythropoietin, and minocycline have a proven track record of translation, the amount of neuroprotection seen with omega-3 PUFA in models of TBI and SCI is comparable. Evidence suggests that DHA and EPA, whether administered orally in acute, systemic, or chronic doses, can help neurological conditions, including neurotrauma. But how exactly this happens is still a mystery (Vreugdenhil et al. 1996; Heurteaux et al. 2004). It is worth noting that in mice brain injuries, TREK-1 ablation does not result in neuroprotection (Namiranian et al. 2011).

The complex family of active metabolites produced by EPA and DHA includes resolvins, docosatrienes, and protectins, among many others. Among these metabolites (Kohli and Levy 2009), a family of DHA derivatives made by macrophages called maresins aids in the resolution of inflammation (Serhan et al. 2009). For instance, resolvin A4 interacts with the GPR32 receptor, whereas resolvin D1 binds to the FPR2/ALX and GPR32 receptors (Krishnamoorthy et al. 2012). One receptor, ChemR23, and another, LTB4, are bound by resolvin E1 (Serhan et al. 2011). Resolvins, docosatrienes, and protectins are a complex family of active metabolites that are produced by EPA and DHA. In the absence of inflammation, these metabolites play an essential role (Bazan 2005).

Resolvins, docosatrienes, and protectins are all members of a vast and complex family of active metabolites produced by EPA and DHA. Metabolites such as maresins, a kind of DHA derivative produced by macrophages, help inflammation resolve (Kohli and Levy 2009) (Serhan et al. 2009). One example is the binding of resolvin D1 to the FPR2/ALX and GPR32 receptors and the interaction of resolvin A4 with the GPR32 receptor (Krishnamoorthy et al. 2012). The ChemR23 and LTB4 receptors are both bound by resolvin E1 (Serhan et al. 2011). When you combine EPA and DHA, you get a complex family of active metabolites called resolvins, docosatrienes, and protectins. These metabolites play an essential role after inflammation decreases (Bazan 2005).

Neuronal cultures from developing hippocampus and cortex reveal that DHA increases the total number of neurites and their maximum length (Cao et al. 2005), similar impact on primary sensory neurons in rats (Robson et al. 2010). An increase in synaptic proteins such as syntaxin-3 after DHA therapy in rats with TBI might be due to the formation of longer neurites (Wu et al. 2011).

Interestingly, compared to control mice, fat-1 animals had a better prognosis after sciatic nerve injury (Gladman et al. 2012). We do not yet know the specific mechanism by which this benefit occurs although it may be because of neuroprotection, enhanced plasticity, or the natural regeneration capacity of tissues rich in omega-3 PUFAs. Lastly, a shift in neurogenesis may underlie the therapeutic benefits of omega-3 PUFA following damage to the central nervous system. The adult hippocampus's neurogenesis is controlled by DHA via GRP40 and, maybe, CREB (Ma et al. 2008).

Traumatic brain injury (TBI) also initiates neurogenesis and endogenous cell proliferation (Kernie and Parent 2010). An increase in neurogenesis is associated with better neurological outcomes (such as cognition), suggesting a possible causal link (Xiong et al. 2008). It is suggested that omega-3 PUFAs have a beneficial effect on the stem cell pool based on the fact that fat-1 mice perform better than their wild-type littermates in neurogenesis and spatial learning (He et al. 2009a). Transgenic mice also outperform their wild-type counterparts in terms of neurogenesis and oligodendrogenesis after a stroke (Hu et al. 2013).

Desai investigated the effect of dietary n-3 PUFA on TBI outcome using rCHIMERA, a modality of mild brain injury that is relevant to clinical practice since it produces persistent gliosis in the brain and impaired memory in mice (Chen et al. 2017). Their results show that rCHIMERA-induced vision impairment and memory loss can be ameliorated by a diet high in n-3 PUFA. This is achieved by elevating brain DHA levels, lowering optic tract (OT) axon degeneration, and inhibiting gliosis in many brain regions (Desai et al. 2021).

In response to brain damage, a diverse population of cells known as activated astrocytes and microglial cells react (Karve et al. 2016). Both reactive astrocytes and activated microglia have a role in the inflammation that follows a traumatic brain injury. Multiple factors, including injuries, phenotypes, the environment, and the passage of time, contribute to or detract from traumatic brain injury pathology (Pekny et al. 2014). In response to stimuli, microglia release both pro- and anti-inflammatory chemicals, including tumor necrosis factor alpha, cyclooxygenase-2, IL-4, IL-10, IL-13, and peroxisome proliferator-activated receptor gamma (Orihuela et al. 2016). However, astrocytes can upregulate immune components and exacerbate injuries, in addition to triggering microglia to produce neurotoxic astrocytes. We expect that reducing glial cell activation after TBI will improve TBI outcome since glial cell activation is associated with persistent damage and persists after TBI (Fehily and Fitzgerald 2017).

Although a recent quantitative MRI study failed to find any noticeable impact of the n-3 PUFA diet on white matter damage following TBI, the observed variation in water maze performance may be explained by the protective effects of increased n-3 PUFA on neuronal integrity of axons in both the OT and the hippocampus (HP) (Reyes et al. 2020). The appropriate n-3 diet considerably reduced the injury-induced decrease in visual evoked potentials (VEP), suggesting that there may not be a memory function-specific difference in water maze performance. If the protective benefits are due to changes in DHA enrichment in cell membranes or the formation of anti-inflammatory DHA metabolites after numerous TBIs, further inquiry can determine the cause (Kuda 2017).

When n-3 PUFA are few, DPAn-6 acts as a stand-in for DHA in the brain, and its ratio to DHA rises (Kim and Spector 2018). When mice are given an acceptable quantity of n-3 PUFA in their food, DPAn-6 is nearly not detected in their brains. However, when mice are given an inadequate amount of n-3 PUFA, it is present in substantial proportions. There is some evidence that human brains contain DPAn-6, which suggests that DHA levels in humans may be suboptimal (Hamazaki et al. 2010). Given that the DHA level in the central nervous system correlates well with that of blood cells, including RBCs, in humans (Guest et al. 2013) and rodents (Ward et al. 1998), the fatty acid compositional changes in RBCs can reflect changes in the brain. Research on the amounts of omega-3 fatty acids in human blood from around the world revealed that Western diets often include low levels of DHA. Several studies have shown that red blood cells typically contain an average of 2.71–7.84% DHA (Stark et al. 2016). Although it is challenging to directly compare humans and experimental animals, it is anticipated that the range of DHA levels (0.4–6.7%) shown in mouse red blood cells in this study would align with the range observed in humans with a diversified diet. Omega-3 polyunsaturated fatty acid (PUFA) supplements may have beneficial effects on humans, such as increasing the levels of n-3 fatty acids in red blood cells (RBCs) and DHA in the brain (Witte et al. 2010).

In the future, studies will show that ῠ-3 can improve neuronal survival, neurovascular regeneration, and cerebral remodeling, which lessens the impact of brain injury from ischemia and reperfusion (Jiang et al. 2019). Additionally, when given to stroke patients after an ischemia-induced ABI, DHA decreased damage severity by reorienting macrophage polarization toward the M2 subtype (Cai et al. 2018). According to the study by Chen et al., ῠ-3 PUFA can reduce inflammation and increase brain damage after experimental TBI by regulating the polarity of microglia. The deacetylation of the High Mobility Group Box 1 (HMGB1)/NF-κB pathway was a part of this molecular process that Silent information regulator sirtuin 1 (SIRT1) protein (SIRT1) helped enhance (Chen et al. 2018). Research conducted by Begum et al. suggests that administering DHA following a traumatic brain injury (TBI) might potentially reduce levels of endogenous stress, aberrant protein accumulation, and neurological deficits (Begum et al. 2014).

Following TBI, omega-3 exhibited a wide array of complex neuroprotective effects. Supplementation with omega-3 polyunsaturated fatty acids enhanced autophagy and decreased neuronal death after TBI (Chen et al. 2018). Omega-3 treatment enhances neurological recovery and reduces the severity of white matter damage caused by experimental traumatic brain injury, according to studies (Pu et al. 2013). According to the in vitro and in vivo evidence provided by Zhu et al., DHA reduces NOX generation in TBI by modulating Nrf2 signaling (Zhu et al. 2020). A short burst of DHA given during the time of acute emergency care after a TBI can enhance long-term neurological outcomes. In addition to the neuroprotective effects, the DHA-derived mediators resolvins and protectins are increased (Thau-Zuchman et al. 2020). It was also noted that DHA administration significantly increased NAD(P)H: quinone oxidoreductase (NQO-1) and heme oxygenase 1 (HO-1) (Zhu et al. 2018). The long-term neurological outcomes of traumatic brain injuries (TBIs) can be improved by administering a single bolus of DHA during the acute emergency intervention period. Additionally, mediators derived from DHA, such as resolvins and protectins, are upregulated, providing additional neuroprotection (Huang et al. 2022).

Aside from lowering neuronal damage caused by hypoglycemia, which was associated with the activation of the downstream NF-κB pathway, ssDHA also decreased the necroptosis pathway, as shown in preliminary research (Chen et al. 2022). Evidence from early studies suggests that DHA can protect neurons from hypoglycemia-induced necroptosis and other forms of cell death. It was discovered that this damage was linked to the activation of the downstream NF-κB pathway (He et al. 2009b; Kaiser et al. 2013). The NF-κB pathways are part of the intricate chemical and process machinery that controls neuroinflammation and necroptosis. The confirmation of ó-3's effect on PPARγ levels has led to a reduction in NF-κB activation, neuroinflammation, and necroptosis. Inhibiting necroptosis signaling, DHA mitigated the effects of tumor necrosis factor (TNF) on cell damage and caspase-3 and caspase-8-induced pathway failure (Xiao et al. 2020).

References

Aki T, Shimada Y, Inagaki K, Higashimoto H, Kawamoto S, Shigeta S et al (1999) Molecular cloning and functional characterization of rat Δ-6 fatty acid desaturase. Biochem Biophys Res Commun 255(3):575–579

Alashmali SM, Hopperton KE, Bazinet RP (2016) Lowering dietary n-6 polyunsaturated fatty acids: interaction with brain arachidonic and docosahexaenoic acids. Curr Opin Lipidol 27(1):54–66

Al-Hilal M, AlSaleh A, Maniou Z, Lewis FJ, Hall WL, Sanders TA et al (2013) Genetic variation at the FADS1-FADS2 gene locus influences delta-5 desaturase activity and LC-PUFA proportions after fish oil supplement [S]. J Lipid Res 54(2):542–551

Askew K, Gomez-Nicola D (2018) A story of birth and death: insights into the formation and dynamics of the microglial population. Brain Behav Immun 69:9–17

Bailes JE, Mills JD (2010) Docosahexaenoic acid reduces traumatic axonal injury in a rodent head injury model. J Neurotrauma 27(9):1617–1624

Bascoul-Colombo C, Guschina IA, Maskrey BH, Good M, O'Donnell VB, Harwood JL (2016) Dietary DHA supplementation causes selective changes in phospholipids from different brain regions in both wild type mice and the Tg2576 mouse model of Alzheimer's disease. Biochim Biophys Acta Mol Cell Biol Lipids 1861(6):524–537

Bazan NG (2005) Neuroprotectin D1 (NPD1): a DHA-derived mediator that protects brain and retina against cell injury-induced oxidative stress. Brain Pathol 15(2):159–166

Bazarian JJ, Zhong J, Blyth B, Zhu T, Kavcic V, Peterson D (2007) Diffusion tensor imaging detects clinically important axonal damage after mild traumatic brain injury: a pilot study. J Neurotrauma 24(9):1447–1459

Bazinet RP, Metherel AH, Chen CT, Shaikh SR, Nadjar A, Joffre C et al (2020) Brain eicosapentaenoic acid metabolism as a lead for novel therapeutics in major depression. Brain Behav Immun 85:21–28

Beck KD, Nguyen HX, Galvan MD, Salazar DL, Woodruff TM, Anderson AJ (2010) Quantitative analysis of cellular inflammation after traumatic spinal cord injury: evidence for a multiphasic inflammatory response in the acute to chronic environment. Brain 133(2):433–447

Begum G, Yan HQ, Li L, Singh A, Dixon CE, Sun D (2014) Docosahexaenoic acid reduces ER stress and abnormal protein accumulation and improves neuronal function following traumatic brain injury. J Neurosci 34(10):3743–3755

Bennett ML, Bennett FC, Liddelow SA, Ajami B, Zamanian JL, Fernhoff NB et al (2016) New tools for studying microglia in the mouse and human CNS. Proc Natl Acad Sci USA 113(12):1738–1746

Blanchard H, Pédrono F, Boulier-Monthéan N, Catheline D, Rioux V, Legrand P (2013) Comparative effects of well-balanced diets enriched in α-linolenic or linoleic acids on LC-PUFA metabolism in rat tissues. Prostaglandins Leukot Essent Fatty Acids 88(5):383–389

Bourre JM, Pascal G, Durand G, Masson M, Dumont O, Piciotti M (1984) Alterations in the fatty acid composition of rat brain cells (neurons, astrocytes, and oligodendrocytes) and of subcellular fractions (myelin and synaptosomes) induced by a diet devoid of n-3 fatty acids. J Neurochem 43(2):342–348

Bradbury EJ (2002) Chondroitinase ABC promotes functional recovery after spinal cord injury. Nature 416:589–590

Brand A, Schonfeld E, Isharel I, Yavin E (2008) Docosahexaenoic acid-dependent iron accumulation in oligodendroglia cells protects from hydrogen peroxide-induced damage. J Neurochem 105(4):1325–1335

Brenna JT, Diau G-Y (2007) The influence of dietary docosahexaenoic acid and arachidonic acid on central nervous system polyunsaturated fatty acid composition. Prostaglandins Leukot Essent Fatty Acids 77(5–6):247–250

Broadhurst CL, Wang Y, Crawford MA, Cunnane SC, Parkington JE, Schmidt WF (2002) Brain-specific lipids from marine, lacustrine, or terrestrial food resources: potential impact on early African Homo sapiens. Comp Biochem Physiol B Biochem Mol Biol 131(4):653–673

Cai W, Liu S, Hu M, Sun X, Qiu W, Zheng S et al (2018) Post-stroke DHA treatment protects against acute ischemic brain injury by skewing macrophage polarity toward the M2 phenotype. Transl Stroke Res 9:669–680

Calder PC (2002) Dietary modification of inflammation with lipids. Proc Nutr Soc 61(3):345–358

Calder PC (2007) Immunomodulation by omega-3 fatty acids. Prostaglandins Leukot Essent Fatty Acids 77(5–6):327–335

Cao D, Xue R, Xu J, Liu Z (2005) Effects of docosahexaenoic acid on the survival and neurite outgrowth of rat cortical neurons in primary cultures. J Nutr Biochem 16(9):538–546

Cao D, Kevala K, Kim J, Moon HS, Jun SB, Lovinger D et al (2009) Docosahexaenoic acid promotes hippocampal neuronal development and synaptic function. J Neurochem 111(2):510–521

Champeil-Potokar G, Chaumontet C, Guesnet P, Lavialle M, Denis I (2006) Docosahexaenoic acid (22: 6n-3) enrichment of membrane phospholipids increases gap junction coupling capacity in cultured astrocytes. Eur J Neurosci 24(11):3084–3090

Chen CT, Bazinet RP (2015) β-Oxidation and rapid metabolism, but not uptake regulate brain eicosapentaenoic acid levels. Prostaglandins Leukot Essent Fatty Acids 92:33–40

Chen CT, Liu Z, Ouellet M, Calon F, Bazinet RP (2009) Rapid β-oxidation of eicosapentaenoic acid in mouse brain: an in situ study. Prostaglandins Leukot Essent Fatty Acids 80(2–3):157–163

Chen H, Desai A, Kim H-Y (2017) Repetitive closed-head impact model of engineered rotational acceleration induces long-term cognitive impairments with persistent astrogliosis and microgliosis in mice. J Neurotrauma 34(14):2291–2302

Chen X, Chen C, Fan S, Wu S, Yang F, Fang Z et al (2018) Omega-3 polyunsaturated fatty acid attenuates the inflammatory response by modulating microglia polarization through SIRT1-mediated deacetylation of the HMGB1/NF-κB pathway following experimental traumatic brain injury. J Neuroinflammation 15(1):1–15

Chen J, Li M, Liu Z, Wang Y, Xiong K (2022) Molecular mechanisms of neuronal death in brain injury after subarachnoid hemorrhage. Front Cell Neurosci 16:1025708

Cho HP, Nakamura M, Clarke SD (1999) Cloning, expression, and fatty acid regulation of the human Δ-5 desaturase. J Biol Chem 274(52):37335–37339

Chung W-L, Chen J-J, Su H-M (2008) Fish oil supplementation of control and (n-3) fatty acid-deficient male rats enhances reference and working memory performance and increases brain regional docosahexaenoic acid levels. J Nutr 138(6):1165–1171

Clandinin M, Chappell J, Leong S, Heim T, Swyer P, Chance G (1980) Intrauterine fatty acid accretion rates in human brain: implications for fatty acid requirements. Early Hum Dev 4(2):121–129

Cohen SR, Bernsohn J (1973) Incorporation of 1-14C labeled fatty acids into isolated neuronal soma, astroglia and oligodendroglia from calf brain. Brain Res J 60(2):521–525

Coti Bertrand P, O'Kusky JR, Innis SM (2006) Maternal dietary (n-3) fatty acid deficiency alters neurogenesis in the embryonic rat brain. J Nutr 136(6):1570–1575

Czapski GA, Gajkowska B, Strosznajder JB (2010) Systemic administration of lipopolysaccharide induces molecular and morphological alterations in the hippocampus. Brain Res J 1356:85–94

Das M, Das S (2019) Docosahexaenoic acid (DHA) induced morphological differentiation of astrocytes is associated with transcriptional upregulation and endocytosis of β2-AR. Mol Neurobiol 56(4):2685–2702

Delpech J-C, Thomazeau A, Madore C, Bosch-Bouju C, Larrieu T, Lacabanne C et al (2015a) Dietary n-3 PUFAs deficiency increases vulnerability to inflammation-induced spatial memory impairment. Neuropsychopharmacology 40(12):2774–2787

Delpech J-C, Madore C, Joffre C, Aubert A, Kang JX, Nadjar A et al (2015b) Transgenic increase in n-3/n-6 fatty acid ratio protects against cognitive deficits induced by an immune challenge through decrease of neuroinflammation. Neuropsychopharmacology 40(3):525–536

DeMar JC Jr, Ma K, Bell JM, Rapoport SI (2004) Half-lives of docosahexaenoic acid in rat brain phospholipids are prolonged by 15 weeks of nutritional deprivation of n-3 polyunsaturated fatty acids. J Neurochem 91(5):1125–1137

Desai A, Chen H, Kevala K, Kim H-Y (2021) Higher n-3 polyunsaturated fatty acid diet improves long-term neuropathological and functional outcome after repeated mild traumatic brain injury. J Neurotrauma 38(18):2622–2632

Dhopeshwarkar GA, Subramanian C (1976) Intracranial conversion of linoleic acid to arachidonic acid: evidence for lack of Δ8 desaturase in the brain. J Neurochem 26(6):1175–1179

Di Miceli M, Bosch-Bouju C, Layé S (2020) PUFA and their derivatives in neurotransmission and synapses: a new hallmark of synaptopathies. Proc Nutr Soc 79(4):388–403

Donnelly DJ, Popovich PG (2008) Inflammation and its role in neuroprotection, axonal regeneration and functional recovery after spinal cord injury. Exp Neurol 209(2):378–388

Dubois J, Dehaene-Lambertz G, Kulikova S, Poupon C, Hüppi PS, Hertz-Pannier L (2014) The early development of brain white matter: a review of imaging studies in fetuses, newborns and infants. Neurosci J 276:48–71

Dyall S, Michael-Titus A (2008) Neurological benefits of omega-3 fatty acids. NeuroMolecular Med 10:219–235

Farooqui AA, Horrocks LA, Farooqui T (2007) Modulation of inflammation in brain: a matter of fat. J Neurochem 101(3):577–599
Faulkner JR, Herrmann JE, Woo MJ, Tansey KE, Doan NB, Sofroniew MV (2004) Reactive astrocytes protect tissue and preserve function after spinal cord injury. J Neurosci 24(9):2143–2155
Fehily B, Fitzgerald M (2017) Repeated mild traumatic brain injury: potential mechanisms of damage. Cell Transpl 26(7):1131–1155
Felix M-S, Popa N, Djelloul M, Boucraut J, Gauthier P, Bauer S et al (2012) Alteration of forebrain neurogenesis after cervical spinal cord injury in the adult rat. Front Neurosci 6:45
Fleming JC, Norenberg MD, Ramsay DA, Dekaban GA, Marcillo AE, Saenz AD et al (2006) The cellular inflammatory response in human spinal cords after injury. Brain 129(12):3249–3269
Franceschini M, Bazo HC, Lauretani F, Agosti M, Cristina PM (2011) Age influences rehabilitative outcomes in patients with spinal cord injury (SCI). Aging Clin Exp Res 23:202–208
Frost F, Roach MJ, Kushner I, Schreiber P (2005) Inflammatory C-reactive protein and cytokine levels in asymptomatic people with chronic spinal cord injury. Arch Phys Med Rehabil 86(2):312–317
Frugier T, Morganti-Kossmann MC, O'Reilly D, McLean CA (2010) In situ detection of inflammatory mediators in post mortem human brain tissue after traumatic injury. J Neurotrauma 27(3):497–507
Giedd JN, Blumenthal J, Jeffries NO, Castellanos FX, Liu H, Zijdenbos A et al (1999) Brain development during childhood and adolescence: a longitudinal MRI study. Nat Neurosci 2(10):861–863
Gilroy DW, Edin ML, De Maeyer RP, Bystrom J, Newson J, Lih FB et al (2016) CYP450-derived oxylipins mediate inflammatory resolution. Proc Natl Acad Sci USA 113(23):3240–3249
Gladman SJ, Huang W, Lim S-N, Dyall SC, Boddy S, Kang JX et al (2012) Improved outcome after peripheral nerve injury in mice with increased levels of endogenous omega-3 polyunsaturated fatty acids. J Neurosci 32(2):563–571
Green P, Glozman S, Kamensky B, Yavin E (1999) Developmental changes in rat brain membrane lipids and fatty acids: the preferential prenatal accumulation of docosahexaenoic acid. J Lipid Res 40(5):960–966
Green JT, Liu Z, Bazinet RP (2010) Brain phospholipid arachidonic acid half-lives are not altered following 15 weeks of N-3 polyunsaturated fatty acid adequate or deprived diet. J Lipid Res 51(3):535–543
Greenhalgh AD, David S, Bennett FC (2020) Immune cell regulation of glia during CNS injury and disease. Nat Rev Neurosci 21(3):139–152
Guest J, Garg M, Bilgin A, Grant R (2013) Relationship between central and peripheral fatty acids in humans. Lipids Health Dis 12(1):1–8
Haggarty P (2010) Fatty acid supply to the human fetus. Annu Rev Nutr 30:237–255
Hall JC, Priestley JV, Perry VH, Michael-Titus AT (2012) Docosahexaenoic acid, but not eicosapentaenoic acid, reduces the early inflammatory response following compression spinal cord injury in the rat. J Neurochem 121(5):738–750
Hamazaki K, Choi KH, Kim H-Y (2010) Phospholipid profile in the postmortem hippocampus of patients with schizophrenia and bipolar disorder: no changes in docosahexaenoic acid species. J Psychiatr Res 44(11):688–693
Hamazaki K, Maekawa M, Toyota T, Dean B, Hamazaki T, Yoshikawa T (2015) Fatty acid composition of the postmortem prefrontal cortex of patients with schizophrenia, bipolar disorder, and major depressive disorder. Psychiatry Res 227(2–3):353–359
Harel T, Quek DQ, Wong BH, Cazenave-Gassiot A, Wenk MR, Fan H et al (2018) Homozygous mutation in MFSD2A, encoding a lysolipid transporter for docosahexanoic acid, is associated with microcephaly and hypomyelination. J Neurogenet 19:227–235
Harris NG, Nogueira MS, Verley DR, Sutton RL (2013) Chondroitinase enhances cortical map plasticity and increases functionally active sprouting axons after brain injury. J Neurotrauma 30(14):1257–1269

He C, Qu X, Cui L, Wang J, Kang JX (2009a) Improved spatial learning performance of fat-1 mice is associated with enhanced neurogenesis and neuritogenesis by docosahexaenoic acid. Proc Natl Acad Sci USA 106(27):11370–11375

He S, Wang L, Miao L, Wang T, Du F, Zhao L et al (2009b) Receptor interacting protein kinase-3 determines cellular necrotic response to TNF-α. Cell J 137(6):1100–1111

Heurteaux C, Guy N, Laigle C, Blondeau N, Duprat F, Mazzuca M et al (2004) TREK-1, a K+ channel involved in neuroprotection and general anesthesia. EMBO J 23(13):2684–2695

Hoeffel G, Ginhoux F (2018) Fetal monocytes and the origins of tissue-resident macrophages. Cell Immunol 330:5–15

Hu X, Zhang F, Leak R, Zhang W, Iwai M, Stetler RA et al (2013) Transgenic overproduction of omega-3 polyunsaturated fatty acids provides neuroprotection and enhances endogenous neurogenesis after stroke. Curr Mol Med 13(9):1465–1473

Huang L, Zhou Y, Gou Z-X, Zhang F, Lu L-Q (2022) Docosahexaenoic acid reduces hypoglycemia-induced neuronal necroptosis via the peroxisome proliferator-activated receptor γ/nuclear factor-κB pathway. Brain Res J 1774:147708

Hurlbert RJ, Hadley MN, Walters BC, Aarabi B, Dhall SS, Gelb DE et al (2013) Pharmacological therapy for acute spinal cord injury. Neurosurgery 72:93–105

Igarashi M, Santos RA, Cohen-Cory S (2015) Impact of maternal n-3 polyunsaturated fatty acid deficiency on dendritic arbor morphology and connectivity of developing Xenopus laevis central neurons in vivo. J Neurosci 35(15):6079–6092

Innis SM, de la Presa OS (2001) Dietary fatty acid composition in pregnancy alters neurite membrane fatty acids and dopamine in newborn rat brain. J Nutr 131(1):118–122

Jiang X, Suenaga J, Pu H, Wei Z, Smith AD, Hu X et al (2019) Post-stroke administration of omega-3 polyunsaturated fatty acids promotes neurovascular restoration after ischemic stroke in mice: efficacy declines with aging. Neurobiol Dis 126:62–75

Joardar A, Sen AK, Das S (2006) Docosahexaenoic acid facilitates cell maturation and β-adrenergic transmission in astrocytes. J Lipid Res 47(3):571–581

Joffre C, Grégoire S, De Smedt V, Acar N, Bretillon L, Nadjar A et al (2016a) Prostaglandins leukot essent fatty acids. Trends Pharmacol Sci. Prostaglandins Leukot Essent Fatty Acids 1-10

Joffre C, Dinel A-L, Aubert A, Fressange-Mazda C, Le Ruyet P, Layé S (2016b) Impact of lactobacillus fermentum and dairy lipids in the maternal diet on the fatty acid composition of pups' brain and peripheral tissues. Prostaglandins Leukot Essent Fatty Acids 115:24–34

Kaiser WJ, Sridharan H, Huang C, Mandal P, Upton JW, Gough PJ et al (2013) Toll-like receptor 3-mediated necrosis via TRIF, RIP3, and MLKL. J Biol Chem 288(43):31268–31279

Karve IP, Taylor JM, Crack PJ (2016) The contribution of astrocytes and microglia to traumatic brain injury. Br J Pharmacol 173(4):692–702

Kernie SG, Parent JM (2010) Forebrain neurogenesis after focal ischemic and traumatic brain injury. Neurobiol Dis 37(2):267–274

Kim H-Y, Spector AA (2018) N-Docosahexaenoylethanolamine: a neurotrophic and neuroprotective metabolite of docosahexaenoic acid. Mol Asp Med 64:34–44

Kitson AP, Metherel AH, Chen CT, Domenichiello AF, Trépanier M-O, Berger A et al (2016) Effect of dietary docosahexaenoic acid (DHA) in phospholipids or triglycerides on brain DHA uptake and accretion. J Nutr Biochem 33:91–102

Kohli P, Levy BD (2009) Resolvins and protectins: mediating solutions to inflammation. Br J Pharmacol 158(4):960–971

Krishnamoorthy S, Recchiuti A, Chiang N, Fredman G, Serhan CN (2012) Resolvin D1 receptor stereoselectivity and regulation of inflammation and proresolving microRNAs. Am J Pathol 180(5):2018–2027

Kuda O (2017) Bioactive metabolites of docosahexaenoic acid. Biochimie 136:12–20

Kuipers RS, Luxwolda MF, Offringa PJ, Boersma ER, Dijck-Brouwer DJ, Muskiet FA (2012) Fetal intrauterine whole body linoleic, arachidonic and docosahexaenoic acid contents and accretion rates. Prostaglandins Leukot Essent Fatty Acids 86(1–2):13–20

Lacombe RS, Giuliano V, Colombo SM, Arts MT, Bazinet RP (2017) Compound-specific isotope analysis resolves the dietary origin of docosahexaenoic acid in the mouse brain [S]. J Lipid Res 58(10):2071–2081
Lapillonne A, Moltu SJ (2016) Long-chain polyunsaturated fatty acids and clinical outcomes of preterm infants. Ann Nutr Metab 69(Suppl. 1):35–44
Lauritzen LA, Hansen H, Jørgensen M, Michaelsen K (2001) The essentiality of long chain n-3 fatty acids in relation to development and function of the brain and retina. Prog Lipid Res 40(1–2):1–94
Laye S, Nadjar A, Joffre C, Bazinet RP (2018) Anti-inflammatory effects of omega-3 fatty acids in the brain: physiological mechanisms and relevance to pharmacology. Pharmacol Rev 70(1):12–38
Lenz KM, Nelson LH (2018) Microglia and beyond: innate immune cells as regulators of brain development and behavioral function. Front Immunol 9:698
Leonard AE, Kelder B, Bobik EG, Chuang L-T, Parker-Barnes JM, Thurmond JM et al (2000) cDNA cloning and characterization of human Δ5-desaturase involved in the biosynthesis of arachidonic acid. Biochem J 347(3):719–724
Lewinska A, Zebrowski J, Duda M, Gorka A, Wnuk M (2015) Fatty acid profile and biological activities of linseed and rapeseed oils. Molecules 20(12):22872–22880
Liu D, Xu G-Y, Pan E, McAdoo D (1999) Neurotoxicity of glutamate at the concentration released upon spinal cord injury. Neurosci J 93(4):1383–1389
Loane DJ, Faden AI (2010) Neuroprotection for traumatic brain injury: translational challenges and emerging therapeutic strategies. Trends Pharmacol Sci 31(12):596–604
López-Vales R, Redensek A, Skinner TA, Rathore KI, Ghasemlou N, Wojewodka G et al (2010) Fenretinide promotes functional recovery and tissue protection after spinal cord contusion injury in mice. J Neurosci 30(9):3220–3226
Lucas M, Asselin G, Mérette C, Poulin M-J, Dodin S (2009) Validation of an FFQ for evaluation of EPA and DHA intake. Public Health Nutr 12(10):1783–1790
Ma D, Lu L, Boneva NB, Warashina S, Kaplamadzhiev DB, Mori Y et al (2008) Expression of free fatty acid receptor GPR40 in the neurogenic niche of adult monkey hippocampus. Hippocampus 18(3):326–333
Madore C, Nadjar A, Delpech J-C, Séré A, Aubert A, Portal C et al (2014) Nutritional n-3 PUFAs deficiency during perinatal periods alters brain innate immune system and neuronal plasticity-associated genes. Brain Behav Immun 41:22–31
Madore C, Leyrolle Q, Lacabanne C, Benmamar-Badel A, Joffre C, Nadjar A et al (2016) Neuroinflammation in autism: plausible role of maternal inflammation, dietary omega 3, and microbiota. Neural Plast 2016:3597209
Madore C, Leyrolle Q, Morel L, Rossitto M, Greenhalgh A, Delpech J et al (2020) Essential omega-3 fatty acids tune microglial phagocytosis of synaptic elements in the mouse developing brain. Nat Commun 11(1):6133
Mann R, Schaefer C, Sadosky A, Bergstrom F, Baik R, Parsons B et al (2013) Burden of spinal cord injury-related neuropathic pain in the United States: retrospective chart review and cross-sectional survey. Spinal Cord 51(7):564–570
Manni MM, Tiberti ML, Pagnotta S, Barelli H, Gautier R, Antonny B (2018) Acyl chain asymmetry and polyunsaturation of brain phospholipids facilitate membrane vesiculation without leakage. Elife 7:34394
Martin RE (1998) Docosahexaenoic acid decreases phospholipase A2 activity in the neurites/nerve growth cones of PC12 cells. J Neurosci 54(6):805–813
Martin RE, Bazan NG (1992) Changing fatty acid content of growth cone lipids prior to synaptogenesis. J Neurochem 59(1):318–325
Martinez M (1991) Developmental profiles of polyunsaturated fatty acids in the brain of normal infants and patients with peroxisomal diseases: severe deficiency of docosahexaenoic acid in Zellweger's and pseudo-Zellweger's syndromes. World Rev Nutr Diet 66:87

Mass E, Ballesteros I, Farlik M, Halbritter F, Günther P, Crozet L et al (2016) Specification of tissue-resident macrophages during organogenesis. Science 353(6304):4238
McNamara RK, Carlson SE (2006) Role of omega-3 fatty acids in brain development and function: potential implications for the pathogenesis and prevention of psychopathology. Prostaglandins Leukot Essent Fatty Acids 75(4–5):329–349
McNamara RK, Able J, Jandacek R, Rider T, Tso P (2009) Inbred C57BL/6J and DBA/2J mouse strains exhibit constitutive differences in regional brain fatty acid composition. J Lipids 44:1–8
Metherel AH, Kitson AP, Domenichiello AF, Lacombe RS, Hopperton KE, Trépanier M-O et al (2017) Docosahexaenoic acid (DHA) accretion in the placenta but not the fetus is matched by plasma unesterified DHA uptake rates in pregnant Long Evans rats. Placenta 58:90–97
Meyer BJ, Mann NJ, Lewis JL, Milligan GC, Sinclair AJ, Howe PR (2003) Dietary intakes and food sources of omega-6 and omega-3 polyunsaturated fatty acids. J Lipids 38(4):391–398
Mills JD, Bailes JE, Sedney CL, Hutchins H, Sears B (2011a) Omega-3 fatty acid supplementation and reduction of traumatic axonal injury in a rodent head injury model. J Neurosurg 114(1):77–84
Mills JD, Hadley K, Bailes JE (2011b) Dietary supplementation with the omega-3 fatty acid docosahexaenoic acid in traumatic brain injury. J Neurosurg 68(2):474–481
Morand O, Chanez C, Masson M, Dumont O, Flexor M, Baumann N et al (1982) Alteration in fatty acid composition of neurons, astrocytes, oligodendrocytes, myelin and synaptosomes in intrauterine malnutrition in rat. Ann Nutr Metab 26(2):111–120
Moriguchi T, Loewke J, Garrison M, Catalan JN, Salem N (2001) Reversal of docosahexaenoic acid deficiency in the rat brain, retina, liver, and serum. J Lipid Res 42(3):419–427
Muc M, Kreiner-Møller E, Larsen JM, Birch S, Brix S, Bisgaard H et al (2015) Maternal fatty acid desaturase genotype correlates with infant immune responses at 6 months. Br J Nutr 114(6):891–898
Myer D, Gurkoff G, Lee S, Hovda D, Sofroniew M (2006) Essential protective roles of reactive astrocytes in traumatic brain injury. Brain 129(10):2761–2772
Namiranian K, Brink CD, Goodman JC, Robertson CS, Bryan RM Jr (2011) Traumatic brain injury in mice lacking the K channel, TREK-1. J Cereb Blood Flow Metab 31(3):1–6
Obata, Nagakura, Masaki, Maekawa, Yamashita (1999) Eicosapentaenoic acid inhibits prostaglandin D2 generation by inhibiting cyclo-oxygenase-2 in cultured human mast cells. Clin Exp Allergy 29(8):1129–1135
Orihuela R, McPherson CA, Harry GJ (2016) Microglial M1/M2 polarization and metabolic states. Br J Pharmacol 173(4):649–665
Paolicelli RC, Gross CT (2011) Microglia in development: linking brain wiring to brain environment. Neuron Glia Biol 7(1):77–83
Pekny M, Wilhelmsson U, Pekna M (2014) The dual role of astrocyte activation and reactive gliosis. Neurosci Lett 565:30–38
Percy P, Percy A, Vilbergsson G, Månsson J-E (1996) Polyunsaturated fatty acid accretion in first-and second-trimester human fetal brain: lack of correlation with levels in paired placental samples. Mol Med 59(1):38–43
Pilitsis JG, Coplin WM, O'Regan MH, Wellwood JM, Diaz FG, Fairfax MR et al (2003) Free fatty acids in cerebrospinal fluids from patients with traumatic brain injury. Neurosci Lett 349(2):136–138
Plourde M, Cunnane SC (2007) Extremely limited synthesis of long chain polyunsaturates in adults: implications for their dietary essentiality and use as supplements. Appl Physiol Nutr Metab 32(4):619–634
Poitelon Y, Kopec AM, Belin S (2020) Myelin fat facts: an overview of lipids and fatty acid metabolism. Cell J 9(4):812
Priestley JV, Michael-Titus AT, Tetzlaff W (2012) Limiting spinal cord injury by pharmacological intervention. Handb Clin Neurol 109:463–484
Profyris C, Cheema SS, Zang D, Azari MF, Boyle K, Petratos S (2004) Degenerative and regenerative mechanisms governing spinal cord injury. Neurobiol Dis 15(3):415–436

Pu H, Guo Y, Zhang W, Huang L, Wang G, Liou AK et al (2013) Omega-3 polyunsaturated fatty acid supplementation improves neurologic recovery and attenuates white matter injury after experimental traumatic brain injury. J Cereb Blood Flow Metab 33(9):1474–1484

Rao J, Ertley R, DeMar J, Rapoport S, Bazinet R, Lee H (2007) Dietary n-3 PUFA deprivation alters expression of enzymes of the arachidonic and docosahexaenoic acid cascades in rat frontal cortex. Mol Neurobiol 12(2):151–157

Rapoport SI, Rao JS, Igarashi M (2007) Brain metabolism of nutritionally essential polyunsaturated fatty acids depends on both the diet and the liver. Prostaglandins Leukot Essent Fatty Acids 77(5–6):251–261

Rey C, Nadjar A, Joffre F, Amadieu C, Aubert A, Vaysse C et al (2018) Maternal n-3 polyunsaturated fatty acid dietary supply modulates microglia lipid content in the offspring. Prostaglandins Leukot Essent Fatty Acids 133:1–7

Rey C, Delpech J, Madore C, Nadjar A, Greenhalgh A, Amadieu C et al (2019) Dietary n-3 long chain PUFA supplementation promotes a pro-resolving oxylipin profile in the brain. Brain Behav Immun 76:17–27

Reyes LD, Haight T, Desai A, Chen H, Bosomtwi A, Korotcov A et al (2020) Investigation of the effect of dietary intake of omega-3 polyunsaturated fatty acids on trauma-induced white matter injury with quantitative diffusion MRI in mice. J Neurosci 98(11):2232–2244

Risdall JE, Menon DK (2011) Traumatic brain injury. Philos Trans R Soc Lond Ser B Biol Sci 366(1562):241–250

Robson LG, Dyall SC, Sidloff D, Michael-Titus AT (2010) Omega-3 polyunsaturated fatty acids increase the neurite outgrowth of rat sensory neurones throughout development and in aged animals. Neurobiol Aging 31(4):678–687

Rohrbough J, Rushton E, Palanker L, Woodruff E, Matthies HJ, Acharya U et al (2004) Ceramidase regulates synaptic vesicle exocytosis and trafficking. J Neurosci 24(36):7789–7803

Roozenbeek B, Maas AI, Menon DK (2013) Changing patterns in the epidemiology of traumatic brain injury. Nat Rev Neurol 9(4):231–236

Sakayori N, Kikkawa T, Tokuda H, Kiryu E, Yoshizaki K, Kawashima H et al (2016) Maternal dietary imbalance between omega-6 and omega-3 polyunsaturated fatty acids impairs neocortical development via epoxy metabolites. Stem Cell Res J 34(2):470–482

Sanders T, Naismith D (1980) The metabolism of α-linolenic acid by the foetal rat. Br J Nutr 44(2):205–208

Sanders T, Rana SK (1987) Comparison of the metabolism of linoleic and linolenic acids in the fetal rat. Ann Nutr Metab 31(6):349–353

Sastry PS (1985) Lipids of nervous tissue: composition and metabolism. Prog Lipid Res 24(2):69–176

Scott BL, Bazan NG (1989) Membrane docosahexaenoate is supplied to the developing brain and retina by the liver. Proc Natl Acad Sci USA 86(8):2903–2907

Senter HJ, Venes JL (1979) Loss of autoregulation and posttraumatic ischemia following experimental spinal cord trauma. J Neurosurg 50(2):198–206

Serhan CN, Chiang N (2013) Resolution phase lipid mediators of inflammation: agonists of resolution. Curr Opin Pharmacol 13(4):632–640

Serhan CN, Chiang N, Van Dyke TE (2008) Resolving inflammation: dual anti-inflammatory and pro-resolution lipid mediators. Nat Rev Immunol 8(5):349–361

Serhan CN, Yang R, Martinod K, Kasuga K, Pillai PS, Porter TF et al (2009) Maresins: novel macrophage mediators with potent antiinflammatory and proresolving actions. J Exp Med 206(1):15–23

Serhan CN, Krishnamoorthy S, Recchiuti A, Chiang N (2011) Novel anti-inflammatory-pro-resolving mediators and their receptors. Curr Top Med Chem 11(6):629–647

Shalini S-M, Ho CF-Y, Ng Y-K, Tong J-X, Ong E-S, Herr DR et al (2018) Distribution of Alox15 in the rat brain and its role in prefrontal cortical resolvin D1 formation and spatial working memory. Mol Neurobiol 55:1537–1550

Shin SS, Dixon CE (2011) Oral fish oil restores striatal dopamine release after traumatic brain injury. Neurosci Lett 496(3):168–171

Shinjyo N, Piscitelli F, Verde R, Di Marzo V (2013) Impact of omega-6 polyunsaturated fatty acid supplementation and γ-aminobutyric acid on astrogliogenesis through the endocannabinoid system. J Neurosci Res 91(7):943–953

Sidhu VK, Huang BX, Kim H-Y (2011) Effects of docosahexaenoic acid on mouse brain synaptic plasma membrane proteome analyzed by mass spectrometry and 16O/18O labeling. J Proteome Res 10(12):5472–5480

Simopoulos AP (2002) The importance of the ratio of omega-6/omega-3 essential fatty acids. Biomed Pharmacother 56(8):365–379

Simopoulos AP (2011) Evolutionary aspects of diet: the omega-6/omega-3 ratio and the brain. Mol Neurobiol 44(2):203–215

Sinclair AT, Crawford M (1972) The incorporation of linolenic acid and docosahexaenoic acid into liver and brain lipids of developing rats. FEBS Lett 26(1–2):127–129

Sprecher H (2000) Metabolism of highly unsaturated n-3 and n-6 fatty acids. Biochim Biophys Acta Mol Cell Biol 1486(2–3):219–231

Stark KD, Van Elswyk ME, Higgins MR, Weatherford CA, Salem N Jr (2016) Global survey of the omega-3 fatty acids, docosahexaenoic acid and eicosapentaenoic acid in the blood stream of healthy adults. Prog Lipid Res 63:132–152

Strobel C, Jahreis G, Kuhnt K (2012) Survey of n-3 and n-6 polyunsaturated fatty acids in fish and fish products. Lipids Health Dis 11:1–10

Sun L, Xu Y-W, Han J, Liang H, Wang N, Cheng Y (2015) 12/15-Lipoxygenase metabolites of arachidonic acid activate PPARγ: a possible neuroprotective effect in ischemic brain. J Lipid Res 56(3):502–514

Suzuki H, Manabe S, Wada O, Crawford M (1997) Rapid incorporation of docosahexaenoic acid from dietary sources into brain microsomal, synaptosomal and mitochondrial membranes in adult mice. Int J Vitam Nutr Res 67(4):272–278

Taha AY, Cheon Y, Faurot KF, MacIntosh B, Majchrzak-Hong SF, Mann JD et al (2014) Dietary omega-6 fatty acid lowering increases bioavailability of omega-3 polyunsaturated fatty acids in human plasma lipid pools. Prostaglandins Leukot Essent Fatty Acids 90(5):151–157

Thau-Zuchman O, Ingram R, Harvey GG, Cooke T, Palmas F, Pallier PN et al (2020) A single injection of docosahexaenoic acid induces a pro-resolving lipid mediator profile in the injured tissue and a long-lasting reduction in neurological deficit after traumatic brain injury in mice. J Neurotrauma 37(1):66–79

Thion MS, Ginhoux F, Garel S (2018) Microglia and early brain development: an intimate journey. Science 362(6411):185–189

Vreugdenhil M, Bruehl C, Voskuyl R, Kang J, Leaf A, Wadman W (1996) Polyunsaturated fatty acids modulate sodium and calcium currents in CA1 neurons. Proc Natl Acad Sci USA 93(22):12559–12563

Wang Y, Botolin D, Christian B, Busik J, Xu J, Jump DB (2005) Tissue-specific, nutritional, and developmental regulation of rat fatty acid elongases. J Lipid Res 46(4):706–715

Ward G, Huang Y, Bobik E, Xing H, Mutsaers L, Auestad N et al (1998) Long-chain polyunsaturated fatty acid levels in formulae influence deposition of docosahexaenoic acid and arachidonic acid in brain and red blood cells of artificially reared neonatal rats. J Nutr 128(12):2473–2487

Witte TR, Salazar AJ, Ballester OF, Hardman WE (2010) RBC and WBC fatty acid composition following consumption of an omega 3 supplement: lessons for future clinical trials. Lipids Health Dis 9:1–7

Wong CT, Ussyshkin N, Ahmad E, Rai-Bhogal R, Li H, Crawford DA (2016) Prostaglandin E2 promotes neural proliferation and differentiation and regulates Wnt target gene expression. J Neurosci Res 94(8):759–775

Wu A, Ying Z, Gomez-Pinilla F (2004) Dietary omega-3 fatty acids normalize BDNF levels, reduce oxidative damage, and counteract learning disability after traumatic brain injury in rats. J Neurotrauma 21(10):1457–1467

Wu A, Ying Z, Gomez-Pinilla F (2011) The salutary effects of DHA dietary supplementation on cognition, neuroplasticity, and membrane homeostasis after brain trauma. J Neurotrauma 28(10):2113–2122
Wu A, Ying Z, Gomez-Pinilla F (2013) Exercise facilitates the action of dietary DHA on functional recovery after brain trauma. Neurosci J 248:655–663
Xiao K, Xu Q, Liu C, He P, Qin Q, Zhu H et al (2020) Docosahexaenoic acid alleviates cell injury and improves barrier function by suppressing necroptosis signalling in TNF-α-challenged porcine intestinal epithelial cells. J Innate Immun 26(8):653–665
Xiong Y, Lu D, Qu C, Goussev A, Schallert T, Mahmood A et al (2008) Effects of erythropoietin on reducing brain damage and improving functional outcome after traumatic brain injury in mice. J Neurosurg 109(3):510–521
Yavin E, Himovichi E, Eilam R (2009) Delayed cell migration in the developing rat brain following maternal omega 3 alpha linolenic acid dietary deficiency. Neurosci J 162(4):1011–1022
Yigitkanli K, Zheng Y, Pekcec A, Lo EH, van Leyen K (2017) Increased 12/15-lipoxygenase leads to widespread brain injury following global cerebral ischemia. Transl Stroke Res 8:194–202
Yurko-Mauro K, Kralovec J, Bailey-Hall E, Smeberg V, Stark JG, Salem N (2015) Similar eicosapentaenoic acid and docosahexaenoic acid plasma levels achieved with fish oil or krill oil in a randomized double-blind four-week bioavailability study. Lipids Health Dis 14:1–9
Zhu W, Ding Y, Kong W, Li T, Chen H (2018) Docosahexaenoic acid (DHA) provides neuroprotection in traumatic brain injury models via activating Nrf2-ARE signaling. J Inflamm 41:1182–1193
Zhu W, Cui G, Li T, Chen H, Zhu J, Ding Y et al (2020) Docosahexaenoic acid protects traumatic brain injury by regulating NOX 2 generation via Nrf2 signaling pathway. Neurochem Res 45:1839–1850
Ziegler AB, Ménagé C, Grégoire S, Garcia T, Ferveur J-F, Bretillon L et al (2015) Lack of dietary polyunsaturated fatty acids causes synapse dysfunction in the drosophila visual system. PLoS One 10(8):135353

Chapter 12
Polyphenols and TBI

Hagar F. Forsan

Abstract Traumatic brain injury (TBI) is a reason for disability and death in teenagers and the youthful adult global population. Although increasing findings and progress in our comprehension of TBI's intricate pathophysiology, the principal processes remain unknown. Whereas an early brain insult causes rapid and irreversible main brain damage, after a secondary brain injury, the healing processes take months or years to complete, allowing for therapeutic approaches. So much study has been devoted to identifying druggable targets implicated in these mechanisms. Despite many years of excellent preclinical trials and extremely promising outcomes, when these medications were brought to clinics, they showed, at most, small therapeutic benefits but frequently no effects or very harsh detrimental effects in those with traumatic brain injuries. This reality has underlined the necessity for creative techniques that can respond to the TBI's complexity and address TBI symptoms on numerous levels. The latest research clearly implies that nutritional treatments may present a special opportunity to enhance TBI healing mechanisms. Dietary polyphenols, a large family of chemicals common in vegetables and fruits, have appeared as prospective agents for usage in TBI settings in recent years due to their demonstrated pleiotropic effects.

Keywords TBI · Polyphenols · Polyphenolic acids · Flavonoids · Oxidative stress

Abbreviations

AQP4	Aquaporin-4
BBB	Blood-brain barrier
Bcl2	B-cell lymphoma 2

H. F. Forsan (✉)
Animal Production Research Institute (APRI), Agricultural Research Center (ARC), Dokki, Giza, Egypt
e-mail: hagarfathy@pg.cu.edu.eg

W. Mohamed (ed.), *Nutrition and Traumatic Brain Injury (TBI)*, Nutritional Neurosciences, https://doi.org/10.1007/978-981-97-6341-2_12

BDNF	Brain-derived neurotrophic factor
CCI	Controlled cortical impact
EC	Catechin, (−)-epicatechin
EGCG	Epigallocatechin-3-gallate
MSCs	Mesenchymal stem cells
mNSS	Modified neurological severity score
NF-B	Nuclear factor kappa
PA	Phenolic acid
QUIN	Quinolinic acid
SOD	Superoxide dismutase
TBI	Traumatic brain injury
TLR4	Toll-like receptor 4
UGT	UDP-glucuronosyltransferase

12.1 Introduction

The majority of polyphenolic antioxidant substances come from plants. Polyphenolic compounds constitute a wide class of phenolic chemicals (de Andrade Teles et al. 2018; Forsan and Awad 2023; Ren et al. 2003). Flavonoids are a diverse group of plant compounds with known anti-inflammatory and antioxidant properties. They have been shown to affect various signaling pathways in the body, including those involving the nuclear factor kappa B (NF-κB) pathway.

The majority of naturally occurring polyphenols come from plants, which produce them either as a defense mechanism against pathogens or as part of regular plant tissue growth (Obrenovich et al. 2010; Scalbert et al. 2005). Numerous polyphenols have a variety of health advantages and make up the biggest amounts of antioxidants in the average person's diet. It is one of the greatest diverse antioxidants found in nature, with approximately 8000 polyphenols known to exist. Since free radicals play a significant part in chronic diseases like diabetes and Alzheimer's disease. Polyphenolic compounds are important free radical scavengers. Polyphenolic phytochemicals are present in a range of plant sources, including black soybeans. Fruits, cereals, nuts, vegetables, and seeds are only a few of the naturally occurring foods that contain polyphenols (Ragusa et al. 2017; Thomasset et al. 2007). Flavonoids, tannins, polyphenolic acids, polyphenols, and lignans are all types of polyphenolic compounds that occur naturally in plants and have been studied for their health benefits. All these compounds are prevalent in nature and contribute to the health benefits associated with plant-based diets. They are involved in various biological activities that can impact inflammation, oxidative stress, and overall health. Secondary metabolites, also known as human lignans, are formed by the metabolism of plant-based lignans. Certain lignans are able to pass through the blood-brain barrier (BBB) and serve as disease neuroprotectors, including autism spectrum disorders and AD (Serra et al. 2019). Treatments for crippling neurological conditions may benefit from new treatment techniques based on the gut-brain axis. The production of therapeutically important polyphenolic substances associated with a higher BBB is influenced by dietary polyphenolic substances in the gut flora. For instance, the naturally occurring polyphenolic chemical curcumin, even though its capacity to traverse the blood-brain barrier is constrained, protects against Alzheimer's disease because of its effectiveness in binding amyloid and its antioxidant qualities. The current working theory for curcumin's neuroprotective impact is that it interacts with the gut-microbiota-brain axis to affect the central nervous system indirectly (Di Meo et al. 2019). Through the metabolism of the gut microbes, curcumin may be changed into more potent neuroprotective compounds. Alternately, the gut-brain axis has an indirect impact on the gut microbial pathways of metabolism, which makes it favorably adapted to provide neuroprotection (Zam 2018).

TBI is described as mechanical and external forces causing damage to the brain both structurally and functionally, with two stages of injury (main and secondary) (Liu et al. 2018). It is a particularly typical neurological condition globally and the primary issue contributing to death and disability in people under 40 (Khellaf et al. 2019).

TBI, additionally known as acquired intracranial injury, is a type of brain injury. An outside factor, like a blow, bump, or shock to the head, as well as a fast and serious knock by something, or a deep piercing of the tissues of the brain, causes it (Huynh et al. 2020). The most common causes are sports, transportation, violence, accidents, and construction.

Furthermore, approximately 288,000 TBI hospitalizations occur each year, with males accounting for 78.8% (B. Lee et al. 2019; Sivandzade et al. 2021). TBI is most common in older persons above the age of 75. As a result, TBI causes substantial financial and spiritual suffering in society and families (Diaz et al. 2019). TBI is characterized according to several factors, such as injury-related physiological reactions, mechanism of injury, location, severity, and type of injury (Werhane et al. 2017). TBI symptoms include cognitive decline, behavioral impairment, affective abnormalities, perceptual loss, and communication difficulties and differ based on brain injury severity (Yeates et al. 2012).

TBI pathophysiological includes both the first injury, brought on by outside factors, and the second injury, brought on by additional harm to both brain tissue and cells (Lorente et al. 2015). Physical stresses on the brain can result in both localized and diffuse damage. According to emerging research, people with moderate or severe traumatic brain injury (TBI) can experience localized damage, though the extent and nature of this damage can vary widely depending on the specifics of the injury (Vos 2011).

Most significantly, secondary brain damage is caused by molecular, physiological, and biochemical events that occur after the initial brain damage (Khatri et al. 2018). Mechanical studies show that inflammation, oxidative stress, glutamate excitotoxicity, DNA damage, neurotrophic impairment, BBB disruption, and mitochondrial dysfunction complement activation all play a role in TBI progression and pathology (Ng and Lee 2019).

An increasing amount of research suggests that progressively aberrant proteins or molecules are markers intimately linked to TBI, which contributes to

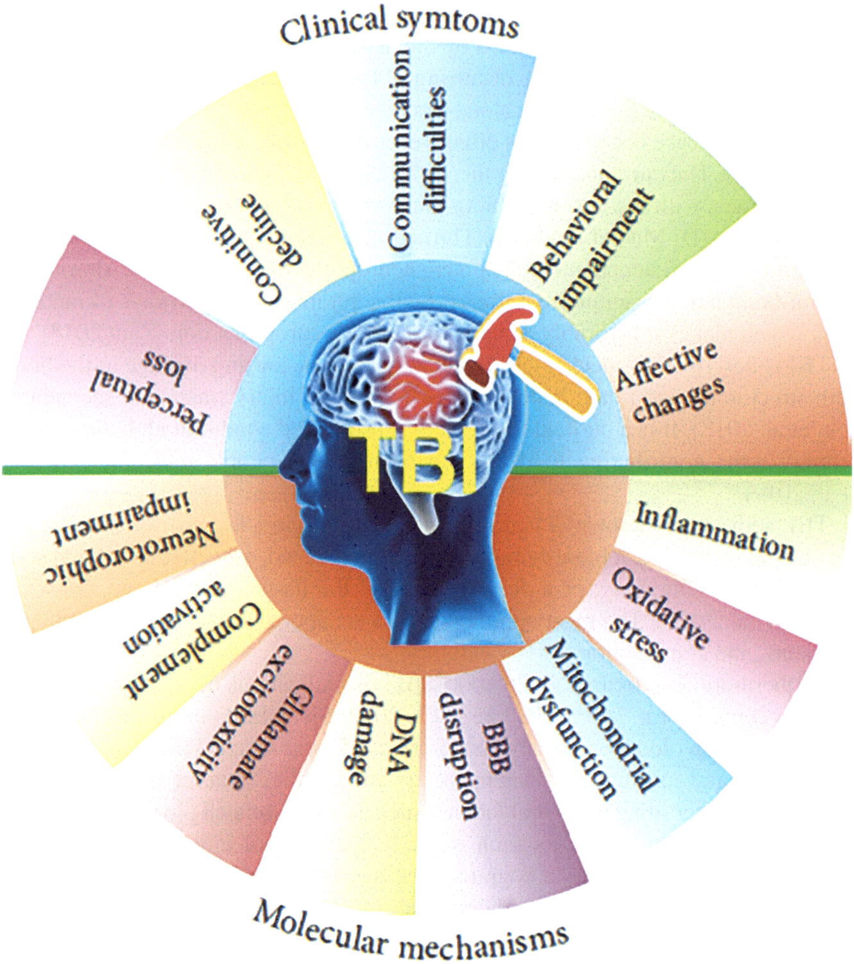

Fig. 12.1 Relationship between sleep, diet, biological factors, and disease (Adapted with permission from Wu et al. (2022). Copyright 2023, Hindawi)

understanding the mechanism of TBI (O'Brien et al. 2020). Early structural injury biomarkers, such as ubiquitin carboxyl-terminal hydrolase L1, blood–nerve glial acidic protein, S100B protein in CSF, and Tau, for example, can help determine whether a head scan needs to be performed after a TBI (Suehiro et al. 2019). Hyperbaric oxygen therapy, hypertonic, hyperventilation therapy pharmaceutical and biological therapies, and noninvasive brain stimulation are all used to treat TBI (Parizad et al. 2021). Most importantly, integrating novel biological gene reagents and stem cell reagents with pharmaceutical treatment preparations has been shown to minimize TBI adverse effects and mortality (Zibara et al. 2019) (Fig. 12.1).

12.1.1 Brain damage produced by an external impact

TBI-related clinical symptoms include affective abnormalities, cognitive decline, perceptual loss, behavioral impairment, and communication difficulties. TBI's molecular processes include oxidative stress, inflammation, BBB disruption, mitochondrial dysfunction, glutamate excitotoxicity, DNA damage, neurotrophic impairment, and complement activation. Treatment with stem cells involves both transplanting stem cells back into the body and stimulating endogenous stem cells using drugs or environmental stimulants. Numerous investigations have demonstrated that some medications can increase stem cells' survivability as well as their efficacy. For example, 24 h after a TBI, an intravenous infusion of mesenchymal stem cells (MSCs) with simvastatin could help in the recovery of the modified neurological severity score (mNSS) (Mahmood et al. 2008).Furthermore, calpain antagonists were administered 30 min after TBI, followed by an MSC transplant 24 h later that could lower pro-inflammatory cytokines near lesions, increase MSC survival, and improve mNSS (Hu et al. 2019). Moreover, minocycline pre-treatment for a day of implanted stem cells may protect from the damaging effects of ischemia via increasing Nrf2 nuclear translocation and raising the downstream protein expression (Kanninen et al. 2015). Because of this, a complete understanding of the mechanism underlying TBI and the adoption of targeted strategies for precise intervention will aid in the restoration of post-activity of the traumatic brain while also minimizing the emergence and progression of problems, effectively opening up an opportunity for effective TBI therapy (Tal et al. 2017).

12.2 Oxidative Stress Role of TBI

An imbalance between free radicals and antioxidants, resulting from environmental factors like pollution and radiation, as well as lifestyle and eating habits choices, is a significant component of the onset of illness (Ayoub et al. 2021; Tal et al. 2017). Nitric oxide, superoxide, and hydroxyl are examples of free radicals or compounds that have a number of electrons that are not paired (de Almeida et al. 2020). It is commonly recognized that oxidative stress contributes to the etiology of certain illnesses, such as hypertension, neurodegeneration, and other conditions that influence CNS (Loessner et al. 1988; Suzuki et al. 2010). Although numerous free radicals are produced during normal metabolic processes, the body's cells may create antioxidants (Li et al. 2021).

Many investigations have revealed that excessive levels of free radicals damage biological elements such as DNA. This imbalance is eventually upset, leading to long-term oxidative stress (Philippe et al. 2010). However, in other circumstances, such as physiologic adaptableness, oxidative stress, and the controller of intracellular sign transmission, can be advantageous (Mukhopadhyay et al. 2015). Therefore, an imbalance between the oxidative and antioxidant systems may be a more accurate definition of oxidative stress than either alone. The application of oxidative stress indicators to evaluate disease pathophysiology and the effectiveness of therapy is currently on the rise (Nishikawa et al. 2020). Lipid peroxides, such as malondialdehyde (Tun and Kang 2017) and 4-hydroxynonenal (4-HNE), are markers of lipid oxidative damage (Gu et al. 2019). Two indicators of DNA oxidative damage are thymine glycol (TG) and 8-oxo guanine (8-oxoG) (Wessels et al. 2014). Furthermore, several amino acids and proteins, such as hydroxyleucine, nitrotyrosine, carbonyl protein, and dehydrovaline, oxidize and produce a range of compounds that are recognized as oxidative stress markers. Among them, clinical investigations identified lipid peroxide as being one of a number of important indicators.

Furthermore, oxidative stress is a factor. Signaling transduction is crucial in regulating various biological processes through complex pathways involving protein kinase activation and transcription factors. The intricate network of signaling pathways involving protein kinases and transcription factors orchestrates multiple biological processes, ensuring that cells respond appropriately to internal and external cues. This regulation is vital for maintaining cellular homeostasis and overall organismal health (Glotov et al. 2015; Tao et al. 2017). Nuclear factors, such as activator protein-1, nuclear factor kappa (NF-B), and gene transcription factors, detect oxidation and reduction cycles to alleviate oxidative stress (Tao et al. 2017). Furthermore, the active oxygen species production affects NF-B activation, affecting pro-inflammatory reactions in a variety of illnesses, including neurodegenerative disease TBI and spinal cord damage and eases (Cannizzaro et al. 2017).

As a result, oxidative stress has an important role in the pathogenesis of CNS diseases. Over months to years, the ensuing secondary brain damage progressively manifests itself and has a considerable impact on TBI treatment and prognosis despite the fact that TBI causes abrupt and irreversible primary damage to the brain parenchyma (Zhang et al. 2018). As a result, treatment approaches are critical for secondary brain injury. So far, several hallmarks of delayed secondary CNS damage have been identified, including oxidative stress, axon degeneration, glial cell overactivation, and mitochondrial failure (Ng and Lee 2019).

Recent evidence points to the pathogenesis and progression of TBI being significantly influenced by oxidative stress (Zhang et al. 2018). Generally speaking, oxidative stress is brought on or accompanied by a number of molecular pathways, including neutrophil activation, neuroexcitation pathway activation, and mitochondrial failure (P. Chen et al. 2019). In a fluid percussion TBI model, superoxide radical levels in brain microvessels rise right away after damage, but oxygen radical scavengers such as catalase and superoxide dismutase (SOD) drastically lower superoxide radical levels and partially repair the brain injury (Kontos and Povlishock 1986).

As a result of the damaged mitochondria, oxidized evacuated hemoglobin, and arachidonic acid cascade as well as the reaction of enzymes or auto-oxidation of biogenic amine neurotransmitters, a significant amount of superoxide radicals are created in the initial few minutes or hours after brain injury (Angeloni et al. 2015).

Following this, neutrophils and macrophages enter, over-activating the microglia and aiding in the production of superoxide radicals (Abbasi-Kangevari et al. 2019; Jarrahi et al. 2020). Additionally, iron overload is a result of the production of numerous lipid peroxidation and hydroxyl radicals that eventually cause oxidative damage and neuronal ferry ptosis. These conditions significantly worsen the pathogenesis of TBI from immunosuppression, the perspectives of brain plasticity, and cerebral blood flow (Tang et al. 2020).

12.3 Dietary Poly Phenols in the Management and Prevention of TBI

Nutritional methods cover a wide spectrum of molecules, including minerals and vitamins, Nutritional methods encompass a broad spectrum of molecules that contribute to health, including minerals, vitamins, and dietary polyphenols.polyphenols represent a significant area of interest in nutrition due to their diverse biological effects and potential therapeutic benefits, particularly in the context of TBI and other health conditions. Because of their diverse pleiotropic effects, a major range of compounds have shown potential as TBI agents in recent years. A wide range of chemical compounds, traditionally derived from nature, are referred to as polyphenols (Fig. 12.2). Large amounts of polyphenols can be found in tea, spices, vegetables, and fruits. Polyphenols are widely present in plant-based foods and everyday beverages, making them a significant source of these bioactive compounds in our diets.The near-universal presence of polyphenols in these everyday foods and drinks

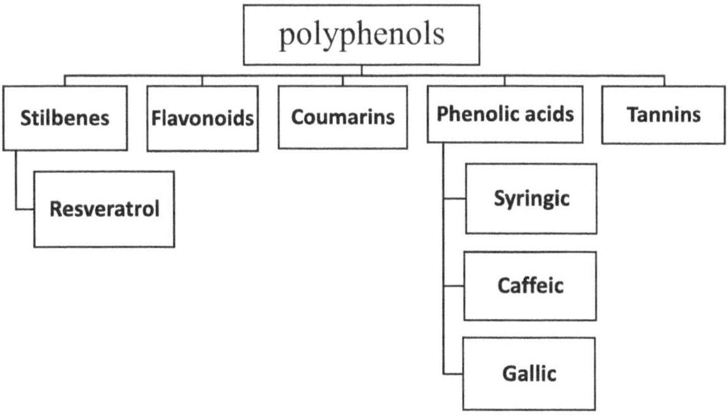

Fig. 12.2 The five primary classes of polyphenols (Adapted with permission from Carecho et al. (2023). Copyright 2023, MDPI)

provides a consistent and accessible way to include these beneficial compounds in the diet. Their diverse range of bioactive effects makes them an important component of a healthy eating pattern (Godos et al. 2017; Taguchi et al. 2020). Dietary polyphenols undergo degradation after consumption, primarily due to the breakdown of their glycoside group in the stomach. Both the large and small intestines continue to absorb them (Del Rio et al. 2013).

Polyphenols are metabolized by gut microbiota into a wide variety of smaller metabolites in the stomach and intestines. This process contributes to the overall health benefits associated with polyphenol consumption, highlighting the important role of gut microbiota in the metabolism and efficacy of dietary polyphenols (Carregosa et al. 2019, 2022). In total, certain polyphenols have the ability to get beyond the barrier of the epithelium and get into the liver, where they undergo phase II converting, including glucuronide, conjugation of sulfate, glycine, and methyl previously being discharged into the bloodstream (Carregosa et al. 2022; Del Rio et al. 2013). Notably, polyphenol byproducts have a greater ability to be ingested and reach intended organs, such as the brain, and are present in the bloodstream at significantly greater levels than their parent counterparts (Carecho et al. 2020). Thus, a lot of the systemic effects of meals high in polyphenol may actually be caused by polyphenol metabolite.

12.4 Flavanols

Catechins, EGCG, and epicatechin are among the flavanols studied in TBI models for their diverse effects on TBI-related pathogenic features. They exhibit a pleiotropic impact by addressing oxidative stress, inflammation, and neuronal damage, which are key factors in TBI. Following a TBI, both catechin and EC therapy reduced brain tissue deterioration and the onset of cerebral edema considerably (Jiang et al. 2017).

As the number of neurons dying by apoptosis in the vicinity of the injury decreases, EC and EGCG have both demonstrated notable neuroprotective benefits (Itoh et al. 2013). Furthermore, EGCG has been shown to upregulate the quantity of neural stem cells or cells that are nestin-positive, in the vicinity of the lesion (Itoh et al. 2012).

Research demonstrates that administering EGCG treatment greatly increases the proportion of neural cells in the vicinity of the injury that are positive for B-cell lymphoma 2 (Bcl2) (Itoh et al. 2013). But it is important to remember that its impacts on further components of secondary brain damage can involve, though not as strongly, the death of neural cells. This is because the effect was shown by Itoh et al.' (2013) investigation. In other words, it has been shown that the three flavanols utilized in the TBI paradigm are potent neuroinflammation and oxidative stress regulators.

12.4.1 Catechins

Low molecular weight phenolic chemicals known as flavonoids are frequently found in foods like green tea. Over 70% of green tea's total leaf polyphenol content is made up of various catechins, such as EGCG, epicatechin gallate, epicatechin, and epigallocatechin, among others (Bae et al. 2020). Monomeric flavonoids like catechins and epicatechins are an excellent choice for avoiding neuroinflammation because of their ability to function as antioxidants and scavenger free radicals as well as their high bioavailability. Studies have shown that catechins can prevent age-related cognitive deterioration and are known to enhance cognitive performance.

Treatment with catechins helps inhibit the formation of induced nitric oxide synthase, and consequently, NO, in activated microglial cells. Additionally, it lessens CNS inflammation also the creation of other pro-inflammatory markers (Farkhondeh et al. 2020). Glutathione S transferase and heme oxygenase 1 (HO-1), among other antioxidant response element genes, are produced when phosphorylated Nrf-2, a transcription factor that oxidative stress requires, translocates to the nucleus. Catechins cause Nrf-2 to become activated, which reduces oxidative stress and eventually prevents the activation of genes linked to inflammation (Du et al. 2018). Catechins have also been shown to inhibit the NF-B inflammatory pathway.

Due to the suppression of TNF-a and IL-6 synthesis, STAT-3 signaling is inhibited, which decreases downstream pro-inflammatory genes' expression, including COX-2 (Farkhondeh et al. 2020). Peroxisome proliferation activated receptor (Kruger et al. 2014), another transcription factor, also stimulates the microglia's production of pro-inflammatory cytokines. Pioglitazone and rosiglitazone, Peroxisome Proliferator-Activated Receptor (PPAR) agonists, decreased amyloid ß plaque development and enhanced cognition in test animals with an animal model of AD (Khan et al. 2019). Furthermore, catechins can prevent hypoxia-induced autophagy in microglia by preserving the mTOR and Akt phosphorylation levels.

Catechins and epicatechins are indeed used in research to evaluate their protective effects against lipid peroxidation and oxidative stress, particularly in the context of neurotoxicity induced by quinolinic acid in rat brain slices. In rat brain (striatum) slices, catechins and epicatechins are employed to evaluate their antioxidant and anti-lipid peroxidation activities against quinolinic acid (QUIN)-induced neurotoxicity (Daniel 2019). QUIN produces NMDA receptor-mediated excitotoxicity that activates persuaded nitric oxide synthase to produce NO and calcium-driven ROS. This results in neuronal death and oxidative damage. QUIN-induced oxidative stress is prevented by catechin and epicatechin, as seen by the oxidized glutathione (GSSG) levels and Glutathione (GSH).

According to reports, when compared to the control, oxidative harm done to lipids in QUIN-treated slices increased by 65%; however, when epicatechin and catechin were added, this decreased to 51% and 53% in QUIN-treated slices. Additionally, QUIN treatment lowers GSH levels by 35% compared to control. Epicatechin and catechin administration can similarly reduce GSSH levels that were raised by QUIN therapy. Research is also being conducted on catechin's effects on chemotherapy-induced cognitive impairment, sometimes referred to as "chemo brain" or "chemo fog." Chemotherapy damages DNA double strands in brain cells, which explains why cancer is linked to cognitive impairment. The neuronal death and activation of p53-mediated autophagy leads to cognitive impairment. The authors used green tea catechin hydrate at doses of 50, 100, and 200 mg/kg of body weight.

To cause chemo brain, Doxorubicin (DOX) was given in 10 cycles at a dose of 2.5 mg/kg over a period of 50 days. Animals given only DOX injections were unable to distinguish between novel and old objects, but after receiving catechins (100 mg/kg), rats showed increased cognitive function and spent more time in the presence of unknown objects. In rats given with DOX + catechin, catechins have been shown to reduce acetylcholinesterase expression in the hippocampus and frontal cortex although they enhance it in animals injected with DOX alone. This might account for catechins' capacity to enhance cognitive function and suggests a future use for catechin as an adjuvant in cancer treatment to prevent chemo brain (Cheruku et al. 2018). However, when taken orally, catechins are frequently excreted from the body within 4–6 h, necessitating a reuptake of the substance every 4–5 h (Janle et al. 2008).

Catechins are being investigated to learn more about how they help reverse the consequences of TBI. They are primarily concerned with addressing the motor dysfunction and cognitive impairment brought on by blood-brain barrier damage and tight junction protein depletion. In order to investigate the neuroprotective properties of catechins, a 2017 study (Jiang et al. 2017) employed adult male Sprague-Dawley rats to inflict a controlled cortical impact (CCI) damage of about 3.2-mm depth. The foot fault test was used in the study to evaluate the motor functioning abilities of TBI rats, while the Morris water maze test was used to conduct cognitive tests. Evans blue evaluation, reverse transcription PCR, and western blot were performed next to examine blood-brain barrier permeability, as well as the expression levels of tight junction proteins and pro-inflammatory indicators.

The motor and cognitive abilities of test rats were dramatically enhanced by a 20-mg-per-kg therapy of catechins from 24 h after TBI to 28 days after the injury in contrast to the models of TBI that were not treated. Animals that received catechins with TBI had reduced edema as shown by smaller brain infarcts and lower levels of brain water. The reduced spreading of the marker in the ipsilateral region of the brain in comparison to the TBI model rats in Evans test results further demonstrated that catechins might prevent blood-brain barrier damage.

This was further demonstrated by the fact that catechin therapy elevated the expression of the tight junction proteins occludin and zonula occluden in TBI rats. When compared to untreated controls, catechins also enhanced the expression of the anti-inflammatory marker arginase 1 and decreased the production of pro-inflammatory markers such IL-1ß, iNOS, and IL-6 in the ipsilateral regions of cortex tissue.

The second leading cause of death, cerebral ischemia, or ischemic stroke is frequently seen after TBI and is brought on by the blockage of an artery delivering blood to the brain. In addition to providing the brain cells with a basic hypoxic and nutrient-deprived state, ischemic stroke weakens the blood vessels, which increases the risk of hemorrhage within a week or two. The function of microglia is frequently viewed as having two negative effects when there is cerebral ischemia. Using hypoxic, BV2 microglial cell lines and reperfusion circumstances can be replicated to comprehend the microglial and neuronal ischemia processes at the molecular level of cell senescence (C.-M. Chen et al. 2016).

Microglial cell death was caused by the suppression of the mTOR/Akt signaling pathways, which led to autophagy, when serum-starved for 1 h and then cultured for 6 h in 1% oxygen and under no-growth conditions for 24 h. Microglia cell apoptosis was reduced and mTOR/Akt was activated when treated with 10–250 M catechin.

12.4.2 Epigallocatechin-3-Gallate (EGCG)

According to Cheng and colleagues' research, EC defends the brain from TBI by lowering iron depositing, activating the Nrf2 pathway, and suppressing the production of the HO-1 protein. In this model of TBI, one of the two responses may indicate an Nrf2-independent mechanism (Cheng et al. 2016), indicating that both Nrf2-dependent and Nrf2-independent pathways are involved in the safety provided by EC. It is noteworthy that following a TBI, all three of the flavanols investigated in the TBI paradigm dramatically enhanced neurological results.

12.5 Flavanones

The only flavanones examined in the TBI paradigm and used in treatment were hesperidin and naringenin albeit using different methods of experimentation. Since hesperidin was studied using a dietary method, it is likely that its metabolites will enter circulation and subsequently the brain. In reality, with an oral dose of 50 mg/kg of hesperidin by weight of the body in people, levels of hesperidin aglycone and hesperidin were expected to be less than 0.02 mg/L in plasma (Boonpawa et al. 2017). However, the naringenin experiment prevented this compound's gut metabolic by considering intraperitoneal (IP) pharmacological delivery. Oral hesperidin therapy has been demonstrated to significantly diminish symptoms associated with depression in rats with mild TBI. According to research by Kosari-Nasab et al., hesperidin decreased the levels of IL-1 and Malondialdehyde (MDA) and increased the generation of brain-derived neurotrophic factor (BDNF) in the hippocampus, suggesting that the observed antidepressant-like effect of hesperidin may be regulated, at least in part, by decreased damage and neurological inflammation from oxidation and enhanced BDNF generation in the hippocampus.

Brain edema has also been demonstrated to be effectively treated with naringenin-administered intraperitoneal at both tested levels (50 and 100 mg/kg), with the effects of the 100 mg/kg dose being more pronounced. The mNSS improved with naringenin (100 mg/kg) at 3, 5, and 7 days after TBI, naringenin (100 mg/kg) improved mNSS scores at 3, 5, and 7 days after TBI, while naringenin (50 mg/kg) only showed improvement at 3 days, it reflects that the higher dose has a more prolonged effect on neurological recovery compared to the lower dose, according to research (Deng et al. 2021). Naringenin, at a dose of 100 mg/kg, inhibited the apoptotic and neuronal cell death caused by TBI at the molecular level by upregulating antiapoptotic (Bcl2) and downregulating pro-apoptotic (Bcl2-associated X [Bax] and caspase-3) proteins (Deng et al. 2021).

12.6 Flavonols

Thus far, four flavonols have been studied in the context of TBI: quercetin, fisetin, rutin, and kaempferol. (Song et al. 2021). While fisetin, rutin, and quercetin have reduced neurodegeneration and apoptosis in the peri-lesioned region, as evidenced by an increase in the total number of living neurons, a decrease in active caspase-3 levels, and a decrease in the number of apoptotic cells, fisetin has been shown to preserve the authenticity of the BBB while decreasing the size of brain lesions (Yang et al. 2014). Quercetin's neuroprotective mechanisms are believed to involve the activation of the PI3K/Akt pathway, increased phosphorylation of Akt and ERK1, and modulation of apoptosis-related proteins like Bax and Bcl2. These actions help promote cell survival and reduce neuronal damage (Yang et al. 2014).

Furthermore, LC3/NeuN double labeling showed that quercetin inhibited neuronal autophagy in the hippocampus. Macroscopic brain edema as determined by the brain content of water was effectively lowered by quercetin, fisetin, and rutin (Song et al. 2021). Additionally, these three compounds significantly enhanced neurological, motor, sensory–motor, and cognitive functioning, all of which had been significantly impaired after TBI (Yang et al. 2014). TBI research on flavonols is limited to their effects when used after the injury. Consequently, there is no comparative data on how flavonols might work as a preventive measure or as part of a long-term treatment regimen. Regarding the administration method, it is worth noting that a boost in TBI outcomes is described regardless of the compound delivering route (oral versus IP). Because flavonol bioavailability is severely restricted by oral gavage, the two primary flavonols and their derivatives may be important to the observed effects.

12.7 Flavones

So far, the only flavones used in TBI settings have been luteolin, one of the more prevalent dietary flavones, and chrysin. Absorb and metabolic investigations revealed that both flavones had comparatively poor bioavailability, which could explain why most luteolin research considered IP delivery. Luteolin, administered intravenously both before and after TBI during the induction process, was useful for reducing brain inflammation, as evidenced by lower TNF- and IL-1 mRNA and protein levels, as well as lower NF-kB p65 nuclear accumulation (Sawmiller et al. 2014). Within 24 h after the TBI, luteolin recovered levels of MDA and glutathione peroxidase (GPx) function. Luteolin decreases neuronal cell damage and death 24 h after TBI and increases beclin1 and LC3 levels, implying increased autophagy (Xu et al. 2014). The TBI trials discussed herein used luteolin either pre- or post-TBI medication, via the IP route of treatment. The primary stages of secondary brain injury were significantly impacted by each pre-treatment and following treatment (Sawmiller et al. 2014; Xu et al. 2014). However, no investigations involving ongoing therapy have been conducted to the best of our understanding.

12.8 Isoflavone

So far, just one study has examined the impact of isoflavone in the TBI model. Soltani and coworkers found that genistein in a low dose (15 mg/kg) administered after a TBI can reduce brain edema, as measured by a substantial decrease in brain levels of water. Genistein also lowered intracranial pressure while improving BBB integrity. Furthermore, with regard to functioning, genistein increased motor function (Soltani et al. 2015).

12.9 Curcuminoids

Curcumin was thoroughly studied in TBI environments as an example polyphenol of the curcuminoid group. It has been linked to a considerable decrease in IL-1, TNF-mRNA, IL-18, IL-6, and levels of protein, as well as a reduction in astrocytic and microglial activity followed by TBI (Dong et al. 2018).

Curcumin management could enhance TBI effects by reducing the acute activation of macrophages and microglia and causing neuronal death in TBI microglia/macrophages through a mechanism linking the toll-like receptor 4 (TLR4)/MyD88/NF-B signaling process.

Furthermore, curcumin reduced the levels of MDA, the primary result of polyunsaturated fatty acid peroxidation (Dong et al. 2018). Wu and colleagues found that dietary curcumin administration significantly decreased damage caused by oxidative stress and restored aberrant levels of BDNF, synapsin 1, and Cyclic AMP Response Element-Binding Protein (CREB) following TBI (A. Wu et al. 2006). Curcumin intervention has also been shown to increase the amounts of molecules that regulate energy balance after TBI by increasing the expression of ubiquitous mitochondrial creatine kinase, AMP-activated protein kinase, and cytochrome c oxidase II (COX-II).

Dong et al. showed that following injury, curcumin treatment decreased the proportion of apoptotic neurons in the injured cortex, decreased the levels of cleaved caspase-3, and elevated the expression of Bcl2 protein (Dong et al. 2018).

These effects were linked to increased Nrf2-enhancing nuclear transfer and increased levels of antioxidant enzymes in the Nrf2-ARE pathway, including NQO-1, HO-1, glutamate-cysteine ligase modifier subunit, and glutamate-cysteine ligase catalytic subunit, suggesting that the Nrf2-ARE pathway plays a role in the achievement of curcumin effects (Dong et al. 2018).

In both investigated concentrations, the animal death rate following a TBI was considerably decreased by administering 50 and 100 mg/kg of curcumin 5 days before the TBI. On the other hand, lesion volume was reduced in a dose-related way, with 100 mg/kg showing greater effectiveness. After damage, administering curcumin significantly decreased cerebral edema (Dong et al. 2018).

Curcumin altered the creation of aquaporin-4 (AQP4), an astrocytic water channel linked to the emergence of cellular edema after head trauma, indicating a possible mechanism via which curcumin inhibits the growth of brain edema following traumatic brain injury. Studies on the functional results after traumatic brain injury have demonstrated that curcumin protects against impairment to the reflex, sensory, cognitive, and motor systems (Soltani et al. 2015).

Despite the curcumin usage differences between research studies regarding the administration method (IP or oral) and regimen (one or more treatments before and after the TBI), outcomes in all of the experiments demonstrated that it was able to decrease neurodegeneration, oxidative stress, and neuroinflammation and improve functions after the TBI. Two investigations (A. Wu et al. 2006) used a continuous treatment with curcumin (4-week pre-treatment and 1-week post-treatment), with curcumin included in the animal chow, and found that curcumin had a strong favorable impact on TBI outcome. However, just one research compared pre- and post-curcumin treatment, determining that pre-treatment processes were more effective than following treatment for decreasing brain edema.

12.10 Stilbenes

While stilbenes are a large subfamily of polyphenolic substances, resveratrol has been only one of them employed in TBI research to our knowledge. However, substantial proof indicates that resveratrol can raise the amount of the anti-inflammatory cytokine IL-10 in damaged brain tissue and has anti-inflammatory qualities while simultaneously decreasing brain levels of IL-1, IL-6, IL-4, IL-12, TNF-, NF-kB p65, and high mobility group box 1 and greatly reducing the activity of microglial cells after TBI (Gatson et al. 2013). There are polyphenol effects described in TBI models. Meanwhile, it is impossible to stop the TBI original stress-causing; the majority of research is primarily concerned with preventing further insults that result from the formation of the damage. The secondary pathogenic cascades in TBI cannot yet be targeted with evidence-based treatments (Lucke-Wold et al. 2018). Recent studies have focused a great deal of attention on the benefits of nutritional supplements and TBI recovery (Lucke-Wold et al. 2018; Sharma et al. 2020). Given the recent surge in scientific interest in resveratrol, this paper focuses on the anti-inflammatory, anti-apoptotic, and antioxidant properties of tannins, resveratrol, and catechins in preclinical models that illustrate the core processes of TBI neuroprotection.

12.10.1 Resveratrol

Resveratrol is widely distributed in plants, such as *Polygonum* sp., grapes, peanuts, and herbs (Burns et al. 2002). Pterostilbene and resveratrol are biochemically related and exhibit many of the same pharmacological properties (Kumar et al. 2013). Because of its lipophilicity and dimethoxy structure, the analog's increased cell membrane permeability makes resveratrol a more effective medicinal drug. However, the phenolic hydroxyl moiety may be susceptible to the glucuronidation and sulfation process because it has not been methylated (Burns et al. 2002; Wang and Sang 2018). This is because it has a moderate level of metabolic stability and offers more conjugating sites than its analog, pterostilbene (Burns et al. 2002; Wang and Sang 2018). However, there is a severe lack of in-depth knowledge regarding the pharmacokinetics and metabolism of resveratrol.

The UDP-glucuronosyltransferase (UGT) family of enzymes catalyzes the conversion of resveratrol to its primary glucuronide metabolites, resveratrol 3-O-glucuronide, and resveratrol 4'-O-glucuronide (Aumont et al. 2001). Most UGT family enzymes are found in high amounts in human hepatic microsomes, which (Aumont et al. 2001) polyphenols from the body are salvation. Resveratrol-3-O-sulfate and resveratrol-4'-O-sulfate are the two primary salvation end products of resveratrol. Human sulfotransferases convert resveratrol to variable degrees, according to a study (Miksits et al. 2005).

Resveratrol's bioavailability is significantly influenced by the route of administration (Walle 2011). Healthy individuals were given radio-labeled resveratrol intravenously and orally, and it was discovered that about 70% of the substance is absorbed orally (Walle et al. 2004). Due to rapid and thorough biotransformation, resveratrol alone has a bioavailability that is almost negligible. As a result, the amount of unmodified Resveratrol that is present in the systemic circulation is extremely little. One hour after taking the oral dose, the plasma's total radioactivity peaked, and it peaked again 6 h later. The first hour saw a sharp decline in the total radioactivity concentrations of plasma, and the recovery of radioactivity in urine varied from 40% to 80%, suggesting a rapid and extensive dispersion phase.

Many clinical trials used an oral dose of 25 mg of resveratrol to evaluate its bioavailability and kinetics, with unchanged resveratrol concentrations ranging from 1 to 5 ng/mL (L. Almeida et al. 2009; Walle et al. 2004). Up to 530 ng/mL of unmodified resveratrol was observed at high dosages (up to 5 g). Resveratrol-3-O-sulfate had a concentration that was more than 15 times higher than resveratrol. A clinical investigation with fewer than 50 healthy participants revealed that resveratrol doses of 0.5–5.0 g for 4 weeks are safe and beneficial, with the exception of moderate gastrointestinal problems at larger doses (Brown et al. 2010). In a related investigation, rats were given pure resveratrol for 90 days at a dose of 700 mg/kg body weight without experiencing any negative side effects (Williams et al. 2009). Resveratrol has a low bioavailability although some preclinical investigations have shown that it can pass BBB (Walle 2011). After intraperitoneal delivery, significant quantities of resveratrol were found in the brain, with a peak concentration occurring after 4 h.

Any brain injury has a long-term effect known as behavioral impairment. The behavioral abnormalities that result from TBI-induced brain damage are most frequently observed (Singleton et al. 2010; Sönmez et al. 2007). These deficits include the loss of visuospatial memory and cognition, anxiety, aggression, and reduced locomotion. The use of resveratrol in treating TBI-related behavioral impairments is supported by growing data (Salberg et al. 2017). Intraperitoneal injection of 100 mg/kg resveratrol ameliorated anxiolytic behavior and repaired locomotor and mental disorders in Westar Albino rats with infantile TBI. Resveratrol at dosages of 25 and 50 mg/kg similarly demonstrated reduced anxiolytic behavior and improved cognitive function in the thoracic blast-induced TBI mice (Cong et al. 2021).

It has been demonstrated that resveratrol therapy can lessen hippocampus neuron loss while maintaining them. In the weight loss and CCI rat model of TBI, intraperitoneal administration of 100 mg/kg resveratrol prevented the loss of hippocampal neurons (CA1, CA2, CA3, and DG areas of the hippocampus) (Singleton et al. 2010; Sönmez et al. 2007). Similarly, after receiving doses of 50 mg/kg/day and 100 mg/kg/day resveratrol, Feeney's falling weight model of TBI mice showed preserved CA1, CA3, and CA4 areas of the hippocampus (Atalay et al. 2017). The loss of neurons in TBI is also correlated with decreased expression of proteins that are abundantly expressed in the presynaptic vesicles. In the cerebral hippocampus of weight-dropping rodents, intraperitoneal injection of 100 mg/kg resveratrol was observed to upregulate the expression of presynaptic proteins, synaptophysin, and PS-95. In the cerebral hippocampus of the weight drop rodent model of TBI, intraperitoneal treatment of 100 mg/kg resveratrol was observed to upregulate the expression of presynaptic proteins, synaptophysin, and PS-95, indicating the restoration of synaptic loss (Feng et al. 2016). The inability of wounded axons to regenerate is one of the main obstacles to TBI recovery. It is interesting to note that in a cellular model, exposure to 5-M resveratrol marginally encouraged neurite regrowth among damaged cerebral granular neurons (Shanan et al. 2019).

Resveratrol shows neuroprotection against the primary damage of TBI, further illuminating the likely mechanism of action. Numerous research conducted in a preclinical setting also suggest that it has the ability to influence key elements of secondary injury following a TBI, such as oxidative stress, neuroinflammation, and apoptosis.

12.11 Phenolic Acids

Phenolic acids (PA) are a large subclass of polyphenolic chemicals. So far, various PAs have been tried in TBI applications, and all of them have shown highly effective impacts on oxidative stress, neurodegenerative processes, and neuroinflammation in addition to functional results afterward TBI. A significant characteristic of this class of substances is that in addition to their inclusion in the matrix of food, among them are polyphenols and more complex catabolites produced by gut bacteria, including ferulic acid, cinnamic acid, and caffeic acid. Therefore, the effects that this group has reported could be similar to the effects of oral parent polyphenols that undergo gut metabolism. After a TBI, it has been demonstrated that gallic acid dramatically lowers the brain's IL-6, TNF-α, and IL-1β protein levels (Gatson et al. 2013). According to a study utilizing protocatechuic acid, this substance reduced GSH depletion and inhibited microglial activation in damaged brain tissue, which helped prevent subsequent injury (S. H. Lee et al. 2017). Surprisingly, using an immediate ferulic acid infusion after TBI increased SOD and GSH levels and decreased neuronal apoptosis. Following brain damage, protocatechuic acid and ferulic acid were shown to restore SOD and GSH levels, whereas ferulic and gallic markedly lower levels of MDA in the tissue of the brain (S. H. Lee et al. 2017). According to a study conducted by Nasution and colleagues, lowering the level of F2-IsoPs in their serum and administering caffeic acid phenethyl ester (CAPE), a small chemical derivative of caffeic acid, to injured animals decrease oxidative damage (Nasution et al. 2020). Following the TBI, therapy with CAPE totally recovered the levels of GPx, SOD, and MDA (Kerman et al. 2012). Protocatechuic acid was able to significantly lessen dendritic loss in the cortex and neurodegeneration and the hippocampal CA1 and CA3 regions when delivered right after TBI. Additionally, it has been demonstrated that ferulic acid dramatically reduces the death of neuronal cells in the prefrontal cortex and the hippocampal CA1, CA2, CA3, and DG regions in rats who have had TBI. Furthermore, by raising synaptophysin protein levels, and synaptic spine density. Synaptophysin is a protein found in the presynaptic vesicles of neurons, and it is commonly used as a marker for synaptic density and function. Synaptic spine density refers to the number of dendritic spines per unit length of a neuron's dendrite, which is an important indicator of synaptic connectivity and plasticity. Research on various neuroprotective agents, including polyphenols like CAPE (Caffeic Acid Phenethyl Ester), often focuses on their effects on these synaptic markers. Increasing synaptophysin levels and synaptic spine density are considered positive indicators of improved synaptic function and plasticity. CAPE and similar compounds may support these aspects of neuronal health, contributing to their potential benefits in cognitive and motor function. However, while these findings are promising, ongoing research is essential to fully understand their implications and mechanisms in both animal models and human studies. It has been shown that cinnamic acid can repair synaptic deficiencies (Guo et al. 2019).

According to Guo and associates, cinnamic acid accomplishes these advantageous effects by inhibiting HDAC2 activity by fostering the expression of miR-455-3p, a miRNA that controls HDAC2 at the post-transcriptional level (Guo et al. 2019).

Cinnamic acid treatment has been demonstrated to considerably diminish brain edema (Guo et al. 2019), whereas just one injection of CAPE therapy substantially decreased the volume of the cortical infarction in rats. According to a research carried out by Nasution and colleagues, treating with CAPE lowers blood levels of AQP4 protein and AQP4 mRNA expression in the brain, which may be the mechanism by which CAPE relieves brain edema. Through the elevation of occludin-5 levels, post-injury CAPE injection also improved BBB integrity and lowered TBI-associated vascular dysfunction.

Vestibulomotor/motor performance memory and learning were improved by CAPE treatment. Additionally, astrocytes, a kind of brain glial cell required for neuronal survival following TBI, are said to be protected by resveratrol. Reactive oxygen species levels rise as a result of glutamate excitotoxicity brought on by TBI (Bunney et al. 2017). An increase in ROS levels activates GSK-3ß and causes autophagy (Mancinelli et al. 2017), which causes neurons and glial cells to die.

To construct the majority of in vitro TBI models, astrocyte cultures are treated with glutamate (1–10 mM) prior to investigating the potential involvement of resveratrol in TBI (Lin et al. 2014).

When glutamate and resveratrol (1–5 M) are combined for 48 h, resveratrol increases tyr216 phosphorylation of GSK-3ß while decreasing ser9 phosphorylation, decreasing ROS levels, and boosting cell viability. In vivo models of TBI, 100 mg/kg of resveratrol administration also revealed a decrease in markers of autophagy. In vivo models of TBI, 100 mg/kg of resveratrol administration furthermore demonstrated a decline in autophagy markers. Upon treatment, it also repairs mitochondrial damage.

12.12 Ellagitannins

Ellagic acid, the sole representative of the subfamily of ellagitannins in the TBI approach, has been demonstrated to dramatically lower the levels of TNF-α, IL-1β, and IL-6 protein in the brain after a TBI (Mashhadizadeh et al. 2017). It decreases BBB absorptivity following a TBI, according to studies by Mashhadizadeh et al. (2017).

Ellagic acid was also successful in improving neurological function, electrophysiological parameters, and memory and cognition in regimens of application—as well as before and after therapy (Mashhadizadeh et al. 2017).

12.13 Tannins

Complex tannins, condensed tannins, and hydrolyzable tannins are the three main categories of tannins. GalloTannins and ellagiTannins are subcategories of hydrolyzable tannins. Nuts, tea, maize, and other foods commonly contain complex tannins. They are created when a catechin molecule is coupled with an ellagitannin or gallotannin. Thus, complex tannins got their name. Condensed tannins are strong antioxidant molecules that are widely distributed in foods including black tea, nuts, and fruits because they contain the flavan-3-ol component. Gallic acid is a constituent of the PA that makes up hydrolyzable tannins. They are abundant in fruits like grapes and bananas. Tannins are more readily absorbed when they are conjugated to proteins and polymers rather than being taken in their pure state due to their larger molecular weight. The gut bacteria convert tannins such as ellagitannins into ellagic acid and gallotannins into gallic acid metabolites (Marín Fernández et al. 2015).

Utilizing a streptozotocin-induced cognitive paradigm, the effect of tannins on neuroinflammation was recently investigated. Wistar male rats were gavage-fed 30 mg/kg of tannic acid for 21 days, and 3 mg/kg of streptozotocin was administered intravenously to create an AD model of neuroinflammation. In comparison to the streptozotocin-treated group, rats given tannic acid treatment showed notably reduced pro-inflammatory cytokine levels and enhanced mental agility in an inhibition avoidance test (Gerzson et al. 2020).

Punicalagin, an ellagitannin, was additionally investigated for its capacity to inhibit neuroinflammation. Punicalagin pre-treatment reduced the amounts of pro-inflammatory indicators in the main rat activation of microglia paradigm treated with LPS, including prostaglandin E2, IL-6, and TNF. When 5–40 were exposed to punicalagin, there was a decrease in NF-B expression, phosphorylation of I-B kinase, I-B, and p65, and TLR-4-mediated activation of TRAF 6, which is what causes NF-B activation (Olajide et al. 2014). Tannic acid, like punicalagin, has been found to reduce neurological inflammation in LPS-stimulated BV2 microglia cells by reducing the expression of NF-B.

In this study, tannic acid was used at 10, 25, and 50 M concentrations (Y. Wu et al. 2019). Neuroinflammation caused by the production of amyloid ß plaques in PSAPP mice was examined after a 6-month gavage feeding of 30 mg/kg tannic acid. Measurements of astrocytosis and reactive microgliosis based on Iba1 expression and GFAP suggested that rats fed tannic acid had less neuroinflammation than controls. Two pro-inflammatory cytokines, IL-1ß and TNF-a, were also expressed in the brain but much less so than in controls (Mori et al. 2012).

A fairly recent study used CCI injury in Westar rats to examine the role of tannic acids in TBI. Tannic acid was administered to the animals at doses of 50 mg/kg body weight 30 min prior to the cortical injury as well as 6 and 18 h later. Tannic acid most likely uses the PGC-1a/Nrf2/HO-1 signaling pathway to carry out its neuroprotective activity, according to the researchers (Salman et al. 2020). The TBI model was examined, and it was then given 50 mg/kg of tannic acid. We looked at things like apoptotic signal activation, lipid peroxidation, impairments in motor function, and pro-inflammatory marker expression.

The expression of the oxidative stress-protective enzymes CAT, GST, GSH, SOD, and GPx, increased; however, the lipid peroxidation indicator 4-HNE dropped. Furthermore, there was decreased expression of GFAP and cleaved caspase 3, which indicates less reactive gliosis and apoptosis, both of which cause neuronal death. Reduced expression of the pro-inflammatory cytokines TNFa and IL-1ß further supported the finding that the brain's water content had decreased.

12.14 Conclusion

Polyphenols are phytonutrients that have been found to be among the most likely to be bioactive molecules thus far. The pleiotropic nature of polyphenols—their ability to influence multiple biological pathways—suggests that incorporating them into TBI (traumatic brain injury) care might be beneficial. All of the findings in this chapter, irrespective of animal species, animal strain, animal sex, or animal age, as well as the TBI models and utilization route applied, demonstrate the enormous ability of polyphenols to effectively fight the TBI's subsequent damage cascade at various stages. Most notably, when animals were given polyphenols, a large number of studies revealed major impacts on the results after TBI in regard to better neurological scores and motor and cognitive function. However, these concerns must be resolved in the planned preclinical research in order to give more data and provide the groundwork for future clinical trials with polyphenols.

References

Abbasi-Kangevari M, Ghamari S-H, Safaeinejad F, Bahrami S, Niknejad H (2019) Potential therapeutic features of human amniotic mesenchymal stem cells in multiple sclerosis: immunomodulation, inflammation suppression, angiogenesis promotion, oxidative stress inhibition, neurogenesis induction, MMPs regulation, and remyelination stimulation. Front Immunol 10:238

Almeida L, Vaz-da-Silva M, Falcão A, Soares E, Costa R, Loureiro AI et al (2009) Pharmacokinetic and safety profile of trans-resveratrol in a rising multiple-dose study in healthy volunteers. Mol Nutr Food Res 53(S1):S7–S15

Angeloni C, Prata C, Vieceli Dalla Sega F, Piperno R, Hrelia S (2015) Traumatic brain injury and NADPH oxidase: a deep relationship. Oxidative Med Cell Longev 2015:370312

Atalay T, Gülşen İ, Çölçimen N, Alp H, Sosuncu E, Alaca I et al (2017) Resveratrol treatment prevents hippocampal neurodegeneration in a rodent model of traumatic brain injury. Turk Neurosurg 27(6):924

Aumont V, Krisa S, Battaglia E, Netter P, Richard T, Mérillon J-M et al (2001) Regioselective and stereospecific glucuronidation of trans-and cis-resveratrol in human. Arch Biochem Biophys 393(2):281–289

Ayoub IM, Korinek M, El-Shazly M, Wetterauer B, El-Beshbishy HA, Hwang T-L et al (2021) Anti-allergic, anti-inflammatory, and anti-hyperglycemic activity of Chasmanthe aethiopica leaf extract and its profiling using LC/MS and GLC/MS. Plan Theory 10(6):1118

Bae J, Kim N, Shin Y, Kim S-Y, Kim Y-J (2020) Activity of catechins and their applications. Biomed Dermatol 4:1–10

Boonpawa R, Spenkelink A, Punt A, Rietjens IM (2017) Physiologically based kinetic modeling of hesperidin metabolism and its use to predict in vivo effective doses in humans. Mol Nutr Food Res 61(8):1600894

Brown VA, Patel KR, Viskaduraki M, Crowell JA, Perloff M, Booth TD et al (2010) Repeat dose study of the cancer chemopreventive agent resveratrol in healthy volunteers: safety, pharmacokinetics, and effect on the insulin-like growth factor axis. Cancer Res 70(22):9003–9011

Bunney P, Zink A, Holm A, Billington C, Kotz C (2017) Orexin activation counteracts decreases in nonexercise activity thermogenesis (NEAT) caused by high-fat diet. Physiol Behav 176:139–148

Burns J, Yokota T, Ashihara H, Lean ME, Crozier A (2002) Plant foods and herbal sources of resveratrol. J Agric Food Chem 50(11):3337–3340

Cannizzaro L, Rossoni G, Savi F, Altomare A, Marinello C, Saethang T et al (2017) Regulatory landscape of AGE-RAGE-oxidative stress axis and its modulation by PPARγ activation in high fructose diet-induced metabolic syndrome. Nutr Metab 14:1–13

Carecho R, Carregosa D, Dos Santos CN (2020) Low molecular weight (poly) phenol metabolites across the blood-brain barrier: the underexplored journey. Brain Plast 6(2):193–214

Carecho R, Carregosa D, Ratilal BO, Figueira I, Ávila-Gálvez MA, Dos Santos CN, Loncarevic-Vasiljkovic N (2023) Dietary (poly) phenols in traumatic brain injury. Int J Mol Sci 24(10):8908

Carregosa D, Carecho R, Figueira IS, Santos CN (2019) Low-molecular weight metabolites from polyphenols as effectors for attenuating neuroinflammation. J Agric Food Chem 68(7):1790–1807

Carregosa D, Pinto C, Ávila-Gálvez MÁ, Bastos P, Berry D, Santos CN (2022) A look beyond dietary (poly) phenols: the low molecular weight phenolic metabolites and their concentrations in human circulation. Compr Rev Food Sci Food Saf 21(5):3931–3962

Chen C-M, Wu C-T, Yang T-H, Chang Y-A, Sheu M-L, Liu SH (2016) Green tea catechin prevents hypoxia/reperfusion-evoked oxidative stress-regulated autophagy-activated apoptosis and cell death in microglial cells. J Agric Food Chem 64(20):4078–4085

Chen P, Totten M, Zhang Z, Bucinca H, Erikson K, Santamaría A et al (2019) Iron and manganese-related CNS toxicity: mechanisms, diagnosis and treatment. Expert Rev Neurother 19(3):243–260

Cheng T, Wang W, Li Q, Han X, Xing J, Qi C et al (2016) Cerebroprotection of flavanol (−)-epicatechin after traumatic brain injury via Nrf2-dependent and-independent pathways. Free Radic Biol Med 92:15–28

Cheruku SP, Ramalingayya GV, Chamallamudi MR, Biswas S, Nandakumar K, Nampoothiri M et al (2018) Catechin ameliorates doxorubicin-induced neuronal cytotoxicity in in vitro and episodic memory deficit in in vivo in Wistar rats. Cytotechnology 70:245–259

Cong P, Wang T, Tong C, Liu Y, Shi L, Mao S et al (2021) Resveratrol ameliorates thoracic blast exposure-induced inflammation, endoplasmic reticulum stress and apoptosis in the brain through the Nrf2/Keap1 and NF-κB signaling pathway. Injury 52(10):2795–2802

Daniel SDÁ (2019) Comparative effects of catechin, epicatechin and N-Ω-nitroarginine on quinolinic acid-induced oxidative stress in rat striatum slices. Biomed Pharmacother 78:210–215

de Almeida AJPO, de Almeida Rezende MS, Dantas SH, de Lima Silva S, de Oliveira JCPL, Alves RMFR et al (2020) Unveiling the role of inflammation and oxidative stress on age-related cardiovascular diseases. Oxidative Med Cell Longev 2020:1954398

de Andrade Teles RB, Diniz TC, Costa Pinto TC, de Oliveira Junior RG, Gama de Silva M, de Lavor ÉM et al (2018) Flavonoids as therapeutic agents in Alzheimer's and Parkinson's diseases: a systematic review of preclinical evidences. Oxidative Med Cell Longev 2018:7043213

Del Rio D, Rodriguez-Mateos A, Spencer JP, Tognolini M, Borges G, Crozier A (2013) Dietary (poly) phenolics in human health: structures, bioavailability, and evidence of protective effects against chronic diseases. Antioxid Redox Signal 18(14):1818–1892

Deng C, Yi R, Fei M, Li T, Han Y, Wang H (2021) Naringenin attenuates endoplasmic reticulum stress, reduces apoptosis, and improves functional recovery in experimental traumatic brain injury. Brain Res 1769:147591

Di Meo F, Margarucci S, Galderisi U, Crispi S, Peluso G (2019) Curcumin, gut microbiota, and neuroprotection. Nutrients 11(10):2426

Diaz B, Elkbuli A, Wobig R, McKenney K, Jaguan D, Boneva D et al (2019) Subarachnoid versus nonsubarachnoid traumatic brain injuries: the impact of decision-making on patient safety. J Emerg Trauma Shock 12(3):173

Dong W, Yang B, Wang L, Li B, Guo X, Zhang M et al (2018) Curcumin plays neuroprotective roles against traumatic brain injury partly via Nrf2 signaling. Toxicol Appl Pharmacol 346:28–36

Du X, Yu J, Sun X, Qu S, Zhang H, Hu M et al (2018) Impact of epigallocatechin-3-gallate on expression of nuclear factor erythroid 2-related factor 2 and γ-glutamyl cysteine synthetase genes in oxidative stress-induced mouse renal tubular epithelial cells. Mol Med Rep 17(6):7952–7958

Farkhondeh T, Pourbagher-Shahri AM, Ashrafizadeh M, Folgado SL, Rajabpour-Sanati A, Khazdair MR, Samarghandian S (2020) 绿茶儿茶素和小胶质细胞功能. 中国神经再生研究(英文版) 15(10):1792

Feng Y, Cui Y, Gao JL, Li R, Jiang XH, Tian YX et al (2016) Neuroprotective effects of resveratrol against traumatic brain injury in rats: involvement of synaptic proteins and neuronal autophagy. Mol Med Rep 13(6):5248–5254

Forsan HF, Awad SS (2023) Cyanidin: advances on resources, biosynthetic pathway, bioavailability, bioactivity, and pharmacology. In: Xiao J (ed) Handbook of dietary flavonoids. Springer International Publishing, Cham, pp 1–50

Gatson JW, Liu M-M, Abdelfattah K, Wigginton JG, Smith S, Wolf S, Minei JP (2013) Resveratrol decreases inflammation in the brain of mice with mild traumatic brain injury. J Trauma Acute Care Surg 74(2):470–475

Gerzson MF, Bona NP, Soares MS, Teixeira FC, Rahmeier FL, Carvalho FB et al (2020) Tannic acid ameliorates STZ-induced Alzheimer's disease-like impairment of memory, neuroinflammation, neuronal death and modulates Akt expression. Neurotox Res 37:1009–1017

Glotov AS, Tiys ES, Vashukova ES, Pakin VS, Demenkov PS, Saik OV et al (2015) Molecular association of pathogenetic contributors to pre-eclampsia (pre-eclampsia associome). BMC Syst Biol 9(2):1–12

Godos J, Rapisarda G, Marventano S, Galvano F, Mistretta A, Grosso G (2017) Association between polyphenol intake and adherence to the Mediterranean diet in Sicily, southern Italy. NFS J 8:1–7

Gu Y-H, Yamashita T, Yamamoto H, Matsuo T, Washino N, Song J-H, Kang K-M (2019) Plant enzymes decrease prostate cancer cell numbers and increase TNF-α in vivo: a possible role in immunostimulatory activity. Int J Food Sci 2019:8103480

Guo S, Zhen Y, Zhu Z, Zhou G, Zheng X (2019) Cinnamic acid rescues behavioral deficits in a mouse model of traumatic brain injury by targeting miR-455-3p/HDAC2. Life Sci 235:116819

Hu J, Chen L, Huang X, Wu K, Ding S, Wang W et al (2019) Calpain inhibitor MDL28170 improves the transplantation-mediated therapeutic effect of bone marrow-derived mesenchymal stem cells following traumatic brain injury. Stem Cell Res Ther 10(1):1–13

Huynh LM, Burns MP, Taub DD, Blackman MR, Zhou J (2020) Chronic neurobehavioral impairments and decreased hippocampal expression of genes important for brain glucose utilization in a mouse model of mild TBI. Front Endocrinol 11:556380

Itoh T, Imano M, Nishida S, Tsubaki M, Mizuguchi N, Hashimoto S et al (2012) (−)-Epigallocatechin-3-gallate increases the number of neural stem cells around the damaged area after rat traumatic brain injury. J Neural Transm 119:877–890

Itoh T, Tabuchi M, Mizuguchi N, Imano M, Tsubaki M, Nishida S et al (2013) Neuroprotective effect of (−)-epigallocatechin-3-gallate in rats when administered pre-or post-traumatic brain injury. J Neural Transm 120:767–783

Janle EM, Morré DM, Morré DJ, Zhou Q, Zhu Y (2008) Pharmacokinetics of green tea catechins in extract and sustained-release preparations. J Diet Suppl 5(3):248–263

Jarrahi A, Braun M, Ahluwalia M, Gupta R, Wilson M, Munie S (2020) Revisiting traumatic brain injury: from molecular mechanisms to therapeutic interventions. Biomedicines 8:389

Jiang Z, Zhang J, Cai Y, Huang J, You L (2017) Catechin attenuates traumatic brain injury-induced blood–brain barrier damage and improves longer-term neurological outcomes in rats. Exp Physiol 102(10):1269–1277

Kanninen KM, Pomeshchik Y, Leinonen H, Malm T, Koistinaho J, Levonen A-L (2015) Applications of the Keap1–Nrf2 system for gene and cell therapy. Free Radic Biol Med 88:350–361
Kerman M, Kanter M, Coşkun KK, Erboga M, Gurel A (2012) Neuroprotective effects of caffeic acid phenethyl ester on experimental traumatic brain injury in rats. J Mol Histol 43:49–57
Khan MA, Alam Q, Haque A, Ashafaq M, Khan MJ, Ashraf GM, Ahmad M (2019) Current progress on peroxisome proliferator-activated receptor gamma agonist as an emerging therapeutic approach for the treatment of Alzheimer's disease: an update. Curr Neuropharmacol 17(3):232–246
Khatri N, Thakur M, Pareek V, Kumar S, Sharma S, Datusalia AK (2018) Oxidative stress: major threat in traumatic brain injury. CNS Neurol Disorders Drug Targets 17(9):689–695
Khellaf A, Khan DZ, Helmy A (2019) Recent advances in traumatic brain injury. J Neurol 266:2878–2889
Kontos HA, Povlishock JT (1986) Oxygen radicals in brain injury. Cent Nerv Syst Trauma 3(4):257–263
Kruger AK, Reither EN, Peppard PE, Krueger PM, Hale L (2014) Do sleep-deprived adolescents make less-healthy food choices? Br J Nutr 111(10):1898–1904
Kumar A, Negi G, Sharma S (2013) Neuroprotection by resveratrol in diabetic neuropathy: concepts & mechanisms. Curr Med Chem 20(36):4640–4645
Lee SH, Choi BY, Lee SH, Kho AR, Jeong JH, Hong DK, Suh SW (2017) Administration of protocatechuic acid reduces traumatic brain injury-induced neuronal death. Int J Mol Sci 18(12):2510
Lee B, Leem J, Kim H, Jo H-G, Yoon S-H, Shin A et al (2019) Herbal medicine for acute management and rehabilitation of traumatic brain injury: a protocol for a systematic review. Medicine 98(3):e14145
Li G, Xiang S, Pan Y, Long X, Cheng Y, Han L, Zhao X (2021) Effects of cold-pressing and hydrodistillation on the active non-volatile components in lemon essential oil and the effects of the resulting oils on aging-related oxidative stress in mice. Front Nutr 8:689094
Lin C, Chen T, Yang L, Shih C (2014) Resveratrol protects astrocytes against traumatic brain injury through inhibiting apoptotic and autophagic cell death. Cell Death Dis 5(3):e1147–e1147
Liu W, Chen Y, Meng J, Wu M, Bi F, Chang C et al (2018) Ablation of caspase-1 protects against TBI-induced pyroptosis in vitro and in vivo. J Neuroinflammation 15(1):1–16
Loessner B, Bullock S, Rose SP (1988) 411B: a monoclonal postsynaptic marker for modulations of synaptic connectivity in the rat brain. J Neurochem 51(2):385–390
Lorente L, Martín MM, Almeida T, Abreu-González P, Ramos L, Argueso M et al (2015) Total antioxidant capacity is associated with mortality of patients with severe traumatic brain injury. BMC Neurol 15:1–7
Lucke-Wold BP, Logsdon AF, Nguyen L, Eltanahay A, Turner RC, Bonasso P et al (2018) Supplements, nutrition, and alternative therapies for the treatment of traumatic brain injury. Nutr Neurosci 21(2):79–91
Mahmood A, Goussev A, Lu D, Qu C, Xiong Y, Kazmi H, Chopp M (2008) Long-lasting benefits after treatment of traumatic brain injury (TBI) in rats with combination therapy of marrow stromal cells (MSCs) and simvastatin. J Neurotrauma 25(12):1441–1447
Mancinelli R, Carpino G, Petrungaro S, Mammola CL, Tomaipitinca L, Filippini A et al (2017) Multifaceted roles of GSK-3 in cancer and autophagy-related diseases. Oxidative Med Cell Longev 2017:4629495
Marín Fernández L, Miguélez González EM, Villar Granja CJ, Lombó Brugos F (2015) Bioavailability of dietary polyphenols and gut microbiota metabolism: antimicrobial properties. Biomed Res Int 2015:905215
Mashhadizadeh S, Farbood Y, Dianat M, Khodadadi A, Sarkaki A (2017) Therapeutic effects of ellagic acid on memory, hippocampus electrophysiology deficits, and elevated TNF-α level in brain due to experimental traumatic brain injury. Iran J Basic Med Sci 20(4):399

Miksits M, Maier-Salamon A, Aust S, Thalhammer T, Reznicek G, Kunert O et al (2005) Sulfation of resveratrol in human liver: evidence of a major role for the sulfotransferases SULT1A1 and SULT1E1. Xenobiotica 35(12):1101–1119

Mori T, Rezai-Zadeh K, Koyama N, Arendash GW, Yamaguchi H, Kakuda N et al (2012) Tannic acid is a natural β-secretase inhibitor that prevents cognitive impairment and mitigates Alzheimer-like pathology in transgenic mice. J Biol Chem 287(9):6912–6927

Mukhopadhyay B, Gongopadhyay AN, Rani A, Gavel R, Mishra SP (2015) Free radicals and antioxidants status in neonates with congenital malformation. J Indian Assoc Pediatr Surg 20(4):179

Nasution RA, Islam AA, Hatta M, Turchan A, Faruk M (2020) Role of CAPE in reducing oxidative stress in animal models with traumatic brain injury. Ann Med Surg 57:118–122

Ng SY, Lee AYW (2019) Traumatic brain injuries: pathophysiology and potential therapeutic targets. Front Cell Neurosci 13:528

Nishikawa S, Inoue Y, Hori Y, Miyajima C, Morishita D, Ohoka N et al (2020) Anti-inflammatory activity of kurarinone involves induction of HO-1 via the KEAP1/Nrf2 pathway. Antioxidants 9(9):842

O'Brien WT, Pham L, Symons GF, Monif M, Shultz SR, McDonald SJ (2020) The NLRP3 inflammasome in traumatic brain injury: potential as a biomarker and therapeutic target. J Neuroinflammation 17:1–12

Obrenovich ME, Nair NG, Beyaz A, Aliev G, Reddy VP (2010) The role of polyphenolic antioxidants in health, disease, and aging. Rejuvenation Res 13(6):631–643

Olajide OA, Kumar A, Velagapudi R, Okorji UP, Fiebich BL (2014) Punicalagin inhibits neuroinflammation in LPS-activated rat primary microglia. Mol Nutr Food Res 58(9):1843–1851

Parizad N, Hajimohammadi K, Goli R (2021) Surgical debridement, maggot therapy, negative pressure wound therapy, and silver foam dressing revive hope for patients with diabetic foot ulcer: a case report. Int J Surg Case Rep 82:105931

Philippe BA, Karine N, Barthélemy AK, Noél ZG, David NG, Joseph DA, Hosttetmann K (2010) Bio-guided isolation of antioxidant compounds from Chrysophyllum perpulchrum, a plant used in the Ivory Coast pharmacopeia. Molecules 15(9):6386–6398

Ragusa A, Centonze C, Grasso ME, Latronico MF, Mastrangelo PF, Fanizzi FP, Maffia M (2017) Composition and statistical analysis of biophenols in Apulian Italian EVOOs. Food Secur 6(10):90

Ren W, Qiao Z, Wang H, Zhu L, Zhang L (2003) Flavonoids: promising anticancer agents. Med Res Rev 23(4):519–534

Salberg S, Yamakawa G, Christensen J, Kolb B, Mychasiuk R (2017) Assessment of a nutritional supplement containing resveratrol, prebiotic fiber, and omega-3 fatty acids for the prevention and treatment of mild traumatic brain injury in rats. Neuroscience 365:146–157

Salman M, Tabassum H, Parvez S (2020) Tannic acid provides neuroprotective effects against traumatic brain injury through the PGC-1α/Nrf2/HO-1 pathway. Mol Neurobiol 57(6):2870–2885

Sawmiller D, Li S, Shahaduzzaman M, Smith AJ, Obregon D, Giunta B et al (2014) Luteolin reduces Alzheimer's disease pathologies induced by traumatic brain injury. Int J Mol Sci 15(1):895–904

Scalbert A, Johnson IT, Saltmarsh M (2005) Polyphenols: antioxidants and beyond. Am J Clin Nutr 81(1):215S–217S

Serra D, Almeida LM, Dinis TC (2019) Polyphenols as food bioactive compounds in the context of Autism Spectrum Disorders: a critical mini-review. Neurosci Biobehav Rev 102:290–298

Shanan N, GhasemiGharagoz A, Abdel-Kader R, Breitinger H-G (2019) The effect of Pyrroloquinoline quinone and resveratrol on the survival and regeneration of cerebellar granular neurons. Neurosci Lett 694:192–197

Sharma S, Kumar A, Choudhary A, Sharma S, Khurana L, Sharma N et al (2020) Neuroprotective role of oral vitamin D supplementation on consciousness and inflammatory biomarkers in determining severity outcome in acute traumatic brain injury patients: a double-blind randomized clinical trial. Clin Drug Investig 40:327–334

Singleton RH, Yan HQ, Fellows-Mayle W, Dixon CE (2010) Resveratrol attenuates behavioral impairments and reduces cortical and hippocampal loss in a rat controlled cortical impact model of traumatic brain injury. J Neurotrauma 27(6):1091–1099
Sivandzade F, Alqahtani F, Cucullo L (2021) Impact of chronic smoking on traumatic brain microvascular injury: an in vitro study. J Cell Mol Med 25(15):7122–7134
Soltani Z, Khaksari M, Jafari E, Iranpour M, Shahrokhi N (2015) Is genistein neuroprotective in traumatic brain injury? Physiol Behav 152:26–31
Song J, Du G, Wu H, Gao X, Yang Z, Liu B, Cui S (2021) Protective effects of quercetin on traumatic brain injury induced inflammation and oxidative stress in cortex through activating Nrf2/HO-1 pathway. Restor Neurol Neurosci 39(1):73–84
Sönmez Ü, Sönmez A, Erbil G, Tekmen I, Baykara B (2007) Neuroprotective effects of resveratrol against traumatic brain injury in immature rats. Neurosci Lett 420(2):133–137
Suehiro E, Fujiyama Y, Kiyohira M, Motoki Y, Nojima J, Suzuki M (2019) Probability of soluble tissue factor release lead to the elevation of D-dimer as a biomarker for traumatic brain injury. Neurol Med Chir 59(2):63–67
Suzuki H, Matsuzaki J, Hibi T (2010) Ghrelin and oxidative stress in gastrointestinal tract. J Clin Biochem Nutr 48(2):122–125
Taguchi C, Kishimoto Y, Fukushima Y, Kondo K, Yamakawa M, Wada K, Nagata C (2020) Dietary intake of total polyphenols and the risk of all-cause and specific-cause mortality in Japanese adults: the Takayama study. Eur J Nutr 59:1263–1271
Tal S, Hadanny A, Sasson E, Suzin G, Efrati S (2017) Hyperbaric oxygen therapy an induce angiogenesis and regeneration of nerve fibers in traumatic brain injury patients. Front Hum Neurosci 11:508
Tang S, Gao P, Chen H, Zhou X, Ou Y, He Y (2020) The role of iron, its metabolism and ferroptosis in traumatic brain injury. Front Cell Neurosci 14:590789
Tao Y, Wang Y, Ma Z, Wang L, Qin L, Wang L et al (2017) Subretinal delivery of erythropoietin alleviates the N-methyl-N-nitrosourea-induced photoreceptor degeneration and visual functional impairments: an in vivo and ex vivo study. Drug Deliv 24(1):1273–1283
Thomasset SC, Berry DP, Garcea G, Marczylo T, Steward WP, Gescher AJ (2007) Dietary polyphenolic phytochemicals—promising cancer chemopreventive agents in humans? A review of their clinical properties. Int J Cancer 120(3):451–458
Tun T, Kang Y-S (2017) Imperatorin is transported through blood-brain barrier by carrier-mediated transporters. Biomol Ther 25(4):441
Vos PE (2011) Biomarkers of focal and diffuse traumatic brain injury. Crit Care 15(4):1–2
Walle T (2011) Bioavailability of resveratrol. Ann N Y Acad Sci 1215(1):9–15
Walle T, Hsieh F, DeLegge MH, Oatis JE, Walle UK (2004) High absorption but very low bioavailability of oral resveratrol in humans. Drug Metab Dispos 32(12):1377–1382
Wang P, Sang S (2018) Metabolism and pharmacokinetics of resveratrol and pterostilbene. Biofactors 44(1):16–25
Werhane ML, Evangelista ND, Clark AL, Sorg SF, Bangen KJ, Tran M et al (2017) Pathological vascular and inflammatory biomarkers of acute-and chronic-phase traumatic brain injury. Concussion 2(1):CNC30
Wessels M, Leyhausen G, Volk J, Geurtsen W (2014) Oxidative stress is responsible for genotoxicity of camphorquinone in primary human gingival fibroblasts. Clin Oral Investig 18:1705–1710
Williams LD, Burdock GA, Edwards JA, Beck M, Bausch J (2009) Safety studies conducted on high-purity trans-resveratrol in experimental animals. Food Chem Toxicol 47(9).2170–2182
Wu A, Ying Z, Gomez-Pinilla F (2006) Dietary curcumin counteracts the outcome of traumatic brain injury on oxidative stress, synaptic plasticity, and cognition. Exp Neurol 197(2):309–317
Wu Y, Zhong L, Yu Z, Qi J (2019) Anti-neuroinflammatory effects of tannic acid against lipopolysaccharide-induced BV2 microglial cells via inhibition of NF-κB activation. Drug Dev Res 80(2):262–268

Wu A-G, Yong Y-Y, Pan Y-R, Zhang L, Wu J-M, Zhang Y et al (2022) Targeting Nrf2-mediated oxidative stress response in traumatic brain injury: therapeutic perspectives of phytochemicals. Oxidative Med Cell Longev 2022:1015791

Xu J, Wang H, Lu X, Ding K, Zhang L, He J et al (2014) Posttraumatic administration of luteolin protects mice from traumatic brain injury: implication of autophagy and inflammation. Brain Res 1582:237–246

Yang T, Kong B, Gu J-W, Kuang Y-Q, Cheng L, Yang W-T et al (2014) Anti-apoptotic and anti-oxidative roles of quercetin after traumatic brain injury. Cell Mol Neurobiol 34:797–804

Yeates KO, Taylor HG, Rusin J, Bangert B, Dietrich A, Nuss K, Wright M (2012) Premorbid child and family functioning as predictors of post-concussive symptoms in children with mild traumatic brain injuries. Int J Dev Neurosci 30(3):231–237

Zam W (2018) Gut microbiota as a prospective therapeutic target for curcumin: a review of mutual influence. J Nutr Metab 2018:1367984

Zhang H-M, Chen W, Liu R-N, Zhao Y (2018) Notch inhibitor can attenuate apparent diffusion coefficient and improve neurological function through downregulating NOX2-ROS in severe traumatic brain injury. Drug Des Devel Ther 12:3847–3854

Zibara K, Ballout N, Mondello S, Karnib N, Ramadan N, Omais S et al (2019) Combination of drug and stem cells neurotherapy: potential interventions in neurotrauma and traumatic brain injury. Neuropharmacology 145:177–198

Chapter 13
The Relation Between Vitamin D and Traumatic Brain Injury: Possible Mechanisms and Effect on the Severity and Progression

Nada Ibrahim Hendi, **Mohamed Khaled, and Nour Shaheen**

Abstract Vitamin D is a fat-soluble, neuroactive secosteroid. It not only regulates calcium homeostasis but also plays an important role in other body functions such as neuronal function. The neuroprotective effect of vitamin D is not fully understood. However, several pathways were suggested. Vitamin D has an antioxidant effect through the upregulation of antioxidant enzymes and downregulation of the culprit cellular elements responsible for oxidative stress such as the M1 microglia. Vitamin D also plays a role in neuromodulation due to the expression of alpha 1 hydroxylase enzyme in multiple immune cells such as B cells, T cells, dendritic cells, monocytes, and macrophages. Moreover, its interaction with nerve growth factors and calcium regulation plays an important role in preserving neuronal function and preventing apoptosis. The relation between vitamin D and traumatic brain injury is complicated and not fully understood. It was hypothesized that they have a bidirectional relationship, meaning that traumatic brain injury may lead to exhaustion of the body's vitamin D storage. Moreover, the level of vitamin D correlates with the severity of traumatic brain injury, with a worse prognosis associated with vitamin D deficiency. Several studies found that patients with TBI had either vitamin D insufficiency or deficiency. Additionally, previous studies reported that TBI patients who had vitamin D supplementation eventually had better prognosis and better management of the long-term sequelae. In this chapter, we aim to summarize the possible mechanisms of vitamin D neuroprotection, the possible role of vitamin D in TBI, and describe the relationship between vitamin D and the severity and progression of TBI.

Keywords vitamin D · TBI · Neuroprotection

N. I. Hendi (✉)
Faculty of Medicine, Ain Shams University, Cairo, Egypt

M. Khaled · N. Shaheen
Faculty of Medicine, Alexandria University, Alexandria, Egypt

W. Mohamed (ed.), *Nutrition and Traumatic Brain Injury (TBI)*, Nutritional Neurosciences, https://doi.org/10.1007/978-981-97-6341-2_13

13.1 Overview of Vitamin D (Its Metabolism and Role in the Central Nervous System)

Vitamin D is a neuroactive steroid-related hormone that is necessary for bone formation and calcium homeostasis. It is a ligand that signals via nuclear receptors (Carlberg 2019). For humans, sunlight (80–90%) and dietary intake (10–20%) are the main sources of vitamin D (Mithal et al. 2009). In the skin, with the help of ultraviolet B radiation, 7-dehydrocholesterol is photolyzed to precalciferol (Norman 1998) and then undergoes two sequential hydroxylation reactions: first in the liver and then in the kidneys. However, the second step can occur to a lesser extent in extrarenal organs. The CYP450-dependent enzyme 25-hydroxylase, sometimes called CYP2R1, rapidly hydroxylates cholecalciferol in the liver, producing calcidiol (25-hydroxyvitamin D) (Jones 2008). Then, under the influence of reduced calcium level and the subsequently elevated parathyroid hormone (PTH), the second hydroxylation reaction takes place resulting in the production of 1,25 dihydroxycholecalciferol (1,25(OH)2D), the active form of vitamin D which is also named calcitriol, through 1α-hydroxylation of 25(OH)D by the mitochondrial 1- α hydroxylase enzyme in the cells of the proximal convoluted tubules (Shaheen et al. 2023; Tanaka and DeLuca 1983). Although vitamin D has long been known for its critical function in maintaining bone health, it has also gained notice for its pleiotropic effects on a wide spectrum of physiological processes, including those related to the central nervous system (Groves et al. 2014; Wrzosek et al. 2013). Regulation of apoptotic activity ; immunomodulation as found in dendritic cells, macrophages, and lymphocytes ; and neuroprotective activities are among the non-calcemic actions (Yuan et al. 2018). Therefore, vitamin D was suggested to influence the treatment of traumatic brain injury (TBI), particularly the secondary damage caused by abnormal neuronal function, cerebral edema, and ischemia (Lee et al. 2019b). Moreover, it has been postulated that vitamin D is a key marker for determining the degree of neurophysiological and cognitive impairment. Additionally, its paucity confers a negative decline in the overall quality of life and rate of recovery (Jamall et al. 2016).

This chapter aims to explain the mechanisms of neuroprotection of vitamin D, demonstrate the correlation between vitamin D levels and TBI status, and discuss the effect of vitamin D supplementation for secondary traumatic brain injury (TBI) and primary TBI rehabilitation.

13.2 Vitamin D's Role in Neuroprotection

The contribution of vitamin D to neuron development and function was first suggested when the vitamin D receptor (VDR) and the 1- α hydroxylase enzyme were found in the neurons, hippocampus, and glial cells in the brain (Langub et al. 2001). It is unclear how vitamin D causes neuroprotection; however, antioxidant pathways, calcium regulation in neurons, immunomodulation, neurotrophin expression, glutamate, and other pathways may be involved.

13.2.1 Antioxidant Mechanisms

In an inflammatory environment, several cells are frequently activated, including microglia and astrocytes that produce reactive oxygen species (ROS) and pro-inflammatory cytokines. The brain, more than any other organ in the body, is particularly susceptible to ROS-induced damage due to its high oxygen demand (X. Wang and Michaelis 2010). Vitamin D has been shown to reduce oxidative stress in the central nervous tissue by reducing lipid hydroxides and protein carbonyls and restoring antioxidant enzymes to normal levels in chronic inflammatory conditions (Haider et al. 2011). Likewise, vitamin D supplementation was found to reduce oxidative stress and enhance antioxidant biomarkers (Ali et al. 2021; Cojic et al. 2021; Moslemi et al. 2022; Sepidarkish et al. 2019). In addition to reducing the population of M1 microglia, which is a major contributor to the increase in nitric oxide and reactive oxygen intermediates, activated microglia subjected to vitamin D3 showed lower expression of pro-inflammatory cytokines (Boontanrart et al. 2016). Furthermore, 1,25(OH)2D3 has been shown to increase the nuclear factor E2-related factor 2 (Nrf2) expression, which upregulates the expression of antioxidant enzymes through binding to the antioxidant response element in the nucleus (Lin et al. 2022; L. Wang et al. 2010). Vitamin D inhibits nitric oxide synthase (iNOS), an enzyme activated in a hypoxic environment that induces oxidative damage (Dursun et al. 2014). It also increases Klotho, a phosphate- regulating hormone, that has a function in the signaling system that includes antioxidant production (Haussler et al. 2021; Razzaque 2012). In vitro, studies on rats' neurons revealed that 1,25(OH)2D3 raises the amounts of glutathione in these cells which is one of the main antioxidants that is delivered to nerve cells by astrocytes (Shinpo et al. 2000).

13.3 Immunomodulatory and Anti-inflammatory Mechanisms

Immune cells, including B and T lymphocytes, phagocytes such as macrophages, dendritic cells, and monocytes, were detected to contain the enzyme 1-α-hydroxylase that activates vitamin D and expresses its receptor VDR (Sassi et al. 2018; Szymczak and Pawliczak 2016; Veldman et al. 2000). Thus, it has been suggested that the immune system can produce active vitamin D and trigger intracrine or paracrine reactions (Hewison 1992). Studies show the effects of vitamin D are driven by the vitamin D–vitamin D receptor (VDR) interaction. When vitamin D binds to its receptor, it interacts with the retinoid X receptor (RXR) leading to its dimerization and transportation to the nucleus to bind with specific sequences of DNA (vitamin D response elements [VDREs]) in the promoter domains of target genes to induce their production (S. Joshi et al. 2011).

13.3.1 Innate Immunity

In monocytes, activation of toll-like receptor (TLR) 1 and 2 in combination with vitamin D stimulates the expression of CAMP, a gene that encodes the cathelicidin antimicrobial peptide (LL-37) in humans, which is a transcriptional target for the 1,25$(OH)_2$D-VDR complex (Gombart et al. 2005). Additionally, 1,25 (OH) 2D3-mediated VDR activity has been found to interact with the TLR-induced interleukin -1 beta (IL-1β) signaling pathway to increase the expression of the antimicrobial peptide DEFB4 in monocytes (Liu et al. 2009). It also stimulates innate antimicrobial effector responses by activating antibacterial autophagy (Shin et al. 2010). Furthermore, 1,25-(OH)2D3 has indirect anti-inflammatory effects on macrophages by increasing IL-10 and decreasing inflammatory stimuli (e.g., IL-1β, IL-6, TNF-α, RANKL, and COX-2) (Colotta et al. 2017). The inhibition of inflammatory cytokines is mediated by the increase of mitogen-activated protein kinase (MAPK) phosphatase (MKP)-1 by 1,25-(OH)2D3, resulting in suppression of lipopolysaccharide-induced p38 activation (Zhang et al. 2012a).

In neutrophils, studies showed 1,25(OH)2D-induced LL37 and VDR expression; however, unlike monocytes, there is no clear evidence that they express a functional CYP27B1 enzyme (Wang et al. 2004). Vitamin D has been found to decrease the production of neutrophil extracellular traps (NETs), which in turn decreases the risk of autoimmunity as well as the defense against invasive pathogens (Agraz-Cibrian et al. 2019).

Natural killer cells, like monocytes and neutrophils, release antimicrobial peptides, including cathelicidin and α-defensin, and express VDR and hydroxylases, indicating that 1,25(OH)2D3 must play a role in the formation of NK cells, as well as in the development, differentiation, and toxicity of these cells (Höck 2014; Joshi et al. 2014).

In keratinocytes, 1,25(OH)2D3 enhances TLR2/1 and CAMP expression, leading to enhanced antimicrobial activity (Schauber et al. 2008).

Vitamin D synthesis and VDR expression in dendritic cells (DCs) play a critical role in the autoregulation and development of human monocyte-derived dendritic cells (Martin Hewison et al. 2003). Additionally, 1,25D3 mDCs demonstrated a significant ability to inhibit the development and activation of aggressive autoreactive $CD4^+$ T cells. Moreover, it increased $CD25^+Foxp3^+$ regulatory T cells (a subset of Tregs) and stimulated intracellular IL-10 production by T cells , which helps in immune downregulation (Ferreira et al. 2014).

Vitamin D influences gut microbiome composition as studies have reported that VDR polymorphisms are significant contributors to microbiota variation and that gut microbiome can be changed by exposure to vitamin D (Bashir et al. 2016; Luthold et al. 2017; Wang et al. 2016; Wu et al. 2011). Additionally, the intestinal epithelium integrity is protected from bacterial lipopolysaccharide damage by vitamin D signals (Lee et al. 2019a).

13.3.2 Adaptive Immunity

Naïve T and B cells harbor vitamin D receptors to a small extent; however, when activated, VDR and 1-α-hydroxylase expression is induced which, in turn, helps in the regulation and suppression of the adaptive response (Chun et al. 2014). Over 500 vitamin D –responsive genes are subject to regulation by the upregulation of VDR expression that follows T and B cell activation and proliferation (Prietl et al. 2013).

13.3.2.1 T cells

Th2 and Th10 are generally anti-inflammatory, while Th1 and Th17 are pro-inflammatory. By inhibiting IL-12 synthesis, which is necessary for Th1 development, and IL-23 and IL-6, which are necessary for Th17 development, 1,25(OH)2D prevents differentiation of Th1 and Th17 T cells. As a result, Th1 cells produce less IFNγ and IL-2, while Th17 cells produce less IL-17 (Daniel et al. 2008), further restricting the recruitment of T lymphocytes (a function of IFNγ) and their proliferation (a function of IL-2). As IL-12 levels drop, Th2 cells proliferate and release IL-4, IL-5, and IL-13, cytokines that shift the balance of Th1/Th2 from a Th1 cell–dominated pattern to one that is dominated by Th2 cells (Baeke et al. 2010; Cantorna and Waddell 2014; Smolders et al. 2009). On the other hand, the count of $CD4^+$/$CD25^+$ regulatory T cells (Treg) is increased by 1,25(OH)2D (Prietl et al. 2014). Through the production of IL-10, Tregs modulate the proliferation of the other T cell subtypes. This nuance regulation in turn helps in immune tolerance which also probably has a major impact on stopping the cytokine storm. Additionally, vitamin D can control some pro-inflammatory cytokines directly on a genomic basis. TNFα is an example; its promoter has a vitamin D response element (VDRE), which allows 1,25(OH)2D attached to the VDR to suppress TNFα production. Another example is IL-2; the Nuclear factor of activated T-cells (NFAT1)-activator protein 1 (AP1) enhancer motif is the first place where VDR and NFAT1 compete for binding, and then VDR binds to c-Jun. This competition results in suppression of IL-2 and granulocyte-macrophage colony-stimulating factor (GM-CSF) expression (Nagpal et al. 2005).

13.3.2.2 B cells

1,25(OH)2D3 inhibits B lymphocytes directly and indirectly. It directly affects human B cells, resulting in the inhibition of IgE production; however, the peripheral B lymphocyte proliferation is not affected maybe through reduced activation of NF-κB (p50 and p65) (Heine et al. 2002). Indirectly, it inhibits plasma cell differentiation through changes in CD4 cell signaling and suppression as mentioned before.

13.4 Calcium Regulation

Normally, calcium is essential for many intracellular processes through multiple pathways. It is involved in alterations of gene expression, neurotransmitter release, neuronal excitability, synaptic plasticity, and apoptosis. Physiologically, Ca^{2+} is responsible for a wide spectrum of neuronal cell functions that are disrupted in dysregulated calcium homeostasis (Burgalossi et al. 2012; Nikoletopoulou and Tavernarakis 2012; Zhang et al. 2009). Pathological elevations of intracellular Ca^{2+} levels can trigger a number of detrimental processes that will ultimately lead to the cell's death through apoptosis, including changes to mitochondrial processes and the production of free radicals (Young 1992) since several brain areas, including the hippocampus, have been found to contain vitamin D receptors. It is believed that vitamin D regulates Ca^{2+}-mediated activities in the brain (Langub et al. 2001). One of the pathways of the neuroprotective functions of vitamin D is through the regulation of the L-type voltage-sensitive Ca^{2+} channels. Vitamin D not only downregulates their expression but also regulates the opening of L-type calcium channels through a variety of kinase cascades and enzyme activity in the cerebral cortex (Zanatta et al. 2012). Another mechanism of vitamin D neuroprotection is through calcium-binding proteins such as calbindin-D28K, parvalbumin, and calretinin. They are believed to serve as calcium buffers maintaining calcium homeostasis intracellularly. Vitamin D is associated with the expression and regulation of these proteins (Alexianu et al. 1998; Halhali et al. 2010; Oveland et al. 2018).

13.5 Neurotrophin Expression

A family of growth factor proteins known as neurotrophins stimulates the growth, survival, and normal function of neuronal and glial cells in addition to the growth and differentiation of neuronal stem cells (NSCs), thus inducing protective and restorative effects. The main members of the neurotrophin family are nerve growth factor (NGF), brain-derived neurotrophic factor (BDNF), neurotrophin-3 (NT-3), neurotrophin-4 (NT-4), and glial cell line–derived neurotrophic factor (GDNF) (Skaper 2018). They are secreted by a variety of cell types, including neurons, neuronal stem cells, oligodendrocytes, M2 microglia, and astrocytes (Shirazi et al. 2015; Traiffort et al. 2020).

13.5.1 Vitamin D and NGF

Several in vitro studies have shown the relation between vitamin D and NGF. Vitamin D upregulated the expression of NGF in hippocampal neurons as well as cortical neurons not transferred to the basal anterior brain (Brown et al. 2003; Gezen-Ak et al. 2014; Neveu et al. 1994; Wion et al. 1991).

13.5.2 Vitamin D and BDNF

The effect of vitamin D on BDNF levels is still controversial. On one hand, vitamin D3 supplementation has been shown in experimental rat models to activate BDNF expression pathways, which in turn raises serum BDNF levels (Eyles et al. 2003; Khairy and Attia 2019; Sanchez et al. 2002). Furthermore, vitamin D3 may regulate apoptosis and age-related pro-inflammatory states indirectly by stimulating the production of BDNF, according to earlier research on animals (Briones and Darwish 2012; Nagatsu and Sawada 2005). On the other hand, there is little trial evidence regarding the potential of vitamin D3 and omega-3s to change blood BDNF in a sample of the older adult population (Gravesteijn et al. 2022). Similarly, Vyas et al. (2023) found that over two years, regular vitamin D3 and/or omega-3 supplementation did not change serum BDNF.

13.5.3 Vitamin D and GDNF

Current evidence suggests that the 1,25(OH)2D3-VDR pathway is important in stimulating the production of GDNF, which supports neuronal processes in the brain including neurite branching and survival (Airaksinen et al. 1999; Chen et al. 2006; Duarte et al. 2012; Gordon 2009). Interestingly, research has demonstrated that reduced GDNF expression occurs in the brain in association with vitamin D deficiency (Eyles et al. 2003).

13.6 Tolerance to Glutamate Toxicity

A high concentration of the excitatory neurotransmitter glutamate is the main mediator of excitotoxicity, which is defined by the overexcitation of neurons that results in neuronal cell death. This mechanism raises the neurological system's oxidative stress and causes glutamate receptors, including the NMDA receptor, to remain activated for a longer period. This increases the amount of calcium that enters the postsynaptic neuron and ultimately causes apoptosis (Olney 1990). Furthermore, glutamate directly and significantly mediates inflammatory reactions, as it was found to affect immune response through human T lymphocyte modulation (Ganor and Levite 2014). Vitamin D has the potential to minimize excitotoxicity through L-type calcium channel modulation, which is believed to regulate NMDA receptors (Brewer et al. 2001). In addition, in cultured cortical neurons receiving long-term vitamin D3 treatment, neuronal cell death caused by short glutamate exposure was substantially prevented (Taniura et al. 2006).

13.6.1 Neuroprotective Role of Vitamin D in Traumatic Brain Injury

The pathophysiology of TBI is divided into two categories: (1) the initial acute event, which is a mechanical damage that primarily occurs at the time of injury and may result in intracranial or extracranial consequences such as hypoxia, subarachnoid, or epidural hemorrhage and (2) secondary damage following the primary injury that entails vascular dysfunction, neuronal cell demise, blood- brain barrier (BBB) dysfunction, glutamatergic excitotoxicity, neuroinflammation, and calcium overload. This long-term inflammation causes neuronal morphological and functional changes precipitating chronic TBI aftereffects, including cognitive and mental illnesses, sensory-motor dysfunction, and psychiatric problems. These secondary injuries have highlighted the need for neuroprotective interventions, especially during the period of evolution of secondary injury. Vitamin D which has a nonclassical role involving the modulation of multiple neuroinflammatory processes is of significant use in the management of the long-term sequelae of TBI secondary injury. The following points summarize the mechanisms of injury in TBI and the role of vitamin D in alleviating them.

13.6.1.1 Microglia Activation

Damage-associated molecular patterns (DAMPs), released from damaged cells as inflammatory stimuli and other extracellular signals originating from damaged neurons, can cause microglia to polarize to M1- and M2-like activation states (Zhang et al. 2012b). Both of these states play opposing roles in tissue healing and neurodegeneration. The M1 phenotype activated by IFN-γ, IL-1α, IL-6, and TNF-α contributes to neurodegeneration causing long-term neurological deficits, oxidative stress, and chronic neuroinflammation as they release pro-inflammatory cytokines and free radicals that damage the ability to heal itself. On the other hand, when the M2-like microglia are stimulated, they emit neurotrophic factors, proteases, and anti-inflammatory cytokines. M2-like microglia are involved in CNS remodeling and help with immunosuppression and the reversal of M1-mediated neuroinflammation (Hu et al. 2015). A microglial transition toward the M2 phenotype is under the control of vitamin D through suppression of the TLR4/MyD88/NF-κB signaling pathways (Jiang et al. 2022).

13.6.1.2 Astrocyte Activation

Normally, astrocytes maintain the metabolism of neurons, buffer potassium, and glutamate and secrete gliotransmitters, which modulate synapses. Astrocyte reactivity and its interaction with microglial cells are induced by the initial mechanical injury of TBI through the mechanosensitive ion channels they possess (Bowman et al. 1992) and stretch-sensitive cation channels that have a role in calcium influx after exposure to glutamate, causing cell death (Floyd et al. 2005). Activated astrocytes start to proliferate and increase in size to form a glial scar to isolate the damaged area from its surroundings. This process is called astrogliosis, and it is a normal protective mechanism unless it is an excessive or prolonged activation. The prolonged astrogliosis would exacerbate the negative outcome due to the release of neurotoxic pro-inflammatory cytokines and ROS eventually resulting in the spread of injury to distal parts, excitotoxicity, edema, and seizures (Mora and Chapouly 2023). Vitamin D3 inhibited astrocyte over-activation, protecting hippocampus neurons. Animals given intraperitoneal doses of 25(OH)D3 were found to have lower astrocyte activation and astrocyte concentration showing reduced levels of GFAP, an intermediate filament highly expressed in reactive astrocytes proximal to CNS traumatic lesions (Liddelow and Barres 2017; Nystad et al. 2014; Zou et al. 2022).

13.6.1.3 Mitochondrial Dysfunction

One of the main factors contributing to mitochondrial damage is energy depletion, which is an outcome of impaired mitochondria. Acidosis and the loss of cellular energy arise from the change of aerobic metabolism to anaerobic metabolism in the injured brain due to insufficient oxygenation. Mitochondrial damage will be severe due to excitotoxicity, excess Ca^{2+}, and ROS overproduction that occur after initial injury and energy failure (Rockswold et al. 2007). Gene activation that occurs when 1,25(OH)2D interacts with its receptor is critical for respiration, maintaining the integrity of the mitochondria, and numerous other physiological processes. Furthermore, increased mitochondrial respiration and the resulting damage as well as an excess of oxygen- free radicles (ROS) that can harm cells and DNA are prevented by the vitamin D signaling pathway (Bouillon and Verstuyf 2013; Ricca et al. 2018; Ryan et al. 2016).

13.6.1.4 Oxidative Stress and Calcium Homeostasis

The role of growth factors, chemokines, and cytokines in the pathophysiology of TBI is being highlighted in an increasing number of researches. Large amounts of cytokines, such as IL-1β, TNF-α, and transforming growth factor-beta (TGF-β), are produced in the acute post-injury period. This, in turn, exacerbates the brain trauma condition by increasing oxidative stress and matrix metalloproteinases (MMPs) (Hans et al. 1999; Lucas et al. 2006; Werner and Engelhard 2007). A disruption of the blood-brain barrier and the entry of inflammatory cells into the brain is caused by the posttraumatic inflammatory response. Additionally, the pathophysiology of TBI and other forms of brain damage has been linked to changes in intracellular calcium concentration ($[Ca^{2+}]_i$) and Ca^{2+} homeostatic mechanisms. These alterations to Ca^{2+} homeostasis are believed to promote changes in neuroplasticity in neurons following brain damage. This eventually leads to excessive stimulation of the glutamate receptors which is a crucial cause of excitotoxicity (Delorenzo et al. 2005; Sun et al. 2008; Weber 2012). Vitamin D's role in ameliorating these injuries is discussed earlier.

13.6.1.5 Blood- Brain Barrier (BBB) Dysfunction

After traumatic brain injury (TBI), glial cell activation and neuroinflammation occur as a secondary damage response that might compromise the blood-brain barrier (BBB). Neutrophils, chemokines, pro-inflammatory mediators, and other pro-inflammatory mediators are recruited and upregulated during this reaction. It has also been shown that pericytes, a particular kind of BBB cells, act as inflammatory mediators after traumatic brain injury. Microglia and other innate immune cells release inflammatory chemicals that lead to BBB disruption and neuroinflammation. Increased BBB permeability results from reactive oxygen species released by microglia, neurons, and astrocytes that stimulate the glial release of cytokines and chemokines. These cytokines induce downstream pathways that have an effect on the BBB's tight junction proteins' integrity (Cash and Theus 2020). Owing to vitamin D's anti-inflammatory and immunomodulatory effects, studies have shown its efficacy in maintaining BBB integrity. Additionally, vitamin D was found to contribute to the modulation of junction protein and inhibition of nuclear factor kappa B (NFκB) activation, a protein mediating inflammatory response (Takahashi et al. 2017; Won et al. 2015).

13.6.1.6 Neurodegenerative Depositions

Among the most prevalent, persistent, and incapacitating alterations in post-traumatic patients are cognitive ones. Learning and memory of new information are frequently affected. Other thinking abilities that are frequently impacted include the ability to focus, arrange ideas, create efficient plans, and make wise decisions. Imaging studies have shown evidence that neurodegeneration following TBI has several characteristics of Alzheimer's disease. After TBI, cerebral atrophy is localized rather than widespread. Additionally, TBI led to neuronal degeneration and gradual atrophy in the brain tissue in rodent models, and these effects persisted for at least a year following the original injury (Bramlett and Dietrich 2002; Shively et al. 2012). A number of proteins related to neurodegenerative diseases in humans have also been shown to increase in mice after an experimental TBI. Following TBI, the amyloid precursor protein increases instantaneously, and weeks and months later, β-amyloid peptide levels up (Iwata et al. 2002). Preclinical and experimental evidence demonstrated a connection between vitamin D levels and cognitive performance (Briones and Darwish 2012; Latimer et al. 2014). Moreover, research on humans demonstrates a substantial link between cognitive decline or dementia and low levels of circulating 25 hydroxy vitamin D (Etgen et al. 2012).

13.6.2 Correlation Between Vitamin D Status and TBI Severity

The direction of causality of vitamin D level and TBI is not fully understood. Vitamin D insufficiency or deficiency was found to be more prevalent in patients with TBI than the normal population. This has raised the assumption that TBI may affect vitamin D levels. It was assumed that vitamin D and TBI status have a complex, bidirectional relationship (Toman et al. 2017). On one hand, pre-morbid vitamin D deficiency makes it more difficult for the body to counteract the neuroinflammatory process or to enhance the neuroprotective or neuroplastic response to injury, which can result in a more serious brain injury. Moreover, studies have demonstrated a negative correlation between vitamin D levels and the severity of TBI (Table 13.1). On the other hand, a more serious injury would hypothetically exhaust the body's supplies of vitamin D, increasing the likelihood of shortage. Additionally, variations in vitamin D levels can have a multifactorial contribution, including altered hormonal regulation (Karaca et al. 2016), altered lifestyle and dietary habits due to chronic fatigue, neuroendocrinal changes (Behan et al. 2008; Ghaffarzadegan et al. 2016; Xydakis et al. 2008) , and metabolic abnormalities following TBI (Lai et al. 2022).

Table 13.1 Prevalence of vitamin D deficiency in patients after TBI

Study ID	Sample size	Severity	Vitamin D paucity percentage	
			Vitamin D insufficient	Vitamin D deficient
Anon (n.d.)	90	Moderate to severe	0	58 (65%)
Socci et al. (2014)	46	Moderate to severe	0	34 (73.9%)
Jamall et al. (2016)	353	Moderate to severe	163 (74.5%)	119 (72.6%)
Toman et al. (2017)	124	Mild to severe	29 (23%)	42 (34%)

13.6.3 Vitamin D Supplementation for Improving Clinical Outcomes in TBI

Given the neuroprotective role of vitamin D and the correlation between vitamin D status and TBI severity, collective evidence from various studies supports the idea that giving vitamin D supplements improves TBI severity outcomes and helps in the management of long-lasting sequelae (Table 13.2).

13.7 Conclusion

Vitamin D's effect on traumatic brain injury (TBI) patients is still not fully understood. The possible relationship between vitamin D levels and TBI outcomes has been the subject of numerous research, which have shed light on the vitamin's effects on neurological function, rehabilitation, and long-term prognosis. Research shows that following a traumatic brain injury, neuroprotective properties may be of significant role by sufficient amounts of vitamin D. Due to its anti-inflammatory and immunomodulatory characteristics, as well as maintaining calcium homeostasis, vitamin D may be able to reduce secondary injury following traumatic brain injury. Moreover, it has been frequently observed the high prevalence of vitamin D deficiency in TBI patients, which raises the idea regarding the possible advantages of vitamin D adjunctive treatment in enhancing clinical outcomes.

The more we learn about the intricate correlation between vitamin D and traumatic brain injury, the more obvious it is that patient care must be provided accordingly. In addition to traditional therapies, ensuring sufficient vitamin D levels may be beneficial to maximize recovery and reduce long-standing neurological impairments in TBI patients. Thorough, well-designed clinical trials are necessary in the future to provide more precise recommendations for vitamin D supplementation in the treatment of traumatic brain injury. Moreover, further investigation may reveal more complex pathways through which vitamin D affects the damaged brain, opening the door to focused therapies and individualized care strategies.

Table 13.2 Role of vitamin D supplementation in TBI outcomes

Study ID	Sample size	Vitamin D status	GCS score	Vitamin D treatment	Associations	Outcome	Follow-up
Matthews et al. (2013)	3	17–29 ng/mL	3–9	Vitamin D3 (50,000 IU), progesterone (20 mg), omega-3 fatty acids (2 grams (Loveza), and enteral glutamine (20 grams) up to 6 weeks	Temporal subdural hematomas (SDH), left frontal subarachnoid hemorrhage (SAH), diffuse axonal injury (DAI), memory deficient	Gas chromatography (GCS): 12–15, extubate: 9–12 days, improved memory	3 months
Aminmansour et al. (2012)	20	10–29 ng/mL	3–8	Progesterone (1 mg/kg), vitamin D (5 μg/kg) up to 5 days	Brain trauma and DAI	Glasgow Outcome Scale (GOS): 7–15, improved	3 months
Karsy et al. (2019)	267	8–20 ng/mL	3–8	Vitamin D (54,000 IU) at once	Neurocritical admitted patient	GCS: 13, ICU and hospital stay: not significant	Till discharge
Sharma et al. (2020)	35	8–58 ng/mL	4–12	Vitamin D (120,000 IU) at once	Temporal SDH, left frontal SAH, DAI, memory deficient	GCS increased by 3.86 units, significant decrease in ICU stay and mechanical ventilation	Till 14 days

References

Agraz-Cibrian JM, Giraldo DM, Urcuqui-Inchima S (2019) 1,25-Dihydroxyvitamin D3 induces formation of neutrophil extracellular trap-like structures and modulates the transcription of genes whose products are neutrophil extracellular trap-associated proteins: a pilot study. Steroids 141:14–22. https://doi.org/10.1016/j.steroids.2018.11.001

Airaksinen MS, Titievsky A, Saarma M (1999) GDNF family neurotrophic factor signaling: four masters, one servant? Mol Cell Neurosci 13(5):313–325. https://doi.org/10.1006/mcne.1999.0754

Alexianu ME, Robbins E, Carswell S, Appel SH (1998) 1Alpha, 25 dihydroxyvitamin D3-dependent up-regulation of calcium-binding proteins in motoneuron cells. J Neurosci Res 51(1):58–66. https://doi.org/10.1002/(SICI)1097-4547(19980101)51:1<58::AID-JNR6>3.0.CO;2-K

Ali A, Shah SA, Zaman N, Uddin MN, Khan W, Ali A, Riaz M, Kamil A (2021) Vitamin D exerts neuroprotection via SIRT1/nrf-2/ NF-kB signaling pathways against D-galactose-induced memory impairment in adult mice. Neurochem Int 142:104893. https://doi.org/10.1016/j.neuint.2020.104893

Aminmansour B, Nikbakht H, Ghorbani A, Rezvani M, Rahmani P, Torkashvand M, Nourian M, Moradi M (2012) Comparison of the administration of progesterone versus progesterone and vitamin D in improvement of outcomes in patients with traumatic brain injury: a randomized clinical trial with placebo group. Adv Biomed Res 1:58. https://doi.org/10.4103/2277-9175.100176

Baeke F, Takiishi T, Korf H, Gysemans C, Mathieu C (2010) Vitamin D: modulator of the immune system. Curr Opin Pharmacol 10(4):482–496. https://doi.org/10.1016/j.coph.2010.04.001

Bashir M, Prietl B, Tauschmann M, Mautner SI, Kump PK, Treiber G, Wurm P, Gorkiewicz G, Högenauer C, Pieber TR (2016) Effects of high doses of vitamin D3 on mucosa-associated gut microbiome vary between regions of the human gastrointestinal tract. Eur J Nutr 55(4):1479–1489. https://doi.org/10.1007/s00394-015-0966-2

Behan LA, Phillips J, Thompson CJ, Agha A (2008) Neuroendocrine disorders after traumatic brain injury. J Neurol Neurosurg Psychiatry 79(7):753–759. https://doi.org/10.1136/jnnp.2007.132837

Boontanrart M, Hall SD, Spanier JA, Hayes CE, Olson JK (2016) Vitamin D3 alters microglia immune activation by an IL-10 dependent SOCS3 mechanism. J Neuroimmunol 292:126–136. https://doi.org/10.1016/j.jneuroim.2016.01.015

Bouillon R, Verstuyf A (2013) Vitamin D, mitochondria, and muscle. J Clin Endocrinol Metab 98(3):961–963. https://doi.org/10.1210/jc.2013-1352

Bowman CL, Ding JP, Sachs F, Sokabe M (1992) Mechanotransducing ion channels in astrocytes. Brain Res 584(1–2):272–286. https://doi.org/10.1016/0006-8993(92)90906-p

Bramlett HM, Dietrich WD (2002) Quantitative structural changes in white and gray matter 1 year following traumatic brain injury in rats. Acta Neuropathol 103(6):607–614. https://doi.org/10.1007/s00401-001-0510-8

Brewer LD, Thibault V, Chen KC, Langub MC, Landfield PW, Porter NM (2001) Vitamin D hormone confers neuroprotection in parallel with downregulation of L-type calcium channel expression in hippocampal neurons. J Neurosci 21(1):98–108. https://doi.org/10.1523/JNEUROSCI.21-01-00098.2001

Briones TL, Darwish H (2012) Vitamin D mitigates age-related cognitive decline through the modulation of pro-inflammatory state and decrease in amyloid burden. J Neuroinflammation 9:244. https://doi.org/10.1186/1742-2094-9-244

Brown J, Bianco JI, McGrath JJ, Eyles DW (2003) 1,25-dihydroxyvitamin D3 induces nerve growth factor, promotes neurite outgrowth and inhibits mitosis in embryonic rat hippocampal neurons. Neurosci Lett 343(2):139–143. https://doi.org/10.1016/s0304-3940(03)00303-3

Burgalossi A, Jung S, Man KM, Nair R, Jockusch WJ, Wojcik SM, Brose N, Rhee J-S (2012) Analysis of neurotransmitter release mechanisms by photolysis of caged Ca^{2+} in an autaptic neuron culture system. Nat Protoc 7(7):1351–1365. https://doi.org/10.1038/nprot.2012.074

Cantorna MT, Waddell A (2014) The vitamin D receptor turns off chronically activated T cells. Ann N Y Acad Sci 1317:70–75. https://doi.org/10.1111/nyas.12408

Carlberg C (2019) Vitamin D: A micronutrient regulating genes. Curr Pharm Des 25(15):1740–1746. https://doi.org/10.2174/1381612825666190705193227

Cash A, Theus MH (2020) Mechanisms of blood-brain barrier dysfunction in traumatic brain injury. Int J Mol Sci 21(9). https://doi.org/10.3390/ijms21093344

Chen PS, Peng GS, Li G, Yang S, Wu X, Wang CC, Wilson B, Lu RB, Gean PW, Chuang DM, Hong JS (2006) Valproate protects dopaminergic neurons in midbrain neuron/glia cultures by stimulating the release of neurotrophic factors from astrocytes. Mol Psychiatry 11(12):1116–1125. https://doi.org/10.1038/sj.mp.4001893

Chun RF, Liu PT, Modlin RL, Adams JS, Hewison M (2014) Impact of vitamin D on immune function: lessons learned from genome-wide analysis. Front Physiol 5:151. https://doi.org/10.3389/fphys.2014.00151

Cojic M, Kocic R, Klisic A, Kocic G (2021) The effects of Vitamin D supplementation on metabolic and oxidative stress markers in patients with type 2 diabetes: a 6-month follow up randomized controlled study. Front Endocrinol 12:610893. https://doi.org/10.3389/fendo.2021.610893

Colotta F, Jansson B, Bonelli F (2017) Modulation of inflammatory and immune responses by vitamin D. J Autoimmun 85:78–97. https://doi.org/10.1016/j.jaut.2017.07.007

Daniel C, Sartory NA, Zahn N, Radeke HH, Stein JM (2008) Immune modulatory treatment of trinitrobenzene sulfonic acid colitis with calcitriol is associated with a change of a T helper (Th) 1/Th17 to a Th2 and regulatory T cell profile. J Pharmacol Exp Ther 324(1):23–33. https://doi.org/10.1124/jpet.107.127209

Delorenzo RJ, Sun DA, Deshpande LS (2005) Cellular mechanisms underlying acquired epilepsy: the calcium hypothesis of the induction and maintenance of epilepsy. Pharmacol Ther 105(3):229–266. https://doi.org/10.1016/j.pharmthera.2004.10.004

Duarte EP, Curcio M, Canzoniero LM, Duarte CB (2012) Neuroprotection by GDNF in the ischemic brain. Growth Factors 30(4):242–257. https://doi.org/10.3109/08977194.2012.691478

Dursun E, Gezen-Ak D, Yilmazer S (2014) The influence of vitamin D treatment on the inducible nitric oxide synthase (INOS) expression in primary hippocampal neurons. Noro Psikiyatri Arsivi 51(2):163–168. https://doi.org/10.4274/npa.y7089

Etgen T, Sander D, Bickel H, Sander K, Förstl H (2012) Vitamin D deficiency, cognitive impairment and dementia: a systematic review and meta-analysis. Dement Geriatr Cogn Disord 33(5):297–305. https://doi.org/10.1159/000339702

Eyles D, Brown J, Mackay-Sim A, McGrath J, Feron F (2003) Vitamin D3 and brain development. Neuroscience 118(3):641–653. https://doi.org/10.1016/s0306-4522(03)00040-x

Faculty of Medicine, Alexandria University, Alexandria, Egypt, Shahee N, Ramadan A, Patel MA, Parikh K, Gupta V, Kumawat S, Labieb F, Kaur P, Patel P, Mohammad N, Anamika FNU, Mahmood T, Gupta S, Jain R (2023) Vitamin D deficiency and heart health: a narrative review. J Nutr Res 11(2):47–52. https://doi.org/10.55289/jnutres/v11i2_23.16

Ferreira GB, Gysemans CA, Demengeot J, da Cunha JPMCM, Vanherwegen A-S, Overbergh L, Van Belle TL, Pauwels F, Verstuyf A, Korf H, Mathieu C (2014) 1,25-Dihydroxyvitamin D3 promotes tolerogenic dendritic cells with functional migratory properties in NOD mice. J Immunol 192(9):4210–4220. https://doi.org/10.4049/jimmunol.1302350

Floyd CL, Gorin FA, Lyeth BG (2005) Mechanical strain injury increases intracellular sodium and reverses Na+/Ca2+ exchange in cortical astrocytes. Glia 51(1):35–46. https://doi.org/10.1002/glia.20183

Ganor Y, Levite M (2014) The neurotransmitter glutamate and human T cells: glutamate receptors and glutamate-induced direct and potent effects on normal human T cells, cancerous human leukemia and lymphoma T cells, and autoimmune human T cells. J Neural Transm 121(8):983–1006. https://doi.org/10.1007/s00702-014-1167-5

Gezen-Ak D, Dursun E, Yilmazer S (2014) The effect of vitamin D treatment on nerve growth factor (NGF) release from hippocampal neurons. Noro Psikiyatri Arsivi 51(2):157–162. https://doi.org/10.4274/npa.y7076

Ghaffarzadegan N, Ebrahimvandi A, Jalali MS (2016) A dynamic model of post-traumatic stress disorder for military personnel and veterans. PLoS One 11(10):e0161405. https://doi.org/10.1371/journal.pone.0161405

Gombart AF, Borregaard N, Koeffler HP (2005) Human cathelicidin antimicrobial peptide (CAMP) gene is a direct target of the vitamin D receptor and is strongly up-regulated in myeloid cells by 1,25-dihydroxyvitamin D3. FASEB J 19(9):1067–1077. https://doi.org/10.1096/fj.04-3284com

Gordon T (2009) The role of neurotrophic factors in nerve regeneration. Neurosurg Focus 26(2):E3. https://doi.org/10.3171/FOC.2009.26.2.E3

Gravesteijn E, Mensink RP, Plat J (2022) Effects of nutritional interventions on BDNF concentrations in humans: a systematic review. Nutr Neurosci 25(7):1425–1436. https://doi.org/10.1080/1028415X.2020.1865758

Groves NJ, McGrath JJ, Burne THJ (2014) Vitamin D as a neurosteroid affecting the developing and adult brain. Annu Rev Nutr 34:117–141. https://doi.org/10.1146/annurev-nutr-071813-105557

Haider L, Fischer MT, Frischer JM, Bauer J, Höftberger R, Botond G, Esterbauer H, Binder CJ, Witztum JL, Lassmann H (2011) Oxidative damage in multiple sclerosis lesions. Brain A J Neurol 134(Pt 7):1914–1924. https://doi.org/10.1093/brain/awr128

Halhali A, Figueras AG, Díaz L, Avila E, Barrera D, Hernández G, Larrea F (2010) Effects of calcitriol on calbindins gene expression and lipid peroxidation in human placenta. J Steroid Biochem Mol Biol 121(1–2):448–451. https://doi.org/10.1016/j.jsbmb.2010.03.008

Hans VH, Kossmann T, Joller H, Otto V, Morganti-Kossmann MC (1999) Interleukin-6 and its soluble receptor in serum and cerebrospinal fluid after cerebral trauma. Neuroreport 10(2):409–412. https://doi.org/10.1097/00001756-199902050-00036

Haussler MR, Livingston S, Sabir ZL, Haussler CA, Jurutka PW (2021) Vitamin D Receptor Mediates a Myriad of Biological Actions Dependent on Its 1,25-Dihydroxyvitamin D Ligand: Distinct Regulatory Themes Revealed by Induction of Klotho and Fibroblast Growth Factor-23. JBMR Plus 5(1):e10432. https://doi.org/10.1002/jbm4.10432

Heine G, Anton K, Henz BM, Worm M (2002) 1α,25-dihydroxyvitamin D3 inhibits anti-CD40 plus IL-4-mediated IgE production in vitro. Eur J Immunol. 2002 Dec;32(12):3395–404.

Hewison M (1992) Vitamin D and the immune system. J Endocrinol 132(2):173–175. https://doi.org/10.1677/joe.0.1320173

Hewison M, Freeman L, Hughes SV, Evans KN, Bland R, Eliopoulos AG, Kilby MD, Moss PAH, Chakraverty R (2003) Differential regulation of vitamin D receptor and its ligand in human monocyte-derived dendritic cells. J Immunol 170(11):5382–5390. https://doi.org/10.4049/jimmunol.170.11.5382

Höck AD (2014) Review: Vitamin D3 deficiency results in dysfunctions of immunity with severe fatigue and depression in a variety of diseases. In Vivo 28(1):133–145

Hu X, Leak RK, Shi Y, Suenaga J, Gao Y, Zheng P, Chen J (2015) Microglial and macrophage polarization—new prospects for brain repair. Nat Rev Neurol 11(1):56–64. https://doi.org/10.1038/nrneurol.2014.207

Iwata A, Chen X-H, McIntosh TK, Browne KD, Smith DH (2002) Long-term accumulation of amyloid-beta in axons following brain trauma without persistent upregulation of amyloid precursor protein genes. J Neuropathol Exp Neurol 61(12):1056–1068. https://doi.org/10.1093/jnen/61.12.1056

Jamall OA, Feeney C, Zaw-Linn J, Malik A, Niemi MEK, Tenorio-Jimenez C, Ham TE, Jilka SR, Jenkins PO, Scott G, Li LM, Gorgoraptis N, Baxter D, Sharp DJ, Goldstone AP (2016) Prevalence and correlates of vitamin D deficiency in adults after traumatic brain injury. Clin Endocrinol 85(4):636–644. https://doi.org/10.1111/cen.13045

Jiang H, Yang X, Wang Y, Zhou C (2022) Vitamin D Protects against traumatic brain injury via modulating TLR4/MyD88/NF-κB pathway-mediated microglial polarization and neuroinflammation. Biomed Res Int 2022:3363036. https://doi.org/10.1155/2022/3363036

Jones G (2008) Pharmacokinetics of vitamin D toxicity. Am J Clin Nutr 88(2):582S–586S. https://doi.org/10.1093/ajcn/88.2.582S

Joshi L, Ponnana M, Penmetsa SR, Nallari P, Valluri V, Gaddam S (2014) Serum vitamin D levels and VDR polymorphisms (BsmI and FokI) in patients and their household contacts susceptible to tuberculosis. Scand J Immunol 79(2):113–119. https://doi.org/10.1111/sji.12127

Joshi S, Pantalena L-C, Liu XK, Gaffen SL, Liu H, Rohowsky-Kochan C, Ichiyama K, Yoshimura A, Steinman L, Christakos S, Youssef S (2011) 1,25-dihydroxyvitamin D(3) ameliorates Th17 autoimmunity via transcriptional modulation of interleukin-17A. Mol Cell Biol 31(17):3653–3669. https://doi.org/10.1128/MCB.05020-11

Karaca Z, Tanrıverdi F, Ünlühızarcı K, Kelestimur F (2016) GH and pituitary hormone alterations after traumatic brain injury. Prog Mol Biol Transl Sci 138:167–191. https://doi.org/10.1016/bs.pmbts.2015.10.010

Karsy M, Guan J, Eli I, Brock AA, Menacho ST, Park MS (2019) The effect of supplementation of vitamin D in neurocritical care patients: RandomizEd Clinical TrIal oF hYpovitaminosis D (RECTIFY). J Neurosurg 1–10. https://doi.org/10.3171/2018.11.JNS182713

Khairy EY, Attia MM (2019) Protective effects of vitamin D on neurophysiologic alterations in brain aging: role of brain-derived neurotrophic factor (BDNF). Nutr Neurosci 1–10. https://doi.org/10.1080/1028415X.2019.1665854

Lai J-Q, Shi Y-C, Lin S, Chen X-R (2022) Metabolic disorders on cognitive dysfunction after traumatic brain injury. Trends Endocrinol Metab 33(7):451–462. https://doi.org/10.1016/j.tem.2022.04.003

Langub MC, Herman JP, Malluche HH, Koszewski NJ (2001) Evidence of functional vitamin D receptors in rat hippocampus. Neuroscience 104(1):49–56. https://doi.org/10.1016/s0306-4522(01)00049-5

Latimer CS, Brewer LD, Searcy JL, Chen K-C, Popović J, Kraner SD, Thibault O, Blalock EM, Landfield PW, Porter NM (2014) Vitamin D prevents cognitive decline and enhances hippocampal synaptic function in aging rats. Proc Natl Acad Sci USA 111(41):E4359–E4366. https://doi.org/10.1073/pnas.1404477111

Lee C, Lau E, Chusilp S, Filler R, Li B, Zhu H, Yamoto M, Pierro A (2019a) Protective effects of vitamin D against injury in intestinal epithelium. Pediatr Surg Int 35(12):1395–1401. https://doi.org/10.1007/s00383-019-04586-y

Lee JM, Jeong SW, Kim MY, Park JB, Kim MS (2019b) The effect of Vitamin D supplementation in patients with acute traumatic brain injury. World Neurosurg 126:e1421–e1426. https://doi.org/10.1016/j.wneu.2019.02.244

Liddelow SA, Barres BA (2017) Reactive astrocytes: production, function, and therapeutic potential. Immunity 46(6):957–967. https://doi.org/10.1016/j.immuni.2017.06.006

Lin J, Niu Z, Xue Y, Gao J, Zhang M, Li M, Peng Y, Zhang S, Li W, Zhang Q, Li X (2022) Chronic vitamin D3 supplementation alleviates cognition impairment via inhibition of oxidative stress regulated by PI3K/AKT/Nrf2 in APP/PS1 transgenic mice. Neurosci Lett 783:136725. https://doi.org/10.1016/j.neulet.2022.136725

Liu PT, Schenk M, Walker VP, Dempsey PW, Kanchanapoomi M, Wheelwright M, Vazirnia A, Zhang X, Steinmeyer A, Zügel U, Hollis BW, Cheng G, Modlin RL (2009) Convergence of IL-1beta and VDR activation pathways in human TLR2/1-induced antimicrobial responses. PLoS One 4(6):e5810. https://doi.org/10.1371/journal.pone.0005810

Lucas S-M, Rothwell NJ, Gibson RM (2006) The role of inflammation in CNS injury and disease. Br J Pharmacol 147(Suppl 1):S232–S240. https://doi.org/10.1038/sj.bjp.0706400

Luthold RV, Fernandes GR, Franco-de-Moraes AC, Folchetti LGD, Ferreira SRG (2017) Gut microbiota interactions with the immunomodulatory role of vitamin D in normal individuals. Metab Clin Exp 69:76–86. https://doi.org/10.1016/j.metabol.2017.01.007

Matthews LR, Danner OK, Ahmed YA, Dennis-Griggs DM, Frederick A, Clark C, Moore R, DuMornay W, Childs EW, Wilson KL (2013) Combination therapy with vitamin D3, progesterone, omega-3 fatty acids and glutamine reverses coma and improves clinical outcomes in patients with severe traumatic brain injuries: a case series. Int J of Case Rep Images 4(3):143. https://doi.org/10.5348/ijcri-2013-03-281-CS-2

Mithal A, Wahl DA, Bonjour JP, Burckhardt P, Dawson-Hughes B, Eisman JA, El-Hajj Fuleihan G, Josse RG, Lips P, Morales-Torres J, IOF Committee of Scientific Advisors (CSA) Nutrition Working Group (2009) Global vitamin D status and determinants of hypovitaminosis D. Osteoporos Int 20(11):1807–1820. https://doi.org/10.1007/s00198-009-0954-6

Mora P, Chapouly C (2023) Astrogliosis in multiple sclerosis and neuro-inflammation: what role for the notch pathway? Front Immunol 14:1254586. https://doi.org/10.3389/fimmu.2023.1254586

Moslemi E, Musazadeh V, Kavyani Z, Naghsh N, Shoura SMS, Dehghan P (2022) Efficacy of vitamin D supplementation as an adjunct therapy for improving inflammatory and oxidative stress biomarkers: An umbrella meta-analysis. Pharmacol Res 186:106484. https://doi.org/10.1016/j.phrs.2022.106484

Nagatsu T, Sawada M (2005) Inflammatory process in Parkinson's disease: role for cytokines. Curr Pharm Des 11(8):999–1016. https://doi.org/10.2174/1381612053381620

Nagpal S, Na S, Rathnachalam R (2005) Noncalcemic actions of vitamin D receptor ligands. Endocr Rev 26(5):662–687. https://doi.org/10.1210/er.2004-0002

Neveu I, Naveilhan P, Jehan F, Baudet C, Wion D, De Luca HF, Brachet P (1994) 1,25-dihydroxyvitamin D3 regulates the synthesis of nerve growth factor in primary cultures of glial cells. Brain Res Mol Brain Res 24(1–4):70–76. https://doi.org/10.1016/0169-328x(94)90119-8

Nikoletopoulou V, Tavernarakis N (2012) Calcium homeostasis in aging neurons. Front Genet 3:200. https://doi.org/10.3389/fgene.2012.00200

Norman AW (1998) Sunlight, season, skin pigmentation, vitamin D, and 25-hydroxyvitamin D: integral components of the vitamin D endocrine system. Am J Clin Nutr 67(6):1108–1110. https://doi.org/10.1093/ajcn/67.6.1108

Nystad AE, Wergeland S, Aksnes L, Myhr K-M, Bø L, Torkildsen O (2014) Effect of high-dose 1.25 dihydroxyvitamin D3 on remyelination in the cuprizone model. Acta Pathologica, Microbiologica, et Immunologica Scandinavica 122(12):1178–1186. https://doi.org/10.1111/apm.12281

Olney JW (1990) Excitotoxicity: an overview. Canada Diseases Weekly Report = Rapport Hebdomadaire Des Maladies Au Canada 16(Suppl 1E):47–57. discussion 57

Oveland E, Nystad A, Berven F, Myhr K-M, Torkildsen Ø, Wergeland S (2018) 1,25-Dihydroxyvitamin-D3 induces brain proteomic changes in cuprizone mice during remyelination involving calcium proteins. Neurochem Int 112:267–277. https://doi.org/10.1016/j.neuint.2017.08.008

Prietl B, Treiber G, Pieber TR, Amrein K (2013) Vitamin D and immune function. Nutrients 5(7):2502–2521. https://doi.org/10.3390/nu5072502

Prietl B, Treiber G, Mader JK, Hoeller E, Wolf M, Pilz S, Graninger WB, Obermayer-Pietsch BM, Pieber TR (2014) High-dose cholecalciferol supplementation significantly increases peripheral $CD4^+$ Tregs in healthy adults without negatively affecting the frequency of other immune cells. Eur J Nutr 53(3):751–759. https://doi.org/10.1007/s00394-013-0579-6

Razzaque MS (2012) FGF23, klotho and vitamin D interactions: what have we learned from in vivo mouse genetics studies? Adv Exp Med Biol 728:84–91. https://doi.org/10.1007/978-1-4614-0887-1_5

Ricca C, Aillon A, Bergandi L, Alotto D, Castagnoli C, Silvagno F (2018) Vitamin D receptor is necessary for mitochondrial function and cell health. Int J Mol Sci 19(6). https://doi.org/10.3390/ijms19061672

Rockswold SB, Rockswold GL, Defillo A (2007) Hyperbaric oxygen in traumatic brain injury. Neurol Res 29(2):162–172. https://doi.org/10.1179/016164107X181798

Ryan ZC, Craig TA, Folmes CD, Wang X, Lanza IR, Schaible NS, Salisbury JL, Nair KS, Terzic A, Sieck GC, Kumar R (2016) 1α,25-Dihydroxyvitamin D3 regulates mitochondrial oxygen consumption and dynamics in human skeletal muscle cells. J Biol Chem 291(3):1514–1528. https://doi.org/10.1074/jbc.M115.684399

Sanchez B, Lopez-Martin E, Segura C, Labandeira-Garcia JL, Perez-Fernandez R (2002) 1,25-Dihydroxyvitamin D(3) increases striatal GDNF mRNA and protein expression in adult rats. Brain Res Mol Brain Res 108(1–2):143–146. https://doi.org/10.1016/s0169-328x(02)00545-4

Sassi F, Tamone C, D'Amelio P (2018) Vitamin D: nutrient, hormone, and immunomodulator. Nutrients 10(11). https://doi.org/10.3390/nu10111656

Schauber J, Oda Y, Büchau AS, Yun Q-C, Steinmeyer A, Zügel U, Bikle DD, Gallo RL (2008) Histone acetylation in keratinocytes enables control of the expression of cathelicidin and CD14 by 1,25-dihydroxyvitamin D3. J Investig Dermatol 128(4):816–824. https://doi.org/10.1038/sj.jid.5701102

Sepidarkish M, Farsi F, Akbari-Fakhrabadi M, Namazi N, Almasi-Hashiani A, Maleki Hagiagha A, Heshmati J (2019) The effect of vitamin D supplementation on oxidative stress parameters: a systematic review and meta-analysis of clinical trials. Pharmacol Res 139:141–152. https://doi.org/10.1016/j.phrs.2018.11.011

Sharma S, Kumar A, Choudhary A, Sharma S, Khurana L, Sharma N, Kumar V, Bisht A (2020) Neuroprotective role of oral Vitamin D supplementation on consciousness and inflammatory biomarkers in determining severity outcome in acute traumatic brain injury patients: a double-blind randomized clinical trial. Clin Drug Investig 40(4):327–334. https://doi.org/10.1007/s40261-020-00896-5

Shin D-M, Yuk J-M, Lee H-M, Lee S-H, Son JW, Harding CV, Kim J-M, Modlin RL, Jo E-K (2010) Mycobacterial lipoprotein activates autophagy via TLR2/1/CD14 and a functional vitamin D receptor signalling. Cell Microbiol 12(11):1648–1665. https://doi.org/10.1111/j.1462-5822.2010.01497.x

Shinpo K, Kikuchi S, Sasaki H, Moriwaka F, Tashiro K (2000) Effect of 1,25-dihydroxyvitamin D(3) on cultured mesencephalic dopaminergic neurons to the combined toxicity caused by L-buthionine sulfoximine and 1-methyl-4-phenylpyridine. J Neurosci Res 62(3):374–382. https://doi.org/10.1002/1097-4547(20001101)62:3<374::AID-JNR7>3.0.CO;2-7

Shirazi HA, Rasouli J, Ciric B, Rostami A, Zhang G-X (2015) 1,25-Dihydroxyvitamin D3 enhances neural stem cell proliferation and oligodendrocyte differentiation. Exp Mol Pathol 98(2):240–245. https://doi.org/10.1016/j.yexmp.2015.02.004

Shively S, Scher AI, Perl DP, Diaz-Arrastia R (2012) Dementia resulting from traumatic brain injury: what is the pathology? Arch Neurol 69(10):1245–1251. https://doi.org/10.1001/archneurol.2011.3747

Skaper SD (2018) Neurotrophic factors: an overview. Methods Mol Biol 1727:1–17. https://doi.org/10.1007/978-1-4939-7571-6_1

Smolders J, Thewissen M, Peelen E, Menheere P, Tervaert JWC, Damoiseaux J, Hupperts R (2009) Vitamin D status is positively correlated with regulatory T cell function in patients with multiple sclerosis. PLoS One 4(8):e6635. https://doi.org/10.1371/journal.pone.0006635

Socci FI, Di Valvasone S, Cecchi A, Bonizzoli M, Terreni A, Peris A (2014) Vitamin D level could affect the recovery rate in traumatic brain injury: a retrospective study. Crit Care 18(Suppl 1):P471. https://doi.org/10.1186/cc13661

Sun DA, Deshpande LS, Sombati S, Baranova A, Wilson MS, Hamm RJ, DeLorenzo RJ (2008) Traumatic brain injury causes a long-lasting calcium (Ca2+)-plateau of elevated intracellular Ca levels and altered Ca2+ homeostatic mechanisms in hippocampal neurons surviving brain injury. Eur J Neurosci 27(7):1659–1672. https://doi.org/10.1111/j.1460-9568.2008.06156.x

Szymczak I, Pawliczak R (2016) The active metabolite of vitamin D3 as a potential immunomodulator. Scand J Immunol 83(2):83–91. https://doi.org/10.1111/sji.12403

Takahashi S, Maeda T, Sano Y, Nishihara H, Takeshita Y, Shimizu F, Kanda T (2017) Active form of vitamin D directly protects the blood–brain barrier in multiple sclerosis. Clin Exp Neuroimmunol 8(3):244–254. https://doi.org/10.1111/cen3.12398

Tanaka Y, DeLuca HF (1983) Stimulation of 1,25-dihydroxyvitamin D3 production by 1,25-dihydroxyvitamin D3 in the hypocalcaemic rat. Biochem J 214(3):893–897. https://doi.org/10.1042/bj2140893

Taniura H, Ito M, Sanada N, Kuramoto N, Ohno Y, Nakamichi N, Yoneda Y (2006) Chronic vitamin D3 treatment protects against neurotoxicity by glutamate in association with upregulation of vitamin D receptor mRNA expression in cultured rat cortical neurons. J Neurosci Res 83(7):1179–1189. https://doi.org/10.1002/jnr.20824

Toman E, Bishop JRB, Davies DJ, Su Z, Criseno S, Mason A, Toogood AA, Belli A (2017) Vitamin D deficiency in traumatic brain injury and its relationship with severity of injury and quality of life: a prospective, observational study. J Neurotrauma 34(7):1448–1456. https://doi.org/10.1089/neu.2016.4494

Traiffort E, Kassoussi A, Zahaf A, Laouarem Y (2020) Astrocytes and microglia as major players of myelin production in normal and pathological conditions. Front Cell Neurosci 14:79. https://doi.org/10.3389/fncel.2020.00079

Veldman CM, Cantorna MT, DeLuca HF (2000) Expression of 1,25-dihydroxyvitamin D(3) receptor in the immune system. Arch Biochem Biophys 374(2):334–338. https://doi.org/10.1006/abbi.1999.1605

Vitamin D deficiency linked to chronic fatigue in brain-injured patients. (n.d.). Retrieved January 29, 2024, from https://www.healio.com/news/endocrinology/20120325/vitamin-d-deficiency-linked-to-chronic-fatigue-in-brain-injured-patients

Vyas CM, Mischoulon D, Chang G, Reynolds CF, Cook NR, Weinberg A, Copeland T, Bubes V, Bradwin G, Lee I-M, Buring JE, Mora S, Rifai N, Manson JE, Okereke OI (2023) Relation of serum BDNF to major depression and exploration of mechanistic roles of serum BDNF in a study of vitamin D3 and omega-3 supplements for late-life depression prevention. J Psychiatr Res 163:357–364. https://doi.org/10.1016/j.jpsychires.2023.05.069

Wang J, Thingholm LB, Skiecevičienė J, Rausch P, Kummen M, Hov JR, Degenhardt F, Heinsen F-A, Rühlemann MC, Szymczak S, Holm K, Esko T, Sun J, Pricop-Jeckstadt M, Al-Dury S, Bohov P, Bethune J, Sommer F, Ellinghaus D et al (2016) Genome-wide association analysis identifies variation in vitamin D receptor and other host factors influencing the gut microbiota. Nat Genet 48(11):1396–1406. https://doi.org/10.1038/ng.3695

Wang L, Lewis T, Zhang Y-L, Khodier C, Magesh S, Chen L, Inoyama D, Chen Y, Zhen J, Hu L, Beamer LJ, Faloon PW, Dandapani S, Perez JR, Munoz B, Palmer M, Schreiber S (2010) The identification and characterization of non-reactive inhibitor of Keap1-Nrf2 interaction through HTS using a fluorescence polarization assay. In: Probe reports from the NIH molecular libraries program. National Center for Biotechnology Information (US)

Wang T-T, Nestel FP, Bourdeau V, Nagai Y, Wang Q, Liao J, Tavera-Mendoza L, Lin R, Hanrahan JW, Mader S, White JH (2004) Cutting edge: 1,25-dihydroxyvitamin D3 is a direct inducer of antimicrobial peptide gene expression. J Immunol 173(5):2909–2912. https://doi.org/10.4049/jimmunol.173.5.2909

Wang X, Michaelis EK (2010) Selective neuronal vulnerability to oxidative stress in the brain. Front Aging Neurosci 2:12. https://doi.org/10.3389/fnagi.2010.00012

Weber JT (2012) Altered calcium signaling following traumatic brain injury. Front Pharmacol 3:60. https://doi.org/10.3389/fphar.2012.00060

Werner C, Engelhard K (2007) Pathophysiology of traumatic brain injury. Br J Anaesth 99(1):4–9. https://doi.org/10.1093/bja/aem131

Wion D, MacGrogan D, Neveu I, Jehan F, Houlgatte R, Brachet P (1991) 1,25-Dihydroxyvitamin D3 is a potent inducer of nerve growth factor synthesis. J Neurosci Res 28(1):110–114. https://doi.org/10.1002/jnr.490280111

Won S, Sayeed I, Peterson BL, Wali B, Kahn JS, Stein DG (2015) Vitamin D prevents hypoxia/reoxygenation-induced blood-brain barrier disruption via vitamin D receptor-mediated NF-kB signaling pathways. PLoS One 10(3):e0122821. https://doi.org/10.1371/journal.pone.0122821

Wrzosek M, Łukaszkiewicz J, Wrzosek M, Jakubczyk A, Matsumoto H, Piątkiewicz P, Radziwoń-Zaleska M, Wojnar M, Nowicka G (2013) Vitamin D and the central nervous system. Pharmacol Rep 65(2):271–278. https://doi.org/10.1016/s1734-1140(13)71003-x

Wu GD, Chen J, Hoffmann C, Bittinger K, Chen Y-Y, Keilbaugh SA, Bewtra M, Knights D, Walters WA, Knight R, Sinha R, Gilroy E, Gupta K, Baldassano R, Nessel L, Li H, Bushman FD, Lewis JD (2011) Linking long-term dietary patterns with gut microbial enterotypes. Science 334(6052):105–108. https://doi.org/10.1126/science.1208344

Xydakis MS, Robbins AS, Grant GA (2008) Mild traumatic brain injury in U.S. soldiers returning from Iraq. N Engl J Med 358(20):2177.; author reply 2179. https://doi.org/10.1056/NEJMc086083

Young W (1992) Role of calcium in central nervous system injuries. J Neurotrauma 9(Suppl 1):S9–S25

Yuan J, Guo X, Liu Z, Zhao X, Feng Y, Song S, Cui C, Jiang P (2018) Vitamin D receptor activation influences the ERK pathway and protects against neurological deficits and neuronal death. Int J Mol Med 41(1):364–372. https://doi.org/10.3892/ijmm.2017.3249

Zanatta L, Goulart PB, Gonçalves R, Pierozan P, Winkelmann-Duarte EC, Woehl VM, Pessoa-Pureur R, Silva FRMB, Zamoner A (2012) 1α,25-dihydroxyvitamin D(3) mechanism of action: modulation of L-type calcium channels leading to calcium uptake and intermediate filament phosphorylation in cerebral cortex of young rats. Biochim Biophys Acta 1823(10):1708–1719. https://doi.org/10.1016/j.bbamcr.2012.06.023

Zhang S-J, Zou M, Lu L, Lau D, Ditzel DAW, Delucinge-Vivier C, Aso Y, Descombes P, Bading H (2009) Nuclear calcium signaling controls expression of a large gene pool: identification of a gene program for acquired neuroprotection induced by synaptic activity. PLoS Genet 5(8):e1000604. https://doi.org/10.1371/journal.pgen.1000604

Zhang Y, Leung DYM, Richers BN, Liu Y, Remigio LK, Riches DW, Goleva E (2012a) Vitamin D inhibits monocyte/macrophage proinflammatory cytokine production by targeting MAPK phosphatase-1. J Immunol 188(5):2127–2135. https://doi.org/10.4049/jimmunol.1102412

Zhang Z, Zhang Z-Y, Wu Y, Schluesener HJ (2012b) Immunolocalization of Toll-like receptors 2 and 4 as well as their endogenous ligand, heat shock protein 70, in rat traumatic brain injury. Neuroimmunomodulation 19(1):10–19. https://doi.org/10.1159/000326771

Zou Y, Mu M, Zhang S, Li C, Tian K, Li Z, Li B, Wang W, Cao H, Sun Q, Chen H, Ge D, Tao H, Tao X (2022) Vitamin D3 suppresses astrocyte activation and ameliorates coal dust-induced mood disorders in mice. J Affect Disord *303*:138–147. https://doi.org/10.1016/j.jad.2022.02.026

Chapter 14
Zinc and TBI: From Bench to Bedside

Basma Emad Aboulhoda and Enas El Saftawy

Abstract Zinc plays a crucial role in neuronal homeostasis due to its extensive physiological distribution in the neural tissue and the abundant metalloproteins and metalloenzymes. Neuronal zinc is involved in regulating neuronal signaling, microglial activation, cell growth, neurogenesis, and antioxidant functions. Zinc supplementation in mild, moderate, and severe traumatic brain injury (TBI) appeared to improve TBI-associated critical conditions such as hypoxia, altered consciousness level, and multiple organ failure owing to the effect of zinc on the endothelium, the immune reaction, and the triglyceride level. Zinc also improves the post-TBI sequelae (e.g., dementia and epilepsy) owing to its effect on the synthesis of the neurons and microtubules. Success factors that enhance the efficacy of zinc treatment involve the administered dose, the onset time of treatment, the duration of zinc intake, the zinc state (adequacy and deficiency), and the route of zinc administration.

Keywords Zinc · Brain · Trauma · Protective · Neuron · TBI

14.1 Introduction

Zinc is a trace transition element and an essential biofactor that sustains health in the human central nervous system. Zinc is considered second to iron posing a pivotal role in multiple neuronal functions (Gower-Winter and Levenson 2012).

B. E. Aboulhoda (✉)
Department of Anatomy and Embryology, Faculty of Medicine, Cairo University, Cairo, Egypt
e-mail: basma.emad@kasralainy.edu.eg

E. El Saftawy
Department of Medical Parasitology, Faculty of Medicine, Cairo University, Cairo, Egypt

Department of Medical Parasitology, Armed Forces College of Medicine, Cairo, Egypt

W. Mohamed (ed.), *Nutrition and Traumatic Brain Injury (TBI)*, Nutritional Neurosciences, https://doi.org/10.1007/978-981-97-6341-2_14

Site: With a concentration of roughly 150 mol/L, the brain has the most significant Zn^{2+} levels in the body. Zn^{2+} concentrates in the neuronal cells present in the cerebral cortex, thalamus, olfactory cortex, amygdala, and hippocampus (Frederickson et al. 2000). About 80–95% of Zn^{2+} is tightly bound to proteins either metalloenzymes or metalloproteins, while the remaining 20% exists in the glutamatergic nerve endings as free ionic (labile/chelate form) in the presynaptic vesicles (Kumar et al. 2021).

Zinc plays a crucial catalytic, structural, and regulatory role in traumatic brain injury (TBI). Yet it remains a research target to explore. An estimated 3000 human zinc metalloproteins have catalytic and structural activities, which equates to around one out of every ten proteins being zinc proteins (Andreini et al. 2006). This wide range of functions highlights the crucial involvement of zinc in cell biology (Maret 2017). In this section, the neuroprotective role of zinc is displayed regarding the following points:

1. The protective functions of neuronal zinc
2. Zinc in mild, moderate, and severe focal brain injury
3. Zinc supplementations and TBI-associated critical conditions (hypoxia and altered conscious level)
4. Factors that affect the protective efficacy of zinc
5. Zinc guarding against post-TBI sequelae

14.1.1 The Protective Functions of Neuronal Zinc

14.1.1.1 Neuronal Signals

Zinc is an essential factor for neurogenesis, the transmission of electrical signals, and synaptic plasticity (Kumar et al. 2021). Zinc is also involved in other functions such as the processing of somatosensory signals and precise perception of nerve stimuli (Patrick and Dyck 2018; Krall et al. 2021). It also contributes to CNS functions, e.g., learning and memory. Zn^{2+} is concentrated in the vesicles of the glutamatergic neurons; thus, upon neuronal excitation, Zn^{2+} is released from synaptic vesicles into synaptic clefts in consort with glutamate (Kawahara et al. 2018). Subsequently, several receptors become activated such as γ-aminobutyric acid type A (GABA-A), N-methyl-d-aspartate (NMDA), α-amino-3-hydroxy-5-methyl-4-isoxazole propionic acid (AMPA) receptors, voltage-gated ion channels, and glycine receptors (Smart et al. 2004). Zinc is also essential for the long-term potentiation of the synaptic junctions in the region of the cortico-amygdala in the brain (Kodirov et al. 2006). Takeda et al. (2013) showed that in mossy nerve fibers, zinc is pivotal to conserving the long-term potentiation (LTP) necessary for increasing the synaptic strength. Zn^{2+} is also involved in the neuronal response to the auditory and olfactory stimuli as the dorsal cochlear nucleus and olfactory bulb are enriched with the glutamatergic neurons (Barr and Burdette 2017). Indeed, zinc is determined to induce pro-survival and pro-death signaling cascades in the neurons (Krall et al. 2021). Furthermore, being a redox-inert metal, zinc functions as an antioxidant element by catalyzing copper/zinc superoxide dismutase enzyme, shielding protein sulfhydryl groups, and increasing the production of metallothionein.

14.1.1.2 Activation of Microglia

Microglia are one of the CNS-specific immune cells. After brain injury, the microglia transform into an "ameboid" form, produce matrix metalloproteinases, and release pro-inflammatory factors and reactive oxygen species (ROS). In this instance, zinc is liberated by neurons to trigger microglia, in a nuclear factor kappa B (NF-κB)–dependent pathway. In this way, the immune functions of zinc contribute to neuropathy and trigger inflammatory cytokines.

14.1.1.3 Regulation of Proteins

Zinc is considered a crucial factor in all cellular structures, where the nucleus, nucleolus, and chromosomes all contain zinc. The RNA, DNA, and ribosomal structures are also all stabilized by zinc. Reverse transcriptase, transcription factor IIIA, RNA polymerase, and other DNA or RNA synthesis enzymes are connected to zinc metalloenzymes. Zinc in DNA serves as a cofactor for numerous enzymes with a direct role in the synthesis of macromolecules, regulation of signal transmission, transcription of genes, and intracellular transportation of proteins (Choi et al. 2020). Yet maintaining zinc homeostatic level is important since the pathological levels of zinc intensely stimulate atypical aggregation of Tau protein which subsequently compromises the intrinsic tube-like skeleton of neurons (Chowdhury and Lu 2022).

It is worth mentioning that the last consequences of spinal cord injury (SCI) that substantially entail functional impairment and long-term disability are glial scarring and irreversible neuronal loss. Previous studies have demonstrated that ectopic zinc finger protein 521 (Zf-p521) expression triggers the generation of induced neural stem cells (iNSCs) from fibroblasts and astrocytes. Zf-p521 was determined as more efficient in reprogramming cultured astrocytes compared with sex-determining region Y-box 2 *(SOX2)*. Zf-p521 is capable of transforming local astrocytes into neurons via a progenitor stage in the damaged spinal cord of an adult rat. This therapy improves the rats' functional abilities as measured by the Basso, Beattie, and Bresnahan locomotor rating scale. These findings suggest that in situ reprogramming could generate induced neural stem cells or neurons from endogenous astrocytes as a viable method for SCI healing (Zarei-Kheirabadi et al. 2019).

14.1.1.4 Antioxidant Properties

Zinc regulates redox homeostasis and acts as a molecular signal in immune cells (Prasad 2008; Wessels et al. 2017). Zinc also controls a number of transcription factors involved in the signaling pathways for inflammatory cytokines and adhesion molecules. In zinc deficiency, injured brains suffer profound gliosis in the areas adjacent to the lesion. In the CNS, metallothioneins (I and II) which are chief zinc-binding proteins markedly increase by TBI and influence the antioxidant defenses and neuronal homeostasis. Also, in TBI, increased oxidative stress is dominated by malondialdehyde, protein tyrosine nitration, inducible nitric oxide synthase, and NF-κB which are potentiated by a deficiency of zinc ions (Penkowa et al. 2001). Likewise, zinc finger protein A-20 (or TNF-α-induced protein 3) exerts negative feedback on the NF-kB signaling pathway via disrupting IkB kinase beta (IKKβ), thus regulating the inflammatory homeostasis (Zhan et al. 2016).

14.1.1.5 Cell Growth and Neurogenesis

Enhancement of neurogenesis following traumatic brain injury has been proposed by several authors. Prior work demonstrated that zinc chelation reduced TBI-induced progenitor cell proliferation and neurogenesis. This highlights the role of zinc as an essential component to control hippocampus neurogenesis following TBI (Choi et al. 2014). Zinc deficiency reduces the capacity of neuronal cell proliferation to half (Corniola et al. 2008). Yet dietary zinc following deficiency showed that p53 gene expression becomes altered, thus regulating the G-1 and S phases of the neuronal cell cycle (Gao et al. 2009). The role of zinc ions in neurogenesis might be attributed to the following mechanisms:

1. The zinc finger domain interconnects cysteine and histidine residues thus constructing a polypeptide chain in the neurons (Vallee and Auld 1995).
2. Zinc regulates cell division growth factors: chiefly nerve growth factor (NGF) and insulin-like growth factor-1 (IGF-1).

The somatotropic axes GH and IGF-1 alter deeply and show different changing levels between the acute and late phases of TBI (Mangiola et al. 2015). IGF-1 is integrated into a diversity of cellular processes necessary for CNS repair after TBI, e.g., amino acids activation, cell cycle regulation, and glucose uptake (De Meyts et al. 1994). Yet reduced IGF-1 levels are ultimately related to the low concentrations of serum zinc (Dorup et al. 1991). NGF is a group of secreted growth factors necessary for survival and plasticity of neuronal populations in the central and peripheral nervous systems (Freeman et al. 2004). The absence of NGF in the neurons is deleterious for protein synthesis (Freeman et al. 2004; Kristiansen and Ham 2014; Zhu et al. 2018). TBI induces the secretion of multiple hormones involving growth hormone as a normal reaction to stress. In addition, Kizhakke Madathil et al. (2010) deduced that TBI induces an acute and brief rise in the IGF-1 and its signaling pathway in the post-traumatic period. Therefore, IGF-1 is suggested as an important factor to trigger repair and induce endogenous neuroprotection. Indeed, zinc is a masterpiece micronutrient that stimulates cell proliferation by GH and IGF-1 through the pituitary axis (Cook et al. 2008). In addition, zinc constitutes the 7S molecule of the nerve growth factor.

3. Zinc is indispensable to induce the cell cycle at the G0/G1 phase.

Neuronal apoptosis is often detected in acute traumatic injury to the CNS, which is usually related to the blockade of the cell cycle at the G1 and GS phases (Wang et al. 2009). Chuang et al. (2017) showed that the zinc finger protein Znf179 is upregulated under the triggering effect of NGF Sp1 transcriptional activity. Cope et al. (2016) used an animal model of TBI to investigate the influence of zinc supplementation on the proliferating cells in the hippocampus. Zinc supplementation showed a robust impact on the neuronal precursor cells (Wang et al. 2009). Prior studies also showed that zinc promotes cell proliferation via its inducing effect on the IGF-1 and pituitary growth hormone. IGF-1 is a chief stimulator of the protein kinase B that subsequently promotes the progression of G1/S and G2/M phases. This explains how zinc is intricately involved in the activation of neuronal cell growth and proliferation (Beyersmann and Haase 2001; Lopez-Pajares et al. 2008).

- The nuclear factor kappa B (NF-κB) and nuclear factor of activated T cells (NFAT) are transcription factors necessary for cell proliferation, differentiation, and synaptic plasticity (Mattson and Meffert 2006). Nevertheless, NF-κB and NFAT–DNA binding in the brain was found to be affected in patients with a low-zinc diet compared with their healthy controls (Aimo et al. 2010). Additionally, severe and marginal zinc deficiency in the diet is associated with lower DNA binding factor of NF-κB and NFAT (Aimo et al. 2010). Thus, the proliferation of the neural progenitor cells in the neurogenic areas, e.g., the subventricular zone and hippocampus, is deprived of the triggering role of the NFAT (Serrano-Perez et al. 2015). Also, the immune responses related to NF-κB become reduced (Karin, 1999).
- Low levels of zinc in serum are also associated with lower expression of the doublecortin (DCX) (Rice et al. 2003). Noteworthy, DCX is a microtubule-associated protein necessary for neuroprotection, neuronal differentiation, and reorganization of neuronal connection after TBI. Therefore, Chiaretti et al. (2008) deduced that DCX possesses a chief role in the prognosis of TBI patients.
- -TBI induces Matrix metalloproteinases (MMPs) which are "zinc-dependent proteolytic enzymes," necessary for active remodeling and cleaving of the extracellular matrix (ECM) to pave the way for migration of the cells. Physiologically, this appears vital for cell growth and healing of wounds. Nevertheless, the pathophysiology of TBI involves activation of the (MMPs) in the blood-brain barrier (BBB), thus degrading the componential elements of its ECM and tight junction proteins causing disruption of the BBB. Impaired BBB leaks blood to the cerebral parenchyma causing edema and microhemorrhages. Moreover, MMPs were reported to disrupt neurogenesis and angiogenesis and contribute to cerebral inflammation and apoptosis. Despite the aforementioned role of MMPs in the pathophysiology of the TBI, Nosrati et al. (2019) speculated that almost 62% of the involved TBI studies (29 out of 47) showed that zinc possesses a repressive influence on the production and activity of the MMPs. The same authors concluded that the suppressive or activator influence of zinc on MMPs relies chiefly on the physiological case of the manipulated cells or animals, the dose of zinc supplementation, and the duration of zinc treatment.

Alteration in neuronal zinc homeostasis also aggravates the brain pathological changes associated with TBI and chronic traumatic encephalopathy. The extreme conditions, whether zinc overload or deficit, intensify brain injury and induce a detrimental effect on the immune reactions. Therefore, preserving a reasonable level of zinc is essential to sustain its distinctive physiological functions.

14.1.2 Zinc in Mild, Moderate, and Severe Focal Brain Injury

In a prior study, patients with moderate-to-severe TBI were randomly assigned to either zinc-adequate or zinc-supplemented therapy for three months in order to evaluate the hypothesis that zinc supplementation could enhance the patient outcome after TBI. After one month, mortality in the zinc-supplemented group was found to be half (12%) the zinc-adequate group (26%). In addition, in the zinc-supplemented group, functional outcomes assessed using the Glasgow Coma Scale exhibited a significant improvement as early as two weeks and persisted during the 3 months of the research (Young et al. 1996). Another double-blind controlled study on 100 cases with severe TBI showed the enhancement effect of zinc supplementation (120 mg zinc/day) with better sequential organ failure assessment ($p < 0.05$), improved Glasgow Coma Scale ($p < 0.05$), and shorter hospital stay ($p < 0.05$) (Khazdouz et al. 2018).

In mild repetitive TBI, a study on the murine models showed that the prophylactic supplementation of zinc modulates ionic zinc in the CA3 region of the hippocampus while the neurological recovery was partially recovered (Neely et al. 2022). Mild TBI has been exhibited as a "signature wound" among military personnel during their wartime activities. Long-term cognitive and behavioral disorders, involving an irregular circadian rhythm, are brought on by repetitive mild TBI. Zinc was suggested to lower the deficits in behavior associated with mild TBI. Additionally, zinc was reported to induce phosphorylation *of the Tropomyosin receptor kinase B* in the hippocampus, signifying a neuroprotective role of zinc in stressed animals (Doherty 2021).

14.1.3 Zinc Supplementations in Hypoxia and Disturbed Conscious Level Associated TBI

14.1.3.1 Hypoxia

Following severe TBI, hypoxic episodes frequently occur. The majority of hypoxic episodes are observed when the mean arterial pressure and cerebral perfusion pressure are within the normal range (Chang et al. 2009). Numerous metabolic changes take place during cerebral hypoxic ischemia, the most evident of which is the production of oxygen-free radicals leading to lipid peroxidation, protein oxidation, and

DNA damage (Chan et al. 1995). Increasing cerebral superoxide dismutase (SOD) by pharmaceutical therapies was found to reduce brain edema, infarction, and focal cerebral ischemia brought on by TBI. In animal models, it has been demonstrated that CuZn/SOD, a naturally occurring antioxidant enzyme, reduces the degree of infarction brought on by TBI (Chan et al. 1993) by scavenging the free radicals of brain neurons by oxidizing superoxide anions into hydrogen peroxide. Dimitrova et al. (2005) reported that the activity of Cu/ZnSOD varies relative to the dietary zinc. Zinc supplementation also reduces serum triglycerides in rat models. The pathogenic cascade following hypoxic-ischemic encephalopathy is significantly influenced by neuroinflammation, which also acts as a mediator for secondary neuronal death. GPR39, a Zn^{2+}-sensing G-protein-coupled receptor on the plasma membrane, has shown significant anti-inflammatory properties (Xie et al. 2021). Through a ZnR/GPR39-dependent pathway, extracellular Zn^{2+} regulates endothelial cell function. Therefore, ZnR/GPR39 might be contemplated as a therapeutic target to control endothelial function (Zhu et al. 2018).

14.1.3.2 Conscious Disturbance

The imbalance of zinc in TBI has been related to disorders of consciousness. Therefore, zinc supplementation is regarded as a promising therapeutic line (Squitti et al. 2023). Patients with severe closed-head injuries showed better healing from their neurological injuries after receiving zinc supplements (Young et al. 1996). Similarly, in a double-blind clinical research, 100 serious head injury patients between the ages of 18 and 65 were randomly assigned to receive either a placebo or 120 mg of zinc via a nasogastric tube for 15 days in order to examine the impact of zinc supplementation on their clinical outcomes. Glasgow Outcome Scale (GOS), mortality rate on day 28, length of stay (LOS), and plasma zinc assessments on days 1, 7, and 16 were compared between groups. The baseline data between groups did not differ significantly (all p values >0.05). On days 7 and 16, the zinc group had a significantly greater mean plasma zinc concentration than the placebo group (p value = 0.001). By day 16, the zinc-supplemented group showed lower levels of inflammatory markers (erythrocyte sedimentation rate and C-reactive protein) and GOS. In the zinc versus placebo groups, the LOS was reduced (52 vs. 65 days; p-value = 0.043), and the mortality rate on day 28 also decreased (17% vs. 22%, p value = 0.507) (Khazdouz et al. 2018). Xu et al. (2015) sought to investigate the correlation between the Glasgow Coma Scale, the injury severity score, and serum levels of trace elements in 126 adult cases with severe TBI. According to the findings, all patients had acute trace element deficiency syndrome following severe TBI, and there was a significant association between injury severity score and serum levels of Zn, Fe, and Mg. Accordingly, zinc was introduced as an essential metal that monitoring and supplementation will be a good strategy in severe brain trauma. Along the same line, Lin et al. (1992) conducted a prospective longitudinal study to evaluate zinc and CRP in serum and urine in 24 head trauma patients. Two groups of cases were assembled: group I, which included 14 patients who had Glasgow

Coma Scale scores ranging from 4 to 8, and group II, which included 10 patients with scores ranging from 9 to 13. All patients were observed for up to 20 days following hospital admission in order to comprehend the impact of injury severity on the two acute-phase reactants. From day 1 to 5, the average serum zinc content of all patients dropped, but it then steadily recovered. Group I experienced more severe hypozincemia than did group II. Through the course of the trial, all the patients have shown low zinc levels, and patients in group I excreted more zinc through their urine than did those in group II. The mean peak urinary zinc loss in group I was greater than 5000 micrograms per day (normal value: less than 800 micrograms per day). The theory that acute hypozincemia seen in patients is stress-induced and that zinc activity is a part of altered nutritional and metabolic responses to severe head injury is supported by acute zinc shortage along with the concurrent rise in CRP.

14.1.4 Zinc Guards Against Post-TBI Sequelae

Changes in the status of the neuronal zinc were found to be related to depression, anxiety, dementia, and neurodegeneration. Seriously, altered neuronal zinc in TBI may result in neuronal cell death (Gower-Winter and Levenson 2012; Graham and Sharp 2019).

The scientific community's comprehension of zinc chelation's potential for enhancing outcomes after TBI has fundamentally changed (Domínguez et al. 2003). In TBI models, it was important to consider functional results rather than only cellular, molecular, or histological markers when evaluating the usefulness of zinc chelation as a treatment (not a pretreatment) for TBI. The CA3 region of the hippocampus had fewer injured neurons after TBI treatment with the heavy metal chelating agent Edetate calcium disodium (CaEDTA). Nevertheless, it is worth mentioning that Morris water maze performance was not improved. According to Hellmich et al. (2008), CaEDTA elevated the pro-apoptotic proteins BAX and caspase-3 two weeks after the first injury, indicating that chelation may result in a second wave of injury weeks later. Chelation or blocking of ionic zinc after TBI has detrimental effects in the acute period following injury (within 24 h), increasing both apoptosis and necrosis in the hippocampus (Doering et al. 2010), in part by increasing reactive oxygen species. This was discovered through subsequent in vitro and in vivo work (Li et al. 2010). A single early intraperitoneal zinc injection along with dietary zinc supplementation also dramatically decreased anhedonia (P 0.001). When compared to wounded rats fed a zinc-adequate diet, dietary supplementation after injury, but not zinc injection, significantly (P 0.05) enhanced cognitive behavior as shown by the time spent locating the hidden platform in the Morris water maze test. These findings imply that zinc supplementation may be a useful therapeutic strategy for treating TBI-related behavioral abnormalities like depression and cognitive impairment (Cope et al. 2012).

Zinc regulates the transport of ions necessary for multiple physiological activities in the neuronal cells via G-protein coupled receptor (ZnR/GPR39). This particular extracellular receptor has a long history of being involved in the control of the PI3K and MAPK pathways (Asraf et al. 2014). Additionally, it causes ERK1/2 to become active, which then triggers p21 and cyclin D1 (Hershfinkel et al. 2007). K+/Cl cotransport (KCC) activity is also upregulated by ZnR/GPR39 activation. The mediation of Zn^{2+}-dependent cellular signaling pathways that result in cell survival, migration, and proliferation depends on these ZnR/GPR39 and KCC inputs (Mero et al. 2019). It has also been demonstrated that the ZnR/GPR39 controls the activity of ion transport systems crucial for the physiological function of epithelial and neuronal cells (Zhu et al. 2018) and that the ZnR/GPR39 activation improves the inhibitory tone of neurons. Epilepsy and seizures have been connected to zinc deficiency, which further confirms the significant physiological function of ZnR/GPR39 (Sunuwar et al. 2017).

14.1.5 Factors Affecting the Efficacy of Zinc Supplementation in TBI

14.1.5.1 The Administered Dose

Clinical trials: Zinc showed improvements in depression scores when the patient's medications were supplemented with 25 mg zinc/day (Sowa-Kucma et al. 2008). Also, 120 mg zinc administration via nasogastric tube for 15 days induced better Glasgow Outcome Scale scores (Khazdouz et al. 2018). Another study supplemented zinc (12 mg/day) 72 h after TBI intravenously in the form of zinc sulfate for 15 days. After this initial period, patients were administered enteral zinc gluconate (22 mg/day) for 3 months. Investigations showed that 3 weeks of zinc supplementation at this dose improved visceral prealbumin. Also, one month later, the mortality rates dropped to 12% in the zinc supplementation group (Young et al. 1996).

Animal models: The impact of zinc dose on depression was experimented with a rodent model of bulbectomy and chronic stress. Zinc (30 mg/kg) single injection reduced the immobility period in the forced swim test (Nowak et al. 2003). Another study assessed the adult male rats given a zinc-supplemented (180 ppm) or zinc-adequate (30 ppm) diet for 4 weeks followed by TBI induced by controlled cortical impact. Rats administered a zinc-supplemented diet exhibited substantially lower depression-like behavior as assessed by the "two-bottle saccharin preference test" (Cope et al. 2011).

14.1.5.2 Timing and Duration of Zinc Intake to Achieve Zinc Neuroprotection

Timing of treatment onset: Zinc supplementation can hinder the poor outcomes associated with TBI if given at appropriate timing (Levenson 2020). Prophylactic zinc supplementation for four weeks prior to TBI showed improved cognitive and depression-like behavioral consequences in murine models (Cope et al. 2011).

One-hour after TBI: Young et al. (1996) reported that instant zinc supplementation after severe closed-head injury boosts neurologic recovery. Interestingly, mice subjected to chemical blocking of zinc ions in CNS showed more neuronal damage following TBI compared to wild-type animals (Doering et al. 2010).

Seventy-two-hour after TBI: Intravenous administration of zinc sulfate for 15 days showed improvement in the scores on the Glasgow Coma Scale and enhanced behavioral resilience; besides, patients with unipolar depression had better responses to antidepressant drug therapy (Cope et al. 2012).

Duration of zinc treatment: Patients with serious closed-head injuries showed faster healing from their neurological injuries with zinc supplementation on days 15 and 21 (Young et al. 1996). The weights of adrenal glands after TBI were also dramatically lowered by zinc supplementation. Thus, prolonged zinc supplementation appeared to reduce the physiological reactions to stress. Also, zinc supplementation reversed deficiencies in spatial learning and memory. There was also no noticeable difference between the TBI-zinc-supplemented group and healthy controls (Cope et al. 2011).

14.1.5.3 Zinc Adequacy, Supplementation, and Deficiency

The gap between safe and hazardous zinc tends to be narrow. Zinc adequacy is regarded at a dose of 30 ppm, while zinc supplementation is at 180 ppm. Assessment of zinc nutrition requires a variety of chemical and functional procedures with variable grades of sensitivity and specificity (Maret and Sandstead 2006; Cope et al. 2012).

Zinc concentrations in the urine of a TBI patient exceed that of a healthy person by 14 times denoting significant zinc loss (Cope et al. 2012). Adult TBI patients in a prior prospective cohort study showed that serum zinc levels 24 h after TBI can be identified as: low (less than 80 mcg/dL), low normal (80–100 mcg/dL), high normal (between 100 and 120 mcg/dL), and high (more than 120 mcg/dL). It is noteworthy that deficiency in zinc serum levels is closely related to 6-month mortality and disability in TBI patients (Kim et al. 2022).

In the TBI animal model, diets which are marginally deficient in zinc (5 ppm) were not suggestively related to deterioration in behavioral outcomes involving spatial learning and memory, anxiety, and depressive-like behaviors (Cope et al. 2011). However, there is proof that low zinc shortage is associated with a number of crucial TBI-related processes, such as compromisation of the BBB, inflammation, and angiogenesis (Abdul-Muneer et al. 2016; Scrimgeour et al. 2018). Moreover, zinc chelation has been shown to inhibit TBI-induced neurogenesis, thereby emphasizing the repairing role of zinc (Choi et al. 2014).

14.1.5.4 Route of Zinc Administration

Multiple routes appeared to be successful in delivering zinc in TBI.

- Diet: Cope et al. (2012) used zinc-adequate (30 ppm) and zinc-supplemented (180 ppm) diets in rat models after TBI. Rats in this model showed ameliorated behavioral deficits, e.g., cognitive impairment and depression.
- Intraperitoneal: Since patients with severe TBI do not receive nutrients peroral in the initial 24 h following TBI, Cope et al. (2012) investigated intraperitoneal injection of zinc (30 mg/kg) in a single dose 1 h after injury for zinc replenishment in murine model.
- Intravenous injection: Young et al. (1996) reported the value of intravenous zinc in patients within the initial 72 h of TBI extended for 15 days in combination with entire parenteral nutrition.
- Nasogastric tube: Khazdouz et al. (2018) applied zinc supplementation in patients with TBI using a nasogastric tube for 15 days and recorded improvements in the Glasgow score, organ failure assessment, and inflammatory scores.

14.2 Conclusion

The utmost percentage of neuronal zinc is found tightly bound to metalloproteins and/or metalloenzymes. Disturbance of neuronal zinc homeostasis aggravates the pathologies associated with TBI and chronic traumatic encephalopathy. Thus, it appeared imperative to assess the levels of zinc to retain neuroprotection while avoiding its neuromodulatory and neurotoxic effects. In extreme conditions, either zinc overload or deficiency play a role in aggravating brain injury and immune reactions. Thus, adequate levels of zinc are crucial to preserve the brain's physiological functions. Zinc supplementation in TBI also improves the Glasgow Coma Scale and the sequential post-TBI organ failure. In repetitive mild TBI, e.g., in wartime, zinc seemed to lessen the long-term cognitive and behavioral deficits.

14.3 Perspectives

- More comprehensive studies are urgently required to determine how to properly implement zinc supplementations, particularly in the TBI vulnerable groups in correlation with their age and medical-associated conditions.
- More studies are required to recognize the descriptive molecular role of zinc in different cerebral regions.
- Clinical pharmacies and pharmaceuticals should focus on the production of improved zinc formulas to eliminate the undesired side effects of zinc administration.
- Experimental studies that assess the long-term effects of zinc after TBI along with the studies that investigate the most adequate zinc dosage and duration for amelioration of the post-traumatic sequelae are highly suggested.

- We recommend repetitive tracing of zinc excretion in the urine of patients with TBI with correlation with the Glasgow Coma Scale.
- Comparing zinc supplementation versus zinc chelation at different time intervals post-TBI remains an area for further investigation.

References

Abdul-Muneer PM, Pfister BJ, Haorah J, Chandra N (2016) Role of matrix metalloproteinases in the pathogenesis of traumatic brain injury. Mol Neurobiol 53:6106–6123

Aimo L, Mackenzie GG, Keenan AH, Oteiza PI (2010) Gestational zinc deficiency affects the regulation of transcription factors AP-1, NF-kappaB and NFAT in fetal brain. J Nutr Biochem 21:1069–1075

Andreini C, Banci L, Bertini I, Rosato I (2006) Counting the zinc-proteins encoded in the human genome. Proteome Res 5:196–201

Asraf H, Salomon S, Nevo A, Sekler I, Mayer D, Hershfinkel M (2014) The ZnR/GPR39 interacts with the CaSR to enhance signaling in prostate and salivary epithelia. J Cell Physiol 229:868–877

Barr CA, Burdette SC (2017) The zinc paradigm for metalloneurochemistry. Essays Biochem 61:225–235

Beyersmann D, Haase H (2001) Functions of zinc in signaling, proliferation and differentiation of mammalian cells. Biometals 14:331–341

Chan PH, Kinouchi H, Epstein CJ, Carlson E, Chen SF, Imaizumi S, Yang GY (1993) Role of superoxide dismutase in ischemic brain injury: reduction of edema and infarction in transgenic mice following focal cerebral ischemia. Prog Brain Res 96:97–104

Chan PH, Epstein CJ, Li Y, Huang TT, Carlson E, Kinouchi H, Yang G, Kamii H, Mikawa S, Kondo T, Copin JC, Chen SF, Chan T, Gafni J, Gobbel G, Reola E (1995) Transgenic mice and knockout mutants in the study of oxidative stress in brain injury. J Neurotrauma 12:815–824

Chang JJ, Youn TS, Benson D, Mattick H, Andrade N, Harper CR et al (2009) Physiologic and functional outcome correlates of brain tissue hypoxia in traumatic brain injury. Crit Care Med 37(1):283–290

Chiaretti A, Antonelli A, Genovese O, Pezzotti P, Di Rocco C, Viola L, Riccardi R (2008) Nerve growth factor and doublecortin expression correlates with improved outcome in children with severe traumatic brain injury. J Trauma Acute Care Surg 65(1):80–85

Choi BY, Kim JH, Kim HJ, Lee BE, Kim IY, Sohn M, Suh SW (2014) Zinc chelation reduces traumatic brain injury-induced neurogenesis in the subgranular zone of the hippocampal dentate gyrus. J Trace Elem Med Biol 28(4):474–481

Choi S, Hong DK, Choi BY, Suh SW (2020) Zinc in the brain: friend or foe? Int J Mol Sci 21(23):8941

Chowdhury SR, Lu HP (2022) Unraveling the mechanism of tau protein aggregation in presence of zinc ion: the earliest step of tau aggregation. Chem Phys Impact 4:100060

Chuang JY, Kao TJ, Lin SH, Wu AC, Lee PT, Su TP et al (2017) Specificity protein 1-zinc finger protein 179 pathway is involved in the attenuation of oxidative stress following brain injury. Redox Biol 11:135–143

Cook AM, Peppard A, Magnuson B (2008) Nutrition considerations in traumatic brain injury. Nutr Clin Pract 23(6):608–620

Cope EC, Morris DR, Scrimgeour AG, VanLandingham JW, Levenson CW (2011) Zinc supplementation provides behavioral resiliency in a rat model of traumatic brain injury. Physiol Behav 104:942–947

Cope EC, Morris DR, Levenson CW (2012) Improving treatments and outcomes: an emerging role for zinc in traumatic brain injury. Nutr Rev 70(7):410–413

Cope EC, Morris DR, Gower-Winter SD, Brownstein NC, Levenson CW (2016) Effect of zinc supplementation on neuronal precursor proliferation in the rat hippocampus after traumatic brain injury. Exp Neurol 279:96–103
Corniola RS, Tassabehji NM, Hare J, Sharma G, Levenson CW (2008) Zinc deficiency impairs neuronal precursor cell proliferation and induces apoptosis via p53-mediated mechanisms. Brain Res 1237:52–61
De Meyts P, Wallach B, Christoffersen CT, Urso B, Gronskov K, Latus LJ, Yakushiji F, Ilondo MM, Shymko RM (1994) The insulin-like growth factor-I receptor. Structure, ligand-binding mechanism and signal transduction. Horm Res 42:152–169
Dimitrova AA, Strashimirov DS, Russeva AL, Andreeva-Gateva PA, Lakova ET, Tzachev KN (2005) Effect of zinc on the activity of Cu/Zn superoxide dismutase and lipid profile in Wistar rats. Folia Medica 47(1):42–46
Doering P, Stoltenberg M, Penkowa M, Rungby J, Larsen A, Danscher G (2010) Chemical blocking of zinc ions in CNS increases neuronal damage following traumatic brain injury (TBI) in mice. PloS One 5(4):e10131
Doherty EN (2021). Intranasal zinc remediates daily activity deficits following stress and traumatic brain injury (Doctoral dissertation)
Domínguez MI, Blasco-Ibáñez JM, Crespo C, Marqués-Marí AI, Martínez-Guijarro FJ (2003) Zinc chelation during non-lesioning overexcitation results in neuronal death in the mouse hippocampus. Neuroscience 116:791–806
Dorup I, Flyvbjerg A, Everts ME, Clausen T (1991) Role of insulin-like growth factor-1 and growth hormone in growth inhibition induced by magnesium and zinc deficiencies. Br J Nutr 66:505–521
Frederickson CJ, Suh SW, Silva D, Frederickson CJ, Thompson RB (2000) Importance of zinc in the central nervous system: The zinc-containing neuron. J Nutr 130:1471S–1483S
Freeman RS, Burch RL, Crowder RJ, Lomb DJ, Schoell MC, Straub JA, Xie L (2004) NGF deprivation-induced gene expression: after ten years, where do we stand? Prog Brain Res 146:111–126
Gao HL, Zheng W, Xin N, Chi ZH, Wang ZY, Chen J, Wang ZY (2009) Zinc deficiency reduces neurogenesis accompanied by neuronal apoptosis through caspase-dependent and -independent signaling pathways. Neurotox Res 16:416–425
Gower-Winter SD, Levenson CW (2012) Zinc in the central nervous system: from molecules to behavior. Biofactors 38(3):186–193
Graham NS, Sharp DJ (2019) Understanding neurodegeneration after traumatic brain injury: from mechanisms to clinical trials in dementia. J Neurol Neurosurg Psychiatry 90(11):1221–1233
Hellmich HL, Eidson K, Cowart J, Crookshanks J, Boone DK, Shah S, Uchida T, DeWitt DS, Prough DS (2008) Chelation of neurotoxic zinc levels does not improve neurobehavioral outcome after traumatic brain injury. Neurosci Lett 440:155–159
Hershfinkel M, Silverman WF, Sekler I (2007) The zinc sensing receptor, a link between zinc and cell signaling. Mol Med 13:331–336
Kawahara M, Tanaka KI, Kato-Negishi M (2018) Zinc, carnosine, and neurodegenerative diseases. Nutrients 10:147
Khazdouz M, Mazidi M, Ehsaei MR, Ferns G, Kengne AP, Norouzy AR (2018) Impact of zinc supplementation on the clinical outcomes of patients with severe head trauma: a double-blind randomized clinical trial. J Dietary Suppl 15(1):1–10
Kim KH, Ro YS, Yoon H, Lee SGW, Jung E, Moon SB et al (2022) Serum zinc and long-term prognosis after acute traumatic brain injury with intracranial injury: a multicenter prospective study. J Clin Med 11(21):6496
Kizhakke Madathil S, Evans HN, Saatman KE (2010) Temporal and regional changes in IGF-1/IGF-1R signaling in the mouse brain after traumatic brain injury. J Neurotrauma 27(1):95–107
Kodirov SA, Takizawa S, Joseph J, Kandel ER, Shumyatsky GP, Bolshakov VY (2006) Synaptically released zinc gates long-term potentiation in fear conditioning pathways. Proc Natl Acad Sci USA 103:15218–15223

Krall RF, Tzounopoulos T, Aizenman E (2021) The function and regulation of zinc in the brain. Neuroscience 457:235–258

Kristiansen M, Ham J (2014) Programmed cell death during neuronal development: The sympathetic neuron model. Cell Death Differ 21:1025–1035

Kumar V, Kumar A, Singh K, Avasthi K, Kim JJ (2021) Neurobiology of zinc and its role in neurogenesis. Eur J Nutr 60:55–64

Levenson CW (2020) Zinc and traumatic brain injury: from chelation to supplementation. Med Sci 8(3):36

Li Y, Hawkins BE, DeWitt DS, Prough DS, Maret W (2010) The relationship between transient zinc ion fluctuations and redox signaling in the pathways of secondary cellular injury: relevance to traumatic brain injury. Brain Res 1330:131–141

Lin CN, Howng SL, Hu SH, Huang TJ (1992) Assessments of nutritional status and immunological responses in head trauma: alterations in zinc and C-reactive protein. Gaoxiong yi xue ke xue za zhi= Kaohsiung J Med Sci 8(4):195–201

Lopez-Pajares V, Kim MM, Yuan ZM (2008) Phosphorylation of MDMX mediated by Akt leads to stabilization and induces 14-3-3 binding. J Biol Chem 283:13707–13713

Mangiola A, Vigo V, Anile C, De Bonis P, Marziali G, Lofrese G (2015) Role and importance of IGF-1 in traumatic brain injuries. BioMed Res Int

Maret W (2017) Zinc in cellular regulation: the nature and significance of "zinc signals". Int J Mol Sci 18(11):2285

Maret W, Sandstead HH (2006) Zinc requirements and the risks and benefits of zinc supplementation. J Trace Elem Med Biol 20(1):3–18

Mattson MP, Meffert MK (2006) Roles for NF-kappaB in nerve cell survival, plasticity, and disease. Cell Death Differ 13:852–860

Mero M, Asraf H, Sekler I, Taylor KM, Hershfinkel M (2019) ZnR/GPR39 upregulation of K(+)/Cl(−)-cotransporter 3 in tamoxifen resistant breast cancer cells. Cell Calcium 81:12–20

Neely CLC, Barkey RE, Hernandez CM, Flinn JM (2022) Prophylactic zinc supplementation modulates hippocampal ionic zinc and partially remediates neurological recovery following repetitive mild head injury in mice. Behav Brain Res 430:113918

Nosrati R, Kheirouri S, Ghodsi R, Ojaghi H (2019) The effects of zinc treatment on matrix metalloproteinases: a systematic review. J Trace Elem Med Biol 56:107–115

Nowak G, Szewczyk B, Wieronska JM et al (2003) Antidepressant-like effects of acute and chronic treatment with zinc in forced swim test and olfactory bulbectomy model in rats. Brain Res Bull 61:159–164

Patrick WH, Dyck RH (2018) Signaling by synaptic zinc is required for whisker-mediated, fine texture discrimination. Neuroscience 369:242–247

Penkowa M, Giralt M, Thomsen PS, Carrasco J, Hidalgo J (2001) Zinc or copper deficiency-induced impaired inflammatory response to brain trauma may be caused by the concomitant metallothionein changes. J Neurotrauma 18(4):447–463

Rice AC, Khaldi A, Harvey HB, Salman NJ, White F, Fillmore H, Bullock MR (2003) Proliferation and neuronal differentiation of mitotically active cells following traumatic brain injury. Exp Neurol 183(2):406–417

Scrimgeour AG, Carrigan CT, Condlin ML, Urso ML, Van den Berg RM, van Helden HPM, Montain SJ, Joosen MJA (2018) Dietary zinc modulates matrix metalloproteinases in traumatic brain injury. J. Neurotrauma 35:2495–2506

Serrano-Perez MC, Fernandez M, Neria F, Berjon-Otero M, Doncel-Perez E, Cano E, Tranque P (2015) NFAT transcription factors regulate survival, proliferation, migration, and differentiation of neural precursor cells. Glia 63:987–1004

Smart TG, Hosie AM, Miller PS (2004) Zn^{2+} ions: modulators of excitatory and inhibitory synaptic activity. Neuroscientist 10:432–442

Sowa-Kucma M, Legutko B, Szewczyk B et al (2008) Antidepressant-like activity of zinc: further behavioral and molecular evidence. J Neural Transm. 115:1621–1628

Squitti R, Reale G, Tondolo V, Crescenti D, Bellini S, Moci M et al (2023) Imbalance of essential metals in traumatic brain injury and its possible link with disorders of consciousness. Int J Mol Sci 24(7):6867

Sunuwar L, Gilad D, Hershfinkel M (2017) The zinc sensing receptor, ZnR/GPR39, in health and disease. Front Biosci (Landmark Ed) 22:1469–1492

Takeda A, Nakamura M, Fujii H, Tamano H (2013) Synaptic Zn^{2+} homeostasis and its significance. Metallomics 5:417–423

Vallee BL, Auld DS (1995) Zinc metallochemistry in biochemistry. EXS 73:259–277

Wang W, Bu B, Xie M, Zhang M, Yu Z, Tao D (2009) Neural cell cycle dysregulation and central nervous system diseases. Progr Neurobiol 89(1):1–17

Xie S, Jiang X, Doycheva DM, Shi H, Jin P, Gao L et al (2021) Activation of GPR39 with TC-G 1008 attenuates neuroinflammation via SIRT1/PGC-1α/Nrf2 pathway post-neonatal hypoxic–ischemic injury in rats. J Neuroinflamm 18(1):1–15

Xu G, Hu B, Chen G, Yu X, Luo J, Lv J, Gu J (2015) Analysis of blood trace elements and biochemical indexes levels in severe craniocerebral trauma adults with glasgow coma scale and injury severity score. Biol Trace Elem Res 164:192–197

Young B, Ott L, Kasarskis E, Rapp R, Moles KAY, Dempsey RJ et al (1996) Zinc supplementation is associated with improved neurologic recovery rate and visceral protein levels of patients with severe closed head injury. J Neurotrauma 13(1):25–34

Zarei-Kheirabadi M, Hesaraki M, Kiani S, Baharvand H (2019) In vivo conversion of rat astrocytes into neuronal cells through neural stem cells in injured spinal cord with a single zinc-finger transcription factor. Stem Cell Res Therapy 10:1–19

Zhan J, Qin W, Zhang Y, Jiang J, Ma H, Li Q, Luo Y (2016) Upregulation of neuronal zinc finger protein A20 expression is required for electroacupuncture to attenuate the cerebral inflammatory injury mediated by the nuclear factor-kB signaling pathway in cerebral ischemia/reperfusion rats. J Neuroinflamm 13:1–18

Zhu D, Su Y, Zheng Y, Fu B, Tang L, Qin YX (2018) Zinc regulates vascular endothelial cell activity through zinc-sensing receptor ZnR/GPR39. Am J Physiol-Cell Physiol 314(4):C404–C414

Chapter 15
Polyphenols: The Key to Unlock TBI Rehabilitation

Rabii Hilali, Fatima Rachidi, Hajar Ouaki, Abdelmonaim Azzouz, Youssef Aoulad El Hadj Ali, Dia Eddine Oudghiri, Ibtissam Chakir, Rabia Ajana, El Hassan Sakar, Ayoub Kasrati, Abdelaati El Khiat, and Lahcen Tamegart

R. Hilali · R. Ajana · E. H. Sakar · L. Tamegart (✉)
Biology, Ecology and Health Lab, Department of Biology, Faculty of Science, Abdelmalek Essaadi University, Tetouan, Morocco
e-mail: l.tamegart@uae.ac.ma

F. Rachidi
Laboratory of Biostatistics, Clinical Research and Epidemiology (LBRCE), Faculty of Medicine and Pharmacy, Mohammed V University, Rabat, Morocco

Biology Department, Higher School of Teachers (ENS), Mohammed V University in Rabat, Rabat, Morocco

H. Ouaki
Laboratory of Biostatistics, Clinical Research and Epidemiology (LBRCE), Faculty of Medicine and Pharmacy, Mohammed V University, Rabat, Morocco

A. Azzouz · Y. A. E. H. Ali
Department of Chemistry, Faculty of Science, University of Abdelmalek Essaâdi, Tetouan, Morocco

D. E. Oudghiri
Biology, Ecology and Health Lab, Department of Biology, Faculty of Science, Abdelmalek Essaadi University, Tetouan, Morocco

Biology Higher Institute of Nursing Professions and Health Techniques (ISPITS), Tetouan, Morocco

I. Chakir
Biology, Ecology and Health Lab, Department of Biology, Faculty of Science, Abdelmalek Essaadi University, Tetouan, Morocco

Multidisciplinary Faculty, LPRD, Sultan Moulay Slimane University, Beni-Mellal, Morocco

A. Kasrati
Biology, Ecology and Health Lab, Department of Biology, Faculty of Science, Abdelmalek Essaadi University, Tetouan, Morocco

Department of Pharmacy Faculty of Medicine and Pharmacy, Hassan 1st University, Settat, Morocco

W. Mohamed (ed.), *Nutrition and Traumatic Brain Injury (TBI)*, Nutritional Neurosciences, https://doi.org/10.1007/978-981-97-6341-2_15

Abstract Traumatic brain injury (TBI) is classified as one of the neurological disorders with various alterations, including neuronal and behavioral damages. TBI could be induced by disruption of the normal physiology of the brain following trauma. As a result, various symptoms such as emotional, physical, and cognitive problems take place. Polyphenols, extracted from various natural matrices such as plants, could play a significant role in regulating ion channels, influencing various cellular processes, and then being used as a potential treatment for TBI. In addition, nanomaterials associated with polyphenols with their strong loading capacity could play an important role in TBI treatment.

Keywords Traumatic brain injury · Trauma · Polyphenols · Plant matrices · Nanoparticles · Curcumin

15.1 Introduction

Traumatic brain injury (TBI) is the main cause of injury-related death and disability worldwide, and it is defined as a disruption of the normal physiology of the brain following trauma (Adugna et al. 2022). It could lead to harmful consequences, such as physical, cognitive, and emotional problems (Adugna et al. 2022; Mohammed et al. 2023).

The primary and secondary injuries are the phases of the pathophysiology of traumatic brain injury (TBI). Primary injuries occur when the brain is subjected to an external force that directly affects the vascular system, glial cells, and neuronal tissue according to their physical properties. The development of biological processes is triggered by the first damage, which leads to secondary injuries. These may include cellular malfunction resulting in apoptsis, ischemia, glutamate toxicity, neuroinflammation, edema, and increased blood-brain barrier (BBB) permeability (Freidin et al. 2023).

It is widely evidenced that polyphenols play a significant role in regulating ion channels, impacting various cellular processes. Research conducted by Scholz et al. (2010) showed that flavonoids act on cardiac ion channels and contribute to attenuation of cardiac calcium by inhibiting the opening of calcium channels. Additionally, the studies performed by Kelemen et al. (2007) emphasized the inhibitory effect of epigallocatechin-3-gallate, a green tea flavonoid, on cardiac *hERG* potassium channels.

A. El Khiat
Biology, Ecology and Health Lab, Department of Biology, Faculty of Science, Abdelmalek Essaadi University, Tetouan, Morocco

Higher Institute of Nursing Professions and Health Techniques in Ouarzazate Ministry of Health, Hay Sidi Hssain, Morocco

A recent study conducted by Modzelewska et al. (2021) demonstrated, for the first time, that quercetin induces relaxation of gastric smooth muscle through its action on ATP-sensitive K+ channels, which block Ca^{2+} channels. This reinforces the control of muscle contraction and provides therapeutic implications for the treatment of gastric disorders.

A research carried out by Agrawal et al. (2015) investigated the therapeutic potential of the flavonoid derivative 7,8-DHF (7,8-dihydroxyflavone) in mitigating the pathology associated with traumatic brain injury (TBI). The study presents compelling evidence that the administration of 7,8-DHF effectively attenuates TBI-induced pathology through the activation of TrkB receptors. By targeting the TrkB receptor pathway, 7,8-DHF demonstrates neuroprotective effects, reducing neuronal damage and enhancing recovery following TBI. Through a series of experimental analyses, including histological assessments and functional assays, Agrawal et al. elucidate the mechanisms underlying the beneficial effects of 7,8-DHF, highlighting its potential as a novel therapeutic intervention for TBI (Agrawal et al. 2015).

In a study by Chitturi et al., the beneficial effects of kaempferol following developmental traumatic brain injury (TBI) were investigated, focusing on its mechanism of action in protecting mitochondrial function, oxidative metabolism, and neural viability. This study demonstrates that kaempferol treatment improved tricarboxylic acid cycle flux, maintained mitochondrial functional integrity by decreasing acylcarnitines, and improved neural viability by increasing N-acetyl aspartate levels. These positive outcomes of kaempferol treatment on the metabolic profile corresponded with improvements in sensorimotor behavior (Chitturi et al. 2019).

A research conducted by Deng et al. (2021) provides significant insights into the potential therapeutic effects of naringenin in mitigating the detrimental consequences of traumatic brain injury (TBI). Through a series of experiments conducted on animal models, the researchers demonstrated that naringenin administration effectively alleviated endoplasmic reticulum stress, a key pathological mechanism implicated in TBI progression. Moreover, naringenin treatment resulted in a notable reduction in apoptosis, thus promoting neuronal survival and tissue repair After injury. Importantly, functional assessments revealed improved neurological outcomes in naringenin-treated subjects, suggesting its promising neuroprotective properties as a potential therapeutic intervention for TBI (Deng et al. 2021).

15.2 Pathophysiology of TBI

Primary brain injury from initial damaging force causes early post-injury tissue deformation and disintegration. Through a set of processes like excitotoxicity, depolarization, edema formation, disruption of the blood-brain barrier, calcium homeostasis, and intracranial hemorrhage, these secondary injuries from TBI alter cell performance and propagate damage (Kaur and Sharma 2018). TBI can be classified into three groups based on the distinct physical processes of injury: (i) closed head injury, (ii) penetrating injury, and (iii) explosive blast injury (Ng and Lee 2019).

Besides, aberrant brain movement upon impact causes axonal stretching and disruption of neuronal cell membranes at the molecular level. This ultimately leads to a large-scale ion flow across neuronal membranes, which produces broad waves of depolarization and triggers the symptoms associated with typical concussions (Barkhoudarian et al. 2011). In fact, the arrangement of tissues, such as skin, bone, and deep structures, can alter when an object is placed on the head. A bone fracture, skin laceration, or lifelong disfigurement will result if the adjustment is larger than the tissue's flexibility. The aggravating factor may cause depressed skull fractures and the destruction of underlying tissue, including the dura, brain, and arteries, with higher weights. The main frequent results are an epidural hematoma, subdural hematoma, contusion, or intracerebral hemorrhage. In more severe circumstances, perforation and penetration may happen, particularly when there are small agent sizes and rapid agent speeds. The impact may have contributed to the energy transfer to the brain and skull. This energy load may induce contusions and deform the brain tissues. Shock waves, acceleration/deceleration, and deformation all put energy on the tissue and result in tissue distortion. These have the potential to harm the skull's structures, including bones, blood vessels, and neurological components (Feng et al. 2010).

Whereas secondary damage starts with inflammation and progresses to breakage of the blood-brain barrier (BBB, infiltration of peripheral blood cells, brain edema, and the release of many immune mediators, including interleukins and chemotactic factors, primary injury results in localized brain damage. TBI modifies cell functioning, architecture, and molecular signaling. In addition to causing tissue damage such as bleeding, contusions, and axonal damage, TBI generally disrupts brain physiology, speech, memory, attention, and cognitive functions. Even though the pathophysiology of TBI is well-understood, the underlying processes still need to be evaluated in order to control the repercussions of TBI (Thapa et al. 2021). It is believed that alterations in BBB function that arise following an injury have a role in the degeneration of brain tissue and influence the way neuroprotective medications work. ABT can result in changes in cerebral blood flow (CBF), oxidative stress, immunoreactive cell activation, immune mediator release, disruption of the blood-brain barrier (BBB), damage to the vascular structure, dysfunction of the neurovascular unit (NVU), destruction of the BBB importance of blood-borne variables in gliovascular unit function, alterations in BBB permeability, and post-traumatic edema formation (Chodobski et al. 2011).

Local inflammation is caused by pro-inflammatory substances secreted by microglia and astrocytes, like interleukin-1, interleukin-6, and tumor necrosis factor. Inflammation is exacerbated by these inflammatory substances, which actively attract glial and immune cells. Following an ABT, the BBB allows peripheral immune cells to infiltrate the body and activate immune cells in the central nervous system (CNS). This process is facilitated by a range of inflammatory cytokines. Activated white blood cells can move to the injured brain parenchyma due to the disruption of the NVU and BBB brought on by ABT. This process is further boosted by the increased expression of cell adhesion molecules. ROS and inflammatory chemicals, such as cytokines and chemokines, are produced by activated white blood cells, microglia, and astrocytes. This results in demyelination and the breakdown of the axonal cytoskeleton. Brain edema and elevated intracranial pressure are the results of inflammatory processes throughout time (Candelario-Jalil et al. 2022).

15.2.1 Protein post-translational modifications (PTMs)

(PTMs), is an amino acid chain or terminal covalent enzyme modification. It is a reversible process that can alter target protein association, localization, activity, and function. When PTMs in the cellular milieu are dysfunctional, linked proteins undergo conformational changes, enzyme activity is unbalanced, proteins fold abnormally, and toxic metabolites are produced. These events ultimately culminate in the disease state (Luo et al. 2023). The pathophysiology of disease and cellular metabolism are intimately related. For cells and organs to continue to operate metabolically, mitochondria are necessary. Within the mitochondria exist several reversible post-translational protein modifications that can alter a variety of mitochondrial functions. These modifications are sensitive to redox conditions and nutrition availability, and they can also act through protein–protein interactions (Luo et al. 2023).

15.3 Neuroprotective Properties of Polyphenols

15.3.1 Role of Polyphenols in Ion Channel Regulation

Several studies have reported that numerous ion channels are modulated by curcumin (Zhang et al. 2014). The effect of curcumin administration on Kv1.3 potassium channels was investigated by Lian et al. (2013). The study results suggest that curcumin exerts a direct inhibition on these channels, potentially leading to a blockade of the release of pro-inflammatory cytokines, thereby exerting anti-inflammatory effects. It has also been demonstrated that curcumin inhibits classical ion flux by acting on Ca^{2+} calcium channels, subsequently limiting the excessive release of glutamates and exerting its neuroprotective effect (Lin et al. 2011).

15.3.2 Anti-inflammatory Effects of Polyphenols

Polyphenols are bioactive molecules found in many plant-based foods and are extensively studied for their anti-inflammatory properties. An in vitro study demonstrated that resveratrol, a polyphenol present in red grapes and nuts, is capable of suppressing the production of cytokines such as TNF-α, IL-1α, IL-6, and IL-17, along with a suppressive effect on NF-κB activation (Fuggetta et al. 2016).

Another example of polyphenols with well-documented anti-inflammatory effects is curcumin. The anti-inflammatory properties of curcumin are ascribed to its interaction with cellular signaling pathways involved in inflammation. Curcumin hinders the action of NF-κB transcription factors; blocks the phosphorylation of the JAK/STAT pathway, a key item in the transmission of inflammatory signals; and inhibits MAPK signaling and the arachidonic acid signaling pathway. These mechanisms enable curcumin to attenuate the production of pro-inflammatory cytokines such as TNF-α, IL-6, and IL-8 (Kahkhaie et al. 2019; Yahfoufi et al. 2018).

Polyphenols exert a significant regulatory action on macrophages, immune cells playing a crucial role in the inflammatory response, by inhibiting the enzyme cyclooxygenase 2 (COX-2) and inducible nitric oxide synthase (iNOS), thereby limiting the production of TNF-α and IL-1β and modulating the expression of IL-6 (González et al. 2011). In humans, a decrease in serum levels of inflammatory markers, such as C-reactive protein, IL-6, and IL-12, has been observed under the influence of polyphenols present in extra-virgin olive and blueberries (Fitó et al. 2008; Kolehmainen et al. 2012).

15.3.3 Influence of Polyphenols on Synaptic Plasticity and Neuronal Recovery

In a clinical study, a high intake of polyphenols, especially flavonoids, was associated with improved cognitive functions over a period of 10 years (Letenneur et al. 2007). Additionally, Witte et al. (2014) demonstrated that the administration of resveratrol led to improved memory and functional connectivity of the hippocampus in elderly individuals compared to a placebo treatment. Epidemiological evidence suggests that the consumption of orange juice rich in hesperidin significantly enhances subjective or objective cognitive function in healthy middle-aged adults (Alharbi et al. 2016).

Furthermore, the administration of polyphenols can influence synaptic morphology (Naomi et al. 2023). For example, it has been demonstrated that resveratrol administration led to an increase in the length and number of dendritic spines in pyramidal neurons. These neuronal changes promote neural plasticity and enhance the efficiency of neuronal connections (Monserrat Hernández-Hernández et al. 2016). Trovò et al. (2020) revealed the effectiveness of epigallocatechin gallate in reversing cognitive deficits by improving neuronal maturation and restoring dendritic branching in a model of CDKL5 deficiency disorder (CCD) syndrome.

It has been demonstrated that grape polyphenol treatment also plays a role in preventing the alteration of neuronal plasticity by promoting dendritic branching (Caracci et al. 2020). Treatment with a combination of polyphenols and proanthocyanidins from tea contributes to memory improvement while enhancing synaptic plasticity, as well as the number and shape of dendrites and synapses in the hippocampus (Yang et al. 2022).

15.4 Underlying Molecular Mechanisms

15.4.1 Impact of Polyphenols on Excitotoxicity

Polyphenols appear to be protective against excitotoxicity by modulating several signaling pathways involved in this pathological mechanism.

They block cell death and decrease mitochondrial membrane depolarization by limiting the uncontrolled entry of Ca^{2+} (Shin et al. 2021). Furthermore, polyphenols exert a preventive effect by limiting the overexpression and translocation of pro-apoptotic Bax proteins. These proapoptotic proteins play a crucial role in triggering apoptosis and are often expressed in the cytosol in response to excitotoxic stimuli. Their translocation to the mitochondria triggers mechanisms leading to cell death. Polyphenols, by acting on Bax proteins, preserve cellular integrity and prevent cell death (Arias-Sánchez et al. 2023).

It has also been demonstrated that polyphenols act on caspase enzymes and apoptosis-inducing factors, preventing cellular damage under excitotoxic conditions (Cong et al. 2016; Elseady et al. 2023). Research conducted by Nantacharoen et al. (2022) showed that the extract of *Cleistocalyx nervosum* suppresses the expression of caspases in a model of oxidative damage induced by glutamate.

15.4.2 Effects of Polyphenols on Neuroinflammatory Processes

The effect of polyphenols on the neuroinflammation process has been extensively studied in recent decades. The results of a study conducted by Zhang et al. (2010) showed that polyphenols negatively regulate microglial activation through the inhibition of mitogen-activated protein kinase (MAPK) and NF-κB signaling pathways. Furthermore, it has been demonstrated that these polyphenols also act on pro-inflammatory molecules by inhibiting their activities, such as cyclooxygenases (COX-2) and lipoxygenase (LOX-5) (Rojas-García et al. 2023).

Furthermore, there is evidence that polyphenols modulate neuronal inflammation through their action on the NF-κB signaling pathway. Indeed, the results of a study conducted by Ben Youssef et al. (2021) highlighted the potential of polyphenols present in grape seed extract to inhibit phosphorylation and thus the activation of the NF-κB pathway, which is essential in triggering the inflammatory response in a murine model of Parkinson's disease.

Phochantachinda et al. (2021) reported that the extract of *Emblica officinalis* fruit suppressed neuronal inflammation in an LPS-activated model by decreasing the levels of nitric oxide (NO), prostaglandin E2 (PGE-2), interleukin (IL-6), and TNF-α through the suppression of the ERK and JNK signaling pathways.

On another note, in the context of the neuroinflammatory disease multiple sclerosis, the administration of polyphenols contributes to a notable increase in brain-derived neurotrophic factor (BDNF), which has an immunomodulatory action (Arias-Sánchez et al. 2023).

15.5 Polyphenols in Humans with Traumatic Brain Injury

Recent studies indicate that nutritional interventions may provide a unique opportunity to enhance the repair processes after traumatic brain injury (TBI). Dietary polyphenols, a diverse group of natural compounds widely distributed in fruits, vegetables, grains, nuts, and seeds, have emerged in the past few years as promising agents to be used in TBI settings (Carecho et al. 2023). Due to their proven pleiotropic effects, polyphenols have the potential to protect against the progression of neurodegenerative pathologies (Vauzour 2012). Delivery of nanoparticle-based BBB-targeted drugs combined with polyphenols that are derived from intestinal bacterial metabolism could prove to be effective treatments for various neurological disorders, including traumatic brain injury (Reddy et al. 2020). Numerous preclinical studies undoubtedly showed that polyphenols can reduce brain edema in rats and mice following TBI and dietary polyphenols may provide protection from the negative consequences of concussive brain injury in flies (Trofimova 2020). There are still human studies missing to confirm the same result in TBI patients (Carecho et al. 2023).

15.6 Preclinical and Clinical Studies with Polyphenols in Brain Injury

Brain damage, whether traumatic, ischemic, or neurodegenerative, is a major public health problem (Dams-O'Connor et al. 2014; Hansson 2021; Maas et al. 2022). They can lead to significant cognitive, motor, and sensory deficits, as well as an altered quality of life (Calina et al. 2020; Calvillo and Irimia 2020).

Polyphenols are natural compounds found in many fruits, vegetables, and beverages, such as red wine, green tea, and coffee (Di Lorenzo et al. 2021). They have been shown to have numerous health benefits, including antioxidant, anti-inflammatory, and neuroprotective properties (Bucciantini et al. 2021; Kowalczewski et al. 2020; Silva and Pogačnik 2020).

15.6.1 Findings from Observational Studies

The association between polyphenol consumption and the risk of brain damage has been examined in several observational studies (Bucciantini et al. 2021; Silva and Pogačnik 2020).

Examples of Observational Studies

- This study followed over 70,000 people for 20 years and found that those who consumed the most polyphenols had a 20% lower risk of dementia (Yang et al. 2021).
- This study compared the polyphenol intake of 100 people with Alzheimer's disease with that of 100 healthy people. The results showed that people with Alzheimer's disease consumed less polyphenols than healthy people (Colizzi 2019).
- This study revealed that polyphenol consumption was associated with a 25% reduction in the risk of cerebrovascular lesions (Bayes et al. 2023).

It is important to note that observational studies cannot prove causality. It is therefore possible that other factors, such as lifestyle or overall diet, are responsible for the associations observed between polyphenol consumption and the risk of brain damage (Manikandan et al. 2023; Mayer et al. 2024).

Further studies are needed to confirm these findings and determine the optimal dose and duration of treatment for patients at risk of brain damage (Wang et al. 2023).

15.6.2 Clinical Trials Evaluating the Efficacy of Polyphenols in TBI

Traumatic brain injury (TBI) is a major public health problem, affecting millions of people every year. Traumatic brain injury can lead to significant cognitive, motor, and sensory deficits, as well as impaired quality of life (Erdman et al. 2011; Hyder et al. 2007) (Table 15.1).

Research performed on the efficacy of polyphenols in the treatment of traumatic brain injury (TBI) is at a preliminary but encouraging stage. Several ongoing and completed clinical trials are exploring the potential of different polyphenols, such as resveratrol, curcumin, and green tea, to improve cognitive function and reduce inflammation in TBI patients (Carecho et al. 2023; Erdman et al. 2011).

Polyphenols represent an interesting avenue of research for the treatment of TBI, but further research is needed before they can be recommended as a standard treatment.

Table 15.1 Clinical studies that evaluate the efficacy of polyphenol in TBI

Clinical trials	Objective	Population	Intervention	Results
Pilot study on the safety and efficacy of resveratrol in the treatment of traumatic brain injury (Brown et al. 2024)	Assessing the safety and tolerability of resveratrol for patients with moderate-to-severe traumatic brain injury	Patients between 18 and 65 years of age	Resveratrol 1000 mg daily for 12 weeks	Changes in cognitive function and inflammation
Efficacy and safety of curcumin in the treatment of moderate head trauma (Khayatan et al. 2022)	Evaluating the efficacy and safety of curcumin in the treatment of moderate head trauma	Patients between 18 and 65 years of age	Curcumin 200 mg daily for 5 weeks	Curcumin's effectiveness of cognitive function and inflammation

15.7 Effects of Polyphenols Associated with Nanoparticles in the Treatment of TBI

Various nanomaterials could be used for the treatment of TBI. Nanoparticles (NPs) can be functionalized with polyphenols (e.g., curcumin) (Banazadeh et al. 2023) to protect drug payloads and improve targeted drug delivery to lesion sites and can be coupled with supportive scaffolds to enhance secondary results (Mohammed et al. 2023). In addition, the encapsulation of pharmaceuticals within NPs is one of the advanced approaches which could enhance site-specific delivery and bioavailability (Zhang et al. 2022). In this sense, Narouiepour et al. developed curcumin-load NPs (CM-NPs) to enhance the therapeutic potential of stem cell therapy in the treatment of TBI (Narouiepour et al. 2022). The CM-NPs in this work were fabricated employing a thin-film hydration approach to encapsulate curcumin into the shell of the NPs. The as-prepared CM-NS with a diameter ranging from 60 to 90 nm were administrated orally to animal models of TBI subjected to stem cell implantations in the injury site. The CM-NS demonstrated the ability to support stem cell therapy efficiency by reducing inflammation and brain edema and improving the overall motor activity of animals (Narouiepour et al. 2022). In another work, the curcumin-loaded lysine/poly(ethylene glycol) diacrylate (C/LP) hydrogels were synthesized for TBI (Sun et al. 2023). The C/LP hydrogels were in situ synthesized via Michael addition. The swelling ratio, morphology, and biocompatibility of C/LP hydrogels were evaluated. An in vitro curcumin release profile was also investigated. After traumatic brain injury, C/LP hydrogels were injected into the injured brain. The brain functional recovery was evaluated by animal behavior, magnetic resonance imaging, and immunohistochemistry (Sun et al. 2023). Furthermore, the curcumin-functionalized Fe_3O_4 magnetic NPs also demonstrated a strong anti-neurotoxic activity in the cerebellum of schizophrenic rats (Naserzadeh et al. 2018).

15.8 Conclusions

TBI is known for its disruptive effects on brain functions. More studies are needed to understand clearly the signaling pathways underlying the physiopathology of TBI. Natural bioactive compounds such as curcumin could be used as a potential drug for the treatment of TBI. In addition, the synthesis of nanoparticles associated with polyphenols like curcumin could increase the permeability of the molecule through the cell membrane and then increase the efficacy of the molecule.

References

Adugna DG, Aragie H, Kibret AA, Belay DG (2022) Therapeutic Application of Stem Cells in the Repair of Traumatic Brain Injury. Stem Cells and Cloning: Advances and Applications 15:53–61. https://doi.org/10.2147/SCCAA.S369577

Agrawal R, Noble E, Tyagi E, Zhuang Y, Ying Z, Gomez-Pinilla F (2015) Flavonoid derivative 7, 8-DHF attenuates TBI pathology via TrkB activation. Biochim Biophys Acta (BBA)-Mol Basis Dis 1852:862–872

Alharbi MH, Lamport DJ, Dodd GF, Saunders C, Harkness L, Butler LT, Spencer JPE (2016) Flavonoid-rich orange juice is associated with acute improvements in cognitive function in healthy middle-aged males. Eur J Nutr 55:2021–2029. https://doi.org/10.1007/s00394-015-1016-9

Arias-Sánchez RA, Torner L, Fenton Navarro B (2023) Polyphenols and neurodegenerative diseases: potential effects and mechanisms of neuroprotection. Molecules 28:5415. https://doi.org/10.3390/molecules28145415

Banazadeh M, Behnam B, Ganjooei NA, Gowda BHJ, Kesharwani P, Sahebkar A (2023) Curcumin-based nanomedicines: a promising avenue for brain neoplasm therapy. J Drug Deliv Sci Technol:105040

Barkhoudarian G, Hovda DA, Giza CC (2011) The molecular pathophysiology of concussive brain injury. Clin Sports Med 30:33–48

Bayes J, Bedaso A, Peng W, Adams J, Sibbritt D (2023) The effect of polyphenols in post stroke adults: a systematic review of randomised controlled trials. Clin Nutr ESPEN 54:113–121

Ben Youssef S, Brisson G, Doucet-Beaupré H, Castonguay A-M, Gora C, Amri M, Lévesque M (2021) Neuroprotective benefits of grape seed and skin extract in a mouse model of Parkinson's disease. Nutr Neurosci 24:197–211. https://doi.org/10.1080/1028415X.2019.1616435

Brown K, Theofanous D, Britton RG, Aburido G, Pepper C, Sri Undru S, Howells L (2024) Resveratrol for the management of human health: how far have we come? A systematic review of resveratrol clinical trials to highlight gaps and opportunities. Int J Mol Sci 25:747

Bucciantini M, Leri M, Nardiello P, Casamenti F, Stefani M (2021) Olive polyphenols: antioxidant and anti-inflammatory properties. Antioxidants 10:1044

Calina D, Buga AM, Mitroi M, Buha A, Caruntu C, Scheau C, Bouyahya A, El Omari N, El Menyiy N, Docea AO (2020) The treatment of cognitive, behavioural and motor impairments from brain injury and neurodegenerative diseases through cannabinoid system modulation—evidence from in vivo studies. J Clin Med 9:2395

Calvillo M, Irimia A (2020) Neuroimaging and psychometric assessment of mild cognitive impairment after traumatic brain injury. Front Psychol 11:1423

Candelario-Jalil E, Dijkhuizen RM, Magnus T (2022) Neuroinflammation, stroke, blood-brain barrier dysfunction, and imaging modalities. Stroke 53:1473–1486

Caracci F, Harary J, Simkovic S, Pasinetti GM (2020) Grape-derived polyphenols ameliorate stress-induced depression by regulating synaptic plasticity. J Agric Food Chem 68:1808–1815. https://doi.org/10.1021/acs.jafc.9b01970

Carecho R, Carregosa D, Ratilal BO, Figueira I, Ávila-Gálvez MA, Dos Santos CN, Loncarevic-Vasiljkovic N (2023) Dietary (poly) phenols in traumatic brain injury. Int J Mol Sci 24:8908

Chitturi J, Santhakumar V, Kannurpatti SS (2019) Beneficial effects of kaempferol after developmental traumatic brain injury is through protection of mitochondrial function, oxidative metabolism, and neural viability. J Neurotrauma 36:1264–1278

Chodobski A, Zink BJ, Szmydynger-Chodobska J (2011) Blood–brain barrier pathophysiology in traumatic brain injury. Transl Stroke Res. 2:492–516

Colizzi C (2019) The protective effects of polyphenols on Alzheimer's disease: a systematic review. Alzheimer's Dement Transl Res Clin Interv 5:184–196

Cong L, Cao C, Cheng Y, Qin X-Y (2016) Green tea polyphenols attenuated glutamate excitotoxicity via antioxidative and antiapoptotic pathway in the primary cultured cortical neurons. Oxid Med Cell Longev 2016:2050435. https://doi.org/10.1155/2016/2050435

Dams-O'Connor K, Cantor JB, Brown M, Dijkers MP, Spielman LA, Gordon WA (2014) Screening for traumatic brain injury: findings and public health implications. J Head Trauma Rehabil 29:479–489

Deng C, Yi R, Fei M, Li T, Han Y, Wang H (2021) Naringenin attenuates endoplasmic reticulum stress, reduces apoptosis, and improves functional recovery in experimental traumatic brain injury. Brain Res 1769:147591. https://doi.org/10.1016/j.brainres.2021.147591

Di Lorenzo C, Colombo F, Biella S, Stockley C, Restani P (2021) Polyphenols and human health: the role of bioavailability. Nutrients 13:273

Elseady WS, Keshk WA, Negm WA, Elkhalawany W, Elhanafy H, Ibrahim MAA, Radwan DA (2023) Saffron extract attenuates Sofosbuvir-induced retinal neurodegeneration in albino rat. Anat Rec (Hoboken, N.J. 2007) 306:422–436. https://doi.org/10.1002/ar.24942

Erdman J, Oria M, Pillsbury L (2011) Workshop speakers' papers. In: Nutrition and traumatic brain injury: improving acute and subacute health outcomes in military personnel. National Academies Press (US)

Feng Y, Abney TM, Okamoto RJ, Pless RB, Genin GM, Bayly PV (2010) Relative brain displacement and deformation during constrained mild frontal head impact. J R Soc Interf 7:1677–1688

Fitó M, Cladellas M, de la Torre R, Martí J, Muñoz D, Schröder H, Alcántara M, Pujadas-Bastardes M, Marrugat J, López-Sabater MC, Bruguera J, Covas MI (2008) Anti-inflammatory effect of virgin olive oil in stable coronary disease patients: a randomized, crossover, controlled trial. Eur J Clin Nutr 62:570–574. https://doi.org/10.1038/sj.ejcn.1602724

Freidin D, Har-Even M, Rubovitch V, Murray KE, Maggio N, Shavit-Stein E, Keidan L, Citron BA, Pick CG (2023) Cognitive and Cellular Effects of Combined Organophosphate Toxicity and Mild Traumatic Brain Injury. Biomedicines 11:1481. https://doi.org/10.3390/biomedicines11051481

Fuggetta MP, Bordignon V, Cottarelli A, Macchi B, Frezza C, Cordiali-Fei P, Ensoli F, Ciafrè S, Marino-Merlo F, Mastino A, Ravagnan G (2016) Downregulation of proinflammatory cytokines in HTLV-1-infected T cells by Resveratrol. J Exp Clin Cancer Res. 35:118. https://doi.org/10.1186/s13046-016-0398-8

González R, Ballester I, López-Posadas R, Suárez MD, Zarzuelo A, Martínez-Augustin O, Sánchez de Medina F (2011) Effects of flavonoids and other polyphenols on inflammation. Crit Rev Food Sci Nutr 51:331–362. https://doi.org/10.1080/10408390903584094

Hansson O (2021) Biomarkers for neurodegenerative diseases. Nat Med 27:954–963

Hyder AA, Wunderlich CA, Puvanachandra P, Gururaj G, Kobusingye OC (2007) The impact of traumatic brain injuries: a global perspective. Neuro Rehabil 22:341–353

Kahkhaie KR, Mirhosseini A, Aliabadi A, Mohammadi A, Mousavi MJ, Haftcheshmeh SM, Sathyapalan T, Sahebkar A (2019) Curcumin: a modulator of inflammatory signaling pathways in the immune system. Inflammopharmacology 27:885–900. https://doi.org/10.1007/s10787-019-00607-3

Kaur P, Sharma S (2018) Recent advances in pathophysiology of traumatic brain injury. Curr Neuropharmacol 16:1224–1238

Kelemen K, Kiesecker C, Zitron E, Bauer A, Scholz E, Bloehs R, Bloehs R, Thomas D, Greten J, Remppis A, Schoels W, Katus HA, Karle CA (2007) Green tea flavonoid epigallocatechin-3-gallate (EGCG) inhibits cardiac hERG potassium channels. Biochem Biophys Res Commun 364:429–435. https://doi.org/10.1016/j.bbrc.2007.10.001

Khayatan D, Razavi SM, Arab ZN, Niknejad AH, Nouri K, Momtaz S, Gumpricht E, Jamialahmadi T, Abdolghaffari AH, Barreto GE (2022) Protective effects of curcumin against traumatic brain injury. Biomed Pharmacother 154:113621

Kolehmainen M, Mykkänen O, Kirjavainen PV, Leppänen T, Moilanen E, Adriaens M, Laaksonen DE, Hallikainen M, Puupponen-Pimiä R, Pulkkinen L, Mykkänen H, Gylling H, Poutanen K, Törrönen R (2012) Bilberries reduce low-grade inflammation in individuals with features of metabolic syndrome. Mol Nutr Food Res 56:1501–1510. https://doi.org/10.1002/mnfr.201200195

Kowalczewski PŁ, Radzikowska D, Ivanišová E, Szwengiel A, Kačániová M, Sawinska Z (2020) Influence of abiotic stress factors on the antioxidant properties and polyphenols profile composition of green barley (Hordeum vulgare L.). Int J Mol Sci 21:397

Letenneur L, Proust-Lima C, Le Gouge A, Dartigues J, Barberger-Gateau P (2007) Flavonoid intake and cognitive decline over a 10-year period. Am J Epidemiol 165:1364–1371. https://doi.org/10.1093/aje/kwm036

Lian Y-T, Yang X-F, Wang Z-H, Yang Y, Yang Y, Shu Y-W, Cheng L-X, Liu K (2013) Curcumin serves as a human kv1.3 blocker to inhibit effector memory T lymphocyte activities. Phyther Res PTR 27:1321–1327. https://doi.org/10.1002/ptr.4863

Lin TY, Lu CW, Wang C-C, Wang Y-C, Wang S-J (2011) Curcumin inhibits glutamate release in nerve terminals from rat prefrontal cortex: possible relevance to its antidepressant mechanism. Prog Neuropsychopharmacol Biol Psychiatry 35:1785–1793. https://doi.org/10.1016/j.pnpbp.2011.06.012

Luo P, Li L, Huang J, Mao D, Lou S, Ruan J, Chen J, Tang R, Shi Y, Zhou S (2023) The role of SUMOylation in the neurovascular dysfunction after acquired brain injury. Front Pharmacol 14:1125662

Maas AIR, Menon DK, Manley GT, Abrams M, Åkerlund C, Andelic N, Aries M, Bashford T, Bell MJ, Bodien YG (2022) Traumatic brain injury: progress and challenges in prevention, clinical care, and research. Lancet Neurol 21:1004–1060

Manikandan P, Raid A-B, Abdelhadi A, Al Othaim A, Vijayakumar R, Ibrahim R, Aldosari S, Alhussain A, Kannaiyan M, Binsaleh NK (2023) Neuroprotective effect of endophytic fungal antioxidant polyphenols on cerebral ischemic stroke-induced Albino rats; memory impairments, brain damage, and upregulation of metabolic proteins. J King Saud Univ 35:102433

Mayer EA, Horn J, Mayer DE, Randolph E (2024) Dietary polyphenols to maintain healthier brain measures and cognitive function, as mediated by gut microbiota metabolites. In: The gut-brain axis. Elsevier, pp 341–360

Modzelewska B, Drygalski K, Kleszczewski T, Chomentowski A, Koryciński K, Kiełczewska A, Pawłuszewicz P, Razak Hady H (2021) Quercetin relaxes human gastric smooth muscles directly through ATP-sensitive potassium channels and not depending on the nitric oxide pathway. Neurogastroenterol Motil 33:14093. https://doi.org/10.1111/nmo.14093

Mohammed FS, Omay SB, Sheth KN, Zhou J (2023) Nanoparticle-based drug delivery for the treatment of traumatic brain injury. Expert Opin Drug Deliv 20:55–73

Monserrat Hernández-Hernández E, Serrano-García C, Antonio Vázquez-Roque R, Díaz A, Monroy E, Rodríguez-Moreno A, Florán B, Flores G (2016) Chronic administration of resveratrol prevents morphological changes in prefrontal cortex and hippocampus of aged rats. Synapse 70:206–217. https://doi.org/10.1002/syn.21888

Nantacharoen W, Baek SJ, Plaingam W, Charoenkiatkul S, Tencomnao T, Sukprasansap M (2022) Cleistocalyx nervosum var. paniala Berry promotes antioxidant response and suppresses glutamate-induced cell death via SIRT1/Nrf2 survival pathway in hippocampal HT22 neuronal cells. Molecules 27:5813. https://doi.org/10.3390/molecules27185813

Naomi R, Yazid MD, Teoh SH, Balan SS, Shariff H, Kumar J, Bahari H, Embong H (2023) Dietary polyphenols as a protection against cognitive decline: evidence from animal experiments; mechanisms and limitations. Antioxidants 12:1054. https://doi.org/10.3390/antiox12051054

Narouiepour A, Ebrahimzadeh-Bideskan A, Rajabzadeh G, Gorji A, Negah SS (2022) Neural stem cell therapy in conjunction with curcumin loaded in niosomal nanoparticles enhanced recovery from traumatic brain injury. Sci Rep 12:3572

Naserzadeh P, Hafez AA, Abdorahim M, Abdollahifar MA, Shabani R, Peirovi H, Simchi A, Ashtari K (2018) Curcumin loading potentiates the neuroprotective efficacy of Fe3O4 magnetic nanoparticles in cerebellum cells of schizophrenic rats. Biomed Pharmacother 108:1244–1252

Ng SY, Lee AYW (2019) Traumatic brain injuries: pathophysiology and potential therapeutic targets. Front Cell Neurosci 13:528

Phochantachinda S, Chatchaisak D, Temviriyanukul P, Chansawang A, Pitchakarn P, Chantong B (2021) Ethanolic fruit extract of *Emblica officinalis* suppresses neuroinflammation in microglia and promotes neurite outgrowth in Neuro2a cells. Evid-Based Compl Altern Med 2021:e6405987. https://doi.org/10.1155/2021/6405987

Reddy VP, Aryal P, Robinson S, Rafiu R, Obrenovich M, Perry G (2020) Polyphenols in Alzheimer's disease and in the gut–brain axis. Microorganisms 8:199

Rojas-García A, Fernández-Ochoa Á, Cádiz-Gurrea ML, Arráez-Román D, Segura-Carretero A (2023) Neuroprotective effects of agri-food by-products rich in phenolic compounds. Nutrients 15:449. https://doi.org/10.3390/nu15020449

Scholz EP, Zitron E, Katus HA, Karle CA (2010) Cardiovascular Ion Channels as a Molecular Target of Flavonoids. Cardiovasc Ther 28:46–52. https://doi.org/10.1111/j.1755-5922.2010.00212.x

Shin SK, Yoo J-M, Li FY, Baek SY, Kim MR (2021) Mulberry fruit improves memory in scopolamine-treated mice: role of cholinergic function, antioxidant system, and TrkB/Akt signaling. Nutr Neurosci 24:940–950. https://doi.org/10.1080/1028415X.2019.1696613

Silva RFM, Pogačnik L (2020) Polyphenols from food and natural products: Neuroprotection and safety. Antioxidants 9:61

Sun L, Li Q, Guo Y, Wu X, Han M, Zhang L, Wang P (2023) Construction of curcumin-loaded hydrogels for treatment of traumatic brain injury. ACS Appl Polym Mater 5:5783–5793

Thapa K, Khan H, Singh TG, Kaur A (2021) Traumatic brain injury: mechanistic insight on pathophysiology and potential therapeutic targets. J Mol Neurosci 71:1725–1742

Trofimova AD (2020) The effects of a polyphenol-rich diet in a fruit-fly model of traumatic brain injury

Trovò L, Fuchs C, De Rosa R, Barbiero I, Tramarin M, Ciani E, Rusconi L, Kilstrup-Nielsen C (2020) The green tea polyphenol epigallocatechin-3-gallate (EGCG) restores CDKL5-dependent synaptic defects in vitro and in vivo. Neurobiol Dis 138:104791. https://doi.org/10.1016/j.nbd.2020.104791

Vauzour D (2012) Dietary polyphenols as modulators of brain functions: biological actions and molecular mechanisms underpinning their beneficial effects. Oxid Med Cell Longev

Wang W, Liu T, Ding Y, Zhang Y (2023) Effects of polyphenol-rich interventions on sleep disorders: A systematic review and meta-analysis. Curr Res Food Sci 100462

Witte AV, Kerti L, Margulies DS, Flöel A (2014) Effects of resveratrol on memory performance, hippocampal functional connectivity, and glucose metabolism in healthy older adults. J Neurosci 34:7862–7870. https://doi.org/10.1523/JNEUROSCI.0385-14.2014

Yahfoufi N, Alsadi N, Jambi M, Matar C (2018) The immunomodulatory and anti-inflammatory role of polyphenols. Nutrients 10. https://doi.org/10.3390/nu10111618

Yang Q, Zhang Y, Zhang L, Li X, Dong R, Song C, Cheng L, Shi M, Zhao H (2022) Combination of tea polyphenols and proanthocyanidins prevents menopause-related memory decline in rats via increased hippocampal synaptic plasticity by inhibiting p38 MAPK and TNF-α pathway. Nutr Neurosci 25:1909–1927. https://doi.org/10.1080/1028415X.2021.1913929

Yang W, Cui K, Li X, Zhao J, Zeng Z, Song R, Qi X, Xu W (2021) Effect of polyphenols on cognitive function: evidence from population-based studies and clinical trials. J Nutr Health Aging:1–15

Zhang F, Liu J, Shi J-S (2010) Anti-inflammatory activities of resveratrol in the brain: Role of resveratrol in microglial activation. Eur J Pharmacol 636:1–7. https://doi.org/10.1016/j.ejphar.2010.03.043

Zhang P, Li Y, Tang W, Zhao J, Jing L, McHugh KJ (2022) Theranostic nanoparticles with disease-specific administration strategies. Nano Today 42:101335

Zhang X, Chen Q, Wang Y, Peng W, Cai H (2014) Effects of curcumin on ion channels and transporters. Front Physiol 5:94. https://doi.org/10.3389/fphys.2014.00094

Chapter 16
Nutritional Recommendations for TBI Management

Sana El Foutat, Abdelaati El Khiat, Bilal El-Mansoury, Youssef Ait Hamdan, Driss Ait Ali, Kamal Kaoutar, Younsse Korchyou, Sakhr Dadda, Mohamed El Koutbi, Abdelmoujoud El Haboussi, Lahcen Tamegart, Brahim El Houate, and Naima Fdil

S. El Foutat · B. El Houate · N. Fdil
Metabolics Platform, Biochemistry Laboratory, Faculty of Medicine, Cadi Ayyad University, Marrakech, Morocco

A. El Khiat (✉)
Biological and Health Sciences Team, Higher Institute of Nursing Professions and Health Techniques, Ministry of Health, Ouarzazate, Morocco

Laboratory of Clinical and Experimental Neurosciences and Environment, Faculty of Medicine and Pharmacy, Cadi Ayyad University, Marrakech, Morocco

Biological Engineering Laboratory, Faculty of Sciences and Techniques (FST), Sultan Moulay Slimane University, Beni Mellal, Morocco

Laboratory of Anthropogenic, Biotechnology and Health, Team Physiopathology Nutritional, Neurosciences and Toxicology, Faculty of Sciences, Chouaib Doukkali University, El Jadida, Morocco

Interdisciplinary Laboratory in Bio-Resources, Environment and Materials, Higher Normal School, Marrakech, Morocco

Institute of Sports Sciences, Hassan 1er University, Settat, Morocco

B. El-Mansoury
Laboratory of Anthropogenic, Biotechnology and Health, Team Physiopathology Nutritional, Neurosciences and Toxicology, Faculty of Sciences, Chouaib Doukkali University, El Jadida, Morocco

Y. A. Hamdan
Interdisciplinary Laboratory in Bio-Resources, Environment and Materials, Higher Normal School, Marrakech, Morocco

D. A. Ali · K. Kaoutar · Y. Korchyou · M. El Koutbi
Biological and Health Sciences Team, Higher Institute of Nursing Professions and Health Techniques, Ministry of Health, Ouarzazate, Morocco

S. Dadda
Department of Biology, Faculty of Sciences and Techniques, Errachidia, Morocco

A. El Haboussi
Institute of Sports Sciences, Hassan 1er University, Settat, Morocco

W. Mohamed (ed.), *Nutrition and Traumatic Brain Injury (TBI)*, Nutritional Neurosciences, https://doi.org/10.1007/978-981-97-6341-2_16

Abstract This chapter presents evidence-based nutritional recommendations for the management of traumatic brain injury (TBI). TBI poses significant challenges to physiological functioning, including metabolic disturbances and increased energy expenditure. Proper nutrition plays a crucial role in supporting recovery and minimizing secondary complications. Recommendations include optimizing macronutrient intake to meet energy demands, emphasizing the importance of protein for tissue repair, and ensuring adequate intake of micronutrients such as omega-3 fatty acids, antioxidants, and vitamins to mitigate oxidative stress and inflammation. Additionally, hydration status should be closely monitored to prevent dehydration, which can exacerbate cognitive deficits and impair recovery. Tailoring nutritional interventions to individual needs, considering factors such as injury severity, comorbidities, and stage of recovery, is essential for maximizing outcomes in TBI management.

Keywords Traumatic brain injury · Nutritional recommendations · Macronutrients · Protein · Micronutrients · Omega-3 fatty acids · Antioxidants · Vitamins · Recovery · Cognitive deficits · Inflammation · Oxidative stress.

16.1 Introduction

Traumatic brain injury (TBI) carries a mortality rate of around 33%, affecting approximately 1.4 million people each year. Survivors often face significant deficits necessitating extended medical and rehabilitation care (Faul et al. 2007). While proven therapies for improved TBI outcomes are limited, interventions like avoiding hypotension, maintaining cerebral perfusion, and preventing complications like mucosal damage and thromboembolism show promise. Although challenging, providing early and adequate nutrition support to TBI patients may improve their overall clinical course (Taylor et al. 1999).

TBI includes both a primary injury brought on by direct harm to brain and a secondary injury cascade that follows the initial insult. Primary injuries range from concussion to intracranial bleeding, posing challenges in treatment due to the rapidity and mechanism of injury. Primarily, the treatment is to control intracranial hypertension, frequently by performing a craniotomy to promptly remove hematomas and hemorrhages.

L. Tamegart
Laboratory of Clinical and Experimental Neurosciences and Environment, Faculty of Medicine and Pharmacy, Cadi Ayyad University, Marrakech, Morocco

Department of Biology, Faculty of Science, Abdelmalek Essaadi University, Tetouan, Morocco

In contrast, secondary injury stems from an inflammatory cascade triggered after the primary injury. This process is characterized by swelling and death of brain cells and is mediated by inflammatory pathways regulated by complement and cellular immunity, reactive oxygen species, and toxic neurotransmitters such as glutamate.

These mechanisms result in metabolic disturbance, cerebral edema, and ultimately energy-related cell death (Hatton 2001).

After TBI, hypermetabolism and increased catabolism result in conditions such as hyperglycemia, protein wasting, and elevated energy demands, reaching up to 200% of the standard energy requirement (Quintard and Ichai 2019; Tavarez et al. 2021). A study has reported that malnutrition affects 68% of individuals with acute TBI (Krakau et al. 2006). This energy imbalance causes the body's mass to diminish, especially that of the skeletal muscles, which in turn causes a negative nitrogen balance and a rise in morbidity and mortality (Kurtz and Rocha 2020).

Effective and early nutritional therapy is critical in managing TBI, especially in cases of severe severity, as it is recognized to have a pivotal role in preserving cellular homeostasis, reducing mortality, and reducing infectious consequences (Bjørklund and Chirumbolo 2017; Oddo et al. 2008). Research on nutrition in stroke patients has been extensively studied, but comparatively less has been done with TBI patients (Jung et al. 2020; Quintard and Ichai 2019). This chapter provides an overview of dietary recommendations for patients with mild-to-moderate TBI and suggests that nutritional solutions are well-positioned to meet the therapeutic demands in the management of mild-to-moderate TBI in the future based on recently published data. According to these results, focused nutritional therapy may improve functional recovery following TBI. Although extensive evidence evaluations have universally recognized the benefits of nutritional therapy in critical illness, recent data suggests that patients with neurotrauma are not receiving appropriate nutritional treatment (Lee and Oh 2022).

More convincing data could be required to persuade medical professionals to provide neurotrauma patients with better nutritional assistance. Feed intolerance and delayed stomach emptying are thought to be most common in critically sick patients who have been admitted with traumatic brain injury (TMI; up to 80% of cases). The Brain Trauma Foundation Guidelines advise that patients with severe TBI obtain their goal nutrition support by at least day 7 post-injury, even if they do not explicitly specify dietary formulations for enhancing outcomes in mild-to-moderate TBI patients

Furthermore, evidence indicates that early feeding, as opposed to conventional nutritional therapy, lowers morbidity and death. Nevertheless, there is little data for the post-acute rehabilitation phase and insufficient data (Scrimgeour and Condlin 2014).

The goal of this review is to compile the difficulties that arise in clinical practice by incorporating knowledge from previous research and the most recent Clinical Practice Guidelines. The objective is to augment our comprehension of the existing information about the best nutritional therapy for patients with TBI and to pinpoint areas that necessitate future research.

16.2 Theoretical Foundations

16.2.1 Impact of TBI on Overall Health and Metabolism

TBI is a serious public health concern, with an estimated 50 million new cases occurring each year. TBI carries a substantial risk of mortality, accounting for 30%–40% of all injury-related deaths in all age groups (Maas et al. 2017). Moreover, a majority of survivors experience enduring neurobehavioral consequences, such as personality changes, emotional instability, and mental and cognitive deficits. These sequelae impose a considerable burden on patients, their families, and society at large (McAllister 2008).

A complex mechanism resulting from primary and secondary insults is involved in the etiology of TBI (Galgano et al. 2017). At the time of injury, mechanical force causes primary injury, which is followed by a lengthy and delayed secondary injury (McAllister 2008).

Secondary injuries manifest through different pathological mechanisms, such as excitotoxicity, oxidative stress, and cerebral metabolic dysfunction (Fujikawa 2015; Tehse and Taghibiglou 2019), as well as cerebrovascular pathology (Fujikawa 2015; Tehse and Taghibiglou 2019), chronic inflammatory events (Fujikawa 2015; Tehse and Taghibiglou 2019), and mitochondrial dysfunction.

16.2.2 Role of Nutrition in TBI Recovery

Commencing nutritional support promptly after a TBI is essential for counteracting the immediate breakdown of tissues often observed in such cases (Kurtz and Rocha 2020). Early nutrition helps maintain muscle mass, reduces the risk of infection-related complications, facilitates cerebral equilibrium, and enhances functional recovery (Bistrian et al. 2011; Wang et al. 2013). It is optimal to initiate nutritional support within 24–48 h following the injury, with the goal of providing 50% of resting energy expenditure (REE) in the first two weeks after the injury.

Determining the ideal quantities of glucose, protein, and fats required for neuronal survival is crucial when considering the nutritional needs after TBI (Cook et al. 2008). Hyperglycemia and hypoglycemia are potential occurrences in TBI patients, with hyperglycemia being more prevalent after TBI (Pillsbury et al. 2011; Shi et al. 2016). It is believed to be brought on by elevated stress levels and increased insulin resistance.

Poorer outcomes after TBI have been associated with elevated blood glucose levels (Shi et al. 2016). Research has demonstrated that controlling blood glucose levels by insulin therapy might enhance results following TBI. However, intense insulin therapy, in contrast to standard insulin therapy, may worsen outcomes, especially in cases of severe TBI (Singer et al. 2019; Vespa et al. 2012). After a TBI, the recommended ideal blood glucose range is 7–9 mmol/L (108–162 mg/dL).

16.2.3 Assessment of Post-TBI Nutritional Needs

There are specific changes in brain metabolism brought on by TBI stress (Carpenter et al. 2014). TBI patients exhibit an increase in metabolic indicators associated with the glycolysis pathway, as well as elevated extracellular lactate levels, resulting in an elevated extracellular lactate/pyruvate ratio (LPR). The latter is frequently used as an indicator of the overall metabolic state of the brain after TBI, where elevated LPR is often linked to mitochondrial dysfunction or anoxia (Timofeev et al. 2011). Furthermore, acute trauma and hypoxia-ischemia (HI) induce fast energy failure, which is followed by a brief return to normal function and ultimately secondary energy failure. According to Perlman (2006) and Timofeev et al. (2011), the latter finally results in brain damage and poor neurodevelopmental outcomes (Perlman 2006; Timofeev et al. 2011). Moreover, the brain becomes vulnerable following brain damage in infants and children due to persistent metabolic dysregulation and acute abnormalities in energy metabolism, which hinder the brain's capacity to sustain several processes necessary for proper growth (Kendall et al. 2014; Perlman 2006).

Neonatal HI is a major public health problem, affecting between 1.8 and 6 per 1000 live births. This condition, known as hypoxic-ischemic encephalopathy (HIE), results from a reduced supply of oxygen and nutrients to the brain, leading to lactic acidosis and reduced CO_2 elimination (Fatemi et al. 2009). Glycolysis and the conversion of pyruvate to lactate take place in the absence of oxygen, thus explaining the elevation of glutamate and lactate levels observed during cerebral hypoxia in patients with TBI (Sarrafzadeh et al. 2003). Similarly, the pentose phosphate pathway is not oxygen-dependent and may be another source of increased lactate after head injury (Jalloh et al. 2015).

Complications from TBI can cause dysfunction and leave patients with an uncertain outlook. Patients with brain injury have a state of hypermetabolism and hypercatabolism, which increases their energy needs and increases their risk of malnutrition in the absence of adequate nutritional support from the earliest stages (Martínez et al. 2023). The brain, essentially, draws its energy from glucose, requiring around 25% of total body glucose. Unfortunately, the central nervous system has limited glycogen stores; thus, neurons must get consistent, sufficient cerebral blood flow in order to meet their energy needs (Martínez et al. 2023).

In the case of TBI, it is crucial to assess the patient's energy and protein requirements, as this can significantly reduce complication and death rates and shorten ICU length of stay, according to studies (Abdelmalik et al. 2017). Top calculating the optimal nutritional requirements of these patients remains a challenge due to various factors such as weight, drugs, and body temperature. Although the indirect calo rimeter method is often considered the most accurate for determining energy requirements, its use may be limited due to variations related to the patient's condition, such as the use of mechanical ventilators, renal therapy, or movements during physiotherapy (Annette and Wenström 2005). Recent research has highlighted a considerable increase in metabolism in patients with TBI, of up to 250%, resulting in a significant loss of nitrogen. It is recommended that these patients consume at least 15% of their total calories as protein, with an amino acid intake of approximately 2 g/kg ideal body weight, which can be beneficial (Martínez et al. 2023).

Beyond only providing calories, protein has other benefits. They have other benefits such as playing a crucial role in recovery from brain damage, boosting immune function, and maintaining body mass (Lee and Oh 2022). Most individuals with significant brain injury have high protein requirements relative to total calorie needs making it difficult to meet their nutritional needs with standard enteral nutrition. Even with increased protein supplementation, nitrogen excretion continues to increase, and this loss may persist for several weeks. Maintaining an adequate protein balance is therefore crucial but can be difficult, especially in the post-acute phase (Kurtz and Rocha 2020). Thus, protein supplementation can be beneficial in patients with TBI, and regular assessments are essential to ensure adequate protein intake. Determining a patient's nutritional requirements, particularly in intensive care, is a major challenge, as both under- and over-nutrition can result in adverse clinical outcomes. Moreover, This procedure is made more difficult by the varied metabolic response observed in acute brain damage (Poblete et al. 2023). TBI patients often require a significant increase, up to 140% or more, in their basal metabolic expenditure due to increased protein catabolism and increased metabolism (Carney et al. 2017). There are practical difficulties in implementing indirect calorimetry in clinical practice despite the fact that it is still the recommended technique for determining the energy needs of critically ill patients (McClave et al. 2016). In patients with TBI, studies have revealed a macronutrient imbalance, with a tendency toward insufficient energy, protein, and fiber intake, and a predominance of fat intake. Furthermore, energy and protein intake appear to be correlated with the severity of the trauma (Kurtz and Rocha 2020).

16.2.4 Specific Nutrients in TBI Management

TBI management involves a multidisciplinary approach, and maintaining recovery and reducing problems are greatly recovery by nutrition.

A complete approach to TBI therapy is necessary although specific nutrients play a crucial role in the recovery process.

The key to general health and healing is a well-balanced diet that consists of a range of nutrient-rich foods.

In some cases, supplementation may be recommended under the guidance of health-care professionals to address specific nutritional needs.

16.2.5 Omega-3 Fatty Acids and Their Neuroprotective Effects

Omega-3 fatty acids have neuroprotective properties and may help lower inflammation (Kiecolt-Glaser et al. 2011). They are regarded as necessary since the body has to get them from food because it is unable to synthesize them on its own (Saini and

Keum 2018). DHA is a major component of cell membranes in the brain; it helps maintain the structural integrity of cell membranes, ensuring proper fluidity and function (Bourre 2006).

Omega-3 fatty acids are involved in the synthesis and function of neurotransmitters (Dighriri et al. 2022). DHA and EPA possess the most encouraging lab data about their ability to restore neurons in TBI (Hasadsri et al. 2013). It has been reported that omega-3 fatty acids could promote neurogenesis (Crupi et al. 2013; Yu et al. 2021). This is particularly relevant for cognitive function and brain repair (Cutuli et al. 2020; Hu et al. 2013). Omega-3 fatty acids possess antioxidant properties, helping neutralize free radicals that can cause cellular damage. This antioxidant activity may contribute to neuroprotection (Wu et al. 2016). EPA and DHA are mainly found in marine sources; fatty fish and certain algae are rich in these forms of omega-3s. Some of the best sources include fatty fish such as salmon, mackerel, herring, anchovies, sardines fish, and oil supplements algal (Saini and Keum 2018).

16.2.6 Essential Micronutrients (Vitamins and Minerals)

Micronutrients contribute significantly to various aspects of TBI management, influencing cellular repair, inflammation, and overall brain health.

Adequate vitamin D levels are important for overall well-being, including bone healing and immune support during the recovery phase after a TBI (Haar et al. 2016). Recent studies provide clear validation for the importance of screening vitamin D levels in cases of neurological deficits. This screening is essential to mitigate the risk of morbidity associated with neurocognitive disorders (Choudhary et al. 2023).

B vitamins (B6 and B12) are crucial for energy metabolism, production of neurotransmitters, mood regulation, and overall neurological health, making them important during TBI recovery (Bianchi et al. 2021; Haar et al. 2016). Vitamin C's antioxidant properties may help reduce oxidative stress associated with TBI, and its role in collagen formation is important for tissue repair. Vitamin E is another important vitamin to consider for TBI. Surprisingly, vitamin E supplementation improved cognition after repeated concussive brain injury in mice.

It also enhanced learning and memory in a fluid percussion model of TBI and decreased oxidative stress. Many authors propose the potential utilization of this vitamin as a dietary supplement for prevention or as an adjunct to the currently available symptomatic treatments for brain injury and neurodegenerative diseases (Castelli et al. 2020; Choudhary et al. 2023).

Numerous brain molecular pathways involving critical metals are altered by TBI, leading to oxidative stress, inflammation, synaptic dysfunction, neurodegeneration, and neural death (Squitti et al. 2023). Minerals are important for the treatment and recuperation of TBI. The best nutrients for enhancing preclinical and clinical TBI outcomes are magnesium and zinc. Superoxide dismutase cofactor zinc has demonstrated potential as a TBI therapeutic adjunct (Lucke-Wold et al. 2018). Additionally, magnesium has been demonstrated to enhance TBI recovery in preclinical animals

(Haar et al. 2016). According to research, modulating minerals such as zinc in the brain may protect neurons from death after events such as stroke and TBI (Park et al. 2023).

16.2.7 Enteral Feeding Strategies Tailored to TBIs

Patients with traumatic brain injury (TBI) often experience malnutrition, which is associated with higher rates of illness and mortality. This condition leads to extended hospital and intensive care unit stays, as well as a greater vulnerability to infections (Meirelles and de Aguilar-Nascimento 2011). Patient outcomes after TBI are influenced by multiple pathophysiological pathways, including hypermetabolism and hypercatabolism (Jeong et al. 2023). So enteral feeding strategies should be conducted to provide adequate nutrition while minimizing the risk of complications and optimizing prognosis.

Early initiation of enteral feeding, typically within 24–48 h post-injury, is recommended to meet increased metabolic demands and support optimal recovery. According to earlier research, immediate nutrition administration may be crucial since it appears to lessen the hypermetabolic reaction to severe injury. Other research showed that delayed nutrition management for TBI patients could increase the risk of undernutrition, comorbidities, and mortality. However, a meta-analysis confirmed that the risk of pneumonia, feed intolerance, or death within 30 days is not known to be influenced by early versus delayed enteral nutrition. Therefore, more studies should evaluate the best time to start nutrition for patients with brain injuries.

TBI patients often have increased energy and protein requirements (Kochanek et al. 2019; Meirelles and de Aguilar-Nascimento 2011). Demand for energy might reach 200% of the average amount needed (Lee and Oh 2022). Thus, energy requirement formulas should be considered to meet these needs while minimizing the risk of overfeeding. Overestimations of caloric needs could have negative consequences, such as increased carbon dioxide generation that would negatively impact the central nervous system and respiratory system (Lee and Oh 2022). In this vision, prediction equations of Harris–Benedict, Ireton-Jones, and Penn State are frequently used to estimate energy consumption (Jeong et al. 2023). Other authors consider that for critically sick individuals, an elementary weight-based formula (25–30 kcal/kg/day) can be used to calculate their energy requirements (Lee and Oh 2022; McClave et al. 2016).

Nonetheless, indirect calorimetry is recognized as the "gold standard" at this time for figuring out how much energy TBI patients need (Jeong et al. 2023; Poblete et al. 2023). Simultaneously, the estimated daily requirements for protein are 1.2–2.0 g/kg of actual body weight (Lee and Oh 2022).

Enteral nutrition (EN) should be administered increasingly to achieve target intake within 5–7 days (Jeong et al. 2023). An excessively high flow may result in gastric discomfort. Therefore, some strategies are used to enhance gastrointestinal tolerance. The first strategy is to use concentrated enteral products ($\geq$1.5 kcal/ml), which minimizes gastric reflux by offering the necessary calories in a reduced volume (Jeong et al. 2023; Lee and Oh 2022). Another strategy relates to the ergonomic posture of patients with TBI. Raising the head position by 30–45 degrees can decrease the risk of reflux and pneumonia (Kattelmann et al. 2006). Lastly, to increase gastrointestinal motility and EN tolerance, promotility or prokinetic drugs like metoclopramide or erythromycin may be taken into consideration (Casaer and Van den Berghe 2014; Lee and Oh 2022; Tan et al. 2011). Additionally, managing medications that delay gastric emptying can also be beneficial (Casaer and Van den Berghe 2014; Jeong et al. 2023).

Finally, to effectively manage EN, the critical care team, the nutrition support team, and the gastroenterology specialists must work closely together and monitor the process (Jeong et al. 2023). Therefore, nutritional status and response to enteral feeding are essential to guide adjustments in feeding regimens and optimize nutritional support by assessing biomarkers (Poblete et al. 2023) or by weekly monitoring of weight gain or loss (Lee and Oh 2022).

16.2.8 Drug–Nutrient Interactions

When preparing to give EN to patients who have had TBI, drug–nutrition interactions must be taken into account. Enteral formulations for pharmaceuticals are being used more often as EN is preferred over parenteral nutrition (PN). When provided enterally, intravenous drugs frequently exhibit adequate bioavailability, facilitating the switch to enteral formulations after bowel function has been recovered and the patient can tolerate nutrition.

On the other hand, when combined with EN solutions, certain antimicrobials might not absorb well. To guarantee the best possible treatment results, it is imperative to evaluate which drugs work well with enteral feeding. Furthermore, corticosteroids, although beneficial for a number of ailments, can cause insulin resistance, which may result in hyperglycemia. Importantly, it is not advised to use corticosteroids to treat TBI. This emphasizes the significance of selecting drugs carefully, taking into account possible interactions and side effects, especially for patients receiving nutritional support.

Among people with TBI, phenytoin is a well-known illustration of a medication-nutrition interaction, commonly used as an anticonvulsant for preventing and treating posttraumatic epilepsy. Among the formulations utilized in the TBI context are phenytoin sodium salt injection and phenytoin acid suspension. In normal conditions, phenytoin is generally well absorbed, exceeding 80% (Doak et al. 1998). However, when the acid suspension is concurrently administered with enteral nutrition (EN) solutions, absorption may be hindered (Doak et al. 1998). Numerous theories have been put out to explain this interaction, such as medication binding to the nasogastric (NG) tube, cation chelation in the enteral product, or a drop in pH in the NG tube and gastrointestinal tract. To prevent simultaneous administration, some health-care professionals advise holding EN for one to two hours prior to and following each phenytoin dose. Others recommend giving the injectable solution enterally while simultaneously giving EN or varying the amount of phenytoin suspension.

Many studies have attempted to address this controversy, but none have been able to replicate the critical care unit in its entirety, and the results have been inconsistent, leaving the issue unresolved. Using the sodium salt intravenously or retaining EN for an hour prior to and following each dosage of the acid mixture are practical and efficient ways to avoid this interaction (ensuring proper shaking for even particle distribution).

Certain drugs, such as carbamazepine and sucralfate, that are given enterally to patients with TBI may not work well with feeding formulae. The carbamazepine suspension may precipitate as an orange, rubbery substance when mixed with water or other drug diluents. When possible, avoid using carbamazepine and other drugs together. Sucralfate coats gastric ulcers albeit it is rarely used nowadays to prevent mucosal disease associated with stress. Enteral formula's acidic proteins may react with sucralfate, perhaps obstructing the feeding tube. Since the insertion of feeding access carries inherent hazards, losing feeding tube patency can be a major problem while caring for a patient with TBI.

Even though they might not obstruct the tube, some enteral formula–medication incompatibilities can nonetheless impede absorption. Concurrent enteral eating, for example, reduces ciprofloxacin's bioavailability by 44%, which is probably caused by chelation with polyvalent cations (Mimoz et al. 1998). A superinfection with more resistant bacteria or the inability to treat an infection could result from this drop in ciprofloxacin concentrations. Furthermore, certain medications may alter the amounts of electrolytes and minerals by altering their absorption, metabolism, or excretion or by supplementing. Hypernutrition or electrolyte imbalances may be caused by prolonged coadministration.

16.2.9 Challenges in Providing Nutrition to Neurologically Ill Patients

Several challenges are faced in providing nutrition to patients who are neurologically injured. First, because of hemodynamic instability following initial blood loss, uncertainty regarding any intrabdominal injury, and the necessity of anesthetic and surgical intervention, clinicians find it challenging to start early feeding in trauma patients (Ganesh and Ibrahim 2023). Second, brain-injured patients would need a continuous infusion of narcotics and sedatives during cerebral resuscitation to reduce the cerebral metabolic rate of oxygen consumption, alleviate pain, and prevent an increase in intracranial pressure (ICP) (Haddad and Arabi 2012). In addition, patients with TBI who also have seizures should get moderate-to-high levels of sedation (Ganesh and Ibrahim 2023).

Awake TBI patients have further challenges that, especially after being released to the ward, could make it difficult for them to receive enough nutrition, including frequent feeding tube removals, food rejection, and swallowing problems (Chapple et al. 2016). Additionally, it has been reported that fasting during surgical or diagnostic operations regularly and for prolonged periods interrupts nutritional intake (Morgan et al. 2004). Since TBI patients often stay in hospitals for extended periods, persistent nutritional deficiencies during hospital stays may have an effect on the ability of patients' functional recovery (Chapple et al. 2016).

On the other hand, delayed stomach emptying is one of the most common complications reported in patients with head injuries. Patients with moderate-to-severe TBI had a more than twofold increase in the stomach emptying half-life. This condition may last for two weeks after an injury (Lee and Oh 2022). It is believed that this condition is connected to elevated ICP, changed levels of endogenous hormones and chemical mediators in circulation, and dysregulation of the autonomic nervous system (Ganesh and Ibrahim 2023).

It has been reported that nutrition delivered to patients with TBI is typically below the prescribed amounts (Chapple et al. 2018). Indeed, in one study, patients in general hospitals revealed a link between decreased oral intake and higher mortality (Hiesmayr et al. 2009). In the week after being released from invasive mechanical ventilation, calorie and protein intakes did not surpass 55% of predicted requirements, according to a prior study that assessed oral intake in patients recovering from critical illness (Peterson et al. 2010). Otherwise, hypermetabolism resulting from TBI can persist for weeks after recovery from the injury, complicating both the initial hospitalization and stability periods (Borzotta et al. 1994; Deutschman et al. 1986; Pepe and Barba 1999). Sustained immobility from injury or a chronic inflammatory response are two possible reasons for this ongoing increase in metabolism and loss of protein (Cook et al. 2008). Finally, the absence of accurate and up-to-date data on the treatment of TBI based on nutrition is a major challenge for health-care providers in managing patients with severe TBI (Ganesh and Ibrahim 2023).

16.3 Practical Recommendations

TBI is a significant contributing factor to long-term impairment and one of the primary causes of death globally. When it comes to managing TBI, nutritional considerations play a crucial role in supporting the recovery and overall well-being of patients (Lee and Oh 2022). Health professionals involved in the care of TBI patients play a vital role in ensuring that nutritional needs are met to promote optimal outcomes.

Practical recommendations in nutritional management for TBI can help guide health-care providers in making informed decisions regarding diet and supplementation (Schwalfenberg and Genuis 2017). Nutritional guidelines for individuals with TBI are an expanding area of study, with new research continually emerging. Although dietitians are essential for assessing and meeting the nutritional requirements of patients, they are just one part of a multidisciplinary team. The perspectives and attitudes of medical and nursing staff are also likely to be vital in implementing nutrition plans for patients with TBI (Chapple et al. 2018).

16.4 Guidelines for Health-Care Professionals

Generally, after an injury occurs, a series of inflammatory, immune, and metabolic responses are triggered, leading to a state of increased metabolism. Adequate macro- and micronutrient support is needed to sustain this heightened metabolic state and ease the healing process (Smith-Ryan et al. 2020).

TBIs, including repeated sub-concussive injuries, lead to a series of neurological disturbances. Following a TBI, there is damage to axons, neuroinflammation, and imbalances in ion levels (such as Na^+, K^+, and Ca^{2+}). The influx of calcium (Ca^{2+}) results in mitochondrial dysfunction and oxidative stress, which contribute to cellular damage (Barrett et al. 2014). Nutritional interventions show promise in alleviating TBI symptoms, and this field is continuously growing, with new treatments being regularly reported. When considering nutrition for TBI patients, it is advised to take into account conditions and physiological changes linked to TBI, such as dysphagia, and alterations in smell and taste. Additionally, individuals may require assistance in monitoring their food intake (e.g., help with calorie counting) (Driver et al. 2017).

Predicting an individual's dietary requirements during the acute phase of TBI can be difficult due to a variety of factors that might alter REE, such as body temperature, usage of sedatives, mechanical ventilation, and the severity of the injury. Due to this challenge, these patients may not receive enough nutrients. Additionally, edema related to resuscitation in severe TBI cases further complicates the accurate prediction of energy requirements using body weight. In the post-acute phase, it is essential to reassess calorie needs. Caloric demands decrease when mobility is limited as medical conditions stabilize. However, during rehabilitation therapy, calorie needs increase significantly, sometimes by 30–60% compared to control groups (Lee and Oh 2022; McClave et al. 2016; Zasler et al. 2012).

However, due to practical issues like high expenses and the requirement for a qualified professional, this method can be problematic (Kurtz and Rocha 2020; McClave et al. 2016).

The guidelines of the Society of Critical Care Medicine (SCCM) and the American Society for Parenteral and Enteral Nutrition (ASPEN) recommends utilizing a published predictive equation or a simple weight-based calculation (25–30 kcal/kg/d) to assess the energy requirements of critically ill patients in situations when indirect calorimetry is not available (McClave et al. 2016; Lee et al. 2021).

Before the 1980s, the standard method of nutrition therapy involved starting patients on nutrition only once their gastrointestinal system had stabilized or they were able to consume enough food (Twyman 1997). In cases of severe brain injury with an acute catabolic state and other typical pathophysiological characteristics, nutrition therapy must be started as soon as possible in order to support the maintenance of brain metabolic equilibrium, important organ function, and skeletal muscle mass. For the first two weeks following an injury, it is recommended to begin nutrition therapy as soon as possible, ideally within the first 24 h, and to provide more than 50% of REE with 1.0–1.5 g of protein per kilogram of body weight (Bistrian et al. 2011; Härtl et al. 2008; Kurtz and Rocha 2020). It has been recommended to feed patients with severe TBI to attain basal caloric replacement by the fifth day after the injury and, ideally, by the fifth day (level IIA recommendation) in order to lower the risk of death (Carney et al. 2017).

In terms of route of access, EN and PN differ in their effects on energy and protein supply, as well as serum glucose levels during the acute phase response. Both EN and PN can provide increasing calories to brain-injured patients. PN provides a significantly greater amount of nitrogen but also results in greater losses (Lee and Oh 2022). However, nitrogen balance is similar in both types of therapy. Compared to EN, PN leads to greater hyperglycemia. Importantly, the choice of nutrition route does not appear to influence the early inflammatory response or clinical outcomes in these patients (Aula et al. 2011). The Brain Trauma Foundation's latest guidelines recommended early transgastric jejunal feeding to decrease the risk of ventilator-associated pneumonia in severe TBI management (Carney et al. 2017). It has also been validated that early EN, administered within 48 h after injury, is linked to improved survival rates, recovery of Glasgow Coma Scale (GCS) scores, and functional outcomes among patients with TBI, particularly those with a GCS score of 6–8 (Chiang et al. 2012).

Supplementing with 20–30 g of glutamine (Gln) per day is expected to normalize plasma glutamine concentration (Tjäder et al. 2004). The recommended dosage of glutamine is typically between 0.3 and 0.5 g per kilogram of body weight per day (equivalent to 25–50 g per day) for a duration of 1–2 weeks, whether administered enterally (EN) or parenterally (PN) (Heyland and Novak 2001; Quintard and Ichai 2019; Scrimgeour and Condlin 2014). Considering the clinical data suggesting that the route of administration, dose, timing, and severity of TBI may affect outcomes, it is advisable to determine plasma glutamine levels before administering Gln to patients with mild-to-moderate TBI (Scrimgeour and Condlin 2014).

Numerous researches have indicated that taking omega-3 fatty acid supplements lowers inflammation, cell death, and axonal damage in rodent models. While human studies on omega-3 supplementation after TBIs are scarce, it has been reported that adults responded well to a suggested dose of 40 mg/kg/d (4 g for an individual weighing 100 kg) as soon as possible following injury (Barrett et al. 2014; Smith-Ryan et al. 2020). Increasing the consumption of foods high in EPA and DHA, such as fish oil and cold-water fish species, may improve the cognitive and long-term potentiation capacities of neurons (Finnegan et al. 2022; Morley and Banks 2010; Wu et al. 2011).

For glycemic control in patients with TBI, current recommendations advise avoiding excessive hyperglycemia (above 10–11 mmol/l) and maintaining a moderate "permissive" glucose control range of 8–11 mmol/l (Badjatia et al. 2014; Hermanides et al. 2018; Quintard and Ichai 2019).

Protein is vital for brain recovery, immune function, and body mass maintenance in brain-injured patients. Their high protein needs are hard to meet with standard nutrition, and nitrogen loss can persist for weeks. Protein supplementation and regular intake assessments are crucial (Kurtz and Rocha 2020).

According to the ASPEN-SCCM recommendation, if the patient is hemodynamically stable, enteral nutrition (EN) with a high-protein polymeric diet should be started as soon as possible after trauma (McClave et al. 2016). Trauma patients are thought to require 1.2–2.0 g of protein per kilogram of actual body weight per day; individuals who have multiple injuries may require more. Experts recommend that protein make up 15%–20% of total calories for TBI patients, with a daily minimum requirement of 2 g per kilogram of body weight (McClave et al. 2016; Quintard and Ichai 2019).

Moreover, giving 100 mg/kg of curcumin, which is 95% curcuminoids, decreased neuronal apoptosis, lowered oxidized protein levels, and improved cognitive function after concussions (Zhu et al. 2014).

Monitoring zinc status and maintaining it within the recommended range is particularly important after a brain injury to support neurotherapeutic and neuroprotective roles; however, it is crucial not to exceed the tolerable upper intake level (UL) of 40 mg/day to avoid potential adverse effects; for the first 15 days after the injury, a dose of 12 mg of zinc sulfate per day was administered intravenously. After day 15, the dose was switched to an oral dose of 22 mg per day (Clausen et al. 2012; Wessels et al. 2013; Young et al. 2009).

Omega-3 fatty acid and vitamin D deficiencies may compound to impair outcomes for TBI patients. Omega-3 fatty acids and vitamin D should be regularly given to at-risk patients, particularly those with TBI. In general, healthy adults should take 600 IU (15 μg) of vitamin D per day; all adults should be able to tolerate an upper limit of 4000 IU/day (100 μg). When exposure to sunshine is inadequate, individuals should consume ≥800–1000 IU of vitamin D/day from dietary and supplementary sources in order to achieve a blood 25(OH)D level of ≥50 nmol/L (Finnegan et al. 2022; Institute of Medicine 2011; Nagpal et al. 2005; Scrimgeour and Condlin 2014).

The existing research on nutritional therapy following TBI primarily addresses severe, acute cases, limiting its applicability to other subgroups such as individuals with mild or chronic TBI. Additionally, particularly in the elderly and pediatric

Table 16.1 Nutritional recommendations for traumatic brain injury (TBI) patients

Aspect	Recommendation	References
Measuring calories needs	Use indirect calorimetry whenever possible to measure the amount of energy that TBI patients need. If not available, consider using a published predictive equation or a basic weight-based equation (25–30 kcal/kg/d) to measure energy requirements. Reassess calorie needs in the post-acute phase, adjusting for changes in mobility and rehabilitation therapy. After an injury, try to reach basal calorie replacement by the fifth day and ideally by the seventh.	McClave et al. (2016), Zasler et al. (2012), Carney et al. (2017)

(continued)

Table 16.1 (continued)

Aspect	Recommendation	References
Timing of nutrition therapy	Initiate nutrition therapy as soon as possible, ideally within the first 24 h of the injury, and supply 1.0–1.5 g of protein per kilogram of body weight or more than 50% of REE, throughout the first two weeks after the injury.	Bistrian et al. (2011), Härtl et al. (2008), Carney et al. (2017)
Route of access (EN vs. PN)	Consider EN as the preferred route of access, as it provides a similar nitrogen balance to PN but with fewer losses. In the treatment of severe TBI, early transgastric jejunal feeding is advised to reduce the incidence of ventilator-associated pneumonia.	Aula et al. (2011), Carney et al. (2017)
Position and method	Patients should be positioned with their heads raised 45 degrees from the bed to reduce reflux and avoid aspiration pneumonia. To meet calorie requirements and minimize the risk of reflux or intolerance, use concentrated enteral formulas (≥1.5 kcal/mL). Administer EN as a continuous infusion for better nitrogen balance and decreased hypercatabolic response.	Mazaherpur et al. (2016)
Omega-3 fatty acids	To enhance various neuronal functions, such as long-term potentiation and cognition, consider increasing the intake of dietary sources rich in EPA and DHA, such as cold-water fish species and fish oil. An adult's response to an injury was found to benefit from a recommended dose of 40 mg/kg/d (4 g for an individual weighing 100 kg).	Barrett et al. (2014), Finnegan et al. (2022), Morley and Banks (2010), Wu et al. (2011), Smith-Ryan et al. (2020)
L-arginine	Supplement with L-arginine to enhance immune function and protein synthesis, as it serves as a precursor to several biologically active compounds.	Martí i Líndez and Reith (2021)
Zinc	Monitor zinc status and maintain it within the recommended range after a brain injury to support neurotherapeutic and neuroprotective roles. Do not exceed the UL of 40 mg/day to avoid potential adverse effects.	Clausen et al. (2012), Wessels et al. (2013), Young et al. (2009)
Glycemic control	Avoid excessive hyperglycemia (above 10–11 mmol/l) and maintain a moderate "permissive" glucose control range of 8–11 mmol/l.	Badjatia et al. (2014); Quintard and Ichai (2019), Hermanides et al. (2018)
Protein	Provide a high-protein polymeric diet within 24–48 h of trauma, aiming for 1.2–2.0 g per kilogram of actual body weight per day. Protein should comprise 15%–20% of total calories, requiring a minimum of 2 g per kilogram of body weight per day.	McClave et al. (2016), Quintard and Ichai (2019)
Curcumin	Consider administering curcumin at a dosage of 100 mg/kg, which contains 95% curcuminoids, to reduce neuronal apoptosis, decrease levels of oxidized proteins, and enhance cognition following concussive brain injuries.	Zhu et al. (2014)
Vitamin D	To achieve a serum 25-OHD level of ≥50 nmol/L, adults should take ≥800–1000 IU of vitamin D/day from dietary and supplemental sources (Finnegan et al. 2022; Institute of Medicine 2011; Nagpal et al. 2005; Scrimgeour and Condlin 2014).	Finnegan et al. (2022), Institute of Medicine (2011); Nagpal et al. (2005), Scrimgeour and Condlin (2014)

populations, the effects of nutrition on post-TBI functioning and practical methods to lower mortality and morbidity are still unknown.

Therefore, addressing the complex nature of TBI requires a multifaceted approach (Table 16.1).

16.5 Conclusion

The neuroprotective effects of magnesium have been demonstrated in cases of severe traumatic brain injuries (TBIs). This underscores the need for additional research, clinical validation, and the potential of magnesium to enhance outcomes in TBI patients. Enhancing the brain's bioavailability of magnesium opens the possibility of administering a low and safe dosage, showing promise in improving clinical results for TBI patients. Combining magnesium with other pharmacological agents could present a secure and clinically effective neuroprotective regime for TBI treatment.

References

Abdelmalik PA, Dempsey S, Ziai W (2017) Nutritional and bioenergetic considerations in critically ill patients with acute neurological injury. Neurocrit Care 27:276–286

Annette H, Wenström Y (2005) Implementing clinical guidelines for nutrition in a neurosurgical intensive care unit. Nurs Health Sci 7(4):266–272

Aula G, España M, Meirelles J, De Aguilar-Nascimento M, Madrid M, Meirelles J, De Aguilar-Nascimento JE (2011) Enteral or parenteral nutrition in traumatic brain injury: a prospective randomised trial Enteral or parenteral nutrition in traumatic brain injury: aprospective randomised trial. Nutr Hosp 26(5):1120–1124

Badjatia N, Vespa P, Le Roux P, Menon DK, Citerio G, Bader MK, Brophy GM, Diringer MN, Stocchetti N, Videtta W, Armonda R, Boesel J, Chesnut R, Chou S, Claassen J, Czosnyka M, De Georgia M, Figaji A, Fugate J et al (2014) Monitoring Nutrition and Glucose in Acute Brain Injury. Neurocrit Care 21(2):159–167. https://doi.org/10.1007/S12028-014-0036-2/METRICS

Barrett EC, McBurney MI, Ciappio ED (2014) ω-3 Fatty acid supplementation as a potential therapeutic aid for the recovery from mild traumatic brain injury/concussion. Adv Nutr 5(3):268–277. https://doi.org/10.3945/AN.113.005280

Bianchi VE, Herrera PF, Laura R (2021) Effect of nutrition on neurodegenerative diseases. A systematic review. Nutr Neurosci 24(10):810–834

Bistrian BR, Askew W, Erdman JW Jr, Oria MP (2011) Nutrition and traumatic brain injury: a perspective from the institute of medicine report. J Parenter Enter Nutr 35(5):556–559

Bjørklund G, Chirumbolo S (2017) Role of oxidative stress and antioxidants in daily nutrition and human health. Nutrition 33:311–321

Borzotta AP, Pennings J, Papasadero B, Paxton J, Mardesic S, Borzotta R, Parrott A, Bledsoe F (1994) Enteral versus parenteral nutrition after severe closed head injury. J Trauma Acute Care Surg 37(3):459–468

Bourre J-M (2006) Effects of nutrients (in food) on the structure and function of the nervous system: update on dietary requirements for brain. Part 1: micronutrients. J Nutr Health Aging 10(5):377

Carney N, Totten AM, O'Reilly C, Ullman JS, Hawryluk GWJ, Bell MJ, Bratton SL, Chesnut R, Harris OA, Kissoon N, Rubiano AM, Shutter L, Tasker RC, Vavilala MS, Wilberger J, Wright DW, Ghajar J (2017) Guidelines for the management of severe traumatic brain injury, fourth edn. Neurosurgery 80(1):6–15. https://doi.org/10.1227/NEU.0000000000001432

Carpenter KLH, Jalloh I, Gallagher CN, Grice P, Howe DJ, Mason A, Timofeev I, Helmy A, Murphy MP, Menon DK (2014) 13C-labelled microdialysis studies of cerebral metabolism in TBI patients. Eur J Pharm Sci 57:87–97

Casaer MP, Van den Berghe G (2014) Nutrition in the acute phase of critical illness. N Engl J Med 370(13):1227–1236. https://doi.org/10.1056/NEJMra1304623

Castelli V, Melani F, Ferri C, d'Angelo M, Catanesi M, Grassi D, Benedetti E, Giordano A, Cimini A, Desideri G (2020) Neuroprotective activities of bacopa, lycopene, astaxanthin, and vitamin B12 combination on oxidative stress-dependent neuronal death. J Cell Biochem 121(12):4862–4869

Chapple L-AS, Deane AM, Heyland DK, Lange K, Kranz AJ, Williams LT, Chapman MJ (2016) Energy and protein deficits throughout hospitalization in patients admitted with a traumatic brain injury. Clin Nutr 35(6):1315–1322. https://doi.org/10.1016/j.clnu.2016.02.009

Chapple LA, Chapman M, Shalit N, Udy A, Deane A, Williams L (2018) Barriers to nutrition intervention for patients with a traumatic brain injury: views and attitudes of medical and nursing practitioners in the acute care setting. J Parenter Enter Nutr 42(2):318–326. https://doi.org/10.1177/0148607116687498

Chiang YH, Chao DP, Chu SF, Lin HW, Huang SY, Yeh YS, Lui TN, Binns CW, Chiu WT (2012) Early enteral nutrition and clinical outcomes of severe traumatic brain injury patients in acute stage: a multi-center cohort study. Https://HomeLiebertpubCom/Neu 29(1):75–80. https://doi.org/10.1089/NEU.2011.1801

Choudhary A, Kumar A, Sharma R, Khurana L, Jain S, Sharma S, Kumar D, Sharma S (2023) Optimal vitamin D level ameliorates neurological outcome and quality of life after traumatic brain injury: a clinical perspective. Int J Neurosci 133(4):417–425

Clausen F, Marklund N, Lewén A, Enblad P, Basu S, Hillered L (2012) Interstitial F2-Isoprostane 8-Iso-PGF2α as a biomarker of oxidative stress after severe human traumatic brain injury. Https://HomeLiebertpubCom/Neu 29(5):766–775. https://doi.org/10.1089/NEU.2011.1754

Cook AM, Peppard A, Magnuson B (2008) Nutrition considerations in traumatic brain injury. Nutr Clin Pract 23(6):608–620. https://doi.org/10.1177/0884533608326060

Crupi R, Marino A, Cuzzocrea S (2013) n-3 fatty acids: role in neurogenesis and neuroplasticity. Curr Med Chem 20(24):2953–2963

Cutuli D, Landolfo E, Nobili A, De Bartolo P, Sacchetti S, Chirico D, Marini F, Pieroni L, Ronci M, D'Amelio M (2020) Behavioral, neuromorphological, and neurobiochemical effects induced by omega-3 fatty acids following basal forebrain cholinergic depletion in aged mice. Alzheimers Res Ther 12(1):1–21

Deutschman CS, Konstantinides FN, Raup S, Thienprasit P, Cerra FB (1986) Physiological and metabolic response to isolated closed-head injury: Part 1: Basal metabolic state: Correlations of metabolic and physiological parameters with fasting and stressed controls. J Neurosurg 64(1):89–98

Dighriri IM, Alsubaie AM, Hakami FM, Hamithi DM, Alshekh MM, Khobrani FA, Dalak FE, Hakami AA, Alsueaadi EH, Alsaawi LS (2022) Effects of omega-3 polyunsaturated fatty acids on brain functions: a systematic review. Cureus 14(10)

Doak KK, Haas CE, Dunnigan KJ, Reiss RA, Reiser JR, Huntress J, Altavela JL (1998) Bioavailability of phenytoin acid and phenytoin sodium with enteral feedings. Pharmacotherapy 18(3):637–645

Driver S, Reynolds M, Kramer K (2017) Modifying an evidence-based lifestyle programme for individuals with traumatic brain injury. Brain Inj 31(12):1612–1616. https://doi.org/10.1080/02699052.2017.1346286

Fatemi A, Wilson MA, Johnston MV (2009) Hypoxic-ischemic encephalopathy in the term infant. Clin Perinatol 36(4):835–858

Faul M, Wald MM, Rutland-Brown W, Sullivent EE, Sattin RW (2007) Using a cost-benefit analysis to estimate outcomes of a clinical treatment guideline: testing the Brain Trauma Foundation guidelines for the treatment of severe traumatic brain injury. J Trauma Acute Care Surg 63(6):1271–1278

Finnegan E, Daly E, Pearce AJ, Ryan L (2022) Nutritional interventions to support acute mTBI recovery. Front Nutr 9:977728. https://doi.org/10.3389/FNUT.2022.977728/BIBTEX

Fujikawa DG (2015) The role of excitotoxic programmed necrosis in acute brain injury. Comput Struct Biotechnol J 13:212–221

Galgano M, Toshkezi G, Qiu X, Russell T, Chin L, Zhao L-R (2017) Traumatic brain injury: current treatment strategies and future endeavors. Cell Transplant 26(7):1118–1130

Ganesh S, Ibrahim NA (2023) Nutrition therapy in critically ill patients with traumatic brain injury. Human Nutr Metab 34(May):200219. https://doi.org/10.1016/j.hnm.2023.200219

Haar CV, Peterson TC, Martens KM, Hoane MR (2016) Vitamins and nutrients as primary treatments in experimental brain injury: Clinical implications for nutraceutical therapies. Brain Res 1640:114–129

Haddad SH, Arabi YM (2012) Critical care management of severe traumatic brain injury in adults. Scand J Trauma Resusc Emerg Med 20(1):1–15

Härtl R, Gerber LM, Ni Q, Ghajar J (2008) Effect of early nutrition on deaths due to severe traumatic brain injury. J Neurosurg 109(1):50–56. https://doi.org/10.3171/JNS/2008/109/7/0050

Hasadsri L, Wang BH, Lee JV, Erdman JW, Llano DA, Barbey AK, Wszalek T, Sharrock MF, Wang H (2013) Omega-3 fatty acids as a putative treatment for traumatic brain injury. J Neurotrauma 30(11):897–906

Hatton J (2001) Pharmacological treatment of traumatic brain injury: a review of agents in development. CNS Drugs 15:553–581

Hermanides J, Plummer MP, Finnis M, Deane AM, Coles JP, Menon DK (2018) Glycaemic control targets after traumatic brain injury: A systematic review and meta-analysis. Crit Care 22(1):1–11. https://doi.org/10.1186/S13054-017-1883-Y/FIGURES/6

Heyland DK, Novak F (2001) Immunonutrition in the critically ill patient: more harm than good? J Parenter Enter Nutr 25(2 Suppl). https://doi.org/10.1177/014860710102500211

Hiesmayr M, Schindler K, Pernicka E, Schuh C, Schoeniger-Hekele A, Bauer P, Laviano A, Lovell AD, Mouhieddine M, Schuetz T (2009) Decreased food intake is a risk factor for mortality in hospitalised patients: the NutritionDay survey 2006. Clin Nutr 28(5):484–491

Hu, X., Zhang, F., Leak, R.K, Zhang, W., Iwai, M., A Stetler, RA, Dai, Y., Zhao, A., Gao, Y., & Chen, J. (2013). Transgenic overproduction of omega-3 polyunsaturated fatty acids provides neuroprotection and enhances endogenous neurogenesis after stroke. Curr Mol Med, 13(9), 1465–1473.

Institute of Medicine, Food and Nutrition Board (2011) Dietary Reference Intakes for Calcium and Vitamin D - Institute of Medicine, Food and Nutrition Board, Committee to Review Dietary Reference Intakes for Vitamin D and Calcium - Google Books

Jalloh I, Carpenter KLH, Grice P, Howe DJ, Mason A, Gallagher CN, Helmy A, Murphy MP, Menon DK, Carpenter TA (2015) Glycolysis and the pentose phosphate pathway after human traumatic brain injury: microdialysis studies using 1, 2-13C2 glucose. J Cereb Blood Flow Metab 35(1):111–120

Jeong H, Kim JH, Choo YH, Kim M, Lee S, Ha EJ, Oh J (2023) Nutrition therapy for patients with traumatic brain injury: a narrative review. Korean J Neurotrauma 19(2):177–184. https://doi.org/10.13004/kjnt.2023.19.e31

Jung HJ, Lee YM, Kim M, Uhm KE, Lee J (2020) Suggested assessments for sarcopenia in patients with stroke who can walk independently. Ann Rehabil Med 44(1):20–37

Kattelmann KK, Hise M, Russell M, Charney P, Stokes M, Compher C (2006) Preliminary evidence for a medical nutrition therapy protocol: enteral feedings for critically ill patients. J Am Diet Assoc 106(8):1226–1241. https://doi.org/10.1016/j.jada.2006.05.320

Kendall GS, Melbourne A, Johnson S, Price D, Bainbridge A, Gunny R, Huertas-Ceballos A, Cady EB, Ourselin S, Marlow N (2014) White matter NAA/Cho and Cho/Cr ratios at MR spectroscopy are predictive of motor outcome in preterm infants. Radiology 271(1):230–238

Kiecolt-Glaser JK, Belury MA, Andridge R, Malarkey WB, Glaser R (2011) Omega-3 supplementation lowers inflammation and anxiety in medical students: a randomized controlled trial. Brain Behav Immun 25(8):1725–1734

Kochanek PM, Tasker RC, Carney N, Totten AM, Adelson PD, Selden NR, Davis-O'Reilly C, Hart EL, Bell MJ, Bratton SL, Grant GA, Kissoon N, Reuter-Rice KE, Vavilala MS, Wainwright MS (2019) Guidelines for the management of pediatric severe traumatic brain injury, third edition: update of the brain trauma foundation guidelines. Pediatr Crit Care Med 20(3S Suppl 1):S1–S82. https://doi.org/10.1097/PCC.0000000000001735

Krakau K, Omne-Ponten M, Karlsson T, Borg J (2006) Metabolism and nutrition in patients with moderate and severe traumatic brain injury: a systematic review. Brain Inj 20(4):345–367

Kurtz P, Rocha EEM (2020) Nutrition therapy, glucose control, and brain metabolism in traumatic brain injury: a multimodal monitoring approach. In: Frontiers in neuroscience, vol 14. Frontiers Media S.A, p 452910. https://doi.org/10.3389/fnins.2020.00190

Lee HY, Oh B-M (2022) Nutrition management in patients with traumatic brain injury: a narrative review. Brain Neurorehabil 15(1)

Lee SJ, Lee HJ, Jung YJ, Han M, Lee SG, Hong SK (2021) Comparison of measured energy expenditure using indirect calorimetry vs predictive equations for liver transplant recipients. J Parenter Enter Nutr 45(4):761–767. https://doi.org/10.1002/JPEN.1932

Lucke-Wold BP, Logsdon AF, Nguyen L, Eltanahay A, Turner RC, Bonasso P, Knotts C, Moeck A, Maroon JC, Bailes JE (2018) Supplements, nutrition, and alternative therapies for the treatment of traumatic brain injury. Nutr Neurosci 21(2):79–91

Maas AIR, Menon DK, Adelson PD, Andelic N, Bell MJ, Belli A, Bragge P, Brazinova A, Büki A, Chesnut RM (2017) Traumatic brain injury: integrated approaches to improve prevention, clinical care, and research. Lancet Neurol 16(12):987–1048

Martí i Líndez, A. A., & Reith, W. (2021) Arginine-dependent immune responses. Cell Mol Life Sci 78(13):5303–5324

Martínez ACC, Sierra LIAO, Galeano AFR, Cundar DAB, Sierra MGO (2023) Key considerations for nutritional management in traumatic brain injury: a narrative review. Roman Neurosurg:292–297

Mazaherpur S, Khatony A, Abdi A, Pasdar Y, Najafi F (2016) The effect of continuous enteral nutrition on nutrition indices, compared to the intermittent and combination enteral nutrition in traumatic brain injury patients. J Clin Diagn Res 10(10):JC01

McAllister TW (2008) Neurobehavioral sequelae of traumatic brain injury: evaluation and management. World Psychiatry 7(1):3

McClave SA, Taylor BE, Martindale RG, Warren MM, Johnson DR, Braunschweig C, McCarthy MS, Davanos E, Rice TW, Cresci GA, Gervasio JM, Sacks GS, Roberts PR, Compher C (2016) Guidelines for the provision and assessment of nutrition support therapy in the adult critically ill patient: society of critical care medicine (SCCM) and American Society for Parenteral and Enteral Nutrition (A.S.P.E.N.). JPEN J Parenter Enteral Nutr 40(2):159–211. https://doi.org/10.1177/0148607115621863

Meirelles CMJ, de Aguilar-Nascimento JE (2011) Enteral or parenteral nutrition in traumatic brain injury: a prospective randomised trial. Nutr Hosp 26(5):1120–1124

Mimoz O, Binter V, Jacolot A, Edouard A, Tod M, Petitjean O, Samii K (1998) Pharmacokinetics and absolute bioavailability of ciprofloxacin administered through a nasogastric tube with continuous enteral feeding to critically ill patients. Intensive Care Med 24:1047–1051

Morgan LM, Dickerson RN, Alexander KH, Brown RO, Minard G (2004) Factors causing interrupted delivery of enteral nutrition in trauma intensive care unit patients. Nutr Clin Pract 19(5):511–517

Morley JE, Banks WA (2010) Lipids and cognition. J Alzheimers Dis 20(3):737–747. https://doi.org/10.3233/JAD-2010-091576

Nagpal S, Na S, Rathnachalam R (2005) Noncalcemic actions of vitamin D receptor ligands. Endocr Rev 26(5):662–687. https://doi.org/10.1210/ER.2004-0002

Oddo M, Schmidt JM, Carrera E, Badjatia N, Connolly ES, Presciutti M, Ostapkovich ND, Levine JM, Le Roux P, Mayer SA (2008) Impact of tight glycemic control on cerebral glucose metabolism after severe brain injury: a microdialysis study. Crit Care Med 36(12):3233–3238

Park MK, Choi BY, Kho AR, Lee SH, Hong DK, Kang BS et al (2023) The protective role of glutathione on zinc-induced neuron death after brain injuries. Int J Mol Sci 24(3):2950

Pepe JL, Barba CA (1999) The metabolic response to acute traumatic brain injury and implications for nutritional support. J Head Trauma Rehabil 14(5):462–474

Perlman JM (2006) Intervention strategies for neonatal hypoxic-ischemic cerebral injury. Clin Ther 28(9):1353–1365

Peterson SJ, Tsai AA, Scala CM, Sowa DC, Sheean PM, Braunschweig CL (2010) Adequacy of oral intake in critically ill patients 1 week after extubation. J Am Diet Assoc 110(3):427–433

Pillsbury L, Oria M, Erdman J (2011) Nutrition and traumatic brain injury: improving acute and subacute health outcomes in military personnel. National Academies Press, Washington DC

Poblete RA, Yaceczko S, Aliakbar R, Saini P, Hazany S, Breit H, Louie SG, Lyden PD, Partikian A (2023) Optimization of nutrition after brain injury: mechanistic and therapeutic considerations. Biomedicine 11(9):2551

Quintard H, Ichai C (2019) Nutritional and metabolic supplementation for the injured brain: an update. Curr Opin Crit Care 25(2):126–131. https://doi.org/10.1097/MCC.0000000000000588

Saini RK, Keum Y-S (2018) Omega-3 and omega-6 polyunsaturated fatty acids: dietary sources, metabolism, and significance—a review. Life Sci 203:255–267

Sarrafzadeh AS, Kiening KL, Callsen TA, Unterberg AW (2003) Metabolic changes during impending and manifest cerebral hypoxia in traumatic brain injury. Br J Neurosurg 17(4):340–346

Schwalfenberg GK, Genuis SJ (2017) The importance of magnesium in clinical healthcare. Scientifica 2017. https://doi.org/10.1155/2017/4179326

Scrimgeour AG, Condlin ML (2014) Nutritional treatment for traumatic brain injury. Https://HomeLiebertpubCom/Neu 31(11):989–999. https://doi.org/10.1089/NEU.2013.3234

Shi J, Dong B, Mao Y, Guan W, Cao J, Zhu R, Wang S (2016) Traumatic brain injury and hyperglycemia, a potentially modifiable risk factor. Oncotarget 7(43):71052

Singer P, Blaser AR, Berger MM, Alhazzani W, Calder PC, Casaer MP, Hiesmayr M, Mayer K, Montejo JC, Pichard C (2019) ESPEN guideline on clinical nutrition in the intensive care unit. Clin Nutr 38(1):48–79

Smith-Ryan AE, Hirsch KR, Saylor HE, Gould LM, Blue MNM (2020) Nutritional considerations and strategies to facilitate injury recovery and rehabilitation. J Athl Train 55(9):918–930. https://doi.org/10.4085/1062-6050-550-19

Squitti R, Reale G, Tondolo V, Crescenti D, Bellini S, Moci M, Caliandro P, Padua L, Rongioletti M (2023) Imbalance of essential metals in traumatic brain injury and its possible link with disorders of consciousness. Int J Mol Sci 24(7):6867

Tan M, Zhu J-C, Yin H-H (2011) Enteral nutrition in patients with severe traumatic brain injury: reasons for intolerance and medical management. Br J Neurosurg 25(1):2–8. https://doi.org/10.3109/02688697.2010.522745

Tavarez T, Roehl K, Koffman L (2021) Nutrition in the neurocritical care unit: a new frontier. Curr Treat Options Neurol 23:1–16

Taylor HG, Yeates KO, Wade SL, Drotar D, Klein SK, Stancin T (1999) Influences on first-year recovery from traumatic brain injury in children. Neuropsychology 13(1):76

Tehse J, Taghibiglou C (2019) The overlooked aspect of excitotoxicity: glutamate-independent excitotoxicity in traumatic brain injuries. Eur J Neurosci 49(9):1157–1170

Timofeev I, Carpenter KLH, Nortje J, Al-Rawi PG, O'Connell MT, Czosnyka M, Smielewski P, Pickard JD, Menon DK, Kirkpatrick PJ (2011) Cerebral extracellular chemistry and outcome following traumatic brain injury: a microdialysis study of 223 patients. Brain 134(2):484–494

Tjäder I, Rooyackers O, Forsberg AM, Vesali RF, Garlick PJ, Wernerman J (2004) Effects on skeletal muscle of intravenous glutamine supplementation to ICU patients. Intensive Care Med 30(2):266–275. https://doi.org/10.1007/S00134-003-2048-9/FIGURES/4

Twyman D (1997) Nutritional management of the critically ill neurologic patient. Crit Care Clin 13(1):39–49. https://doi.org/10.1016/S0749-0704(05)70295-5

Vespa P, McArthur DL, Stein N, Huang S-C, Shao W, Filippou M, Etchepare M, Glenn T, Hovda DA (2012) Tight glycemic control increases metabolic distress in traumatic brain injury: a randomized controlled within-subjects trial. Crit Care Med 40(6):1923–1929

Wang X, Dong Y, Han X, Qi X-Q, Huang C-G, Hou L-J (2013) Nutritional support for patients sustaining traumatic brain injury: a systematic review and meta-analysis of prospective studies. PloS One 8(3):e58838

Wessels I, Haase H, Engelhardt G, Rink L, Uciechowski P (2013) Zinc deficiency induces production of the proinflammatory cytokines IL-1β and TNFα in promyeloid cells via epigenetic and redox-dependent mechanisms. J Nutr Biochem 24(1):289–297. https://doi.org/10.1016/J.JNUTBIO.2012.06.007

Wu A, Ying Z, Gomez-Pinilla F (2011) The salutary effects of DHA dietary supplementation on cognition, neuroplasticity, and membrane homeostasis after brain trauma. Https://HomeLiebertpubCom/Neu 28(10):2113–2122. https://doi.org/10.1089/NEU.2011.1872

Wu Y-Q, Dang R-L, Tang M-M, Cai H-L, Li H-D, Liao D-H, He X, Cao L-J, Xue Y, Jiang P (2016) Long chain omega-3 polyunsaturated fatty acid supplementation alleviates doxorubicin-induced depressive-like behaviors and neurotoxicity in rats: involvement of oxidative stress and neuroinflammation. Nutrients 8(4):243

Young B, Ott L, Kasarskis E, Rapp R, Moles K, Dempsey RJ, Tibbs PA, Kryscio R, McClain C (2009) Zinc supplementation is associated with improved neurologic recovery rate and visceral protein levels of patients with severe closed head injury. Https://HomeLiebertpubCom/Neu 13(1):25–34. https://doi.org/10.1089/NEU.1996.13.25

Yu BS, Sung YJ, Choi HI, Sirohi R, Sim SJ (2021) Concurrent enhancement of CO2 fixation and productivities of omega-3 fatty acids and astaxanthin in Haematococcus pluvialis culture via calcium-mediated homeoviscous adaptation and biomineralization. Bioresour Technol 340:125720

Zasler N, Katz D, Zafonte, and R. D. (2012) Brain injury medicine: principles and practice - google livres, 2nd edn. Demos Medical Publishing

Zhu H-T, Bian C, Yuan J-C, Chu W-H, Xiang X, Chen F, Wang C-S, Feng H, Lin J-K (2014) Curcumin attenuates acute inflammatory injury by inhibiting the TLR4/MyD88/NF-κB signaling pathway in experimental traumatic brain injury. J Neuroinflamm 11(1):1–17. https://doi.org/10.1186/1742-2094-11-59/FIGURES/8

Chapter 17
Nutritional Intervention After Pediatric Traumatic Brain Injury

Jana Jaber, Bouchra Anzieh, Zahraa Awada, Andrea Eid, Ahmad Jaber, and Hiba Hamdar

Abstract Traumatic brain injury (TBI) is a leading cause of long-term physical and psychological disability, and ultimately mortality, in children. In pediatric TBI patients, undernutrition is correlated with higher mortality rates, greater risk of infectious complications, and more severe neurologic outcomes. Nutritional intervention and support are therefore essential in the management of severe TBI since they could promote optimal recovery. This narrative review will provide a comprehensive overview of the metabolic changes, commonly occurring in pediatric TBI patients, in the glucose, lipid, and protein metabolism which further emphasizes the importance of this therapeutic approach. Nutritional interventions in the pediatric care unit (PICU) depend on identifying the patient's most suitable energy goal and the route and timing of nutrient delivery that leads to better outcomes, in addition to balancing these interventions against the risks associated with nutrition treatment. It will also highlight emerging therapies such as specific diets, including macronutrients, vitamins, and mineral supplementations, that intend to reduce neuroinflammation, immunodeficiency, and metabolic crises, by presenting clinical evidence of their usage benefits in neurologic recovery in TBI patients. This review will sum-

J. Jaber · Z. Awada
Lebanese University, Faculty of Medical Sciences, Beirut, Lebanon

Medical Learning Skills Academy, Beirut, Lebanon

B. Anzieh · A. Eid
Medical Learning Skills Academy, Beirut, Lebanon

Holy Spirit University of Kaslik, School of Medicine and Medical Sciences, Jouniyeh, Lebanon

A. Jaber
Medical Learning Skills Academy, Beirut, Lebanon

Gilbert and Rose-Marie Chagoury School of Medicine, Lebanese American University, Byblos, Lebanon, Byblos, Lebanon

H. Hamdar (✉)
Medical Learning Skills Academy, Beirut, Lebanon

Plovdiv Medical University, Plovdiv, Bulgaria

W. Mohamed (ed.), *Nutrition and Traumatic Brain Injury (TBI)*, Nutritional Neurosciences, https://doi.org/10.1007/978-981-97-6341-2_17

marize the currently available literature on nutrition treatment for children with severe traumatic brain injury (sTBI) and highlight on the gaps for future studies.

Keywords Traumatic brain injury · Nutrition · Pediatric intensive care unit · Neurocritical care · Enteral nutrition

17.1 Introduction

Traumatic brain injury (TBI) is defined as an acquired brain injury caused by an external force that results in structural or functional damage to the brain (Popernack et al. 2015). TBI is the primary cause of mortality and disability for children (Dewan et al. 2016). Nearly half a million children visit the emergency department each year for TBI-related injuries, leading to the death of more than 7000 children in the USA, which are mostly reported in children aged 0–4 years and teenagers aged 15–19 years (Kanya Wati 2021; Araki et al. 2017). Compared with adults, children have several anatomical variations that may make them more susceptible to brain injury such as thinner cranial bones, a larger head-to-torso ratio, an underdeveloped immune system, and differences in thermal regulation (Alexiou et al. 2011). TBI can occur as a result of either a primary or secondary injury. The severity of a TBI has been defined as mild, moderate, or severe based on the patient's degree of consciousness as determined by the Glasgow Coma Scale (GCS) (Mckee and Daneshvar 2015). Mild TBI contributes to 70–90% of all TBI-related emergency department visits, while severe TBI accounts for only 3–7% of all cases. However, it has been established that the majority of concussions remain misdiagnosed, and mild pediatric TBI cases are frequently not reported to health services, leading to an underestimation of their true incidence (Kanya Wati 2021; Peterson et al. 2019). The most common causes of pediatric TBI are sporting activities, falls, and motor vehicles (Haydel et al. 2023). In general, pediatric TBI is associated with poor outcomes such as impaired processing speed, attention, memory, and cognitive abilities. The magnitude of these outcomes varies based on the nature, severity, and location of the traumatic injury. In addition to that, the timing of injury during a specific stage of brain development also contributes to the severity of the consequent dysfunctions (Panchenko et al. 2023). Indeed, several studies have shown that the prognosis of pediatric TBI is worse compared to adult TBI with a similar level of severity, as it was associated with more significant edema, larger dysregulation of cerebral blood flow (CBF), greater mortality risk, and more behavioral sequelae (Pop and Badaut 2011). As the physiological processes underlying TBI, such as inflammation and oxidation, are elucidated, it becomes more evident that proper nutrition may have a pivotal role in alleviating the primary, secondary, and long-term consequences of TBI (Institute of Medicine (US) Committee on Nutrition, Trauma, and the Brain 2011a). Furthermore, malnutrition in critically ill people has been associated with endocrine dysfunction, multiorgan failure, decreased immunity, and higher mortality. Furthermore, these dietary deficiencies tend to correlate with poor functional results. Thus, nutritional therapy in critically ill pediatric patients must be optimized to enhance both short- and long-term recovery (Nwafor et al. 2023). Accurate knowledge of the correct dosage, timing, and formulation is paramount to ensure the effectiveness and appropriateness of the therapeutic strategy in pediatric TBI care. This review will focus on explaining the metabolic changes of macronutrients that commonly occur in severe TBI in children and elaborate on the currently acknowledged nutritional interventions that have been implemented to surpass these variations, specifically focusing on the timing, route, and formula of the nutritional supplementation.

17.2 TBI Impacts the Nutritional Status of Children

TBI occurs as a result of primary and secondary injuries. The primary injury occurs at the time of initial insult resulting in focal, multifocal, or diffuse parenchymal and cerebrovascular injury with secondary injury developing over time following the initial impact evolving over the course of minutes to hours to days (Ashwal et al. 2021). Secondary injury has been attributed to a combination of inflammation, ischemia, cerebral edema, excitotoxicity, reactive oxygen species, and immunosuppression (Giza and Hovda 2014). Following TBI, there is an immediate neuroinflammatory response defined by the significant release of proinflammatory cytokines such as IL1, IL6, and TNF alpha from the brain parenchyma of TBI children (Postolache et al. 2020). While these proinflammatory cytokines stimulate and activate lymphocytes in order to promote damaged tissue healing, they can also exaggerate secondary injury processes, metabolic disturbances, and the course of the injury. For example, vascular leakage could lead to an increased lesional and perilesional edema which in turn could trigger decreased diffusion of oxygen and metabolites (Kinoshita 2016). In addition to that, ischemic conditions contribute to cerebral hypoperfusion following TBI. Cerebral blood flow (CBF) insufficiency deprives cells of their metabolic needs, forcing them to convert to anaerobic metabolism. Less energy and higher lactate generation in anaerobic metabolism induce cellular dysfunction and promote an acidic environment (Allen 2016).

These injuries, when combined, enhance the release of catecholamines, corticosteroids, and counter-regulatory hormones, all of which influence cellular metabolism by promoting a hypermetabolic state (Şimşek et al. 2014). Indeed, during the acute phase of TBI, energy requirements can rise to 100%–200% of normal levels, and this can last for weeks to months, depending on the magnitude of the brain injury and level of recovery (Lee and Oh 2022). This hypermetabolic phase is followed by a hypercatabolic phase, which accelerates muscle loss by increasing the mobilization of amino acids from skeletal muscle (Williams et al. 2009). As a consequence of these changes, the glucose, lipid, and protein metabolism are affected by the TBI injury which results in an impairment in the nutritional status of children as shown in Fig. 17.1.

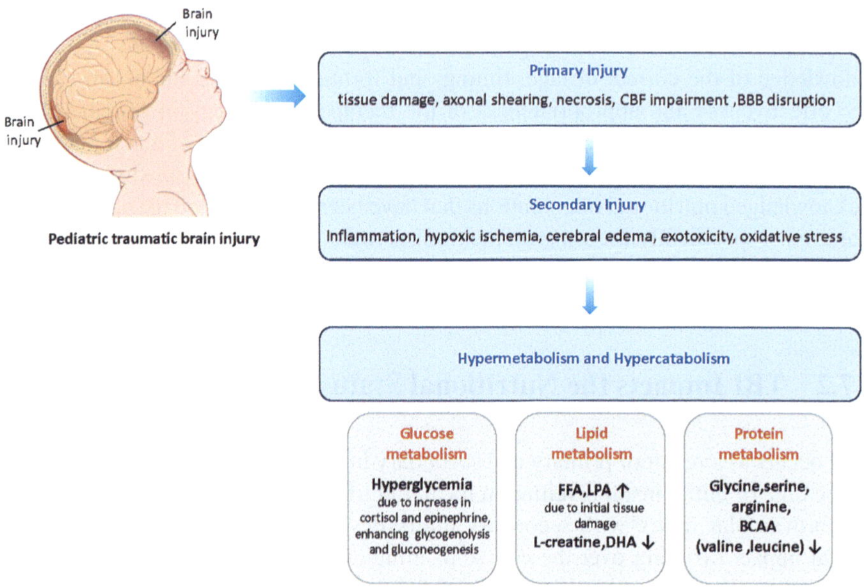

Fig. 17.1 Impact of pediatric traumatic brain injury on metabolism. *CBF* cerebral blood flow, *BBB* blood-brain barrier, *FFA* free fatty acid, and *DHA* docosahexaenoic acid

17.2.1 TBI Affecting Glucose Metabolism

Hyperglycemia is a commonly observed response to the stress of critical illness, including TBI (Mechanick 2006). It results from the changes in glucose metabolism caused by elevated counter-regulatory hormones such as cortisol and epinephrine, which enhance glycogenolysis and gluconeogenesis without a prompt parallel increase in insulin secretion and sensitivity, consequently leading to acute hyperglycemia (Kurtz and Rocha 2020). Hyperglycemia has been reported to be manifested within the initial 12 h after trauma in as much as 44% of pediatric patients with intermediate to severe TBI (Smith et al. 2012). Early hyperglycemia has been reported as associated with higher mortality, extended mechanical ventilation, and hospital length of stay in the pediatric intensive care unit in children with intermediate-to-severe TBI (Chong et al. 2015). Patients with TBI frequently experience hyperglycemia. A study postulated that early hyperglycemia could be a transient condition unrelated to prognosis (Parish and Webb 1988). However, another study demonstrated that severe hyperglycemia (>11 mmol/L) in the initial 12 h following trauma is an independent risk factor for predicting poor prognosis and mortality in children with TBI (Elkon et al. 2014). Importantly, this initial transient increase in glucose uptake is followed by a prolonged period of glucose metabolic depression. This metabolic depression has been established as a hallmark metabolic response after traumatic brain injury (Prins and Matsumoto 2016).

17.2.2 TBI Affecting Lipid Metabolism

Lipids are essential components of the brain because they have significant functions in the composition of plasma membranes and the synthesis of myelin sheath and lipid rafts. When the lipid metabolism is disrupted, cognitive performance could be significantly impaired. TBI impairs the metabolism of not only glucose but also lipids, ketone bodies, and amino acids (Lai et al. 2022). These metabolic changes are especially important in pediatric TBI because the developing brain is highly adapted to produce energy by utilizing alternative substrates for proper brain growth (Bowman et al. 2019). In TBI patients, free fatty acids (FAs) are elevated at 24 and 48 h after brain injury and return to normal values at 96 h. It is possible that such alterations during the acute phase (less than 24 h after injury) indicate initial tissue damage (Phillis and O'Regan 2003). The free FAs are then oxidized, which causes inflammation by triggering apoptosis via caspase-8 and caspase-3, leading to neuronal death and stimulation of pro-inflammatory cytokine release (Bowman et al. 2019). An elevated level of free FA has been linked to worse outcomes in TBI patients (Phillis and O'Regan 2003). Another type of bioactive lipid, lysophosphatidic acid (LPA), has been also found to be elevated in the circulation of TBI patients. While these findings imply that free FA and other bioactive lipids contribute to secondary damage, it can also be hypothesized that the enhancement of mitochondrial fatty acid oxidation may be a possible approach to reduce free fatty acid concentrations in these patients. Furthermore, after a brain injury, there is a reduction in L-carnitine, palmitic acid (PA), docosahexaenoic acid (DHA), and caprylic acid levels, which can negatively impact the injury prognosis (Chitturi et al. 2019). For example, L-carnitine is required for normal mitochondrial activity; it also repairs mitochondrial function and enhances functional recovery after brain damage. The decrease of certain lipids, especially L-creatinine and DHA, in the brain tissue implies that supplementation with this fatty acid may have therapeutic benefits following brain damage (Scafidi et al. 2011).

17.2.3 TBI Affecting Protein Metabolism

During the early post-injury phase, children with moderate-to-severe TBI have an elevated metabolic rate with rapid protein breakdown. Hypercatabolism is demonstrated by degradation of proteins, which is evidenced by significant urinary urea nitrogen excretion. In the absence of dietary intake, this extent of nitrogen loss can lead to a 10% reduction in lean mass in a duration of 7 days (Bagheri and Ekramzadeh 2018). Many amino acids, including glycine and serine, had their metabolism changed as a result of the brain injury (Fedoruk et al. 2023). Glycine is an amino acid abundant in the central nervous system and acts as both an inhibitory and excitatory neurotransmitter (Hajiaghamemar et al. 2020). Serine is another important amino acid in the brain, as it functions as a neuromodulator and neurotransmitter under diverse conditions. Additionally, it has a neuroprotective effect on the CNS by alleviating inflammation through the reduction of pro-inflammatory factors and decreasing neurotoxicity via glycine receptor activation (Sun et al. 2014). A study indicated that plasma glycine levels were significantly decreased in both moderate and severe TBI patients (Dash et al. 2016). Another study showed that serine was downregulated in all TBIs but more promptly in sTBI. As a result, glutamine and serine supplementation may provide protection in TBI patients (Sun et al. 2014). Arginine is another amino acid that is altered by brain damage. In fact, arginine deficiency was linked to poor outcomes in individuals with severe TBI. This is due to its well-known immune-boosting characteristics and critical involvement in the generation of creatine, nitric oxide, and agmatine (Fedoruk et al. 2023; Jeter et al. 2012). Furthermore, two frequent branched-chain amino acids (BCAAs) that should be mentioned are valine and leucine. These amino acids are critical in aromatic amino acid catabolism for the synthesis of brain neurotransmitters and were shown to be depleted following TBI. All of these studies show that TBI has a negative impact on the availability of essential amino acids in patients, implying that nutritional supplementation should be considered (Fedoruk et al. 2023).

17.3 Importance of Proper Nutrition in Promoting Recovery and Rehabilitation after TBI

Current traditional treatment strategies based mainly on pharmacotherapy have shown poor clinical improvement in TBI patients, especially in the improvement of neurological illness. As a result, evidence-based guidelines for the care of severe TBI in children have stressed the need for nutritional non-pharmacological support (Mao et al. 2021; Meinert et al. 2018). Dietary supplementation presents a favorable therapeutic option compared to other treatments in TBI management due to its low-cost noninvasive approach with a reduced risk of unfavorable complications and higher acceptability by the patient's family leading to higher compliance to treatment (Mao et al. 2021). In fact, TBI patients who maintain good nutrition were shown to have a lower risk of immunological dysfunction, infections, and neurological complications and have better prognosis compared to TBI patients with malnutrition (Caliri et al. 2019). This can be explained by the pivotal role of timely nutritional intervention in enhancing cellular homeostasis and mediating the immune and inflammatory cascades that are secondary to initial TBI injury. For example, amino acids, electrolytes, omega-3 fatty acids, and antioxidants have been recommended dietary supplements due to their anti-inflammatory properties which mitigate neuroinflammation and alleviate secondary brain damage, hence lowering the severity and long-term impact of traumatic brain injury (Elliott et al. 2022). Another example is the implementation of hyperosmolar therapy with 3% hypertonic saline which decreases intracranial pressure and has been associated with a positive immune-modulatory effect in severe TBI in infants, children, and adolescents (Sabers et al. 2020). Beyond the strictly caloric value of a TBI patient's diet, nutritional supplementation may present an opportunity to prevent other systemic complications such as gut dysbiosis (Zheng et al. 2022). TBI leads to gut dysbiosis which is marked by a decrease in beneficial bacterial species and a rise in potentially dangerous and toxic bacterial species over time. Based on expanding knowledge of the gut-brain axis, it has been proposed that alterations in gut microbiota might impact post-traumatic neuroinflammation (Elliott et al. 2022; Zheng et al. 2022). Probiotic and prebiotic supplementation was advised due to the prebiotics' immunoregulatory impact in the alteration of the gastrointestinal epigenome and enhancement of permeability, allowing immune cell infiltration (Mao et al. 2021). Probiotic supplementation has been linked to dramatically decreased incidence of illness and mortality (Curtis and Epstein 2014). Therefore, timely and effective nutritional therapy is particularly crucial in the management of TBI to improve patients' prognosis, especially in more severe TBI cases. However, questions remain about the optimal prescription of nutrition for children who have suffered a severe TBI, specifically concerning the most appropriate timing, route, and composition of the nutritional intervention. Dietary interventions in the pediatric population could be also challenging considering the behavioral difficulties and the intolerance of frustration secondary to frontal lobe injuries (Jourdan et al. 2012).

17.4 Nutrition Interventions for Pediatric Severe Traumatic Brain Injury

17.4.1 Assessment of Nutritional Requirements

17.4.1.1 Calculation of Nutritional Requirements

Nutrition treatment in the PICU should be individualized to each child's specific demands since both under- and overfeeding can predispose critically ill children to the risk of malnourishment or metabolic overload (Briassoulis 2014). In pediatric TBI patients, failure to meet energy requirements can result in an extended hospital stay, infections, organ failure, prolonged mechanical ventilation, and elevated mortality rates. Furthermore, in the case of feeding hyperglycemia, hypertriglyceridemia, hepatic steatosis, azotemia, hypercapnia, and an elevated risk of mortality are possible consequences (Delsoglio et al. 2019). Therefore, the initial step in providing patients with proper nutrition is to establish a balanced diet with the goal of preventing the adverse effects of both under or overfeeding the patients (Briassoulis 2014).

However, proper estimation of protein and energy is difficult to achieve. In healthy young children, the nutritional demands vary substantially based on age, growth rate, body composition, present nutritional status, and degree of physical activity (Pontzer et al. 2021). When children become severely ill, the number of factors increases, including fever, septicemia sedative/neuromuscular blockade, medication, and critical illness duration and stage (Larsen et al. 2018). Predictive equations are often used to predict energy expenditure and orient dietary prescription in critically unwell children, but they might overestimate or underestimate accurate metabolic demand. The most commonly used equations are Harris-Benedict, Ireton-Jones, and Penn State predictive equations, listed in Table 17.1 (Lee and Oh 2022; Vasileiou et al. 2020). Harris-Benedict equations estimate energy expenditure based on height, weight, sex, and age (Luy and Dampil 2018). Although this equation is easy to use, is widely known, and provides an approximately accurate computation of basal energy expenditure, it has been proven to be misleading in individuals suffering from trauma, such as a TBI (Donaldson et al. 2000). Pepe et al. reported that estimations of the expenditure of energy applying the Harris-Benedict equation appeared to be only 50% accurate (Pepe and Barba 1999). Penn State equation uses REE calculated by the Harris-Benedict method (HBE), maximum body temperature in the past 24 h (Tmax), and the expired minute volume (VE) (Lee et al. 2021a). These equations that aim to predict REE are valid only in the case of healthy patients. Therefore, in case of an altered REE, variable metabolic needs, or nutritional inability to maintain body weight, these formulas were shown to exhibit limitations in their application (Delsoglio et al. 2019). Importantly, indirect calorimetry (IC) has been found to be the gold standard method to determine energy requirements in critically ill patients. By measuring oxygen consumption (VO_2) and carbon dioxide production (VCO_2), IC aids in the determination of REE as well as the identification of the substrate required. Other features of the IC include personalization of the nutrition support and the evaluation of the metabolic response to nutrition administration (Kagan et al. 2018).

Nutritional monitoring should be an integral part of the care for every pediatric critically ill patient, including TBI patients. It has been shown that during the acute phase of TBI, neuroprotective strategies, including temperature control, use of sedatives, and mechanical ventilation, modify the REE, making it very challenging to predict an individual's nutritional requirements, as presented in Table 17.2 (Michaud et al. 1992; Flower and Hellings 2012; Mrozek et al. 2012). To elaborate, the latest studies on energy expenditure following TBI do not reveal the same extent of hypermetabolism, with certain studies even indicating hypometabolism during the initial post-injury period. These findings support speculation that modern neurocritical

Table 17.1 Common equations for predicting resting energy expenditure

Equations:
Penn State Equation:
REE = (1.1 × value of HBE) + (140 × Tmax) + (32 × VE) – 5340
Ireton-Jones Equations for ventilated patients:
Male REE = 2028 – 11 × age + 5 × weight + 239 × trauma + 804 × burn
Female REE = 1,784 – 11 × age + 5 × weight + 239 × trauma + 804 × burn
Harris-Benedict Equations:
Male REE = 66.47 + 13.75 × weight + 5 × height – 6.755 × age
Female REE = 665.1 + 9.563 × weight + 1.85 × height – 4.676 × age

Table 17.2 Impact of neuroprotective strategies on the resting energy expenditure in TBI patients

Neuroprotective strategy	Impact on resting energy expenditure
Mechanical ventilation (Michaud et al. 1992)	Interrupts the feeding process and can lead to patients not receiving their desired amount of nutrition
Use of sedatives (Flower and Hellings 2012)	Decreases the metabolic rate of critically ill patients and, consequently, REE, while aiming to reduce intracranial pressure and to help patients better tolerate the mechanical ventilation
Temperature control (hypothermia) (Mrozek et al. 2012)	Leads to reduced metabolic rate and lower cerebral metabolic rate of oxygen which is very related to oxygen and glucose consumption and lactate production in neuronal cells

Table 17.3 Common TBI complications that affect nutritional intervention approach

Complication	Impact on nutrition route or initiation in TBI patients
Dysphagia	More than 68% of pediatric TBI patients suffer from impairment of swallowing, either due to sensory problems in the throat or weakening of the muscles, with the average time from injury to oral feeding being 13 weeks post injury. Oral feeding is unsafe because patients' dysphagia leads to a high risk of aspiration as much as 40% of TBI patients with swallowing difficulties experienced aspiration (Bagheri and Ekramzadeh 2018; Morgan et al. 2003; Terré and Mearin 2007).
Gastroparesis (delayed gastric emptying)	As much as half of TBI patients do not tolerate enteral feeding due to gastroparesis, esophageal reflux, and decreased intestinal peristalsis which persists for 2 weeks post injury. This delay of adequate nutrition favors aspiration of gastric contents and pneumonia, thus prolonging hospital stay and increasing the risk of mortality (Mass-Ramirez et al. 2021).

care procedures may be blunting any predicted hypermetabolism. This causes a hypometabolic condition during the first week following injury (Elliott et al. 2022).

17.4.1.2 Determination of the Route of Nutrition

The route of nutritional support is an important consideration following the determination of the total energy requirements in pediatric patients with TBI. Many TBI patients encounter several TBI complications, such as dysphagia and delayed gastric feeding, that affect the choice of the most optimal route or timing of intervention in the patient, as shown in Table 17.3 (Morgan et al. 2003; Terré and Mearin 2007; Mass-Ramirez et al. 2021). For example, many TBI patients face difficulties with swallowing which will prohibit feeding through the oral route. In addition, some patients might depend on mechanical ventilation; hence, the enteral nutritional route is the more favorable route of nutritional support, particularly within the first 24–48 hours after the injury (Cook et al. 2008). EN maintains structural and functional gut integrity, preserves gut microbiota, and permits, through access to the gut, a passageway for immune-modulating substances and is therefore effective in preventing infections and stress ulcers (Kang and Kudsk 2007).

Enteral nutrition is provided by inserting a tube through the nasal passage into the stomach or intestine. If the patient is using a mechanical ventilator, the tube could be inserted through the patient's mouth (Kang and Kudsk 2007). When starting enteral nutrition, several feeding approaches must be addressed. First, the patient's head needs to be elevated by 30–45 degrees off the bed during and after feeding to reduce the risk of aspiration and prevent gastric reflux. Second, in order to be better tolerated by the patient, a more concentrated formula is recommended to provide the quantity of calories required with less volume intake. Third, a gradual increase of EN is necessary to avoid overfeeding in the early phase, especially in patients who are intolerant initially (Nwafor et al. 2023; Kattelmann et al. 2006). In addition to that, prokinetic agents, such as metoclopramide or erythromycin, may be prescribed to improve gastrointestinal motility and alleviate intolerance. Oral naloxone and alvimopan, which are opioid receptor antagonists, are other drugs that can increase EN tolerance. These pharmaceuticals also reduce the risk of ventilator-associated pneumonia (Tan et al. 2011). If prokinetic medications fail to improve bowel motility or are poorly tolerated, as in the case of food intolerance, small bowel feeding should be considered over stomach feeding (Elliott et al. 2022). A randomized controlled study of pediatric patients requiring mechanical ventilation found that individuals receiving gastric EN achieved a lower proportion of their daily caloric target than patients receiving small-bowel EN (Meert et al. 2004). It should be noted that if the necessity for enteral nourishment lasts longer than 2–4 weeks, percutaneous enteral gastrostomy (PEG) could be more favorable (Sundstrøm et al. 2020).

Although early EN is advised for the majority of patients, there is no doubt that PN supplementation is needed if malnutrition persists in order to limit the harmful effects of an unfavorable energy balance (Fivez et al. 2017). Comparing the PN to EN route in TBI patients, a study found that the PN causes more hyperglycemia, but the route had no effect on either the early initial inflammatory response or patient outcomes (Justo Meirelles and de Aguilar-Nascimento 2011). However, it is important to consider that PN is frequently associated with higher rates of infection, immunosuppression, hyperglycemia, hepatic steatosis, diminished gastrointestinal (GI) integrity, and the expression of gut-associated lymphoid tissue (GALT) (Kurtz and Rocha 2020). A three-chamber bag containing glucose, lipids, and amino acids is used to provide the daily needs for parenteral nutrition. It is best to start parenteral feeding with a separate nutritional central venous catheter from those used to deliver drugs or fluids. Furthermore, the nutritional line must be reviewed frequently (Araki et al. 2017).

17.4.1.3 Determination of Timing of Nutrition

Determination of the timing of starting nutritional support depends on the route of administration. For EN, initiation of feeding should be within 48 h of administration in stable patients to avoid adverse effects of delayed feeding (Balakrishnan et al. 2019). Early nutritional treatment after TBI is critical in reversing the acute catabolic state found in TBI patients (Kurtz and Rocha 2020). Early feeding plays a role in the maintenance of muscle mass, reduction of infectious complications, brain homeostasis, and improvement in functional outcomes (Taha et al. 2011). Nutritional support should ideally begin within 24–48 h following a TBI and supply 50% of resting energy expenditure (REE) within the first two weeks. This intervention is crucial in reducing the inflammatory reaction due to TBI and therefore improving the outcome (Nwafor et al. 2023). Hart and collaborators showed that patients who were not fed within 5–7 days after TBI had a two- to fourfold increase in the mortality rate, respectively. In addition, every 10-kcal/kg reduction in caloric intake during the first 5 days was correlated with a 30–40% increase in mortality rates (Härtl et al. 2008). Similar data have been reported for pediatric TBI patients. In a retrospective study of 109 children with sTBI, initiating dietary support during 72 h of admission and reaching target caloric intake by day 7 were associated with a shorter ICU duration of stay and better clinical outcomes at discharge (Taha et al. 2011). Similarly, another study found that patients who received nutritional treatment within 72 h following their injury had reduced mortality rates and better functional outcomes than those who received no nutritional support during the first 7 days (Meinert et al. 2018).

Some variables that might delay the initiation of feeding in critically ill children include illness severity scores, noninvasive and invasive mechanical ventilation, surgical operations within 48–72 h of PICU admission, expected extubation, and high vasoactive levels. Consequently, delaying the start of enteral feeding lowers Glasgow Coma Scale (GCS) scores while increasing Injury Severity Score (ISS). Thus, a bowel regimen is required in these children (Balakrishnan et al. 2019). Although EN is recommended as first-line nutritional therapy for children with TBI, PN is necessary as a supplement to ensure energy balance especially if the patient is in a malnutrition state. The favorable timing of PN feeding is still unknown. While the European Society for Clinical Nutrition and Metabolism (ESPEN) guideline recommends starting PN after 4–7 days, the American Society for Parenteral and Enteral Nutrition-Society of Critical Care Medicine (ASPEN-SCCM) guideline suggests a PN supplementation after 7–10 days (Lee and Oh 2022).

17.4.1.4 Determination of Enteral Formula

Standard Versus Immunization Formula

The primary brain injury is followed by neuroinflammation, free radical production, excitatory toxicity, and oxidative stress (Fesharaki-Zadeh 2022). Given that early initiation of enteral feeding to TBI patients is relatively well-established, the combined administration of immune-modulating nutrient supplementation through the utilization of formulas containing nutrients such as glutamine, omega-3 fatty acids, arginine, and nucleotides has been suggested to enhance neuroprotection from secondary brain insults by resolving the early hyperinflammatory phase while also improving immunological functions, which could significantly favor TBI prognosis

(Razmkon et al. 2011; Bastian and Weimann 2002). In fact, a study assessing the impact of an immunonutrition formulation on inflammatory biomarkers in 36 brain injury patients found that immunonutrition significantly reduced inflammation, decreasing IL-6 serum levels and enhancing antioxidant defense through the endogenous elevation in glutathione levels and improved visceral nutritional status, with an increase in total visceral proteins (Rai et al. 2017). Interestingly, another study demonstrated that severe acute TBI patients who received IEN had greater prealbumin levels, indicating a better nutritional status during hospitalization, and showed a lower incidence of nosocomial infections, specifically bacteremia (Painter et al. 2015). In the pediatric population, however, children given an immune-boosting diet enriched with glutamine, arginine, antioxidants, and omega-3 fatty acids did not show better results than those given a standard formula. The IE formula had no effect on infection rates or bronchoalveolar lavage colonization; however, fewer infectious complications and a marked reduction in gastric colonization were reported in critically ill patients on IE diets (Briassoulis 2014). Although this type of intervention may be safe and practical in children with severe brain damage, it is not linked with any significant benefit above conventional early enteral feeding. As a result, current pediatric guidelines merely state that an immune-enhanced diet is not advised due to a lack of efficacy and research confirming its benefit in this population (Meinert et al. 2018).

Hyperglycemia Control: Tight versus Permissive

When addressing dietary needs following TBI, it is critical to establish the optimum glucose requirement for neuronal survival (Cook et al. 2008). TBI patients may experience hyperglycemia or hypoglycemia (Nwafor et al. 2023). Hyperglycemia is more frequent after a TBI and has been associated with worse outcomes in pediatric patients. In children with TBI, early hyperglycemia correlates with higher mortality, extended mechanical ventilation, and PICU stay (Chong et al. 2015). Prevention of hyperglycemia after brain injury in adults has been proposed to provide significant advantages such as lower vasopressor need for sustaining cerebral perfusion pressure, decreased seizure frequency, and reduced risk of diabetes insipidus (Van den Berghe et al. 2005). However, when compared to normal insulin therapy, intensive insulin therapy had worse post-TBI outcomes, particularly in individuals with severe TBI (Nwafor et al. 2023). In a large international RCT in adults with TBI, patients who received intensive glucose control (target range 80–110 mg/dL) and patients who received conventional glucose control (target range less than 180 mg/dL) had the same long-term outcomes, yet intensive glucose control was associated with a higher incidence of moderate-to-severe hypoglycemia episodes, implying potential harm (NICE-SUGAR Study Investigators for the Australian and New Zealand Intensive Care Society Clinical Trials Group and the Canadian Critical Care Trials Group et al. 2015). Furthermore, using cerebral microdialysis, several studies have found that tight glycemic control is associated with a significant decline in glucose and an elevation in the lactate/pyruvate ratio in the brain, implying that damaged brain tissue is unable to access the glucose required for survival and repair (Vespa et al. 2012). There have been no studies examining glycemic control in children with sTBI (Elliott et al. 2022). One randomized controlled trial of critically ill children (not limited to TBI) found that strict glycemic control improved inflammatory markers and shortened ICU stay, but it significantly raised the risk of hypoglycemia (Vlasselaers et al. 2009). Another recent randomized trial of children admitted to the PICU found no substantial difference in the number of days alive and free of mechanical ventilation after 30 days of the intervention, with a higher frequency of hypoglycemia in the tight glucose control group (Macrae et al. 2014). Both of these studies did not precisely address the subset of TBI children. Research focusing particularly on TBI patients would be able to provide a more definitive response to the question of whether glucose control may influence outcomes in pediatric TBI (Chong et al. 2015).

Lipid Support

The integration of n-3 fatty acid supplementation into normal nutritional support is a distinctive approach for optimizing ICU nutrition and promoting recovery after severe brain damage. EN formulations containing n-3 fatty acids have been established for their anti-inflammatory and pro-immune characteristics (Poblete et al. 2023). Omega-3 polyunsaturated fatty acids (n-3 PUFA), which include eicosapentaenoic acid (EPA) and docosahexaenoic acid (DHA), are essential for proper brain development and function. Omega-3 fatty acids protect neuronal cell membranes, decrease oxidative stress and inflammation, and restore hippocampus lysosome biosynthesis and function (Lai et al. 2022). Several studies have found that n-3 PUFA supplementation may reduce neuroinflammation following brain injury. A recent study of nine individuals with severe TBI showed that omega-3 fatty acid supplementation improved their prognosis (Bailes et al. 2020). Another study, aimed at evaluating the neuroprotective effects of omega 3 PUFA, noticed that omega 3 supplementation resulted in the resolution of mTBI symptoms, such as headache, dizziness, taking longer to think, nausea, fatigue, and phobia, in a median of 12 days, and the complete resolution was noted in most patients within 30 days with no recorded side effects (Galuta et al. 2018). Thus, n-3 PUFA supplementation has been identified as an emerging potential candidate for TBI management, due to its low cost, feasibility, and commercially available feeding formulations which are well-tolerated and safe since they do not cause allergic responses or unfavorable effects. Another advantage is that EPA and DHA, unlike previous pharmaceutical methods, easily penetrate the BBB, providing quick access to wounded brain regions (Poblete et al. 2023). In the general medical and surgical ICU population, an n-3 PUFA-enriched diet decreases nosocomial infection risk and improves hospital outcomes with research required to confirm these findings after TBI (Grau-Carmona et al. 2015).

Protein Support

Protein catabolism is an important contributor to the loss of lean body mass which frequently occurs in TBI patients. It increases significantly during the acute phase of TBI, seems to peak within 8–14 days after injury, and correlates to TBI severity (Young et al. 1992). Nitrogen excretion also increases gradually up to 4 weeks after injury, regardless of supplementation dosage. As a result, it is critical to maintain protein balance through proper supplementation even though balancing nitrogen may be difficult to achieve (Kurtz and Rocha 2020). Protein supplementation must start early, ideally within the first 24 h after injury, and supply more than 50% of resting energy expenditure (REE) with 1.0–1.5 g protein/kg for the following two weeks (McClave et al. 2016). Standard intact protein formulations are tolerated by the majority of TBI patients. Enteral formulas containing partially digested proteins, on the other hand, are recommended over non-digested formulas due to their capacity to improve stomach emptying and digestion, which is typical in TBI patients (Mahapatra 2019). Immunomodulating diets may potentially be a suitable option for addressing this protein supply deficit in TBI patients. Glutamine and arginine are key components of the immunomodulating formula (Painter et al. 2015). Glutamine, an abundant amino acid that is synthesized in muscles for catabolic metabolism, is mostly catabolized more than produced after trauma, sepsis, or major surgery, which can lead to its depletion (Cruzat et al. 2018). A study found that glutamine-enriched enteral meals decreased ICU time, hospitalization time, and ventilator duration among sTBI ICU patients (Xiong and Qian 2021). However, when plasma glutamine levels are normal, large dosages have the potential to be harmful in patients (Heyland et al. 2013). A proper dose of glutamine is between 0.3 and 0.5 g/kg body weight per day (i.e., 25–50 g/day) for 1–2 weeks during EN and PN doses (Quintard and Ichai 2019). L-arginine is required for the production of nitric oxide (NO), creatine, polyamines, homoarginine, and agmatine (Wu et al. 2021). Critical care formulations containing arginine, fish oil, and other antioxidants have been demonstrated to be useful in decreasing infection in patients with brain injuries (Bower et al. 1995). Protein supplements should be monitored and modified during the course of TBI patients. Aggressive supplementation with protein to reduce nitrogen loss and maintain muscle mass is not efficient in reversing TBI's catabolic condition. Clifton and colleagues, who attempted to prevent nitrogen wasting after TBI by increasing protein administration, found that this intervention has resulted in increased protein catabolism, with only 50% of the nitrogen supply retained at high levels of protein intake (Clifton et al. 1986). Adequate water should be given to patients on these high-protein diets to prevent azotemia, especially if they have kidney disease (Cook et al. 2008).

Fluid and Electrolyte Requirements

TBI patients are frequently given intravenous fluid resuscitation, such as hypertonic saline or mannitol, to address the frequent complication of elevated intracranial pressure (ICP) (Fenn 3rd and Sierra 2019). ICP elevation and poor cerebral perfusion, resulting in limited oxygen delivery, are the most prominent causes of oxidative metabolic failure in TBI patients. Therefore, ICP treatment with 3% hypertonic saline or mannitol is critical for improving cerebral metabolism (Cottenceau et al. 2011). In children with TBI, both 3% hypertonic saline and mannitol can successfully lower intracranial pressure, but 3% hypertonic saline has a longer prolonged effect on intracranial pressure and can effectively raise cerebral perfusion pressure (Shi et al. 2020). Interestingly, a meta-analysis study found that neither solution affected oxygen or glucose metabolic rates and that there was no significant difference in neurological results between the two approaches (Cottenceau et al. 2011). Nutritional treatments need to take into account these fluid resuscitation strategies and record both fluid and electrolyte levels to avoid elevated fluid, glucose shifts, and electrolyte derangements, which are also frequent after neurologic injury, including TBI (Kurtz and Rocha 2020). For example, a high protein intake with inadequate water and moderate sodium consumption may be the primary cause of tube feeding syndrome (hypernatremia, azotemia, and dehydration) (Gault et al. 1968). Furthermore, the increased nitrogen levels reported after TBI increase the need for excretion by the kidneys, requiring a larger volume of fluid (Kurtz and Rocha 2020). However, excessive fluids should not be given to TBI patients because they may aggravate cerebral edema and raise the risk of acidosis, hemodilution, and acute respiratory distress syndrome (ARDS) (Lee et al. 2021b). In the case of HTS, intense therapy might cause severe hypernatremia, which can lead to volume overload, heart failure, and pulmonary edema, or it can cause hyperchloremic metabolic acidosis and coagulopathy (Horras et al. 2015). In the case of mannitol, intravenous dextrose must be utilized with caution in the initial phase of TBI because it presents an increased risk of hyperglycemia. Potassium, magnesium, calcium, and phosphate are some of the other electrolytes that must be monitored after a brain injury (Nwafor et al. 2023) (Table 17.4).

17.4.1.5 Vitamins and Mineral Supplementation

Vitamin C

Ascorbic acid (vitamin C) is a potent antioxidant that has significance in immunological functions. Along with vitamin E and glutathione (a tripeptide), ascorbic acid effectively eliminates reactive oxygen species (e.g., superoxides) produced during the generation of adenosine triphosphate (ATP) (Denniss and Barker 2023). Lack of vitamin C stores is a characteristic feature of critical diseases, including TBI (Leichtle et al. 2020). In a small cohort, including TBI and hemorrhagic stroke patients, plasma vitamin C concentrations were shown to be low and adversely

Table 17.4 Supplementation of vitamins and minerals in clinical trials of TBI

Vitamin and mineral	Dosage used	Suggested/proposed mechanism	Findings
Vitamin C	Group A low-dose (500 mg/d IV) for 1 week and group B high-dose (10 g IV on the first and fourth day, then 4 g/d IV for the remaining 3 days)	Antioxidant activity that reduces ROS	The low dose had no impact on the disease prognosis, whereas the high dose decreased perilesional edema. There were no short- and long-term neurological outcomes in both doses (Razmkon et al. 2011).
Vitamin D	A one-time oral dose of 120,000 IU	Neuroprotective effects via amelioration of cytokine level post-traumatic brain insult	A significant improvement in the GCS, reduction in ventilation duration, and decrease of inflammation in patients (Sharma et al. 2020).
	Immediate 100,000 IU IM cholecalciferol and 0.5 mg/day alfacalcidol if oral feeding is feasible	Calcium-ion regulation, immune response, and regulation of free radical production	Improved cognitive results and GOS-E score three months after TBI in patients who first had significant vitamin deficiency (Lee et al. 2019).
Vitamin E	400 IU/d IM for 1 week	Antioxidant activity that reduces ROS and decreases lipid peroxidation rate	Reduced mortality rate and improved functional outcome at discharge, but this effect dissipated throughout the 6-month follow-up (Razmkon et al. 2011).
Magnesium	Unspecified	Neuronal inhibitor, specifically to glutamate, which reduces exotoxicity	Increase in the somatic scores after mild TBI injury (Wahls et al. 2014).
	Continuous infusions of magnesium for 5 days being given to patients within 8 h	No neuroprotective effect of Mg supplementation	No neuroprotective effect was noted, and even mild adverse outcomes such as increased blood pressure were reported in the high magnesium group (Temkin et al. 2007).
Zinc	First group received a standard level (2.5 mg) and the second group high level (12 mg) for 15 days; then both groups 22 mg oral zinc for 3 months	Glutamate-induced excitotoxicity reduction	Zn supplementation enhanced blood prealbumin and retinol-binding protein levels in TBI patients, indicating better visceral protein synthesis. After 2 weeks of supplementation, the GCS score improved (Young et al. 1996).
	120 mg zinc via a nasogastric tube for 15 days	Anti-inflammatory properties, reducing the plasma cytokine concentrations	Improvement in the GOS, SOFA score, and inflammatory markers with a shorter length of hospital stay (Khazdouz et al. 2018).

(continued)

Table 17.4 (continued)

Vitamin and mineral	Dosage used	Suggested/proposed mechanism	Findings
Curcumin	500 mg curcuminoids with 5 mg piperine for 1 week	Anti-inflammatory activity that reduces oxidative damage	Reduction in inflammatory cytokines such as IL-6, TNF-, and MCP-1 [131].
	500 mg/d for one month	Anti-inflammatory activity that reduces oxidative damage	Reduction in leptin serum levels but no change in adiponectin levels, CBC (complete blood count), or blood glucose levels (Shadnoush et al. 2020).

GCS Glasgow Coma Scale, *GOS-E* Glasgow Outcome Scale-Extended, and *SOFA* Sequential Organ Failure Assessment

associated with injury severity (Polidori et al. 2001). As a result, the administration of antioxidants, particularly AA, in critical illness is regarded as an appealing approach for mitigating TBI secondary damage and improving prognosis. In fact, a study examining the impact of AA in TBI patients reported a decrease in the extent of perilesional edema following high vitamin C administration (Razmkon et al. 2011). Increased vitamin C consumption following TBI might therefore minimize the detrimental effects of free radical production and improve cerebral reparative mechanisms (Denniss and Barker 2023).

Vitamin D

Vitamin D is a crucial neuroactive steroid, known for its antioxidant and neuroprotective functions in addition to its classical role in bone metabolism and calcium–phosphate homeostasis (Cui et al. 2019). Vitamin D insufficiency is frequent in TBI patients, and it has been correlated to poor functional outcomes at hospital release as well as mortality within six months after injury in TBI patients with cerebral bleeding or diffuse axonal damage (Jung et al. 2022). As a result, vitamin D administration has been proposed to be neuroprotective by lowering cytokine levels after traumatic brain injury. Several studies have been conducted to assess the effectiveness of vitamin D treatment in individuals with severe TBI. A study of 35 TBI patients found that vitamin D administration had a significant favorable effect on GCS, ventilation duration, and reduction of inflammation in these patients (Sharma et al. 2020). Another study found that providing vitamin D supplements to mild-to-moderate TBI patients who had considerable vitamin D insufficiency during the acute phase of the injury resulted in better cognitive results and GOS-E score three months after TBI (Lee et al. 2019). As a result, vitamin D supplementation may have both short- and long-term favorable benefits on TBI prognosis.

Vitamin E

Vitamin E is a lipid-soluble vitamin and strong antioxidant that suppresses free radical propagation and prevents membrane alterations (Böhm 2018). Deficiency in vitamin E promotes neurodegeneration, notably in peripheral axons and posterior column neurons, resulting in peripheral neuropathy, ataxia, and skeletal myopathy (Herrera and Barbas 2001). Multiple studies have found that plasma vitamin E levels dropped rapidly after TBI, indicating early vitamin E depletion and a demand for additional supplementation (Park et al. 2022). Razmkon et al. studied the impact of vitamin E supplementation on severe brain damage. Vitamin E was found to affect mortality in addition to functional outcomes at discharge. However, this impact seemed to be of a brief duration since it dissipated throughout the 6-month follow-up (Razmkon et al. 2011). More research into the role of vitamin E and associated molecules in TBI is necessary.

Magnesium

Magnesium is required for cell membrane integrity, genetic material stability, and a variety of important enzymatic activities such as glycolysis. It is additionally recognized for its function in limiting the release of excitatory amino acids, especially glutamate, presynaptically, which is a crucial link to the excitotoxic state of secondary brain damage (Lyons and Blackshaw 2018). As a result, magnesium supplementation in TBI patients has been of interest to researchers. Magnesium levels were found to be raised in cerebrospinal fluid (CSF) following TBI, while serum levels were lowered during the first hour after damage and persisted declined for a duration of five days (Kafadar et al. 2007; Nayak et al. 2018). The degree of magnesium deficiency has been correlated with the severity of brain damage (Kafadar et al. 2007). Parenteral delivery of high doses of magnesium in severe TBI has been found to be neuroprotective and aid in the recovery of severe brain injury in patients (Standiford et al. 2018). In particular, magnesium relaxes vascular smooth muscle, leading to vasodilation and improved cerebral blood flow in TBI patients (Institute of Medicine (US) Committee on Nutrition, Trauma, and the Brain 2011b). In fact, Wahls and colleagues concluded that magnesium represented one of the four main vitamins that were associated with higher somatic scores when provided after mild TBI patients (Wahls et al. 2014). However, these findings remain controversial. In a large randomized placebo-controlled trial with over 400 TBI patients, no benefit or even a mild negative impact upon magnesium therapy was detected (Temkin et al. 2007). Due to the major effects, magnesium administration has on numerous bodily systems (particularly cardiovascular), it is crucial to examine the dosage relationship with potential positive effects when utilizing magnesium supplementation (Davies 2018).

Zinc

Following TBI, glutamate gets released from presynaptic terminals, and intracellular zinc accumulates improperly in postsynaptic neurons, leading to excitotoxicity and cell death (Li et al. 2022). TBI patients are noted to have considerably lower blood Zn levels as well as higher urine Zn excretion. Furthermore, urine Zn excretion is correlated with the degree of brain damage, with mean urinary Zn levels in the most severely injured individuals 14 times higher than normal values (Institute of Medicine (US) Committee on Nutrition, Trauma, and the Brain 2011c). TBI additionally contributes to hypoalbuminemia, which has an influence on Zn availability since albumin is a significant serum transport protein for Zn (Scrimgeour and Condlin 2014). Zinc supplementation has been associated with better results in TBI patients. Young and colleagues found that Zn supplementation enhanced blood prealbumin and retinol-binding protein levels in TBI patients, indicating better visceral protein synthesis. After 2 weeks of supplementation, the GCS score improved (Young et al. 1996). In another double-blind controlled trial, zinc treatment resulted in considerably greater plasma zinc concentrations, improved measurements on the Sequential Organ Failure Assessment and Glasgow Coma Scale, and a shorter length of hospital stay (Khazdouz et al. 2018). Furthermore, major depression impacts up to 40% of TBI patients, making it the most prevalent long-term consequence of TBI, and depression is related to a reduced Zn content in peripheral blood (Scrimgeour and Condlin 2014). Zinc involvement in antidepressant medication may be an additional favorable consequence of zinc supplementation in TBI patients, while further research is needed to validate this potential effect (Cope et al. 2012).

Curcumin

Curcumin, a phenolic yellow curry color, has powerful anti-inflammatory and antioxidant properties that may help prevent oxidative damage and cognitive deficiencies associated with aging (Ringman et al. 2005). The potential of curcumin to protect the brain from free radical damage is thought to be many times greater than that of vitamin E (Wu et al. 2006). Zahedi et al. conducted a study with 62 TBI patients to further evaluate its influence on TBI prognosis. One group took curcuminoids in conjunction with piperine (5 mg). They reported a decrease in inflammatory cytokines such as IL-6, TNF-, and MCP-1, indicating that curcumin may benefit TBI patients (Zahedi et al. 2020). In another study, curcuminoids supplementation (500 mg daily) for one month in TBI was related to a significant reduction in leptin serum levels but no change in adiponectin levels, CBC (complete blood count), or blood glucose levels (Shadnoush et al. 2020). These findings imply that curcumin can affect important signaling pathways and molecules that govern neuroinflammation in TBI patients; nevertheless, more research is needed to support its use in TBI patients (Khayatan et al. 2022).

17.5 Monitor Nutrition During PICU Care: The Role of Cerebral Microdialysis in Pediatric TBI

Cerebral microdialysis (CMD) is a valuable monitoring technique used in neuro-critical care to assess brain chemistry and guide treatment in conditions of traumatic brain injury (TBI). It involves the insertion of a microdialysis catheter into brain tissue to continuously sample the interstitial fluid, providing insights into the brain's metabolic state (Zhou and Kalanuria 2018). Microdialysis is commonly employed in tandem with other well-established methods of cerebral monitoring intracranial pressure (ICP), neurophysiological monitoring, positron emission tomography (PET) scans, transcranial Doppler, and brain tissue pO_2 levels (Charalambides et al. 2010). This helps detect brain injury before any further tissue damage and provides the appropriate management plan. Despite that, there is insufficient information about CMD's parameters in children and their probable benefits in TBI management (Thango et al. 2021).

The role of cerebral microdialysis is pivotal as it helps measure neurochemical markers undetected in the systemic circulation due to their inability to traverse the blood-brain barrier. Metabolic decline is commonly observed in TBI patients, which is caused by disturbance of transmembrane ionic gradients, excitotoxicity resulting from calcium and excitatory amino acids, the formation of free radical species, and the initiation of inflammatory cascades. These changes can serve as indicators of metabolic distress and early indications of brain ischemia. With the assistance of CMD, there is accessibility to real-time monitoring of important metabolites such as glucose, lactate, pyruvate, glutamate, and glycerol. These molecules reveal significant information about the brain's principal energy source, anaerobic metabolism, and excitatory amino acids and are thus critical for the identification and prevention of secondary damage following TBI, such as cerebral ischemia (Zhou and Kalanuria 2018; Ketharanathan et al. 2019). The recent consensus guidelines on CMD monitoring recommend primarily using lactate/pyruvate ratio (LPR) and glucose as steps to guide clinical interventions (Oddo and Hutchinson 2018).

For example, glutamate, an excitatory amino acid, is important in the pathophysiology of TBI (Ketharanathan et al. 2019). After being released by excited neurons, an ATP-powered ion cotransport mechanism allows it to be reuptake by astrocyte cells, which maintains its low levels and prevents excitotoxic damage. A cerebral microdialysis would note this increase in glutamate levels, in states of ischemia or hypoxia in TBI patients, due to decreased uptake (Zhou and Kalanuria 2018). Glutamate measurement, particularly through microdialysis, is correlated with mortality rate and 6-month functional outcome, highlighting its importance in the management of TBI (Chamoun et al. 2010). However, in another CMD study that monitored the extracellular glutamate in the brains of children with sTBI, there was no association between glutamate levels and the outcomes upon discharge. The amounts of structural amino acids did not fluctuate in response to glutamate variations, and there was no correlation with the extent of brain lesions on the head CTs, contrary to the adult's observations. This further emphasizes the promising role of

CMD in better understanding the differences in the biochemical response to injury between adults and children. Furthermore, brain glycerol, a component of the phospholipid bilayer cell membrane, is another important parameter that helps evaluate tissue hypoxia in patients. A retrospective cohort study monitored glycerol levels through CMD in children <13 years with severe TBI in parallel to physiological, head CT variables and clinical outcome. CMD glycerol, especially when raised beyond the baseline, was shown to be associated with additional indicators of damage and a significant increase in lesion size on repeat head CT (Thango et al. 2021). It is important to highlight that none of these indicators are totally specific, and their concentration in extracellular fluid varies depending on a variety of factors (Charalambides et al. 2010).

It is important to consider the limitations of CMD use in TBI patients, for better evaluation and understanding of the patient's nutritional status. CMD provides information about chemical variables from a very restricted zone surrounding the catheter, which would limit the generalizability of its findings. It is important to note that energy metabolism varies greatly in different areas of the brain in TBI patients (Zhou and Kalanuria 2018). Accordingly, the CMD probe should be placed in vulnerable brain areas, or in some cases, numerous CMD probes should be placed to monitor various at-risk brain areas. Furthermore, CMD supports sequential ECF sampling, which implies that each sample collection might take up to an hour. As a result, the results should be viewed as averages over the collection time rather than as real-time ECF analyte concentrations (Zhou and Kalanuria 2018; Nordström 2017). Therefore, it is recommended to analyze the CMD values in conjunction with other physiologic measures such as PbTO2, ICP, and MAP in order to obtain more accurate analysis and better evaluation of a patient's condition (Zhou and Kalanuria 2018). More research needs to be done in regard to the benefit of cerebral microdialysis in ICU monitoring in the case of pediatric TBI.

17.6 Conclusion

In conclusion, nutrition improves outcomes for pediatric TBI patients and has been regarded as a very promising therapeutic approach. This narrative review summarizes the current literature about the nutritional interventions in severe TBI pediatric patients concerning the most optimal timing, route, and formula to be provided to these patients hence, limiting the impact of the secondary injury, that commonly follows the initial brain injury, by reducing inflammation and enhancing neuroprotection of the neural cells. Several clinical trials have shown that such an approach carries great potential in the recovery of TBI patients. Further studies are required to specifically address the pediatric population as they have been shown to have several disparities compared to the adult population. Also, most of the macronutrients, vitamins, and mineral studies presented in this review presented the short-term benefit of such interventions. Future studies are needed to examine the long-term impact of adequate nutritional care in adult and pediatric patients with TBI.

Declarations **Funding**: Not applicable

Conflicts of Interest: The authors declare that they have no conflict of interest.

Ethics Approval: Not applicable

Consent to Participate: Not applicable

Consent for Publication: Not applicable

Availability of Data and Material (Data Transparency): Not applicable

Code Availability: Not applicable

Authors' Contributions: Each author has contributed in the same manner to this manuscript.

References

Alexiou GA, Sfakianos G, Prodromou N (2011) Pediatric head trauma. J Emerg Trauma and Shock 4(3):403–408. https://doi.org/10.4103/0974-2700.83872

Allen KA (2016) Pathophysiology and treatment of severe traumatic brain injuries in children. J Neurosci Nurs 48(1):15–E1. https://doi.org/10.1097/JNN.0000000000000176

Araki T, Yokota H, Morita A (2017) Pediatric traumatic brain injury: characteristic features, diagnosis, and management. Neurol Med Chir 57(2):82–93. https://doi.org/10.2176/nmc.ra.2016-0191

Ashwal S, Siebold L, Krueger AC, Wilson CG (2021) Post-traumatic Neuroinflammation: relevance to pediatrics. Pediatr Neurol 122:50–58. https://doi.org/10.1016/j.pediatrneurol.2021.04.010

Bagheri S, Ekramzadeh M (2018) Metabolic profile and nutritional status of traumatic brain injury patients. Int J Nutr Sci 3(4):177–184

Bailes JE, Abusuwwa R, Arshad M, Chowdhry SA, Schleicher D, Hempeck N, Gandhi YN, Jaffa Z, Bokhari F, Karahalios D, Barkley J, Patel V, Sears B (2020) Omega-3 fatty acid supplementation in severe brain trauma: case for a large multicenter trial. J Neurosurg:1–5. Advance online publication. https://doi.org/10.3171/2020.3.JNS20183

Balakrishnan B, Flynn-O'Brien KT, Simpson PM, Dasgupta M, Hanson SJ (2019) Enteral nutrition initiation in children admitted to pediatric intensive care units after traumatic brain injury. Neurocrit Care 30(1):193–200. https://doi.org/10.1007/s12028-018-0597-6

Bastian L, Weimann A (2002) Immunonutrition in patients after multiple trauma. Br J Nutr 87(Suppl 1):S133–S134. https://doi.org/10.1079/bjn2001466

Böhm V (2018) Vitamin E. Antioxidants (Basel, Switzerland) 7(3):44. https://doi.org/10.3390/antiox7030044

Bower RH, Cerra FB, Bershadsky B, Licari JJ, Hoyt DB, Jensen GL, Van Buren CT, Rothkopf MM, Daly JM, Adelsberg BR (1995) Early enteral administration of a formula (impact) supplemented with arginine, nucleotides, and fish oil in intensive care unit patients: results of a multicenter, prospective, randomized, clinical trial. Crit Care Med 23(3):436–449. https://doi.org/10.1097/00003246-199503000-00006

Bowman CE, Scafidi J, Scafidi S (2019) Metabolic perturbations after pediatric TBI: It's not just about glucose. Exp Neurol 316:74–84. https://doi.org/10.1016/j.expneurol.2019.03.018

Briassoulis G (2014) Nutrition monitoring in the PICU. Pediatric Crit Care Med 579–601. https://doi.org/10.1007/978-1-4471-6362-6_42

Caliri S, Andaloro A, Corallo F, Donato A, Marino S, Mantarro C, Terranova A, Bramanti P, Caminiti F, Rifici C (2019) Recovery of malnutrition in a patient with severe brain injury outcomes: a case report. Medicine 98(40):e16755. https://doi.org/10.1097/MD.0000000000016755

Chamoun R, Suki D, Gopinath SP, Goodman JC, Robertson C (2010) Role of extracellular glutamate measured by cerebral microdialysis in severe traumatic brain injury. J Neurosurg 113(3):564–570. https://doi.org/10.3171/2009.12.JNS09689

Charalambides C, Sgouros S, Sakas D (2010) Intracerebral microdialysis in children. Childs Nerv Syst 26(2):215–220. https://doi.org/10.1007/s00381-009-1031-3

Chitturi J, Li Y, Santhakumar V, Kannurpatti SS (2019) Consolidated biochemical profile of subacute stage traumatic brain injury in early development. Front Neurosci 13:431

Chong SL, Harjanto S, Testoni D, Ng ZM, Low CY, Lee KP, Lee JH (2015) Early hyperglycemia in pediatric traumatic brain injury predicts for mortality, prolonged duration of mechanical ventilation, and intensive care stay. Int J Endocrinol 2015:719476. https://doi.org/10.1155/2015/719476

Clifton GL, Robertson CS, Choi SC (1986) Assessment of nutritional requirements of head-injured patients. J Neurosurg 64(6):895–901. https://doi.org/10.3171/jns.1986.64.6.0895

Cook AM, Peppard A, Magnuson B (2008) Nutrition considerations in traumatic brain injury. Nutr Clin Pract 23(6):608–620. https://doi.org/10.1177/0884533608326060

Cope EC, Morris DR, Levenson CW (2012) Improving treatments and outcomes: an emerging role for zinc in traumatic brain injury. Nutr Rev 70(7):410–413. https://doi.org/10.1111/j.1753-4887.2012.00486.x

Cottenceau V, Masson F, Mahamid E, Petit L, Shik V, Sztark F, Zaaroor M, Soustiel JF (2011) Comparison of effects of equiosmolar doses of mannitol and hypertonic saline on cerebral blood flow and metabolism in traumatic brain injury. J Neurotrauma 28(10):2003–2012. https://doi.org/10.1089/neu.2011.1929

Cruzat V, Macedo Rogero M, Noel Keane K, Curi R, Newsholme P (2018) Glutamine: metabolism and immune function, supplementation and clinical translation. Nutrients 10(11):1564. https://doi.org/10.3390/nu10111564

Cui C, Xu P, Li G, Qiao Y, Han W, Geng C, Liao D, Yang M, Chen D, Jiang P (2019) Vitamin D receptor activation regulates microglia polarization and oxidative stress in spontaneously hypertensive rats and angiotensin II-exposed microglial cells: Role of renin-angiotensin system. Redox Biol 26:101295. https://doi.org/10.1016/j.redox.2019.101295

Curtis L, Epstein P (2014) Nutritional treatment for acute and chronic traumatic brain injury patients. J Neurosurg Sci 58(3):151–160

Dash PK, Hergenroeder GW, Jeter CB, Choi HA, Kobori N, Moore AN (2016) Traumatic brain injury alters methionine metabolism: implications for pathophysiology. Front Syst Neurosci 10:36. https://doi.org/10.3389/fnsys.2016.00036

Davies DJ (2018) Supplementary magnesium in traumatic brain injury: where do we go from here? J R Army Med Corps 164(6):397–398. https://doi.org/10.1136/jramc-2018-000985

Delsoglio M, Achamrah N, Berger MM, Pichard C (2019) Indirect calorimetry in clinical practice. J Clin Med 8(9):1387. https://doi.org/10.3390/jcm8091387

Denniss RJ, Barker LA (2023) Brain trauma and the secondary cascade in humans: review of the potential role of vitamins in reparative processes and functional outcome. Behav Sci (Basel, Switzerland) 13(5):388. https://doi.org/10.3390/bs13050388

Dewan MC, Mummareddy N, Wellons JC 3rd, Bonfield CM (2016) Epidemiology of global pediatric traumatic brain injury: qualitative review. World Neurosurg 91:497–509.e1. https://doi.org/10.1016/j.wneu.2016.03.045. Epub 2016 Mar 25. PMID: 27018009

Donaldson J, Borzatta MA, Matossian D (2000) Nutrition strategies in neurotrauma. Crit Care Nurs Clin North Am 12(4):465–475

Elkon B, Cambrin JR, Hirshberg E, Bratton SL (2014) Hyperglycemia: an independent risk factor for poor outcome in children with traumatic brain injury*. Pediatr Crit Care Med 15(7):623–631. https://doi.org/10.1097/PCC.0000000000000170

Elliott E, Shoykhet M, Bell MJ, Wai K (2022) Nutritional support for pediatric severe traumatic brain injury. Front Pediatr 10:904654. https://doi.org/10.3389/fped.2022.904654

Fedoruk RP, Lee CH, Banoei MM, Winston BW (2023) Metabolomics in severe traumatic brain injury: a scoping review. BMC Neurosci 24(1):54. https://doi.org/10.1186/s12868-023-00824-1

Fenn NE 3rd, Sierra CM (2019) Hyperosmolar therapy for severe traumatic brain injury in pediatrics: a review of the literature. J Pediatr Pharmacol Therap 24(6):465–472. https://doi.org/10.5863/1551-6776-24.6.465

Fesharaki-Zadeh A (2022) Oxidative Stress in Traumatic Brain Injury. Int J Mol Sci 23(21):13000. https://doi.org/10.3390/ijms232113000

Fivez T, Kerklaan D, Mesotten D, Verbruggen S, Joosten K, Van den Berghe G (2017) Evidence for the use of parenteral nutrition in the pediatric intensive care unit. Clin Nutr 36(1):218–223., ISSN 0261-5614,. https://doi.org/10.1016/j.clnu.2015.11.004

Flower O, Hellings S (2012) Sedation in traumatic brain injury. Emerg Med Int 2012:637171. https://doi.org/10.1155/2012/637171

Galuta A, Ghinda D, Chen S, Bedaiwy M, Jabri H, Alshardan M, et al (2018) The 3rd joint symposium of the international and national neurotrauma societies and AANS/CNS section on neurotrauma and critical care August 11–16, 2018 Toronto, Canada. J Neurotrauma 16:A-1

Gault MH, Dixon ME, Doyle M, Cohen WM (1968) Hypernatremia, azotemia, and dehydration use to high-protein tube feeding. Ann Intern Med 68(4):778–791. https://doi.org/10.7326/0003-4819-68-4-778

Giza CC, Hovda DA (2014) The new neurometabolic cascade of concussion. Neurosurgery 75 Suppl 4(0 4):S24–S33. https://doi.org/10.1227/NEU.0000000000000505

Grau-Carmona T, Bonet-Saris A, García-de-Lorenzo A, Sánchez-Alvarez C, Rodríguez-Pozo A, Acosta-Escribano J et al (2015) Influence of n-3 polyunsaturated fatty acids enriched lipid emulsions on nosocomial infections and clinical outcomes in critically ill patients: ICU lipids study. Crit Care Med 43(1):31–39

Hajiaghamemar M, Kilbaugh T, Arbogast KB, Master CL, Margulies SS (2020) Using serum amino acids to predict traumatic brain injury: a systematic approach to utilize multiple biomarkers. Int J Mol Sci 21(5):1786. https://doi.org/10.3390/ijms21051786

Härtl R, Gerber LM, Ni Q, Ghajar J (2008) Effect of early nutrition on deaths due to severe traumatic brain injury. J Neurosurg 109(1):50–56. https://doi.org/10.3171/JNS/2008/109/7/0050

Haydel MJ, Weisbrod LJ, Saeed W (2023) Pediatric Head Trauma. [Updated 2022 Nov 12]. In: StatPearls [Internet]. StatPearls Publishing, Treasure Island (FL)

Herrera E, Barbas C (2001) Vitamin E: action, metabolism and perspectives. J Physiol Biochem 57(2):43–56

Heyland D, Muscedere J, Wischmeyer PE, Cook D, Jones G, Albert M, Elke G, Berger MM, Day AG, Canadian Critical Care Trials Group (2013) A randomized trial of glutamine and antioxidants in critically ill patients. N Engl J Med 368(16):1489–1497. https://doi.org/10.1056/NEJMoa1212722

Horras S, Bepko J, Longstreet N (2015) Fluid, electrolyte, and acid–base disorders. In: Paulman P, Taylor R (eds) Family medicine. Springer, Cham. https://doi.org/10.1007/978-1-4939-0779-3_103-1

Institute of Medicine (US) Committee on Nutrition, Trauma, and the Brain (2011a) In: Erdman J, Oria M, Pillsbury L (eds) Nutrition and traumatic brain injury: improving acute and subacute health outcomes in military personnel. National Academies Press (US), Washington (DC), p 17, Summary of Recommendations

Institute of Medicine (US) Committee on Nutrition, Trauma, and the Brain (2011b) In: Erdman J, Oria M, Pillsbury L (eds) Nutrition and traumatic brain injury: improving acute and subacute health outcomes in military personnel. National Academies Press (US), Washington DC, p 12, Magnesium

Institute of Medicine (US) Committee on Nutrition, Trauma, and the Brain (2011c) In: Erdman J, Oria M, Pillsbury L (eds) Nutrition and traumatic brain injury: improving acute and subacute health outcomes in military personnel. National Academies Press (US), Washington DC, p 16, Zinc

Jeter CB, Hergenroeder GW, Ward NH 3rd, Moore AN, Dash PK (2012) Human traumatic brain injury alters circulating L-arginine and its metabolite levels: possible link to cerebral blood flow, extracellular matrix remodeling, and energy status. J Neurotrauma 29(1):119–127. https://doi.org/10.1089/neu.2011.2029

Jourdan C, Brugel D, Hubeaux K, Toure H, Laurent-Vannier A, Chevignard M (2012) Weight gain after childhood traumatic brain injury: a matter of concern. Dev Med Child Neurol 54(7):624–628. https://doi.org/10.1111/j.1469-8749.2012.04291.x

Jung E, Ro YS, Park JH, Moon SB, Lee SGW, Park GJ, Ryu HH, Shin SD (2022) Vitamin D deficiency and prognosis after traumatic brain injury with intracranial injury: a multi-center observational study. J Neurotrauma 39(19-20):1408–1416. https://doi.org/10.1089/neu.2022.0053

Justo Meirelles CM, de Aguilar-Nascimento JE (2011) Enteral or parenteral nutrition in traumatic brain injury: a prospective randomised trial. Nutr Hosp 26(5):1120–1124. https://doi.org/10.1590/S0212-16112011000500030

Kafadar AM, Sanus GZ, Is M, Coskun A, Tanriverdi T, Hanimoglu H, Uzan M (2007) Prolonged elevation of magnesium in the cerebrospinal fluid of patients with severe head injury. Neurol Res 29(8):824–829. https://doi.org/10.1179/016164107X181879

Kagan I, Zusman O, Bendavid I, Theilla M, Cohen J, Singer P (2018) Validation of carbon dioxide production (VCO_2) as a tool to calculate resting energy expenditure (REE) in mechanically ventilated critically ill patients: a retrospective observational study. Crit Care (London, England) 22(1):186. https://doi.org/10.1186/s13054-018-2108-8

Kang W, Kudsk KA (2007) Is there evidence that the gut contributes to mucosal immunity in humans? JPEN J Parenter Enteral Nutr 31(3):246–258. https://doi.org/10.1177/0148607107031003246

Kanya Wati D (2021) Traumatic brain injury in children. IntechOpen. https://doi.org/10.5772/intechopen.96010

Kattelmann KK, Hise M, Russell M, Charney P, Stokes M, Compher C (2006) Preliminary evidence for a medical nutrition therapy protocol: enteral feedings for critically ill patients. J Am Diet Assoc 106(8):1226–1241. https://doi.org/10.1016/j.jada.2006.05.320

Ketharanathan N, Yamamoto Y, Rohlwink UK, Wildschut ED, Mathôt RAA, de Lange ECM, de Wildt SN, Argent AC, Tibboel D, Figaji AA (2019) Combining brain microdialysis and translational pharmacokinetic modeling to predict drug concentrations in pediatric severe traumatic brain injury: the next step toward evidence-based pharmacotherapy? J Neurotrauma 36(1):111–117. https://doi.org/10.1089/neu.2017.5588

Khayatan D, Razavi SM, Arab ZN, Niknejad AH, Nouri K, Momtaz S, Gumpricht E, Jamialahmadi T, Abdolghaffari AH, Barreto GE, Sahebkar A (2022) Protective effects of curcumin against traumatic brain injury. Biomed Pharmacoth 154:113621. https://doi.org/10.1016/j.biopha.2022.113621

Khazdouz M, Mazidi M, Ehsaei MR, Ferns G, Kengne AP, Norouzy AR (2018) Impact of zinc supplementation on the clinical outcomes of patients with severe head trauma: a double-blind randomized clinical trial. J Diet Suppl 15(1):1–10. https://doi.org/10.1080/19390211.2017.1304486

Kinoshita K (2016) Traumatic brain injury: pathophysiology for neurocritical care. J Inten Care 4:29. https://doi.org/10.1186/s40560-016-0138-3

Kurtz P, Rocha EEM (2020) Nutrition therapy, glucose control, and brain metabolism in traumatic brain injury: a multimodal monitoring approach. Front Neurosci 14:190. https://doi.org/10.3389/fnins.2020.00190

Lai JQ, Shi YC, Lin S, Chen XR (2022) Metabolic disorders on cognitive dysfunction after traumatic brain injury. Trends Endocrinol Metab 33(7):451–462. https://doi.org/10.1016/j.tem.2022.04.003

Larsen BMK, Beggs MR, Leong AY, Kang SH, Persad R, Garcia Guerra G (2018) Can energy intake alter clinical and hospital outcomes in PICU? Clin Nutr ESPEN 24:41–46. https://doi.org/10.1016/j.clnesp.2018.02.002

Lee HY, Oh BM (2022) Nutrition management in patients with traumatic brain injury: a narrative review. Brain NeuroRehabil 15(1):e4. https://doi.org/10.12786/bn.2022.15.e4

Lee J, Corl K, Levy MM (2021b) Fluid therapy and acute respiratory distress syndrome. Crit Care Clin 37(4):867–875. https://doi.org/10.1016/j.ccc.2021.05.012

Lee JM, Jeong SW, Kim MY, Park JB, Kim MS (2019) The effect of vitamin D supplementation in patients with acute traumatic brain injury. World Neurosurg 126:e1421–e1426. https://doi.org/10.1016/j.wneu.2019.02.244

Lee SJ, Lee HJ, Jung YJ, Han M, Lee SG, Hong SK (2021a) Comparison of measured energy expenditure using indirect calorimetry vs predictive equations for liver transplant recipients. JPEN J Parenter Enteral Nutr 45(4):761–767. https://doi.org/10.1002/jpen.1932

Leichtle SW, Sarma AK, Strein M, Yajnik V, Rivet D, Sima A, Brophy GM (2020) High-dose intravenous ascorbic acid: ready for prime time in traumatic brain injury? Neurocrit Care 32(1):333–339. https://doi.org/10.1007/s12028-019-00829-x

Li Z, Liu Y, Wei R, Yong VW, Xue M (2022) The important role of zinc in neurological diseases. Biomol Ther 13(1):28. https://doi.org/10.3390/biom13010028

Luy SC, Dampil OA (2018) Comparison of the harris-benedict equation, bioelectrical impedance analysis, and indirect calorimetry for measurement of basal metabolic rate among adult obese Filipino patients with prediabetes or Type 2 diabetes mellitus. J ASEAN Fed Endocr Soc 33(2):152–159. https://doi.org/10.15605/jafes.033.02.07

Lyons MWH, Blackshaw WJ (2018) Does magnesium sulfate have a role in the management of severe traumatic brain injury in civilian and military populations? A systematic review and meta-analysis. J R Army Med Corps 164(6):442–449. https://doi.org/10.1136/jramc-2018-000916

Macrae D, Grieve R, Allen E, Sadique Z, Morris K, Pappachan J, Parslow R, Tasker RC, Elbourne D, CHiP Investigators (2014) A randomized trial of hyperglycemic control in pediatric intensive care. N Engl J Med 370(2):107–118. https://doi.org/10.1056/NEJMoa1302564

Mahapatra A (2019) Role of nutrition in head injury patient. Acta Sci Nutr Health 3(9):111–114

Mao XY, Yin XX, Guan QW, Xia QX, Yang N, Zhou HH, Liu ZQ, Jin WL (2021) Dietary nutrition for neurological disease therapy: Current status and future directions. Pharmacol Ther 226:107861. https://doi.org/10.1016/j.pharmthera.2021.107861

Mass-Ramirez SM, Lozada Marintez ID, Bohorquez-Rivero J, Garcia-Ballestas E, Bustos D, Poveda-Castillo IT, Delgado-Marrugo RD, Daza-Martínez VJ, Sastre-Martínez AD, Ortega-Sierra MG, Janjua T, Bolaño-Romero MP, Moscote-Salazar LR (2021) Gastroparesis as a complication in the patient with traumatic brain injury. Roman Neurosurg 35(3):345–351. https://doi.org/10.33962/roneuro-2021-058

McClave SA, Taylor BE, Martindale RG, Warren MM, Johnson DR, Braunschweig C, McCarthy MS, Davanos E, Rice TW, Cresci GA, Gervasio JM, Sacks GS, Roberts PR, Compher C, Society of Critical Care Medicine, & American Society for Parenteral and Enteral Nutrition (2016) Guidelines for the provision and assessment of nutrition support therapy in the adult critically ill patient: society of critical care medicine (SCCM) and American Society for Parenteral and Enteral Nutrition (A.S.P.E.N.). JPEN J Parenter Enteral Nutr 40(2):159–211. https://doi.org/10.1177/0148607115621863

Mckee AC, Daneshvar DH (2015) The neuropathology of traumatic brain injury. Handb Clin Neurol 127:45–66. https://doi.org/10.1016/B978-0-444-52892-6.00004-0

Mechanick JI (2006) Metabolic mechanisms of stress hyperglycemia. JPEN J Parenter Enteral Nutr 30(2):157–163. https://doi.org/10.1177/0148607106030002157

Meert KL, Daphtary KM, Metheny NA (2004) Gastric vs small-bowel feeding in critically ill children receiving mechanical ventilation: a randomized controlled trial. Chest 126(3):872–878. https://doi.org/10.1378/chest.126.3.872

Meinert E, Bell MJ, Buttram S, Kochanek PM, Balasubramani GK, Wisniewski SR, Adelson PD, Pediatric Traumatic Brain Injury Consortium: Hypothermia Investigators (2018) Initiating nutritional support before 72 hours is associated with favorable outcome after severe traumatic brain injury in children: a secondary analysis of a randomized, controlled trial of

therapeutic hypothermia. Pediatr Crit Care Med 19(4):345–352. https://doi.org/10.1097/PCC.0000000000001471
Michaud LJ, Rivara FP, Grady MS, Reay DT (1992) Predictors of survival and severity of disability after severe brain injury in children. Neurosurgery 31(2):254–264. https://doi.org/10.1227/00006123-199208000-00010
Morgan A, Ward E, Murdoch B, Kennedy B, Murison R (2003) Incidence, characteristics, and predictive factors for Dysphagia after pediatric traumatic brain injury. J Head Trauma Rehabil 18(3):239–251. https://doi.org/10.1097/00001199-200305000-00002
Mrozek S, Vardon F, Geeraerts T (2012) Brain temperature: physiology and pathophysiology after brain injury. Anesthesiol Res Pract 2012:989487. https://doi.org/10.1155/2012/989487
Nayak R, Attry S, Ghosh SN (2018) Serum magnesium as a marker of neurological outcome in severe traumatic brain injury patients. Asian J Neurosurg 13(3):685–688. https://doi.org/10.4103/ajns.AJNS_232_16
NICE-SUGAR Study Investigators for the Australian and New Zealand Intensive Care Society Clinical Trials Group and the Canadian Critical Care Trials Group, Finfer S, Chittock D, Li Y, Foster D, Dhingra V, Bellomo R, Cook D, Dodek P, Hebert P, Henderson W, Heyland D, Higgins A, McArthur C, Mitchell I, Myburgh J, Robinson B, Ronco J (2015) Intensive versus conventional glucose control in critically ill patients with traumatic brain injury: long-term follow-up of a subgroup of patients from the NICE-SUGAR study. Intens Care Med 41(6):1037–1047. https://doi.org/10.1007/s00134-015-3757-6
Nordström CH (2017) Cerebral microdialysis in TBI-limitations and possibilities. Acta Neurochir 159(12):2275–2277. https://doi.org/10.1007/s00701-017-3339-1
Nwafor D, Goeckeritz J, Hasanpour Z, Davidson C, Lucke-Wold B (2023) Nutritional support following traumatic brain injury: a comprehensive review. Exp Res Hypoth Med 8(3):236–247. https://doi.org/10.14218/erhm.2022.00086
Oddo M, Hutchinson PJ (2018) Understanding and monitoring brain injury: the role of cerebral microdialysis. Intensive Care Med 44(11):1945–1948. https://doi.org/10.1007/s00134-017-5031-6
Painter TJ, Rickerds J, Alban RF (2015) Immune enhancing nutrition in traumatic brain injury - a preliminary study. Int J Surg (London, England) 21:70–74. https://doi.org/10.1016/j.ijsu.2015.07.008
Panchenko PE, Hippauf L, Konsman JP, Badaut J (2023) Do astrocytes act as immune cells after pediatric TBI? Neurobiol Dis 185:106231. https://doi.org/10.1016/j.nbd.2023.106231
Parish RA, Webb KS (1988) Hyperglycemia is not a poor prognostic sign in head-injured children. J Trauma 28(4):517–519. https://doi.org/10.1097/00005373-198804000-00017
Park GJ, Ro YS, Yoon H, Lee SGW, Jung E, Moon SB, Kim SC, Shin SD, Pan-Asia Trauma Outcomes Study for Traumatic Brain Injury (PATOS-TBI) Research Network (2022) Serum vitamin E level and functional prognosis after traumatic brain injury with intracranial injury: A multicenter prospective study. Front Neurol 13:1008717. https://doi.org/10.3389/fneur.2022.1008717
Pepe JL, Barba CA (1999) The metabolic response to acute traumatic brain injury and implications for nutritional support. J Head Trauma Rehabil 14(5):462–474. https://doi.org/10.1097/00001199-199910000-00007
Peterson AB, Xu L, Daugherty J, Breiding MJ (2019) Surveillance report of traumatic brain injury-related emergency department visits, hospitalizations, and deaths, United States, 2014
Phillis JW, O'Regan MH (2003) The role of phospholipases, cyclooxygenases, and lipoxygenases in cerebral ischemic/traumatic injuries. Crit Rev™ Neurobiol 15(1)
Poblete RA, Yaceczko S, Aliakbar R, Saini P, Hazany S, Breit H, Louie SG, Lyden PD, Partikian A (2023) Optimization of nutrition after brain injury: mechanistic and therapeutic considerations. Biomedicine 11(9):2551. https://doi.org/10.3390/biomedicines11092551
Polidori MC, Mecocci P, Frei B (2001) Plasma vitamin C levels are decreased and correlated with brain damage in patients with intracranial hemorrhage or head trauma. Stroke 32(4):898–902

Pontzer H, Yamada Y, Sagayama H, Ainslie PN, Andersen LF, Anderson LJ, Arab L, Baddou I, Bedu-Addo K, Blaak EE, Blanc S, Bonomi AG, Bouten CVC, Bovet P, Buchowski MS, Butte NF, Camps SG, Close GL, Cooper JA, Cooper R et al (2021) Daily energy expenditure through the human life course. Science (New York, N.Y.) 373(6556):808–812. https://doi.org/10.1126/science.abe5017

Pop V, Badaut J (2011) A neurovascular perspective for long-term changes after brain trauma. Transl Stroke Res 2(4):533–545. https://doi.org/10.1007/s12975-011-0126-9

Popernack ML, Gray N, Reuter-Rice K (2015) Moderate-to-severe traumatic brain injury in children: complications and rehabilitation strategies. J Pediatr Health Care 29(3):e1–7. https://doi.org/10.1016/j.pedhc.2014.09.003. Epub 2014 Oct 28. PMID: 25449002; PMCID: PMC4409446

Postolache TT, Wadhawan A, Can A, Lowry CA, Woodbury M, Makkar H, Hoisington AJ, Scott AJ, Potocki E, Benros ME, Stiller JW (2020) Inflammation in traumatic brain injury. J Alzheimer's Dis: JAD 74(1):1–28. https://doi.org/10.3233/JAD-191150

Prins ML, Matsumoto J (2016) Metabolic response of pediatric traumatic brain injury. J Child Neurol 31(1):28–34. https://doi.org/10.1177/0883073814549244

Quintard H, Ichai C (2019) Nutritional and metabolic supplementation for the injured brain: an update. Curr Opin Crit Care 25(2):126–131. https://doi.org/10.1097/MCC.0000000000000588

Rai VRH, Phang LF, Sia SF, Amir A, Veerakumaran JS, Kassim MKA, Othman R, Tah PC, Loh PS, Jailani MIO, Ong G (2017) Effects of immunonutrition on biomarkers in traumatic brain injury patients in Malaysia: a prospective randomized controlled trial. BMC Anesthesiol 17(1):81. https://doi.org/10.1186/s12871-017-0369-4

Razmkon A, Sadidi A, Sherafat-Kazemzadeh E, Mehrafshan A, Jamali M, Malekpour B, Saghafinia M (2011) Administration of vitamin C and vitamin E in severe head injury: a randomized double-blind controlled trial. Neurosurgery 58:133–137

Ringman JM, Frautschy SA, Cole GM, Masterman DL, Cummings JL (2005) A potential role of the curry spice curcumin in Alzheimer's disease. Curr Alzheimer Res 2(2):131–136. https://doi.org/10.2174/1567205053585882

Sabers EJ, Reiter PD, Skillman HE, DeMasellis G (2020) Concentrated hypertonic saline in severe pediatric traumatic brain injury. Brain Inj 34(6):828–833. https://doi.org/10.1080/02699052.2020.1752938

Scafidi S, Racz J, Hazelton J, McKenna MC, Fiskum G (2011) Neuroprotection by acetyl-L-carnitine after traumatic injury to the immature rat brain. Dev Neurosci 32(5-6):480–487

Scrimgeour AG, Condlin ML (2014) Nutritional treatment for traumatic brain injury. J Neurotrauma 31(11):989–999. https://doi.org/10.1089/neu.2013.3234

Shadnoush M, Zahedi H, Norouzy A, Sahebkar A, Sadeghi O, Najafi A, Hosseini S, Qorbani M, Ahmadi A, Ardehali SH, Hosseinzadeh-Attar MJ (2020) Effects of supplementation with curcuminoids on serum adipokines in critically ill patients: a randomized double-blind placebo-controlled trial. Phytoth Res: PTR 34(12):3180–3188. https://doi.org/10.1002/ptr.6749

Sharma S, Kumar A, Choudhary A, Sharma S, Khurana L, Sharma N, Kumar V, Bisht A (2020) Neuroprotective role of oral vitamin d supplementation on consciousness and inflammatory biomarkers in determining severity outcome in acute traumatic brain injury patients: a double-blind randomized clinical trial. Clin Drug Investig 40(4):327–334. https://doi.org/10.1007/s40261-020-00896-5

Shi J, Tan L, Ye J, Hu L (2020) Hypertonic saline and mannitol in patients with traumatic brain injury: a systematic and meta-analysis. Medicine 99(35):e21655. https://doi.org/10.1097/MD.0000000000021655

Şimşek T, Şimşek HU, Cantürk NZ (2014) Response to trauma and metabolic changes: post-traumatic metabolism. Ulusal Cerrahi Dergisi 30(3):153–159. https://doi.org/10.5152/UCD.2014.2653

Smith RL, Lin JC, Adelson PD, Kochanek PM, Fink EL, Wisniewski SR, Bayir H, Tyler-Kabara EC, Clark RS, Brown SD, Bell MJ (2012) Relationship between hyperglycemia and outcome

in children with severe traumatic brain injury. Pediatr Crit Care Med 13(1):85–91. https://doi.org/10.1097/PCC.0b013e3182192c30
Standiford L, O'Daniel M, Trigger C et al (2018) A randomized cohort study of the efficacy of PO magnesium in the treatment of acute concussions in adolescents. Am J Emerg Med. https://doi.org/10.1016/j.ajem.2020.05.010
Sun L, Qiang R, Yang Y, Jiang ZL, Wang GH, Zhao GW, Ren TJ, Jiang R, Xu LH (2014) L-serine treatment may improve neurorestoration of rats after permanent focal cerebral ischemia potentially through improvement of neurorepair. PLoS One 9(3):e93405. https://doi.org/10.1371/journal.pone.0093405
Sundstrøm T, Grände PO, Luoto T, Rosenlund C, Undén J, Wester KG (2020) Management of severe traumatic brain injury. Springer, Cham
Taha AA, Badr L, Westlake C, Dee V, Mudit M, Tiras KL (2011) Effect of early nutritional support on intensive care unit length of stay and neurological status at discharge in children with severe traumatic brain injury. J Neurosci Nurs 43(6):291–297. https://doi.org/10.1097/JNN.0b013e318234e9b2
Tan M, Zhu JC, Yin HH (2011) Enteral nutrition in patients with severe traumatic brain injury: reasons for intolerance and medical management. Br J Neurosurg 25(1):2–8. https://doi.org/10.3109/02688697.2010.522745
Temkin NR, Anderson GD, Winn HR, Ellenbogen RG, Britz GW, Schuster J, Lucas T, Newell DW, Mansfield PN, Machamer JE, Barber J, Dikmen SS (2007) Magnesium sulfate for neuroprotection after traumatic brain injury: a randomised controlled trial. Lancet Neurol 6(1):29–38. https://doi.org/10.1016/S1474-4422(06)70630-5
Terré R, Mearin F (2007) Prospective evaluation of oro-pharyngeal dysphagia after severe traumatic brain injury. Brain Inj 21(13-14):1411–1417. https://doi.org/10.1080/02699050701785096
Thango NS, Rohlwink UK, Dlamini L, Tshavhungwe MP, Banderker E, Salie S, Enslin JMN, Figaji AA (2021) Brain interstitial glycerol correlates with evolving brain injury in paediatric traumatic brain injury. Child's Nerv Syst 37(5):1713–1721. https://doi.org/10.1007/s00381-021-05058-2
Van den Berghe G, Schoonheydt K, Becx P, Bruyninckx F, Wouters PJ (2005) Insulin therapy protects the central and peripheral nervous system of intensive care patients. Neurology 64(8):1348–1353. https://doi.org/10.1212/01.WNL.0000158442.08857.FC
Vasileiou G, Qian S, Iyengar R, Mulder MB, Gass LM, Parks J, Pust GD, Rattan R, Lineen E, Byers P, Yeh DD (2020) Use of predictive equations for energy prescription results in inaccurate estimation in trauma patients. Nutr Clin Pract 35(5):927–932. https://doi.org/10.1002/ncp.10372
Vespa P, McArthur DL, Stein N, Huang SC, Shao W, Filippou M, Etchepare M, Glenn T, Hovda DA (2012) Tight glycemic control increases metabolic distress in traumatic brain injury: a randomized controlled within-subjects trial. Crit Care Med 40(6):1923–1929. https://doi.org/10.1097/CCM.0b013e31824e0fcc
Vlasselaers D, Milants I, Desmet L, Wouters PJ, Vanhorebeek I, van den Heuvel I, Mesotten D, Casaer MP, Meyfroidt G, Ingels C, Muller J, Van Cromphaut S, Schetz M, Van den Berghe G (2009) Intensive insulin therapy for patients in paediatric intensive care: a prospective, randomised controlled study. Lancet (London, England) 373(9663):547–556. https://doi.org/10.1016/S0140-6736(09)60044-1]
Wahls T, Rubenstein L, Hall M, Snetselaar L (2014) Assessment of dietary adequacy for important brain micronutrients in patients presenting to a traumatic brain injury clinic for evaluation. Nutr Neurosci 17(6):252–259. https://doi.org/10.1179/1476830513Y.0000000088
Williams FN, Jeschke MG, Chinkes DL, Suman OE, Branski LK, Herndon DN (2009) Modulation of the hypermetabolic response to trauma: temperature, nutrition, and drugs. J Am Coll Surg 208(4):489–502. https://doi.org/10.1016/j.jamcollsurg.2009.01.022
Wu A, Ying Z, Gomez-Pinilla F (2006) Dietary curcumin counteracts the outcome of traumatic brain injury on oxidative stress, synaptic plasticity, and cognition. Exp Neurol 197(2):309–317. https://doi.org/10.1016/j.expneurol.2005.09.004

Wu G, Meininger CJ, McNeal CJ, Bazer FW, Rhoads JM (2021) Role of L-Arginine in nitric oxide synthesis and health in humans. Adv Exp Med Biol 1332:167–187. https://doi.org/10.1007/978-3-030-74180-8_10

Xiong W, Qian K (2021) Low-protein, hypocaloric nutrition with glutamine versus full-feeding in the acute phase in ICU patients with severe traumatic brain injury. Neuropsychiatr Dis Treat 17:703–710. https://doi.org/10.2147/NDT.S296296

Young B, Ott L, Yingling B, McClain C (1992) Nutrition and brain injury. J Neurotrauma 9(Suppl 1):S375–S383

Young B, Ott L, Kasarskis E, Rapp R, Moles K, Dempsey RJ, Tibbs PA, Kryscio R, McClain C (1996) Zinc supplementation is associated with improved neurologic recovery rate and visceral protein levels of patients with severe closed head injury. J Neurotrauma 13(1):25–34. https://doi.org/10.1089/neu.1996.13.25

Zahedi H, Hosseinzadeh-Attar MJ, Sahebkar A, Ranjbar SH, Najafi A, Hosseini S, Qorbani M, Ahmadi A, Ardehali SH, Moravvej H, Pourmand G, Norouzy A, Shadnoush M (2020) Therapeutic effects of supplementation with Curcuminoids in critically ill patients receiving enteral nutrition: a randomized controlled trial protocol. J Diab Metab Disord 19(2):1609–1614. https://doi.org/10.1007/s40200-019-00451-w

Zheng Z, Wang S, Wu C, Cao Y, Gu Q, Zhu Y, Zhang W, Hu W (2022) Gut microbiota dysbiosis after traumatic brain injury contributes to persistent microglial activation associated with upregulated Lyz2 and shifted tryptophan metabolic phenotype. Nutrients 14(17):3467. https://doi.org/10.3390/nu14173467

Zhou T, Kalanuria A (2018) Cerebral Microdialysis in Neurocritical Care. Curr Neurol Neurosci Rep 18(12):101. https://doi.org/10.1007/s11910-018-0915-6

Chapter 18
Nutritional Support for Mental Health in TBI Patients

Jia Whei See

Abstract Many studies have talked about traumatic brain injury (TBI) and how it can affect the mental illness of the affected individual, but nutritional interventions are rarely discussed on how they can assist in alleviating the symptoms of the affected individual. Current reviews have discussed each aspect of dietary interventions such as vitamin E, green tea, curcumin and so on how to alleviate symptoms of mental health illness. For example, depression, anxiety disorder, post-traumatic stress disorder (PTSD) and aggression. In this chapter, we will discuss the relationship between nutritional support for mental health in traumatic brain injury patients and address the gap in knowledge in regard to the aspect. We will also address the future aspects and the limitations of this aspect.

Keywords Mental illness · Mental health · Dietary interventions · Nutritional interventions

18.1 Introduction

Traumatic brain injuries (TBIs) are a grave and mysterious issue. This is due to unrecognised injuries that are excluded from epidemiological data (Bogner et al. 2020). TBI can have different categories graded by its severity which is mild, moderate and severe injury using the Glasgow Coma Scale (Bogner et al. 2020). In those with people suffering from TBI, it was shown that people with TBI, especially mild TBI, have recovered within 2 months (Bogner et al. 2020). But recovery is not absolute into a perfect recovery; some experienced persistent symptoms such as headaches, fatigue, cognitive impairment, personality changes and psychiatric disorders

J. W. See (✉)
General Medicine, Sriwijaya University, Palembang, Indonesia

W. Mohamed (ed.), *Nutrition and Traumatic Brain Injury (TBI)*, Nutritional Neurosciences, https://doi.org/10.1007/978-981-97-6341-2_18

(Bogner et al. 2020). The life quality of these people will be affected, including unemployment, social detachment, family problems and relationship problems (Bogner et al. 2020).

Traumatic brain injury has been shown to have negative effects on cognitive and mental conditions, including major depressive disorder, anxiety disorder and aggressiveness (Duarte et al. 2023). Neuropsychiatric symptoms such as memory and cognitive dysfunction, anxiety, major depressive disorder, social withdrawal and aggressiveness will exist long term after the first brain injury and can last for decades (Duarte et al. 2023). These factors can slow down the recovery and rehabilitation rate (Duarte et al. 2023). People who suffer from this will also suffer from work issues and rely heavily on the health-care system (Duarte et al. 2023). Studies have shown that there is no connection between the gravity of the first injury or with pain and these symptoms (Duarte et al. 2023). In 1848, a man named Phineas P. Gage survived a severe brain trauma from a tamping iron, which led to extreme personality changes (Faria 2013). He was a friendly, cheerful, motivated man who changed into an obstinate, socially uninhibited individual that reflects the frontal lobe syndrome (Faria 2013). This proves that the relationship between traumatic brain injury and mental illness caused by the trauma has been existing for more than decades.

Is there a way to combat these problems faced by these poor people? Most studies encourage patient bedrest or medical or physiotherapy-led therapies in either acute or chronic recovery in TBI patients (Bogner et al. 2020). However, nutrition plays a very important role in preserving neurological function, but there are limited amount of research that emphasises the part of nutrition in the mental health of TBI patients. Even in the 2017 concussion consensus statement, there is no mention of the nutrition role. This chapter will talk about how nutrition can improve recovery for mental health in TBI patients and show some promising results for the correlation between nutrition and mental health in TBI patients.

18.2 Definition

18.2.1 Traumatic Brain Injury (TBI)

TBI is a kind of brain injury where some force externally causes damage to the head (Duarte et al. 2023). Symptoms vary and are categorised based on the severity level. It is divided into mild, moderate and severe depending on how serious the damage is (Duarte et al. 2023).

18.2.2 Psychiatric Illness/Disorder

According to the American Psychiatric Association, psychiatric illness or, in layman's terms, mental illness means a health condition that affects mood, thinking, behaviour or a combination of these (Duarte et al. 2023). Risk factors vary; it could be because of environmental, family, relationships or even neurological causes, depending on each individual (Duarte et al. 2023).

18.2.3 History of Mental Disorders Caused by TBI

Trephination, also called trepanation of the human skull, is one of the ancient documented surgical procedures done by humans (Faria 2013). There is evidence of trephined skulls all over the world, such as from the Old World of Europe or Asia to the New World, especially Peru which is located in South America (Faria 2013). It also evidently happened in the Neolithic age or to the very earliest stages in human history. Although the exact motivations behind this ancient surgery remain unclear, it is clear that it was done by shamans or witch doctors whose aim was to change human behaviour, which paralleled with a speciality called psychosurgery in the middle of the twentieth century (Faria 2013).

It is plausible to suggest that various ailments like uncontrollable migraines, seizures, belief of possession by evil spirits or mental illness, conveyed by unusual behaviour could have been indications for performing surgery by the shaman or medicine man in the earliest stages of human history (Faria 2013). Dr William Osler states that trephination was performed for conditions such as epilepsy, infantile seizures, migraine or other brain diseases which are known to result from confined demons, and with a pierce, it gives an escape route for these spirits (Faria 2013). Once approval is given by the clan, the medicine man in pre-Incan civilisation in Peru would perform an incision in the scalp utilising a surgical instrument and use his knife on the affected individual's skull to alleviate migraines or triggering the releasing of evil spirits which is believed to have possessed the affected individual (Faria 2013). Surprisingly, many people from prehistoric times have survived such gruesome surgeries, as archaeologists observed healing at trephined skulls (Faria 2013). However, it is still unclear whether clinical symptoms have improved or behaviour has changed in these affected individuals (Faria 2013).

In ancient Greece and Rome, they have developed surgical equipment for skull penetration (Faria 2013). The Roman doctors (surgeons) created the terebra serrata, which is rolled between the surgeon's hands to perforate the cranium (Faria 2013). Well-known physicians such as Celsus and Galen have used these instruments (Faria 2013). The terebra has played a role in modern manual burr holes and electric drill neurosurgical procedures (Faria 2013).

In the Middle Age and Renaissance, skull fractions no longer are the limitations to this act but expanded to treat insanity and seizure (Faria 2013). There is an art from this period drawn by Hieronymus Bosch named "The Extraction of the Stone of Madness" that show the intersection of descriptive art and surgical practices (Faria 2013). Fabricius of Aquapendente created the triangular trephine instrument to enter the skull, which is used for elevating depressed skull fractures (Faria 2013). There are variations of Peter Treveris's engraving depicting surgical heightening of skull fractures found in antiquarian medicine-themed books (Faria 2013).

The early nineteenth century saw a shift towards more humane treatment, but institutions still remained less equipped to handle the complexities of mental illness (Faria 2013). Cerebral localisation has been analysed with brain pathology studies and famous studies such as the American Crowbar Case (Faria 2013). In 1848, a man named Phineas P. Gage survived a severe brain trauma from a tamping iron, which led to extreme personality changes (Faria 2013). He was a friendly, cheerful, motivated man who changed into an obstinate, socially uninhibited individual that reflects the frontal lobe syndrome (Faria 2013). Dr John M. Harlow analysed and documented the case thoroughly, observing the behavioural changes after the brain trauma (Faria 2013). In 1860, near San Fransisco, Gage died. Gage's skull and the tamping iron that injured him were preserved and donated to Harvard's Warren Anatomical Museum, where it still remains today (Faria 2013).

In 1888, the first psychosurgery was done in Switzerland by Dr Gottlieb Burckhardt. He removed areas of the cerebral cortex which he believed to have modified the behaviour of the affected individual (Faria 2013). In 1935, the Second International Neurologic Congress in London presented some experiments on chimpanzees done by Dr John F. Fulton, which showed emotional expression loss after the removal of the prefrontal cortex (Faria 2013). Two other doctors, Dr Antonio Egas Moniz and Dr Almeida Lima of Portugal also started to perform frontal leucotomies in the 1930s to treat mental illness, after experimenting on frontal lobe ablation in aggressive chimpanzees (Faria 2013). In the 1940s, Dr Walter Freeman from the United States worked with Dr James W. Watts to study the transorbital lobotomy and observed improvements in patients with mood disorders and schizophrenia (Faria 2013). After World War II, the use of lobotomy and psychosurgery has risen to improve mental illness when medication and psychotherapy were limited (Faria 2013). In the mid-twentieth century, a societal shift towards deinstitutionalisation and the establishment of community mental health centres (Faria 2013). But there are still challenges to combating mental illness, contributing to homelessness, crime and violence (Faria 2013).

18.2.4 Statistics of TBI and Mental Health

Mental health illness is usually found in traumatic brain injury patients. In the initial year after brain trauma, a psychiatric diagnosis is found in minimally 77% of the patients. Psychiatric diagnosis consists of anxiety, mood and substance-use disorders that are not unusual and often exist comorbidly (Duarte et al. 2023). A higher rate of mental health disorders can be affected by unpleasant health results, increased mental health care usage and bad quality of life (Duarte et al. 2023).

18.2.5 Correlation Between TBI Severity and Neuropsychiatric Disorder

It is still unclear how the correlation between the severity of traumatic brain injury and mental health illness outcomes. Few studies have shown that between the level of severity in traumatic brain injury and mental health illness outcomes, the higher the severity level is in a traumatic brain injury that is either moderate or severe, the more unbecoming the prolonged neurocognitive impairment and mental health disorder (Duarte et al. 2023). However, some studies also prove that there is an opposite outcome to this, where mild TBI will have more symptoms compared to moderate-to-severe TBI (Halbauer et al. 2009). There are also few studies that analyse the correlation between the level of severity of TBI and mental health illness results, incorporating grief measures, tiredness, migraine, pain and cognitive concerns (Schwarzbold et al. 2008) Moreover, other studies reported that there may have delayed symptom onset of higher severity TBI in comparison to mild TBI in both military and civilian population (Halbauer et al. 2009).

18.2.6 Effects of TBI-Related Factors Like Location, Extent of Damage and Comorbidities on Mental Health Illness Outcomes

Neuropsychiatric symptoms are a common occurrence in affected individuals who experienced a traumatic brain injury (TBI) (Duarte et al. 2023). TBI will cause direct damage to brain tissue, affecting neural circuits that are responsible for mood regulation, cognition and behaviour, causing mood disorders or symptoms like depression, anxiety, irritability and impulsivity (Duarte et al. 2023). Moreover, TBI will affect the let out of pro-inflammatory cytokines and reactive oxygen species, something which neuroinflammation and oxidative stress can be the consequence of it, along with worsening neuropsychiatric symptoms (Duarte et al. 2023). Differences in neurotransmitter levels following traumatic brain injury can also play a role in the progression of symptoms (Duarte et al. 2023). For example, emotional lability, attention deficits and impaired decision-making. Other than that, the mental and emotional impact of traumatic brain injury, not excluding adjustment to disability, personality change and social isolation, can also trigger or exacerbate neuropsychiatric symptoms (Duarte et al. 2023). Thus, we need to further take in the complex interplay of structural, functional, biochemical and psychosocial factors that are important in precisely diagnosing and effectively controlling mental health illness in one with traumatic brain injury, at the same time enhancing their quality of life and functional outcomes.

18.3 TBI-Related Aftermath

18.3.1 Post-Concussion Syndrome (PCS)

Based on the definition of the World Health Organization, there are three symptoms out of 8, including headache, dizziness, fatigue, irritability, sleep disorder, memory disorder, stress intolerance, mood disorder and alcohol (Halbauer et al. 2009).

The diagnosis of PCS would be complicated, as it is needed to differentiate from behavioural disorders (Halbauer et al. 2009). Furthermore, the mechanism of PCS is still unidentified (Halbauer et al. 2009). Studies show that PCS diagnosis has low specificity to mild TBI, which shows that the cause may not only be from brain injury. PCS also happens in people that are not having TBI, particularly those who have been suffering from psychiatric illnesses (Halbauer et al. 2009). There are several risk factors that indicate to have a higher risk of getting acute PCS, which include having a gender female, higher intelligence quotient (IQ) and pain history (Halbauer et al. 2009). While TBI might cause PCS, it does not impose any role in TBI severity because it might also be caused by other risk factors (Halbauer et al. 2009).

18.3.2 Neuropsychiatric Aftermath

The Diagnostic and Statistical Manual of Mental Disorders-Fourth Edition-Text Revision (DSM-IV-TR) has categorised neuropsychiatric caused by TBI, including delirium, dementia, amnesic disorder, mood disorder, anxiety disorder and psychotic illness caused by TBI (Halbauer et al. 2009). There are reviews that addressed the gap in this issue between TBI and mental health disorders including major depressive disorder, manic disorder, PTSD, psychotic disorder, agitation and aggressiveness (Halbauer et al. 2009).

There are three categories that are significant:

1. Mild neurocognitive impairment (MNI)

   MNI defines cognitive impairment and incorporates the cognitive regions of memory, attention, language, speed of processing, perceptual and motor abilities and executive function (Halbauer et al. 2009). Neuropsychological testing is required for diagnosis (Halbauer et al. 2009). If a solitary cognitive-area dysfunction that is memory-related in area is found in the affected individual, mild cognitive impairment will be the diagnosis (Halbauer et al. 2009). If a non-memory-related and single cognitive-area impairment is found, a single mild non-memory-related cognitive impairment will be the diagnosis (Halbauer et al. 2009).

If over two cognitive areas are impaired, in spite of the classification of the impaired area, the result would be mild neurocognitive impairment (Halbauer et al. 2009).

2. Personality disorder caused by a general medical condition

   Diagnostic indicator for this illness needs at least two areas to be influenced in cognition, affectivity, interpersonal functioning and impulse control (Halbauer et al. 2009). However, it lacks specificity regarding cognitive impairment and may lead to overlooking other mood or affect disorders (Halbauer et al. 2009).

3. Post-concussion disorder (PCD)

   This illness needs 3 of the following symptoms for at least 3 months to secure a diagnosis which is tiredness, sleep disorder, migraine, vertigo or dizziness, mood disorder, self-identity disorder and lack of interest or spontaneous behaviour (Halbauer et al. 2009). However, it must be diagnosed with extra caution as it may be diagnosed wrongly in people who have anxiety or depression (Halbauer et al. 2009).

18.3.3 Clinical Bio Approach to TBI

TBI is a multisystem condition, where it not only affects the brain but also affects the other parts of the organs and systems in the body (Halbauer et al. 2009). The biopsychosocial model, which is often utilised in modern psychiatry, has included biological, psychological and social factors in diagnosis and therapy (Halbauer et al. 2009). By utilising this model in TBI, doctors can analyse dysfunction in aspects of mental, work and social interaction in patients (Halbauer et al. 2009).

TBI can cause symptom clusters, which cause cognitive dysfunctions (memory, attention and language), neurobehavioral disorders (mood, anxiety, etc.), somatosensory disruptions (changes in senses), somatic symptoms such as migraine or chronic pain and substance dependence. Each cluster requires specific management approaches (Halbauer et al. 2009).

The brain can re-establish connections and undergo neuroplasticity, which is very useful in TBI recovery and rehabilitation (Halbauer et al. 2009). However, to restore the function correctly is challenging and may require focused interventions. Genetic factors including the APOE genotype, can also influence neuroplastic processes (Halbauer et al. 2009).

It is very crucial to consider both biological and psychological factors in TBI treatment as well as their effects on social functioning (Halbauer et al. 2009). A holistic approach that addresses the particular damage and patient needs can promote rehabilitation and long-term function (Halbauer et al. 2009).

Neuropsychiatric symptoms post-TBI are classified into (Halbauer et al. 2009):

- Cognitive impairment
- Neuropsychological disorders
- Somatosensory impairment

- Somatic symptoms
- Substance abuse

Increasing levels of cognitive abilities are fixed mental functions determined by frontal lobes, including decision-making, abstract thinking, planning, mental flexibility and judgment (Halbauer et al. 2009). TBI can impact these functions, termed executive functions, affecting abilities like time control, judgement and forming ideas (Halbauer et al. 2009). Cognitive impairment is normal post-TBI, but the outcome and the mechanism are unclear due to classification difficulties (Halbauer et al. 2009). Measurement and quantification are challenging as cognition is not a single entity, and assessment tools constrain conceptualisation (Halbauer et al. 2009). TBI can cause diffuse trauma, complicating evaluation by not respecting neuroanatomical boundaries (Halbauer et al. 2009). Neurocognitive deficits and neurobehavioural disorders vary based on TBI impact sites, force intensity, premorbid intellectual status and subsequent treatment (Halbauer et al. 2009). Cognitive symptoms post-TBI include memory issues, attention deficits, language difficulties, visuospatial cognition impairments, sensory-motor integration problems and executive function deficits (Halbauer et al. 2009). Organising cognitive symptoms post-TBI is complex due to the wide range of potential dysfunctions, impacted by various factors including injury severity and patient characteristics (Halbauer et al. 2009) (Tables 18.1, 18.2 and 18.3).

18.3.4 Mechanism of TBI and Mental Health

Brain metabolic depression refers to a state of reduced metabolic activity in the brain following traumatic brain injury (TBI), observed both in humans and animal models (Halbauer et al. 2009). This phase involves decreased glucose utilisation and energy production in brain cells, persisting for as long as 10 days post injury (Halbauer et al. 2009). During this period, cells are less responsive to normal stimulation, leading to neurobehavioural deficits. The brain becomes more vulnerable to further injury, and a second concussive event during this vulnerable period can prolong the time needed for metabolic normalisation by up to 15 days (Halbauer et al. 2009). Experimental studies in animals have shown similar metabolic depression post-injury, impacting synaptic plasticity and the brain's ability to recover (Halbauer et al. 2009). This is why in order to manage TBI and its recovery, it is important to regulate the brain metabolism and assist with neuronal function (Halbauer et al. 2009). There is an important contributor in maintaining the growth, survival and synaptic plasticity in neurons which is a brain-derived neurotrophic factor (Halbauer et al. 2009). This serves its purpose in cognitive function. There are experimental TBI that has been done, and it resulted in decreasing this contributor which mediates the synaptic plasticity (Halbauer et al. 2009). There is also research that concludes that this factor is significant for our memory and learning abilities (Halbauer

Table 18.1 Psychopharmacological management for cognitive impairment and neuropsychiatric symptoms (Halbauer et al. 2009)

I. Memory dysfunction	• Cholinesterase inhibitors (donepezil, galantamine, rivastigmine) • NMDA receptor modulator (memantine) • Stimulants (methylphenidate) • Neurogenesis stimulation (SSRIs)
II. Attention deficits	• Texas Medication Algorithm Project • Stimulants (methylphenidate, dextroamphetamine) • Noradrenergic reuptake inhibitor (atomoxetine) • Bupropion • Alpha-2 agonists (clonidine) • Other stimulants (modafinil) • Cholinesterase inhibitors (donepezil, galantamine, rivastigmine) • NMDA receptor modulator (memantine) • Noncompetitive NMDA receptor antagonists (amantadine) • Dopamine receptor agonist (bromocriptine)
III. Communication: Linguistics: speech and language (Primary treatment: speech and language therapy)	• Monoaminergic agents (bromocriptine, levodopa, d-amphetamine) • Cholinesterase inhibitors (donepezil, galantamine, rivastigmine) • Nootropic agents (piracetam) • Experimental-repetitive transcortical magnetic stimulation
IV. Communication: Affect recognition and emotion expression	No pharmacological recommendations
V. Visuospatial perception and sensory–motor integration—behavioural training programs	• Neurogenesis promoters (SSRIs, lithium) • Cholinesterase inhibitors (donepezil, galantamine, rivastigmine)
VI. Executive function	• Dopamine receptor agonist (bromocriptine) • Therapies of each cognitive region • Therapies made for attention deficits (see II Attention deficits)

NMDA N-methyl-D-aspartate, *SSRI* selective serotonergic reuptake inhibitor

Table 18.2 Wrap up of pharmacological management contemplation for neuropsychiatric illness (stay away from narcotics, benzodiazepines) (Halbauer et al. 2009)

I. Major depressive disorder	• SSRIs (citalopram, sertraline, paroxetine, fluoxetine) • SNRIs (duloxetine, venlafaxine) • Tricyclics with noradrenergic predominance (nortriptyline, desipramine) • Other antidepressants (mirtazapine, bupropion) • Stimulants (methylphenidate, dextroamphetamines)
II. Manic disorder	• Mood stabilisers (lithium) • Mood stabilisers, anticonvulsants (valproate, carbamazepine, oxcarbazepine, topiramate and lamotragine) • Atypical antipsychotics (aripiprazole, olanzapine, ziprasidone and quetiapine) • Combinations of medications (including low-dose lithium)
III. Post-traumatic stress disorder (PTSD)	• SSRIs (citalopram, sertraline, paroxetine, fluoxetine) • Serotonin—5HT2 agonist—trazodone (especially for sleep) • Noradrenergic alpha-1 blockers—prazosin (especially for sleep) • Noradrenergic beta-blockers centrally acting—propranolol • Atypical neuroleptics with serotonergic effects (quetiapine, olanzapine, aripiprazole and ziprazadone) • Avoid typical neuroleptic agents (haloperidol) • Avoid sedative hypnotics (benzodiazepines)
IV. Psychotic disorder	• Atypical neuroleptics (quetiapine, olanzapine, risperidone, etc.) • Typical neuroleptic agents (haloperidol)
V. Aggressiveness	• Beta-blockers with central effects (propranolol) • Alpha-2 antagonist (clonidine) • Stimulants (methylphenidate, dextroamphetamines) • SSRIs (citalopram, sertraline, paroxetine and fluoxetine) • Tricyclic antidepressants (especially low doses of doxepin) • Mood stabilisers, lithium (low dose, e.g., 300 mg SA at bedtime) • Mood stabilisers, anticonvulsants (valproate, carbamazepine, oxcarbazepine, topiramate and lamotragine) • Atypical antipsychotics (aripiprazole, olanzapine, ziprasidone and quetiapine) • Nonbenzodiazepine anxiolytics
VI. Sexual drive	• Therapy for hypopituitarism • Stay away from pharmacological treatment that lowers libido (pay attention to the side effects of SSRIs)

SA sustained action; *SNRI* serotonin-norepinephrine reuptake inhibitor; *SSRI* selective serotonergic reuptake inhibitor

Table 18.3 Types of neuropsychiatric illness associated with TBI (Halbauer et al. 2009)

Cognitive impairments	Memory Attention Language, facial, affect recognition Visuospatial cognition Executive function
Neurobehavioural disorders	Major depressive disorder Manic syndrome Post-traumatic stress disorder, anxiety disorder Psychosis
Sensory disturbance	Altered vision Faint, deaf (ear function affected) Changed perception of smell Changed perception of taste Touch hypersensitivity
Somatoform disorder	Long-term pain, migraine Loss of sexual interest Tired
Substance dependence	

et al. 2009). Moreover, difficulty in learning can be found in people with certain polymorphisms in this factor's gene (Halbauer et al. 2009).

18.3.5 The Correlation of Brain-Derived Neurotrophic Factor and Energy Metabolism

The brain-derived neurotrophic factor is shown to maintain the metabolism of energy in the brain region specifically in the hippocampus (Halbauer et al. 2009). Moreover, the molecular systems that are important for regulating energy are affected by this factor (Halbauer et al. 2009). Studies have shown that hyperphagia and obesity are found in animals that has a genetic deletion of this factor's gene (Halbauer et al. 2009). The input of this factor is found to lower the body weight, maintain the level of glucose to its normal level, assist with the metabolism of lipids and spike the sensitivity of the hormone insulin (Halbauer et al. 2009). The distribution of BDNF in the brain is most abundant in brain regions that are correlated to cognitive and neuroendocrine regulation, for example, the hippocampus and hypothalamus (Halbauer et al. 2009).

18.3.6 Potential Factors of Nutrition to Lessen Damaging Effects Caused by TBI

Potential factors of nutrition or dietary interventions to lessen damaging effects done by TBI have been shown to influence minimally four primary objectives to moderate cognitive and emotional distress following TBI (Halbauer et al. 2009):

- To decrease energy setback and assist with membrane restoration
- To increase the development of synaptic restructuring
- To promote neuronal signalling
- To trigger a process that can give wide preservation (Halbauer et al. 2009)

18.4 Types of Nutrition Intake That Can Affect Mental Illness Post TBI

18.4.1 Omega-3 Fatty Acids

There are 2 researches that utilised n-3FA supplements in unlike dosages post-traumatic brain injury. One of the studies, each subject was provided with a fish oil supplement that contain 1720 mg eicosapentaenoic acid (EPA) and 1160 mg docosahexaenoic acid (DHA), a multivitamin and a supplement that improves cognitive function on a day-to-day basis (Lee and Oh 2022). The data for the two latter supplements was not reported (Lee and Oh 2022). Each subject was taught ways to improve cognitive health such as smoking discontinuance, regular exercise and substance dependence (Lee and Oh 2022). After intervention, it has shown positive results in terms of cognition, mood, sleep and memory (Lee and Oh 2022). But it is hard to determine whether these results are correlated with the fish oil supplement or the other supplement which the report does not have data (Lee and Oh 2022). In the second study, the subjects were given n-3FA supplements which consist of 1000–1500 mg of EPA and DHA every day accompanied by a Mediterranean-style diet (Grajek et al. 2022). Each subject was also given brain health improvement education. The result has shown good results in terms of cognitive brain function, mood, memory and reaction time (Grajek et al. 2022).

18.4.2 Mediterranean Diet

This diet is well-known to improve health and longevity (Butler and Mörkl 2023). This diet is an all-right and feasible dietary pattern which consists of whole plant-based foods, incorporating fruits, vegetables, legumes, nuts, seeds, herbs, spices, olive oil and whole grains (Butler and Mörkl 2023). Fish, commonly those strong in omega-3 fatty acids and other seafood, dairy and poultry are incorporated in moderation, while red meats and sweets are taken occasionally (Butler and Mörkl 2023). Highly processed foods are not taken in this diet. Although there are no definite studies that directly link mental illness and TBI together, several studies have shown that there are benefits for each to treat mental illness and TBI (Butler and Mörkl 2023). One of the studies has proven that this diet is correlated with a decreased level of depression risk and cognitive impairment (Franco et al. 2023). In people who are diagnosed with depression, they are suggested to switch to this diet for a more effective treatment strategy (Franco et al. 2023). Other than that, the Mediterranean diet is also useful to protect from memory and cognitive decline which works to combat mental health illness caused by TBI (Franco et al. 2023).

18.4.3 Melatonin

Melatonin is a very popular dietary supplement and is often sold by ordinary retail purchases, with no need for a prescription or license and in health food shops globally (Grajek et al. 2022). There are 2 studies that used this supplement to analyse the benefits of improving psychiatric illness post-TBI (Grajek et al. 2022). Each included a grade in TBI severity and had two affected individuals with mTBI (Grajek et al. 2022). Both studies also applied a randomised, double-blind, controlled and crossover design and 10 weeks of follow-up period. The doses used are not the same, one used 5 mg and another study used 2 mg. Both studies have shown a decline in anxiety (Grajek et al. 2022). Results were more significant in one of the studies, whereas the other study has several factors which may contribute to the result and cause a bias. Other than that, both studies did not show any significant outcome for depression (Grajek et al. 2022). But sleep quality was shown to improve using the actigraphy technology after the use of melatonin (Grajek et al. 2022).

18.4.4 Enzogenol

Enzogenol is a natural extract obtained from the bark of Pinus Radiata and contains a high antioxidant and anti-inflammatory attribute (Grajek et al. 2022). There are two studies that used 1000mg of this supplement daily (Grajek et al. 2022). One of the studies shows that there is an increase in cognitive performance after consuming the supplement. But there are no significant findings for episodic memory (Grajek et al. 2022). The effects of therapy were stabilised after 11 weeks, and it was shown that anxiety and depression were significantly improved after the treatment (Grajek et al. 2022). Another study has shown that sleep and mental fatigue were improved significantly (Theadom et al. 2013). Overall majority did not experience any side effects from the supplement, except for a few who complained of visual blurring, sleep disruption and migraines (Theadom et al. 2013).

18.4.5 MLC901 Capsules

This capsule is an antioxidant that contains herbal components that consist of *Radix astragali*, *Prunus persica* and Radix Polygalae (Grajek et al. 2022). This capsule is very common in Chinese medicine practice (Grajek et al. 2022). A study has shown that there is increased performance in executive functioning and complex attention (Grajek et al. 2022). The commitment to the medication remains strong at 6 months, and the therapy effects stay even after the capsule consumption is stopped (Grajek et al. 2022). In this study, subjects were already posting injuries for a few moments; it is a mystery if better results will be shown if the supplements were given at an earlier time (Grajek et al. 2022). Side effects are found in a few subjects like headache, sore tongue and itchiness. But the majority of this capsule was well taken by subjects in the study (Grajek et al. 2022).

18.4.6 Probiotic Supplementation

Studies have found that there is a difference in microbiota which is the bacteria living in the gut of people with psychiatric illnesses like depression and bipolar disorder (Grajek et al. 2022). Particularly there is a decline in bacteria that produce butyrate. In both illnesses, the gut bacteria have different changes (Grajek et al. 2022). In subjects with major depressive disorder, the changes of bacteria in the gut are beta-diversity whereas the other in bipolar disorder, the changes found is a decrease in the overall bacteria is called alpha diversity (Grajek et al. 2022). Some early studies found that gut bacteria can be a predictor of how well someone responds to the treatment for psychiatric illness (Grajek et al. 2022). There are also studies that suggest using blood microbiome for the diagnosis of psychiatric disorders (Grajek et al. 2022). Clinical trials conducted to analyse the use of probiotics on psychiatric disorders have shown positive results (Grajek et al. 2022). In one case report, a subject with bipolar disorder has improved symptoms after receiving a faecal microbiota transplantation (Grajek et al. 2022).

18.4.7 Curcumin

Curcumin is a well-known polyphenol that we can obtain from turmeric for its multiple abilities for brain health, one of which includes regulating energy (Jeong et al. 2023). Another is to help in reducing cerebral oedema, which always results from TBI due to its effects on brain function to hinder the repairing of neurons and its functions in the brain (Jeong et al. 2023). Speed recovery post TBI is found to be contributed by curcumin due to its ability to reduce the dysfunction of metabolism.

In theory, molecular systems such as AMP-activated protein kinase and mitochondrial creatine kinase play a role in maintaining energy metabolism (Jeong et al. 2023). Curcumin can be the catalyst in this process to restore neuronal function and cognitive brain function after a brain trauma injury (Jeong et al. 2023). Furthermore, curcumin's anti-inflammatory characteristics can also lessen cerebral oedema and improve mental illness caused by TBI. This is due to aquaporin-4 expression (which is a contributor to cerebral oedema) and the interleukin-1 beta expression is controlled by curcumin's properties (Jeong et al. 2023). Therefore, curcumin is vital to be implemented into the diet as it helps speed up the recovery of patients post TBI in terms of cognitive health and mental health (Jeong et al. 2023).

18.4.8 Flavonoids

Other than curcumin, there is another polyphenol that helps reduce mental illness post TBI (Jeong et al. 2023). Which is flavonoids an antioxidant that we can obtain from plant-based foods. Flavonoids are quite similar to curcumin in their functions such as anti-inflammatory characteristics that help restore neurons and brain functions (Jeong et al. 2023). Other than that, it also helps maintain signaling pathways in the brain that are crucial for memory and cognitive function, especially post TBI (Jeong et al. 2023). Studies have also shown that there is a decreased risk of stroke and Alzheimer's disease. This highlights flavonoids' abilities to preserve brain function and may also help protect from mental illness post TBI (Jeong et al. 2023). This shows that flavonoids should be included in the diet especially in patients post TBI as it can help increase cognitive function and preserve mental health caused by TBI (Jeong et al. 2023).

18.4.9 Green Tea

Green tea is rich in flavonoids, especially catechins like epigallocatechin gallate that help reduce TBI and overall mental health concerns (Jeong et al. 2023). As discussed earlier, flavonoids are well-known for their antioxidant, anti-inflammatory and neuroprotective characteristics against cognitive decline in ageing (Jeong et al. 2023). Having green tea regularly on a daily basis is found to help protect us from stroke and any cognitive decline in ageing (Jeong et al. 2023). A study conducted on aged mice concluded that green tea lessens hippocampal degeneration caused by ageing, improves memory function and protects from DNA oxidative impact. The benefits above highly suggest implementing green tea in the diet of TBI patients as it helps improve overall mental and cognitive health post-TBI (Jeong et al. 2023).

18.4.10 Resveratrol

Have you ever heard that consuming grapes, red wine and berries can help improve brain health? Well, it is proven in studies that have found that wine consumption is helpful to reduce the risk of Alzheimer's disease (Jeong et al. 2023). This is due to resveratrol, a polyphenol that has the ability to reduce neurodegeneration (Jeong et al. 2023). Moreover, it has the ability to mimic some features of caloric restriction such as increasing sirtuin proteins' production (Jeong et al. 2023). This may increase metabolism and maintain brain functions. Thus, resveratrol is crucial as it is neuroprotective and able to help improve recovery in patients post TBI for their mental and cognitive function (Jeong et al. 2023).

18.4.11 Ketogenic Diet

I believe many of us are familiar with this term, the ketogenic diet. This diet highly consists of high fat, moderate protein and low carbohydrate intake (Jeong et al. 2023). It is widely popular for aiding neurological disorders including TBI. A ketogenic diet uses its alternative energy source in the form of ketone bodies such as D-b-hydrocybutyrate to reduce metabolic dysfunction caused by TBI (Jeong et al. 2023). This shows that this diet is helpful in recovering and also in protecting brain function. Moreover, ketosis occurred that it can decrease the impacts of metabolic stress and excitotoxicity (Jeong et al. 2023). Other than that, restricting calories in diets that are similar to a ketogenic diet is found to increase brain-derived neurotrophic factor which is highly beneficial in neuron and cognitive functions. This diet also helps quicken response time to metabolic issues caused by injuries (Jeong et al. 2023). Thus, a ketogenic diet should be included in the diet of patients post TBI so that mental health concerns and cognitive functions can be improved with a higher success rate (Jeong et al. 2023).

18.4.12 Caloric Restriction and Dietary Strategies

Maintaining metabolic homeostasis is important in the pathophysiology of TBI (Suárez-López et al. 2023). It works by affecting the neuronal plasticity and the brain function (Suárez-López et al. 2023). Dietary interventions that assist with metabolic balance and by changing the caloric load of the diet can provide potential routes for post-traumatic brain injury recovery (Suárez-López et al. 2023). Studies have shown that reducing caloric intake, for example, through intermittent fasting can help take care of neurons in brain regions specifically the hippocampus in opposition to excitotoxicity-induced death (Suárez-López et al. 2023).

Other than that, caloric restriction has also appeared to heighten the degree of brain-derived neurotrophic factor, a protein that is important for neuronal function and cognitive function (Suárez-López et al. 2023). Moreover, the pros of caloric restriction not only work on the brain; studies have shown and proven that there are improvements in recovery of spinal cord injury (Suárez-López et al. 2023). By regulating metabolic pathways and supporting neuronal resilience, we should instil dietary strategies focused on caloric restriction to improve the prognosis of mental health illness caused by TBI patients (Suárez-López et al. 2023).

18.4.13 Vitamin E

Vitamin E is derived from oils, nuts and spinach (Suárez-López et al. 2023). It is proven to be very helpful as a neuroprotective agent against traumatic brain injury (Suárez-López et al. 2023). Not only that, it also contains antioxidative properties (Suárez-López et al. 2023). This allows them to assist in lessening the free radical levels in the brain, therefore helping optimise neuronal function (Suárez-López et al. 2023). Studies have shown that vitamin E supplementation not only encourages memory function in people of old age but is also linked with enhancing neurological performance and brain mitochondrial function that is found in animal models (Suárez-López et al. 2023). By lessening oxidative stress and enhancing neuronal health, vitamin E highlights a feasible dietary intervention to improve cognitive outcomes post trauma brain injury (Suárez-López et al. 2023). Not only that, vitamin E also helps provide benefits for mental health and overall brain function, especially for post brain trauma injury individuals (Suárez-López et al. 2023).

18.4.14 Dietary Strategies and Exercise Synergy for Mental Health Improvement

There are ways to enhance mental health improvement in post traumatic brain injury patients, one way is to combine specific dietary interventions with exercise (Gomez-Pinilla 2011). Exercise is to apply after traumatic brain injury, specifically after dietary adjustments. Studies have demonstrated its benefits on post-traumatic brain injury patients (Gomez-Pinilla 2011). This can work by reinstating the metabolic balance and also by supporting neuronal resilience. These interventions combination have great potential to enhance mental health disorders in affected individuals recovering from traumatic brain injury (Gomez-Pinilla 2011).

18.4.15 DHA and Exercise Interaction

One of the dietary strategies we can talk about is to instil DHA, which is an omega-3 polyunsaturated fatty acid into our dietary plan (Gomez-Pinilla 2011). If DHA is combined with exercise, it allows synergistic effects, especially on cognitive function and synaptic plasticity. This correlation does not only help regulate neuronal signaling but also can enhance BDNF-mediated synaptic plasticity (Gomez-Pinilla 2011). These mechanisms are significant in improving mood disorders, cognitive abilities and holistic mental well-being (Gomez-Pinilla 2011). This shows how important the combination of dietary strategies with exercise is, as it helps regulate and optimise mental health issues and post-traumatic brain injury (Gomez-Pinilla 2011).

18.4.16 Relationship between Diet Strategies and Exercise on Brain Homeostasis

We have already talked about how the combination of diet strategy and exercise works on the recovery of TBI patients (Kurtz and Rocha 2020). How about its influence on brain hemostasis? This combination is proven to correlate with regulating brain health, which is linked with mood and cognitive function (Kurtz and Rocha 2020). This combination's mechanism is by changing gene expression linked with reducing inflammation and reorganising its structure, creating a feasible environment for better mental health well-being (Kurtz and Rocha 2020). Moreover exercising has been shown to lower harmful impacts caused by unhealthy diets on restructuring brain structure and cognitive function (Kurtz and Rocha 2020). This further showcases the benefits of implementing these in the lifestyle of post TBI patients for health improvement (Kurtz and Rocha 2020).

18.4.17 Cell Energy Metabolism and Cognitive Function

Cell energy metabolism is important in the correlation between diet and exercise in patients with post TBI's mental health (Kurtz and Rocha 2020). This is due to metabolic pathways and energy being maintained to preserve cognitive function and to reduce mood disorders (Kurtz and Rocha 2020). Thus, this is crucial as it helps in neuron function and metabolism of energy, furthermore aiding in mental health concerns post-TBI patients (Kurtz and Rocha 2020).

18.4.18 Diet Composition and the Risk of Mood Disorder

One study shows that a healthier diet is important, especially in patients post TBI (Ryan et al. 2023). This healthier diet that is useful for improving mood disorders especially depression and bipolar disorder would be a diet greater intake of fruits, vegetables, legumes, nuts, whole grains and good quality protein sources such as fish or seafood (Ryan et al. 2023). Other study shows that diets higher in processed and free sugar-rich foods can cause sleep quality to deteriorate (Ryan et al. 2023). It is important to avoid diets that are high in saturated fats and sugars, which will aggravate the detrimental effects of traumatic brain injury. Food intake of high-fat, high-sugar diets, similar to "junk food", has been correlated with cognitive impairments and decreased levels of hippocampal BDNF, an important contributor to synaptic plasticity and cognitive abilities (Ryan et al. 2023). Other than that, such diets are detrimental to the outcomes of traumatic brain injury by impairing synaptic plasticity and memory function (Ryan et al. 2023). Nevertheless, there are antioxidant treatments like vitamin E that have been shown to annull these detrimental effects, showcasing the importance of dietary interventions in lessening the cognitive consequences of unhealthy eating habits (Ryan et al. 2023). Thus, to increase mental health in TBI recovery, a healthier lifestyle is vital like implementing exercise with a healthier choice of diet (Ryan et al. 2023).

18.4.19 Glutamate and Depression

High blood glutamate levels are found in patients with post-traumatic brain injury (TBI) depression (Boyko et al. 2023). An important factor that degenerates neurons in brain and triggered by glutamate toxicity is the blood–brain barrier (BBB) permeability (Boyko et al. 2023). Reducing blood glutamate and restoring the blood–brain barrier must be focused on treating and preventing post-TBI depression (Boyko et al. 2023). There are factors that can influence blood glutamate levels, one of them is the dietary aspect where glutamate-rich food contributes significantly (Boyko et al. 2023). Furthermore, chronic BBB disruption post-TBI may cause sustained high blood glutamate levels, increasing neurotoxicity and leading to depression (Boyko et al. 2023). Several strategies to combat this issue such as low-glutamate diets, medications that target glutamate specifically, and supplements to lower blood glutamate levels are suggested to manage post-TBI depression (Boyko et al. 2023).

18.4.20 Malnutrition Caused by TBI and Its Relation with Mental Illness

It is very important for a body to heal especially after severe TBI, with adequate nutrition. Study shows that patients with brain trauma will need minimal 140% of typical metabolic expenditure (Ford and Ouma 2021).75% of TBI patients in ICU emit symptoms of malnutrition at 3 weeks (Ford and Ouma 2021). It is shown that malnutrition is more common in TBI patients with lower GCS scores, leading to poorer prognosis (Ford and Ouma 2021). Lower GCS, total protein and albumin serum levels correlate with malnutrition development. In order for approximate nutritional assessment, BMI can be utilised to measure body fat and it is widely used for nutritional assessment (Ford and Ouma 2021). BMI is usually associated with mid-upper arm circumference (MUAC), especially useful in malnourished adults (Ford and Ouma 2021). Other than that, MUAC is easier to utilise and increasingly valuable in examining nutritional status in acutely ill patients (Ford and Ouma 2021). Hospital malnutrition prevalence is higher in older patients, those with comorbidities, nonmanual workers and prolonged stays. Severe TBIs increase malnutrition risk because of factors such as drowsiness, dysphagia and confusion (Ford and Ouma 2021). It is also shown that malnutrition is correlated to higher infection rates, complications, prolonged hospital stays and increased morbidity and mortality. Studies have shown that if early nutrition is conducted within 5–7 days post-injury, it will reduce 2-week mortality in severe TBI patients (Ford and Ouma 2021). Therefore, it is very important to optimise nutrition intake for TBI patients for a faster recovery (Ford and Ouma 2021).

Although there is currently no research to link between malnutrition caused by TBI and mental illness, many studies have shown that mental illness can be caused by malnutrition itself. One study shows that a greater number of patients with severe mental illness are caused by nutritional issues, while inpatients and outpatients are seen to possess each nutritional need (Risch et al. 2022). This study further analyses and concludes that inpatients' mental illness is more influenced by a decrease in body weight, whereas outpatients are prone to be influenced by increasing in weight (Risch et al. 2022). Although overweight and associated risk factors seem to be the majority, there is still quite a portion of participants at risk of malnutrition, which is 32% of inpatients and 34% of outpatients, which is judged and analysed based on the severity of symptoms and weight loss (Risch et al. 2022).

Another study has also proven this theory, where 507 subjects were participants in the study (Shawel et al. 2023). Anthropometric data was measured using a calibrated weight scale and a height board (Shawel et al. 2023). The study concluded that the percentage of malnourishment among study participants is quite high which is 62.7% (Shawel et al. 2023). There are several factors that are linked with malnourishment which include numerous medications, smoking, the number of meals and taking a formulated diet (Shawel et al. 2023). It is crucial to educate patients and caregivers about nutrition and also to optimise own diet plan according to one's needs (Shawel et al. 2023).

Therefore, there could be a strong relation between mental illness and how will it be affected by malnutrition caused by TBI.

18.5 Conclusion

It is concluded that nutrition in general plays a major role in mental illness that occurs in traumatic brain injury patients. There are several diets that are proven to be beneficial to these concerns, such as including omega-3 fatty acids, the Mediterranean diet, the ketogenic diet, green tea, curcumin, flavonoids, probiotics and supplements. Diet strategy and exercise also have to be incorporated into the lifestyle of the patient in order to facilitate a fast recovery and rehabilitation. Caregivers and patients also have to be given education in these matters in order to better understand this aspect to improve the well-being of the affected individual. This is also to prevent malnourishment as it is also one of the risk factors to affect cognitive and mental health overall. Although this psychiatric illness caused by TBI and TBI itself cannot run away from pharmacological treatment, nutrition can be a supplement in order to treat the mental illness caused by TBI. More research still needs to be done to analyse the relationship and impact of nutrition and how it helps in aiding mental illness caused by traumatic brain injury. The condition of mental health caused by nutrition will not run from nutrition, but mental health caused by TBI also cannot be blamed solely on nutrition.

References

Bogner J, Corrigan JD, Yi H, Singichetti B, Manchester K, Huang L, Yang J (2020) Lifetime history of traumatic brain injury and behavioral health problems in a population-based sample. J Head Trauma Rehabil 35(1):E43–E50. https://doi.org/10.1097/HTR.0000000000000488

Boyko M, Gruenbaum BF, Oleshko A, Merzlikin I, Zlotnik A (2023) Diet's impact on post-traumatic brain injury depression: exploring neurodegeneration, chronic blood-brain barrier destruction, and glutamate neurotoxicity mechanisms. Nutrients 15(21):4681. https://doi.org/10.3390/nu15214681

Butler MI, Mörkl S (2023) The mediterranean diet and mental health. In: Dinan T (ed) Nutritional psychiatry: a primer for clinicians. Cambridge University Press, Cambridge, pp 39–54

Duarte DC, Duarte JC, Ocampo González ÁA, Castillo García JF (2023) Psychiatric disorders in post-traumatic brain injury patients: a scoping review. Heliyon 9(1):e12905. https://doi.org/10.1016/j.heliyon.2023.e12905

Faria MA Jr (2013) Violence, mental illness, and the brain - a brief history of psychosurgery: part 1 - from trephination to lobotomy. Surg Neurol Int 4:49. https://doi.org/10.4103/2152-7806.110146. PMID: 23646259; PMCID: PMC3640229

Ford L, Ouma J (2021) A descriptive study of malnutrition in traumatic brain injury patients. Panam J Trauma Crit Care Emerg Surg 10:107–112. https://doi.org/10.5005/jp-journals-10030-1359

Franco GA, Interdonato L, Cordaro M, Cuzzocrea S, Di Paola R (2023) Bioactive compounds of the mediterranean diet as nutritional support to fight neurodegenerative disease. Int J Mol Sci 24(8):7318. https://doi.org/10.3390/ijms24087318

Gomez-Pinilla F (2011) The combined effects of exercise and foods in preventing neurological and cognitive disorders. Prev Med 52(Suppl 1):S75–S80. https://doi.org/10.1016/j.ypmed.2011.01.023

Grajek M, Krupa-Kotara K, Białek-Dratwa A, Sobczyk K, Grot M, Kowalski O, Staśkiewicz W (2022) Nutrition and mental health: a review of current knowledge about the impact of diet on mental health. Front Nutr 9:943998. https://doi.org/10.3389/fnut.2022.943998

Halbauer JD, Ashford JW, Zeitzer JM, Adamson MM, Lew HL, Yesavage JA (2009) Neuropsychiatric diagnosis and management of chronic sequelae of war-related mild to moderate traumatic brain injury. J Rehabil Res Dev 46(6):757–796. https://doi.org/10.1682/jrrd.2008.08.0119

Jeong H, Kim JH, Choo YH, Kim M, Lee S, Ha EJ, Oh J (2023) Nutrition therapy for patients with traumatic brain injury: a narrative review. Korean J Neurotrauma 19(2):177–184. https://doi.org/10.13004/kjnt.2023.19.e31

Kurtz P, Rocha EEM (2020) Nutrition therapy, glucose control, and brain metabolism in traumatic brain injury: a multimodal monitoring approach. Front Neurosci 14:190. https://doi.org/10.3389/fnins.2020.00190

Lee HY, Oh BM (2022) Nutrition management in patients with traumatic brain injury: a narrative review. Brain NeuroRehabil 15(1):e4. https://doi.org/10.12786/bn.2022.15.e4

Risch L, Hotzy F, Vetter S, Hiller S, Wallimann K, Seifritz E, Mötteli S (2022) Assessment of nutritional status and risk of malnutrition using adapted standard tools in patients with mental illness and in need of intensive psychiatric treatment. Int J Environ Res Public Health 20(1):109. https://doi.org/10.3390/ijerph20010109

Ryan T, Nagle S, Daly E, Pearce AJ, Ryan L (2023) A potential role exists for nutritional interventions in the chronic phase of mild traumatic brain injury, concussion and sports-related concussion: a systematic review. Nutrients 15(17):3726. https://doi.org/10.3390/nu15173726

Schwarzbold M, Diaz A, Martins ET, Rufino A, Amante LN, Thais ME, Quevedo J, Hohl A, Linhares MN, Walz R (2008) Psychiatric disorders and traumatic brain injury. Neuropsychiatr Dis Treat 4(4):797–816. https://doi.org/10.2147/ndt.s2653

Shawel S, Baraki N, Alemeshet Y, Abebe DS, Egata G (2023) Undernutrition and associated factors among adults with mental and neurological disorders in public health hospitals, Eastern Ethiopia, 2019: a cross-sectional study. BMC Psychiatry 23(1):617. https://doi.org/10.1186/s12888-023-05117-9

Suárez-López LM, Bru-Luna LM, Martí-Vilar M (2023) Influence of nutrition on mental health: scoping review. Healthcare 11(15):2183. https://doi.org/10.3390/healthcare11152183

Theadom A, Mahon S, Barker-Collo S, McPherson K, Rush E, Vandal AC, Feigin VL (2013) Enzogenol for cognitive functioning in traumatic brain injury: a pilot placebo-controlled RCT. Eur J Neurol 20(8):1135–1144. https://doi.org/10.1111/ene.12099

Chapter 19
Nutrition and TBI: Synthesizing Insights from Bench to Bedside

Wael Mohamed

Abstract This chapter explores the complex connection between nutrition and traumatic brain injury (TBI), with a focus on the use of tables to aid comprehension. The text thoroughly examines the most recent discoveries from both experimental and clinical investigations, emphasizing the crucial importance of different nutrients and dietary treatments in the management and recovery of traumatic brain injury (TBI). We focus specifically on the influence of macronutrients and micronutrients on key processes such as neuroinflammation, oxidative stress, and neuroplasticity. The chapter highlights the significance of customized nutrition regimens designed specifically for individual TBI patients, including the most suitable time, dose, and ways of delivering nutrients. Moreover, it tackles the pragmatic obstacles involved in converting laboratory findings into clinical practice. The chapter's objective is to provide health-care practitioners, researchers, and policymakers with practical ideas to improve recovery outcomes for patients with traumatic brain injuries, thereby closing the existing gap. The use of various tables helps clarify intricate nutritional concepts and their implementation in TBI therapy, promoting an integrated strategy that gives priority to nutritional issues.

Keywords TBI · Neuroprotective · Supplements · Geriatric · Calories · Recovery · Meals

"Nutrition and TBI: From Bench to Bedside" explores the complex connection between nutrition and traumatic brain injury (TBI) in many areas. The book starts by covering fundamental concepts in basic science and then delves into the molecular processes that explain how diet affects the outcomes of traumatic brain injury

W. Mohamed (✉)
Basic Medical Science Department, Kulliyyah of Medicine, International Islamic University Malaysia (IIUM), Kuantan, Pahang, Malaysia

Clinical Pharmacology Department, Menoufia Medical School, Menoufia University, Shibin el Kom, Egypt

W. Mohamed (ed.), *Nutrition and Traumatic Brain Injury (TBI)*, Nutritional Neurosciences, https://doi.org/10.1007/978-981-97-6341-2_19

(TBI). Subsequently, it smoothly shifts to clinical viewpoints, offering valuable information on how dietary therapies might impact patient recuperation and long-term prognosis. The book facilitates the integration of nutritional therapy into practical applications by bridging the gap between laboratory research and clinical practice. This book provides a thorough and extensive approach that covers scientific, clinical, and translational aspects. It is a definitive resource for researchers, clinicians, and policymakers who want to improve nutritional techniques for managing traumatic brain injury (TBI).

Ultimately, "Nutrition and TBI: From Bench to Bedside" emphasizes the crucial significance of nutrition in the treatment of traumatic brain injury. The book emphasizes the diverse influence of dietary treatments on TBI outcomes, ranging from clarifying molecular pathways to informing therapeutic practice. By combining knowledge from fundamental scientific research, clinical studies, and practical applications, it provides a comprehensive understanding of how diet affects the course of recovery from traumatic brain injury (TBI). Therefore, this book offers a significant guide for improving patient treatment and achieving better results in populations with traumatic brain injuries.

In the future, research on the subject of nutrition and traumatic brain injury (TBI) should prioritize many important areas. Exploring further the molecular processes that govern the connection between diet and TBI will provide a more profound understanding of potential targets for therapy. Furthermore, it is necessary to conduct extensive clinical studies to confirm the effectiveness of certain nutritional therapies in a wide range of traumatic brain injury (TBI) populations. Furthermore, it is crucial to emphasize the translation of research results into clinical practice, namely by concentrating on the creation of standardized procedures for nutritional therapy in patients with traumatic brain injury (TBI). Interdisciplinary cooperation among academics, clinicians, and policymakers will be crucial for promoting innovation and adopting evidence-based dietary solutions to enhance outcomes for persons with TBI.

Gaining a comprehensive understanding of the complex connection between nutrition and traumatic brain injury (TBI) is essential for maximizing patient outcomes and improving the rehabilitation process. The relationship between food and brain health is complex, with nutrition affecting many elements of traumatic brain injury (TBI) development and recovery. This chapter contains a collection of informative tables that examine various aspects of the link between diet and traumatic brain injury (TBI). These tables provide useful information for clinicians, researchers, and policymakers who are interested in understanding the relationship between nutrition and brain damage. They summarize nutritional needs, outline dietary treatments, and explain how these interventions may affect the outcomes of traumatic brain injury.

The tables provided in this chapter provide extensive information on many elements of the correlation between diet and traumatic brain injury (TBI). Tables provide a concise summary of vital nutrients and their functions in the rehabilitation of traumatic brain injury (TBI), emphasizing important dietary elements that are crucial for enhancing brain well-being and supporting the healing processes after an injury. Additionally, they delineated prevalent nutritional insufficiencies reported in individuals with traumatic brain injury (TBI), underscoring the need to treat these deficiencies to enhance results and avert consequences.

Furthermore, several tables classify dietary therapies for TBI care according to their mechanisms of action, which span from neuroprotective techniques to anti-inflammatory and antioxidant approaches. This table provides a thorough foundation for comprehending the many therapeutic possibilities of diet in reducing the pathophysiology associated with traumatic brain injury (TBI) and facilitating the recovery of the nervous system.

In addition, additional tables provide a concise overview of evidence-based dietary recommendations for TBI patients at various phases of rehabilitation, assisting physicians in customizing nutritional therapies to address individual requirements and enhance long-term results. Some tables in the study emphasize emerging patterns and future paths in nutritional research for traumatic brain injury (TBI), providing valuable information on areas of study that show promise and possible opportunities for therapeutic advancements.

These tables are essential tools for health-care workers, researchers, and policymakers interested in managing traumatic brain injury. These tables simplify intricate information, making it easier to understand. They help people make well-informed decisions and contribute to the progress of dietary strategies for TBI care and rehabilitation.

19.1 Neuroprotective Foods Useful in the TBI Diet

Table 19.1 summarizes neuroprotective foods that can be incorporated into the diet to support brain health and potentially aid in the prevention of neurodegenerative diseases. Each food item is listed along with its key properties, including the

Table 19.1 Key prosperities of various neuroprotective foods

Neuroprotective foods	Key properties
Fatty fish (e.g., salmon, mackerel)	Rich source of omega-3 fatty acids, which have anti-inflammatory and neuroprotective effects, supporting brain health and cognitive function
Berries (e.g., blueberries, strawberries)	High in antioxidants, particularly flavonoids and anthocyanins, which can reduce oxidative stress and inflammation in the brain, promoting neuroprotection and cognitive function
Dark leafy greens (e.g., spinach, kale)	Abundant in vitamins (e.g., folate, vitamin K) and minerals (e.g., magnesium, iron) that support brain health and neuroprotection while also providing antioxidants and anti-inflammatory compounds
Nuts and seeds (e.g., walnuts, flaxseeds)	Excellent sources of omega-3 fatty acids, antioxidants, and vitamin E, which can protect against oxidative damage, inflammation, and cognitive decline, supporting brain health and function
Turmeric and curcumin	Containing curcumin, a potent anti-inflammatory and antioxidant compound that may protect against neurodegenerative diseases, reduce brain inflammation, and support cognitive function
Dark chocolate	Rich in flavonoids, particularly cocoa flavanols, which have antioxidant and anti-inflammatory properties, supporting brain health, blood flow, and cognitive function.

presence of antioxidants, anti-inflammatory compounds, omega-3 fatty acids, and other nutrients that contribute to neuroprotection and cognitive function.

19.2 Supplements and Nutraceuticals for TBI Recovery

Table 19.2 provides a summary of supplements and nutraceuticals commonly used in traumatic brain injury (TBI) recovery. Each supplement is listed along with its potential benefits, including anti-inflammatory, antioxidant, neuroprotective, and cognitive-enhancing properties. These supplements may support brain health, promote tissue repair, and aid in cognitive function and mood regulation during the recovery process after TBI.

Table 19.2 Various supplements and nutraceuticals and their potential benefits in TBI

Supplement/nutraceutical	Potential benefits
Omega-3 fatty acids	– Reduce brain inflammation through it anti-inflammatory properties. – Support brain cell membrane integrity and neuroplasticity. – Aid in cognitive function and mood regulation.
Vitamin D	– Modulates inflammation and support neuronal health. – Enhances neurotrophic factors involved in brain repair and recovery. – May help prevent cognitive decline and improve mood.
Antioxidants (e.g., vitamin C, vitamin E, selenium)	– Neutralize free radicals and reduce oxidative stress in the brain. – Protect brain cells from damage and promote tissue repair. – Support overall brain health and cognitive function.
Magnesium	– Regulates neurotransmitter function and synaptic plasticity. – Reduces excitotoxicity and support mitochondrial function. – Aids in relaxation and stress management.
Zinc	– Be essential for neurogenesis and neuronal development. – Modulates immune function and reduce inflammation in the brain. – Supports cognitive function and memory retention.
Curcumin (turmeric extract)	– Has potent anti-inflammatory and antioxidant properties. – Promotes neurogenesis and enhance brain-derived neurotrophic factor (BDNF) levels. – May improve cognitive function and mood in TBI patients.
Acetyl-L-carnitine	– Enhances mitochondrial function and energy production in the brain. – Improves cognitive function and memory recall. – Supports neuronal repair and recovery after brain injury.

Table 19.3 Nutritional support for TBI in specific vulnerable populations

Population	Nutritional considerations
Children and adolescents	– Adequate calorie and nutrient intake for growth and development – Emphasis on nutrient-dense foods to support cognitive function and physical recovery – Monitoring for signs of malnutrition or nutrient deficiencies due to increased nutritional demands during growth
Patients with co-occurring conditions (e.g., diabetes, obesity, cardiovascular disease)	– Individualized meal planning to manage both TBI recovery and concurrent conditions – Emphasis on balanced meals to regulate blood sugar levels and support cardiovascular health – Collaboration with health-care providers to optimize nutrition while managing medications and treatment protocols
Geriatric TBI patients	– Addressing age-related nutritional concerns such as decreased appetite, dental issues, and reduced nutrient absorption – High-protein, nutrient-rich diet to support muscle strength and prevent muscle wasting – Monitoring for dehydration and malnutrition, which may be exacerbated in older adults with TBI

19.3 Nutritional Support for Specific TBI Populations

Table 19.3 summarizes the specific nutritional considerations for various populations affected by traumatic brain injury (TBI). It highlights the importance of tailored nutrition plans to address the unique needs and challenges faced by children, adolescents, patients with co-occurring conditions, and geriatric TBI patients.

19.4 Nutritional Considerations for TBI Recovery

Table 19.4 summarizes the key nutritional considerations for traumatic brain injury (TBI) recovery. It highlights the importance of adequate calorie intake, protein consumption, micronutrient support, hydration, balanced diet, and nutrient timing in promoting healing and rehabilitation following TBI.

Table 19.4 Nutritional consideration for TBI recovery

Nutritional consideration	Key points
Adequate calorie intake	Ensure sufficient energy intake to support metabolic needs and facilitate tissue repair and recovery following traumatic brain injury.
Protein	Promote wound healing, muscle repair, and neurotransmitter synthesis. Aim for adequate protein intake to support tissue regeneration.
Micronutrients	Optimize intake of vitamins and minerals essential for brain health and function, including vitamin D, B vitamins, and antioxidants.
Fluid balance	Maintain hydration to support cognitive function, regulate body temperature, and facilitate nutrient transport and waste removal.
Balanced diet	Emphasize a varied diet rich in fruits, vegetables, whole grains, lean proteins, and healthy fats to provide essential nutrients for recovery.
Nutrient timing	Consider the timing of meals and snacks to optimize energy levels, support physical rehabilitation, and enhance recovery outcomes.

Table 19.5 Recipes examples tailored for TBI recovery

Recipe name	Main ingredients	Key nutrients/health benefits
Brain-boosting smoothie	Blueberries, spinach, banana, almond milk, Greek yogurt, chia seeds	Rich in antioxidants, omega-3 fatty acids, vitamins, and minerals to support brain health and cognitive function
Salmon and quinoa bowl	Grilled salmon, quinoa, mixed greens, avocado, cherry tomatoes, lemon vinaigrette	High in omega-3 fatty acids, protein, fiber, and essential vitamins and minerals for brain repair and inflammation reduction
Turkey and veggie stir-fry	Lean turkey breast, broccoli, bell peppers, carrots, snap peas, brown rice, soy sauce	Providing lean protein, fiber, and an array of vitamins and minerals to support energy levels and cognitive function
Sweet potato and black bean tacos	Roasted sweet potatoes, black beans, whole grain tortillas, avocado, salsa, cilantro	Packed with fiber, antioxidants, and plant-based protein to promote brain health and overall well-being
Spinach and mushroom omelette	Eggs, spinach, mushrooms, onions, feta cheese, olive oil	Containing choline, vitamin K, and other nutrients essential for brain function and memory retention

19.5 Sample Recipes and Meal Plans

These meal plans are designed to provide a balance of macronutrients, vitamins, and minerals to support brain health and aid in the recovery process after traumatic brain injury (TBI). Each recipe and meal include ingredients rich in antioxidants, omega-3 fatty acids, lean protein, and fiber, which are beneficial for TBI rehabilitation.

19.5.1 Nutrient-Dense Recipes Tailored for TBI Recovery (Table 19.5)

Table 19.6 Examples of weekly plans for brain health

Day	Breakfast	Lunch	Dinner	Snacks
Monday	Brain-boosting smoothie	Salmon and quinoa bowl	Turkey and veggie stir-fry	Greek yogurt with berries
Tuesday	Spinach and mushroom omelette	Sweet potato and black bean tacos	Grilled chicken salad	Almonds and apple slices
Wednesday	Overnight oats with berries	Quinoa salad with chickpeas	Baked cod with roasted vegetables	Carrot sticks with hummus
Thursday	Greek yogurt parfait with granola	Turkey and avocado wrap	Lentil soup with whole grain bread	Trail mix (nuts and dried fruits)
Friday	Whole grain pancakes with fruit	Mediterranean pasta salad	Grilled salmon with asparagus	Cottage cheese with pineapple
Saturday	Veggie frittata with whole grain toast	Quinoa and black bean burrito bowl	Chicken stir-fry with brown rice	Yogurt smoothie with spinach
Sunday	Breakfast burritos with eggs and veggies	Chicken caesar salad	Vegetable curry with quinoa	Whole grain crackers with cheese

19.5.2 Weekly Meal Plans to Support Brain Health and Healing (Table 19.6)

19.6 Emerging Trends and Innovations in Nutritional Therapy for TBI

Table 19.7 summarizes emerging trends and innovations in nutritional therapy for traumatic brain injury (TBI). Each trend or innovation is described, highlighting its potential role in enhancing TBI recovery and promoting brain health through personalized nutrition interventions, targeted nutrient delivery, and advanced analytical techniques.

Table 19.7 Emerging trends in nutritional therapy of TBI

Emerging trends and innovations	Description
Ketogenic diet	High-fat, low-carbohydrate diet that may provide neuroprotective benefits and enhance brain metabolism and recovery after TBI
Nutrigenomics	Study of how individual genetic variations influence dietary response and nutrient metabolism, guiding personalized nutrition strategies for TBI patients
Microbiome modulation	Manipulation of gut microbiota through diet or probiotic supplementation to promote gut-brain axis communication and support brain health and recovery
Functional doods and nutraceuticals	Development of specialized food products and supplements fortified with bioactive compounds targeting brain health and neuroprotection in TBI patients
Precision nutrition	Tailoring nutrition interventions based on individual patient characteristics, such as age, sex, genetics, and TBI severity, to optimize recovery outcomes
Brain-targeted Nutrient delivery systems	Innovative formulations and delivery methods to enhance bioavailability and targeted delivery of nutrients and bioactive compounds to the brain for neuroprotection and repair
Artificial intelligence (AI) in dietary planning	Utilization of AI algorithms and machine learning to analyze patient data and dietary patterns, providing personalized nutrition recommendations for TBI rehabilitation

19.7 Different Animal Models Techniques of TBI

Table 19.8 compares different animal models of traumatic brain injury (TBI) in terms of their methods, advantages, and disadvantages. Each model offers distinct approaches to replicate TBI pathophysiology and investigate injury mechanisms, providing valuable insights into TBI pathophysiology, treatment, and prevention strategies. However, each model also has limitations and considerations that should be considered when designing and interpreting experimental studies.

Table 19.8 Available animal models techniques to study TBI

Animal model	Methods	Advantages	Disadvantages
Controlled cortical impact (CCI) model	Controlled mechanical impact to the exposed cortex using a pneumatic or electromagnetic device	– Reproducible injury severity and location. – Allows for precise control over impact parameters. – Mimics focal brain injury with contusion and cortical tissue damage.	– Surgical exposure of the brain, which may induce additional trauma – Lack of rotational forces seen in some clinical TBI cases – Limited representation of diffuse axonal injury (DAI)
Fluid percussion injury (FPI) model	Rapid injection of fluid onto the intact dura mater, inducing transient deformation and pressure changes in the brain	– Mimics the biomechanics of closed head injury and diffuse brain damage. – Adjustable injury severity by varying fluid pressure and duration. – Can model both focal and diffuse brain injury.	– Variable injury outcomes due to differences in injury induction techniques – Producing mixed injury patterns with variable severity – Limited control over injury location and directionality
Weight drop model	Free-falling weight directly impacts the exposed skull, inducing focal or diffuse brain injury depending on impact location and force	– Simple and cost-effective model requiring minimal surgical intervention. – Mimics closed head injury mechanisms and diffuse axonal injury. – Allows for manipulation of injury severity through adjustment of weight and height.	– Lack of precise control over injury parameters such as location and directionality – Variable injury outcomes due to differences in impact force and angle – Not accurately representing rotational forces seen in human TBI
Blast injury model	Exposure of animals to blast waves generated by explosive devices, simulating blast-induced TBI observed in military and civilian settings	– Mimics the complex pathophysiology of blast-related TBI, including primary, secondary, and tertiary injury mechanisms. – Allows for investigation of blast wave effects on the brain and associated systemic responses. – Provides insights into blast injury biomechanics and neurotrauma.	– Technical challenges in generating reproducible blast conditions and controlling injury parameters – Heterogeneity in injury outcomes due to variations in blast wave characteristics and animal susceptibility – Difficulties in separating primary blast effects from secondary injury mechanisms

19.8 Different Animal Models Species of TBI

Table 19.9 compares different animal species used as models of traumatic brain injury (TBI), including rodents (mice, rats), swine (pigs), non-human primates (NHPs), and zebrafish, in terms of their methods, advantages, and disadvantages. Each species offers unique experimental advantages and considerations for studying TBI pathophysiology, treatment, and recovery. Researchers select animal models based on factors such as translational relevance, anatomical similarities, experimental manipulability, and ethical considerations.

Table 19.9 Different animal models species for TBI study

Animal species	Methods	Advantages	Disadvantages
Rodents (mice, rats)	Commonly used for their small size, ease of handling, and relatively low cost. Various injury models, including controlled cortical impact (CCI), fluid percussion injury (FPI), weight drop, and blast can be induced in rodents.	– Well-established models with extensive literature and standardized protocols – Short reproductive cycles and large litter sizes facilitate genetic and pharmacological manipulations – Availability of transgenic and knockout strains for studying specific molecular pathways	– Brain anatomy and physiology that differ from humans, limiting the direct translatability of findings – Small brain size that may affect injury severity and behavioral outcomes – Limited capacity for complex cognitive and behavioral assessments
Swine (pigs)	Larger brains and gyrencephalic cortex like humans, allowing for better modeling of human TBI biomechanics and injury patterns. Various injury methods including CCI, FPI, weight drop, and blast can be applied to pigs.	– Anatomical and physiological similarities to humans enable better translation of findings to clinical TBI – Large size allows for invasive monitoring, neuroimaging, and surgical interventions like human patients – Allows for a more comprehensive assessment of cognitive, motor, and sensory functions	– Higher cost of acquisition, housing, and maintenance compared to rodents – Longer lifespan and slower maturation rate that may limit experimental throughput and longitudinal studies – Limited availability of genetically modified and transgenic strains for mechanistic studies

(continued)

Table 19.9 (continued)

Animal species	Methods	Advantages	Disadvantages
Non-human primates (NHPs)	Closest phylogenetic relatives to humans with similar brain structure, organization, and cognitive abilities. Injury models, including CCI, FPI, and blast can be induced in NHPs.	– A highly translational model for studying TBI pathophysiology, treatment, and rehabilitation – Similarities in brain size, cortical folding, and white matter architecture allow for more accurate modeling of human TBI – Capability for sophisticated behavioral, cognitive, and neuroimaging assessments	– Ethical concerns regarding the use of primates in research and potential welfare issues – High cost of acquisition, housing, and specialized care, limiting accessibility to a few research institutions – Limited availability of genetically modified and transgenic strains for mechanistic studies
Zebrafish	Utilized due to their transparent embryos, allowing for real-time visualization of brain injury and regeneration processes. TBI can be induced using mechanical or chemical methods in zebrafish larvae.	– High fecundity and rapid development enable large-scale genetic and drug screening studies – Ability to manipulate gene expression and study neurodevelopmental processes in vivo – Cost-effective model with low housing and maintenance requirements	– Evolutionary distance from mammals that may limit direct translation of findings to human TBI – Limited behavioral and cognitive assays compared to mammalian models. – Ethical considerations regarding the use of vertebrate animals in research.

19.9 Resources and Tools for Monitoring Nutritional Status and Dietary Intake

Table 19.10 summarizes resources and tools available for monitoring nutritional status and dietary intake. These tools provide valuable information for assessing dietary patterns, nutrient intake, and nutritional adequacy, facilitating personalized nutrition assessment, counseling, and intervention strategies. Researchers, healthcare professionals, and individuals can utilize these resources to evaluate and optimize dietary habits and nutritional status for improved health outcomes.

Table 19.10 Resources and tools for monitoring nutritional status and dietary intake

Resource/tool	Description
Dietary assessment questionnaires	Structured questionnaires or surveys designed to collect information on dietary intake, eating habits, and food preferences. Common examples include food frequency questionnaires (FFQs) and 24-hour dietary recalls.
Food diaries/food records	Written or electronic records kept by individuals to track their food and beverage consumption over a specified period. Allows for detailed assessment of dietary patterns, portion sizes, and nutrient intake.
Nutrient analysis software	Computer programs or applications that analyze dietary intake data to estimate nutrient intake, energy balance, and dietary adequacy. Examples include USDA Food Composition Databases, MyFitnessPal, and Cronometer.
Biochemical biomarkers	Objective measures of nutrient status obtained from biological samples such as blood, urine, or tissue. Biomarkers include serum levels of vitamins, minerals, and metabolic indicators reflecting dietary intake and nutritional status.
Anthropometric measurements	Physical measurements of body composition, including height, weight, body mass index (BMI), waist circumference, and skinfold thickness. These measurements provide insights into nutritional status, growth, and health risks.
Dietary assessment tools for specific populations	Specialized tools and protocols for assessing dietary intake and nutritional status in specific populations such as infants, children, adolescents, elderly individuals, athletes, and individuals with chronic diseases or dietary restrictions.
Mobile applications	Smartphone apps designed to track dietary intake, monitor nutritional status, set dietary goals, and provide personalized nutrition advice. Examples include MyFitnessPal, Lose It!, and MyPlate.
Wearable devices	Wearable technology such as activity trackers and smartwatches equipped with sensors to monitor physical activity, energy expenditure, sleep patterns, and dietary intake. Integrates data for comprehensive health and nutrition assessment.

19.10 Practical Strategies for Caregivers and Health-care Professionals on Nutritional Support of TBI Patients

Table 19.11 summarizes practical strategies for caregivers and health-care professionals to provide nutritional support for traumatic brain injury (TBI) patients. By implementing these strategies, caregivers and health-care teams can optimize the nutritional status and dietary management of TBI patients, promoting recovery, and enhancing overall well-being.

Table 19.11 Practical strategies for caregivers and health-care professionals on nutritional support of TBI patients

Practical strategy	Description
Education on nutritional needs	Provide caregivers and health-care professionals with information on the specific nutritional needs of TBI patients, including energy requirements, macronutrient and micronutrient needs, and dietary modifications for optimal recovery.
Individualized meal planning	Develop individualized meal plans tailored to the patient's nutritional requirements, preferences, and dietary restrictions. Consider factors such as taste changes, swallowing difficulties, and texture modifications to ensure adequate nutrient intake.
Oral supplementation	Recommend oral nutritional supplements or fortified foods to TBI patients with increased nutritional needs, difficulty eating, or inadequate oral intake. Select supplements containing protein, vitamins, minerals, and essential nutrients to support recovery.
Monitoring nutritional status	Regularly assess the nutritional status of TBI patients through dietary intake monitoring, anthropometric measurements, biochemical biomarkers, and clinical evaluations. Use validated assessment tools to identify malnutrition risk and nutrient deficiencies.
Dysphagia management	Implement dysphagia management strategies to address swallowing difficulties commonly seen in TBI patients. Modify food consistency, texture, and presentation to ensure safe swallowing and prevent aspiration pneumonia.
Nutritional counseling	Offer nutritional counseling and guidance to caregivers and TBI patients on healthy eating habits, meal planning, portion control, and food safety practices. Provide practical tips for grocery shopping, meal preparation, and dining out.
Multidisciplinary collaboration	Foster collaboration among health-care professionals, including dietitians, speech therapists, physicians, and rehabilitation specialists, to coordinate comprehensive nutritional care for TBI patients. Incorporate nutritional goals into the overall rehabilitation plan.
Family and caregiver support	Educate family members and caregivers on the importance of nutritional support in TBI recovery and empower them to assist with meal preparation, feeding assistance, and monitoring dietary intake. Provide emotional support and resources to alleviate caregiver's burden.

19.11 Advocacy and Policy Strategies for Integrating Nutrition into Standard TBI Care Practices

Table 19.12 summarizes advocacy and policy strategies for integrating nutrition into standard traumatic brain injury (TBI) care practices. By advocating for policy changes, promoting professional training, increasing research funding, raising public awareness, and enhancing health-care system integration, stakeholders can work together to prioritize and improve nutrition management for TBI patients.

Table 19.12 Advocacy and policy strategies for integrating nutrition into standard TBI care practices

Advocacy and policy strategy	Description
Policy advocacy	Advocate for the inclusion of nutrition assessment, counseling, and intervention services as integral components of TBI rehabilitation programs and health-care policies. Lobby policymakers and health-care decision-makers to prioritize nutrition education, training, and resources for TBI care providers.
Clinical practice guidelines	Develop evidence-based clinical practice guidelines for nutrition management in TBI patients, outlining recommended practices for nutritional assessment, intervention, and monitoring throughout the continuum of care. Disseminate guidelines to health-care professionals and organizations involved in TBI care.
Professional training	Offer continuing education programs, workshops, and training sessions for health-care professionals, including physicians, nurses, dietitians, and therapists, to enhance their knowledge and skills in providing nutritional support for TBI patients. Incorporate nutrition education into medical and allied health curricula.
Research funding	Advocate for increased funding for nutrition-related TBI research initiatives, including studies on dietary interventions, nutritional biomarkers, and the impact of nutrition on TBI outcomes. Support collaborative research efforts to generate evidence for integrating nutrition into TBI care practices.
Public awareness campaigns	Launch public awareness campaigns to highlight the importance of nutrition in TBI recovery and raise awareness about the role of diet and lifestyle factors in brain health and injury prevention. Target caregivers, patients, policymakers, and the general public through multimedia platforms and community outreach initiatives.
Health-care system integration	Advocate for the integration of nutrition services into existing TBI care systems and health-care delivery models, including acute care settings, rehabilitation facilities, outpatient clinics, and community-based programs. Promote interdisciplinary collaboration and care coordination among health-care providers to ensure comprehensive nutritional care for TBI patients.
Legislative support	Work with legislators and policymakers to enact legislation supporting nutrition interventions for TBI patients, such as insurance coverage for medical nutrition therapy, access to specialized foods and supplements, and reimbursement for nutrition counseling services. Collaborate with patient advocacy groups to advocate for patient-centered policies.

19.12 Integrative Approaches and Holistic Management of TBI Rehabilitation

Table 19.13 summarizes integrative approaches and holistic management strategies for traumatic brain injury (TBI) rehabilitation. By adopting a person-centered, multidisciplinary approach and addressing physical, cognitive, emotional, social, and lifestyle factors, health-care professionals can optimize TBI recovery outcomes and promote overall well-being for patients and their families.

Table 19.13 Integrative approaches and holistic management of TBI rehabilitation

Integrative approaches	Description
Multidisciplinary care teams	Implement multidisciplinary care teams comprising physicians, neurologists, neuropsychologists, physical therapists, occupational therapists, speech therapists, dietitians, and social workers to provide comprehensive TBI rehabilitation. Collaborate to address physical, cognitive, emotional, and social aspects of recovery.
Person-centered care	Adopt a person-centered approach to TBI rehabilitation, focusing on individualized treatment plans tailored to the patient's unique needs, preferences, goals, and values. Involve patients and their families in shared decision-making and goal-setting throughout the rehabilitation process.
Functional goal-setting and rehabilitation	Set functional goals for TBI rehabilitation based on the patient's abilities, impairments, and aspirations. Use evidence-based rehabilitation techniques, including task-oriented training, cognitive-behavioral therapy, and environmental modifications, to promote functional independence and community reintegration.
Complementary and alternative therapies	Integrate complementary and alternative therapies such as acupuncture, massage therapy, yoga, meditation, music therapy, and art therapy into TBI rehabilitation programs to address pain management, stress reduction, relaxation, and emotional well-being. Offer modalities that align with patient preferences and cultural beliefs.
Lifestyle interventions and health promotion	Promote healthy lifestyle interventions, including physical activity, exercise, sleep hygiene, stress management, and smoking cessation, to optimize overall health and well-being during TBI recovery. Educate patients and caregivers on the importance of lifestyle factors in brain health and rehabilitation outcomes.
Nutritional support and dietary management	Provide nutritional support and dietary management as integral components of TBI rehabilitation. Assess nutritional status, address malnutrition and swallowing difficulties, and optimize nutrient intake to support brain healing, physical recovery, and cognitive function. Collaborate with dietitians to develop personalized nutrition plans.
Psychosocial support and mental health care	Offer psychosocial support and mental health care to address the emotional and psychological impact of TBI on patients and their families. Provide counseling, psychotherapy, and support groups to help individuals cope with grief, anxiety, depression, and adjustment challenges post-injury.
Community reintegration and long-term support	Facilitate community reintegration and long-term support for TBI survivors transitioning from rehabilitation to independent living. Coordinate vocational rehabilitation, educational resources, housing assistance, transportation services, and peer support networks to promote social participation and quality of life.

The manufacturer's authorised representative in the EU is Springer Nature Customer Service Centre GmbH, Europaplatz 3, 69115 Heidelberg, Germany. If you have any concerns regarding our products, please contact ProductSafety@springernature.com

Printed and bound by CPI Group (UK) Ltd, Croydon, CR0 4YY
01/10/2026
02225859-0015